SIP HANDBOOK

SERVICES, TECHNOLOGIES, AND SECURITY OF SESSION INITIATION PROTOCOL

SIP HANDBOOK

SERVICES, TECHNOLOGIES, AND SECURITY OF SESSION INITIATION PROTOCOL

EDITED BY

SYED A. AHSON
MOHAMMAD ILYAS

CRC Press
Taylor & Francis Group
Boca Raton   London   New York

CRC Press is an imprint of the
Taylor & Francis Group, an **informa** business

CRC Press
Taylor & Francis Group
6000 Broken Sound Parkway NW, Suite 300
Boca Raton, FL 33487-2742

© 2009 by Taylor & Francis Group, LLC
CRC Press is an imprint of Taylor & Francis Group, an Informa business

Library of Congress Cataloging-in-Publication Data

SIP handbook : services, technologies, and security of Session Initiation Protocol
/ edited by Syed A. Ahson and Mohammad Ilyas.
p. cm.
Includes bibliographical references and index.
ISBN-13: 978-1-4200-6603-6
ISBN-10: 1-4200-6603-X
1. Computer network protocols. 2. Internet telephony. 3. Multimedia systems.
I. Ahson, Syed. II. Ilyas, Mohammad, 1953- III. Title.

TK5105.55.S58 2008
004.6'2--dc22   2008021972

**Visit the Taylor & Francis Web site at
http://www.taylorandfrancis.com**

**and the CRC Press Web site at
http://www.crcpress.com**

Contents

Preface   ix

Editors   xi

Contributors   xiii

I   CONCEPTS & SERVICES   1

1   **SIP in Peer-to-Peer Networks**   3
Holger Schmidt and Franz J. Hauck

2   **SIP: Advanced Media Integration**   15
Marisol Hurtado, Andre Rios, Antoni Oller, Jesus Alcober, and Sebastia Sallent

3   **SIP and IPv6 – Migration Considerations, Complications, and Deployment Scenarios**   43
Mohamed Boucadair and Yoann Noisette

4   **SIP Mobility Management**   71
Chung-Ming Huang and Chao-Hsien Lee

5   **SIP Event Notification and Presence Information**   87
Bo Zhao and Chao Liu

6   **Group Conference Management with SIP**   123
Thomas C. Schmidt and Matthias Wählisch

7   **Home Networking System with SIP**   159
Yukikazu Nakamoto and Naoko Kuri

8   **Compliance and Interoperability Testing of SIP Devices**   173
Doris Bao, Luca De Vito, Sergio Rapuano, and Laura Tomaciello

II TECHNOLOGIES   201

9 **P2P SIP: Network Architecture and Resource Location**   203
   Strategy
   Guiran Chang, Chuan Zhu, and Wei Ning

10 **SIP Mobility Technologies and the Use of Persistent**   227
   Identifiers to Improve Inter-Domain Mobility and
   Security
   Henry Jerez, Joud Khoury, and Chaouki Abdallah

11 **SIP and Vertical Handoffs in Heterogeneous Wireless**   253
   Networks
   Jie Zhang, F. Richard Yu, Xiaolei Wang, Henry C. B. Chan,
   and Victor C. M. Leung

12 **NAT Traversal**   277
   Mario Baldi, Fulvio Risso, and Livio Torrero

13 **Multipoint Extensions for SIP**   307
   Sureswaran Ramadass and Omar Abouabdalla

14 **Narrowcasting in SIP: Articulated Privacy Control**   323
   Sabbir Alam, Michael Cohen, Julián Villegas, and Ashir Ahmed

15 **Q-SIP/SDP for QoS-Guaranteed End-to-End**   347
   Real-Time Multimedia Service Provisioning on
   Converged Heterogeneous Wired and Wireless
   Networks
   Young-Tak Kim and Young-Chul Jung

16 **SIP Modeling and Simulation**   373
   Yanlan Ding, GuiPing Su, and Huaxu Wan

17 **SIP-Based Mobility Management and Its Performance**   397
   Evaluation
   Nilanjan Banerjee and Kalyan Basu

III SECURITY   433

18 **SIP Security: Threats, Vulnerabilities and**   435
   Countermeasures
   Dimitris Geneiatakis, Georgios Kambourakis and Costas
   Lambrinoudakis

19 SIP Vulnerabilities for SPIT, SPIT Identification Criteria, Anti-SPIT Mechanisms Evaluation Framework and Legal Issues 457
G. F. Marias, L. Mitrou, M. Theoharidou, J. Soupionis, S. Ehlert, and D. Gritzalis

20 Anonymity in SIP 481
L. Kazatzopoulos, K. Delakourides, and G. F. Marias

21 Secure Intelligent SIP Services 499
Handoura Abdallah

22 SIP Security and Quality of Service Performance 521
Johnson Ihyeh Agbinya

23 A Conceptual Architecture for SPIT Mitigation 563
Yacine Rebahi, Stelios Dritsas, Tudor Golubenco, Benjamin Pannier, and Johan Fredrik Juell

24 Towards a Fraud Detection Framework in VoIP Networks 583
Yacine Rebahi, Thomas Magedanz, and Dorgham Sisalem

Index 601

Preface

Session Initiation Protocol (SIP) is among the most widely adopted protocols for IP telephony. SIP, as presented by the IETF, forms a flexible, comprehensive signaling solution and allows for a lightweight deployment of many, but not all, of the well-known service features. The simpler implementation and open collaboration of SIP make it the signaling protocol of choice for advanced multimedia communications signaling. This fact is evidenced by its acceptance into the 3rd Generation Partnership Project (3GPP) as a signaling protocol for establishing real time multimedia sessions. SIP is continuously gaining in popularity and deployment and has been widely adopted by service providers to provide IP telephony, instant messaging, and other data services. SIP is increasingly gaining in popularity as the next generation multimedia signaling and session establishment protocol. SIP is a key technology which enables this evolution, opening new possibilities to develop multimodal communications. SIP increases the possibilities of developing integrated multimedia services and offers advantages such as independence from the underlying transport network, session establishment with the involvement of multiple devices and the support for mobility.

SIP is a session-layer signaling protocol that is used for establishment, modification, and release of sessions among participants. The session may involve VoIP (Voice over IP), video conference, multimedia distribution, and IMS (IP Multimedia Subsystem) applications. SIP supports five facets of establishing and terminating multimedia communications: user location, user availability, user capabilities, session setup, and session management. The SIP architecture is also proposed as an efficient candidate that can be reused to provide personal, terminal, and session mobility with a readily available infrastructure. SIP infrastructure is extensively deployed all over the Internet and has already been accepted as the signaling protocol of preference for many multimedia frameworks. Recently, SIP has gained widespread acceptance and deployment among wireline service providers for introducing new services such as VoIP, Tele-presence, video conferencing; within the enterprises for Instant Messaging and collaboration; and amongst mobile carriers for push-to-talk services. Industry acceptance of SIP as the protocol of choice for converged communications over IP networks is thus highly likely.

The *SIP Handbook* provides technical information about all aspects of SIP. The areas covered in the handbook range from basic concepts to research grade material including future directions. The *SIP Handbook* captures the current state of IP Multimedia Subsystem technology and serves as a source of comprehensive reference material on this subject. It comprises three sections,

Concepts & Services, Technologies and Security. It has a total of 24 chapters authored by 65 experts from around the world. The targeted audience includes professionals who are designers and/or planners for SIP systems, researchers (faculty members and graduate students), and those who would like to learn about this field.

The handbook is expected to have the following salient features:

- To serve as a single comprehensive source of information and as reference material on SIP technology.

- To deal with an important and timely topic of emerging technology of today, tomorrow, and beyond.

- To present accurate, up-to-date information on a broad range of topics related to SIP technology.

- To present material authored by the experts in the field.

- To present information in an organized and well-structured manner.

Although the handbook is not precisely a textbook, it can certainly be used as a textbook for graduate courses and research-oriented courses that deal with SIP. Any comments from the readers will be highly appreciated.

Many people have contributed to this handbook in their unique ways. The first and the foremost group that deserves immense gratitude is the group of highly talented and skilled researchers who have contributed 24 chapters. All of them have been extremely cooperative and professional. It has also been a pleasure to work with Ms. Nora Konopka, Ms. Jessica Vakili, and Richard Tressider of CRC Press, and we are extremely gratified for their support and professionalism. Our families have extended their unconditional love and strong support throughout this project and they all deserve very special thanks.

Syed Ahson
Redmond, Washington

Mohammad Ilyas
Boca Raton, Florida

Editors

Syed A. Ahson is a senior software design engineer with Microsoft in Redmond, Washington. As part of the Mobile Voice and Partner Services group, he is busy creating new and exciting end-to-end services and applications. Prior to working at Microsoft, he was a senior staff software engineer with Motorola, where he contributed significantly in leading roles towards the creation of several advanced and exciting cellular phones. Dr. Ahson has extensive experience with Wireless Data Protocols, Wireless Data Applications and Cellular Telephony Protocols. Prior to joining Motorola, he was a senior software design engineer with NetSpeak Corporation (now part of Net2Phone), a pioneer in VoIP telephony software.

Dr. Ahson has published more than ten books on emerging technologies such as WiMAX, RFID, Mobile Broadcasting and IP Multimedia Subsystem. His recent books include *IP Multimedia Subsystem Handbook* (2008), and *Handbook of Mobile Broadcasting: DVB-H, DMB, ISDB-T and MediaFLO* (2007). Dr. Ahson has published several research articles and has taught computer engineering courses as adjunct faculty at Florida Atlantic University, Boca Raton, Florida where he introduced a course on Smartphone technology and applications. He received his MS degree in computer engineering in July 1998 at Florida Atlantic University and his B.Sc. degree in electrical engineering from Aligarh University, India, in 1995.

Mohammad Ilyas is associate dean for research and industry relations in the College of Engineering and Computer Science at Florida Atlantic University, Boca Raton, Florida. Previously, he served as chair of the Department of Computer Science and Engineering and interim associate vice president for research and graduate studies. He received his PhD degree from Queen's University in Kingston, Canada. His doctoral research was about switching and flow control techniques in computer communication networks. He received his BSc degree in electrical engineering from the University of Engineering and Technology, Pakistan and his MS degree in electrical and electronic engineering at Shiraz University, Iran.

Dr. Ilyas has conducted successful research in various areas including traffic management and congestion control in broadband/high-speed communication networks, traffic characterization, wireless communication networks, performance modeling, and simulation. He has published over 20 books on emerging technologies, and over 150 research articles. His recent books include *RFID Handbook* (2008), and *WiMAX Handbook* (3 vols.) (2007). He has supervised 11 PhD dissertations and more than 37 MS theses to completion. He has been a consultant to several national and international organizations. Dr. Ilyas is an active participant in several IEEE technical committees and activities and he is a senior member of IEEE and a member of ASEE.

Contributors

Chaouki Abdallah
Electrical and Computer
  Engineering Department
The University of New Mexico
Albuquerque, New Mexico

Handoura Abdallah
École Nationale Supérieure des
  Télécommunications de Bretagne
Brest, France

Omar Abouabdalla
National Advanced IPv6 Center
University of Science, Malaysia
Penang, Malaysia

Johnson Ihyeh Agbinya
School of Computing and
  Communication
University of Technology
Sydney, Australia

Ashir Ahmed
Department of Computer
  Science and Communication
  Engineering
Kyushu University
Fukuoka, Japan

Sabbir Alam
Spatial Media Group
University of Aizu
Aizu-Wakamatsu, Japan

Jesus Alcober
Polytechnic University of
  Catalonia
Barcelona, Spain

Mario Baldi
Department of Control and
  Computer Engineering
Technical University of Turin
Turin, Italy

Nilanjan Banerjee
IBM India Research Lab
New Delhi, India

Doris Bao
Department of Engineering
University of Sannio
Benevento, Italy

Kalyan Basu
Treveni Systems
Plano, Texas

Mohamed Boucadair
CORE/M2V
France Telecom
Caen, France

Henry C.B. Chan
Department of Computing
The Hong Kong Polytechnic
  University
Hong Kong, People's Republic of
  China

Guiran Chang
Computing Center
Northeastern University
Shenyang, People's Republic of
  China

Michael Cohen
Spatial Media Group
University of Aizu
Aizu-Wakamatsu, Japan

K. Delakourides
Athens University of Economics and
  Business
Athens, Greece

Luca De Vito
Department of Engineering
University of Sannio
Benevento, Italy

Yanlan Ding
Graduate University of Chinese
  Academy of Sciences
Beijing, People's Republic of China

Stelios Dritsas
Athens University of Economic and
  Business
Athens, Greece

S. Ehlert
Fraunhofer Institute for Open
  Communication Systems
Berlin, Germany

Dimitris Geneiatakis
Department of Information and
  Communication Systems
  Engineering
University of the Aegean
Karlovassi, Greece

Tudor Golubenco
Iptego
Berlin, Germany

D. Gritzalis
Athens University of Economics and
  Business
Athens, Greece

Franz J. Hauck
Institute of Distributed Systems
Ulm University
Ulm, Germany

Chung-Ming Huang
Department of Computer
  Science and Information
  Engineering
National Cheng Kung University
Tainan City, Taiwan,
  Republic of China

Marisol Hurtado
i2CAT Foundation
Barcelona, Spain

Henry Jerez
Corporation for National Research
  Initiatives
Reston, Virginia

Johan Fredrik Juell
Telio Telecom AS
Oslo, Norway

Young-Chul Jung
Department of Information
  and Communication
  Engineering
Yeungnam University
Gyeongsan, South Korea

Georgios Kambourakis
Department of Information and
  Communication Systems
  Engineering
University of the Aegean
Karlovassi, Greece

L. Kazatzopoulos
Athens University of Economics and
  Business
Athens, Greece

Joud Khoury
Electrical and Computer
  Engineering Department
The University of New Mexico
Albuquerque, New Mexico

Young-Tak Kim
Department of Information
 and Communication
 Engineering
Yeungnam University
Gyeongsan, South Korea

Naoko Kuri
Graduate School of Applied
 Informatics
University of Hyogo
Kakogawa, Japan

Costas Lambrinoudakis
Department of Information and
 Communication Systems
 Engineering
University of the Aegean
Karlovassi, Greece

Chao-Hsien Lee
Department of Medical Information
 Management
Kaohsiung Medical University
Kaohsiung, Taiwan, Republic of
 China

Victor C.M. Leung
Department of Electrical and
 Computer Engineering
The University of British Columbia
Vancouver, British Columbia,
 Canada

Chao Liu
Yahoo! Inc.
Sunnyvale, California

Thomas Magedanz
Technical University of Berlin
Berlin, Germany

G. F. Marias
Athens University of Economics and
 Business
Athens, Greece

L. Mitrou
Department of Information and
 Communication Systems
 Engineering
University of the Aegean
Karlovassi, Greece

Yukikazu Nakamoto
Graduate School of Applied
 Informatics
University of Hyogo
Kakogawa, Japan

Wei Ning
School of Information Science and
 Engineering
Northeastern University
Shenyang, People's Republic of
 China

Yoann Noisette
CORE/M2V
France Telecom
Caen, France

Antoni Oller
Polytechnic University of Catalonia
Barcelona, Spain

Benjamin Pannier
eleven GmbH
Berlin, Germany

Sureswaran Ramadass
National Advanced IPv6 Center
University of Science, Malaysia
Penang, Malaysia

Sergio Rapuano
Department of Engineering
University of Sannio
Benevento, Italy

Yacine Rebahi
Fraunhofer Institute for
 Open Communication Systems
Berlin, Germany

Andre Rios
Polytechnic University of Catalonia
Barcelona, Spain

Fulvio Risso
Department of Control and
  Computer Engineering
Technical University of Turin
Turin, Italy

Sebastia Sallent
i2CAT Foundation
Barcelona, Spain

Holger Schmidt
Institute of Distributed Systems
Ulm University
Ulm, Germany

Thomas C. Schmidt
Department of Computer Sciences
Hamburg University of Applied
  Sciences
Hamburg, Germany

Dorgham Sisalem
Tekelec
Berlin, Germany

J. Soupionis
Athens University of Economics and
  Business
Athens, Greece

GuiPing Su
Graduate University of Chinese
  Academy of Sciences
Beijing, People's Republic of China

M. Theoharidou
Athens University of Economics and
  Business
Athens, Greece

Laura Tomaciello
Department of Engineering
University of Sannio
Benevento, Italy

Livio Torrero
Department of Control and
  Computer Engineering
Technical University of Turin
Turin, Italy

Julián Villegas
Spatial Media Group
University of Aizu
Aizu-Wakamatsu, Japan

Matthias Wählisch
Department of Computer Sciences
Hamburg University of Applied
  Sciences
Hamburg, Germany

Huaxu Wan
School of Information Science and
  Engineering
Graduate University of Chinese
  Academy of Sciences
Beijing, People's Republic of China

Xiaolei Wang
Department of Electrical and
  Computer Engineering
The University of British Columbia
Vancouver, British Columbia,
  Canada

F. Richard Yu
School of Information Technology
Carleton University
Ottawa, Ontario, Canada

Bo Zhao
Department of Computer Science
  and Engineering
The Pennsylvania State University
University Park, Pennsylvania

Jie Zhang
Department of Electrical and
 Computer Engineering
The University of British Columbia
Vancouver, British Columbia,
 Canada

Chuan Zhu
School of Information Science and
 Engineering
Northeastern University
Shenyang, People's Republic of
 China

Part I

CONCEPTS & SERVICES

1

SIP in Peer-to-Peer Networks

Holger Schmidt and Franz J. Hauck
Ulm University

CONTENTS

1.1	Introduction	3
1.2	Background	4
1.3	Integration of P2P Technologies into SIP	6
1.4	Issues in P2P SIP	10
1.5	Field of Application of P2P SIP	10
1.6	Standardization	11
1.7	Conclusion	11
	References	12

1.1   Introduction

Session Initiation Protocol (SIP [1]) has become the quasi-standard for Voice-over-Internet Protocol (VoIP) communications. As an IETF protocol for session signalling in general, SIP is, for instance, adopted for session management in the Third Generation Partnership Project (3GPP) [2], which aims at providing a next generation infrastructure for mobile phone systems. SIP is based on a client–server infrastructure in which *user agents* represent the end-terminals as clients, *proxy servers* handle SIP message routing between the user agents, and *registrar servers* store the client's contact information into a *location service.*

As a result of the client–server infrastructure, SIP networks are in general characterized by high administrative efforts. Therefore, Skype [3] started to integrate peer-to-peer (P2P) mechanisms into VoIP communications, which led to an infrastructure without expensive high-end server machines and with almost no need for client-side configuration. However, as Skype builds on proprietary, obfuscated protocols [4] and telecom companies require standards for interoperability, there are efforts to integrate P2P techniques into SIP (P2P SIP).

There are several approaches to integrate P2P mechanisms into SIP: an extension of the SIP protocol itself (e.g., Bryan et al. [5]), an integration of P2P mechanisms into the SIP message forwarding mechanism or a direct integration of P2P techniques into the user agent (e.g., Schmidt et al. [6]).

The protocol extension leads to changes in the SIP headers to support P2P overlays, which restricts backward compatibility to standard SIP. An integration of P2P mechanisms into the SIP message forwarding mechanism or directly into the SIP user agent entails changes to SIP entities but not to the protocol itself. This maintains compatibility to standard SIP.

In the next section, we provide background on the basic technologies needed for P2P SIP. On this basis, we present possible integration approaches of P2P mechanisms into SIP: either the protocol or standard SIP network entities are extended. Furthermore, we will describe issues that have to be considered in P2P SIP environments. At the end of the chapter, we describe possible fields of application as well as current standardization efforts regarding P2P SIP.

1.2   Background

This section describes fundamental technologies, which are the basis for P2P SIP. First of all, we introduce P2P networking topologies: unstructured and structured topologies. Then, we provide background on standard SIP.

1.2.1   Peer-to-Peer Networks

P2P networks evolved from the client–server paradigm, in which clients consume server resources. However, the client–server paradigm has some drawbacks such as bad scalability, single-point-of-failure on the server side, and unused resources within the network. In P2P networks, peers are equal, i.e., they have client and server roles at the same time. Thus, a crash of part of the P2P system has no severe consequences for the whole system; other parts of the system are able to overtake the crashed part's tasks. P2P systems are separated into two groups: hybrid and pure P2P systems. While hybrid P2P systems have client–server components, pure P2P systems do not have any single-point-of-failure due to client–server components. There are two types of pure P2P systems. These are characterized by the way resources are distributed and localized within the system: they use either a structured or an unstructured topology.

The idea behind an unstructured P2P topology is flooding. In order to propagate a message from one peer to another, the message is forwarded by each peer to every known neighbor peer until the message has reached the target. A popular P2P system that uses an unstructured topology is Gnutella*.

Unstructured P2P systems use many networking resources to propagate messages to all known neighbor peers. This fact led to the development of structured P2P systems, which are usually built on the basis of a distributed hash table (DHT). In such structured P2P systems each node as well as each

---

*http://rfc-gnutella.sourceforge.net/

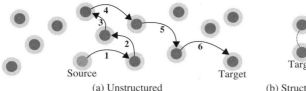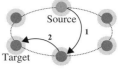

(a) Unstructured   (b) Structured (here: ring-structured)

FIGURE 1.1
Resource location within unstructured and structured P2P topologies.

resource is identified by a hash value acting as the node or resource identifier (ID). The DHT specifies a closeness metric for distributing and locating resources at particular nodes within the network. In order to enable location, resources are stored at the node that has the closest node ID according to the given metric (some systems allow resource replication at nodes close to the resource ID as well). Structured P2P topologies ensure that nodes are aware of nodes with IDs that are close to their own ID according to the given metric. This allows the selective propagation of messages to nodes, which are closer to the target ID. A popular P2P system relying on a structured topology is Chord [7], which uses a ring structure topology ordered by node IDs (see Figure 1.1b).

Figure 1.1 shows the difference of resource location in unstructured and structured P2P topologies. In general, discovery in structured systems leads to fewer messages that have to be propagated. However, maintaining the overlay structure requires additional periodic messages.

1.2.2 SIP Fundamentals

SIP was developed by the Internet Engineering Task Force (IETF) as RFC 3261 [1]. SIP is a text-based protocol for session management in general, which is built upon standard Internet Protocols only. The protocol is predominantly used for multimedia session management, e.g., for establishing and terminating voice over IP (VoIP) sessions.

The SIP RFC specifies a set of entities on the basis of the client–server paradigm (see Figure 1.2). User agents (SIP-UA) send and receive SIP messages to establish, modify, and terminate sessions. End-terminals register their current contact information (i.e., IP address and port of the endpoint) at a registrar server using a REGISTER message. Then, the registrar stores this data in a location service (LOC), which represents some kind of database for contact information of participating user agents within a specific domain. Proxy servers control the SIP network traffic, as shown in Figure 1.2. For establishing a SIP session, the end-terminal sends an INVITE message, which contains the target end-terminal's address as SIP URI (format: username@domainname), to a locally predefined proxy server (Figure 1.2, message 1). The proxy server first checks the location of the target end-terminal. If the target's SIP-URI is within another domain, the message is forwarded to a proxy of the other

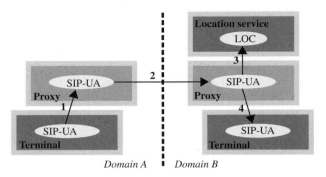

FIGURE 1.2
Message flow of standard SIP session establishment.

domain, which is discovered using the Domain Name System (DNS) (Figure 1.2, message 2). If the target is located within the same domain the proxy is able to locate the target's current contact address using the domain-local location service. With this information, the proxy is able to forward the message to the target user agent (Figure 1.2, Messages 3 and 4).

1.3   Integration of P2P Technologies into SIP

As already mentioned in the introduction, there are several ways to integrate P2P mechanisms into SIP in order to eliminate the need of central servers. These are either an extension of standard SIP or an extension of the SIP entity regarding endpoint location, i.e., the SIP location service. Both mechanisms are described in detail in the next two sections.

1.3.1   Extension of SIP

Bryan et al. developed a SIP extension called *dSIP*, which uses P2P mechanisms [5]. The idea is to maintain a P2P overlay on the basis of extended SIP messages serving as a distributed registrar server for resource location. The protocol extension allows publishing and querying contact information, which is required for routing SIP messages without the need of any central server entities. Therefore, additional SIP headers are introduced but no new SIP methods (i.e., message types) are needed.

dSIP is designed to support various DHT protocols. In order to guarantee minimal compatibility between potential participants at least Chord has to be supported [7, 8]. dSIP peers (i.e., terminals that are compatible with dSIP) have to provide server-like as well as infrastructure-maintaining functionality, i.e., they have to act like registrar and proxy servers at the same time.

In general, P2P infrastructures require mechanisms to store and to retrieve resource information. dSIP uses the SIP REGISTER message for binding SIP resource names (i.e., SIP URIs) to particular contact details. Therefore, each peer and resource has a globally unique identifier (GUID) within the dSIP infrastructure. Resources are stored and located according to the DHT algorithm mapping their GUID onto a specific node GUID that is closest according to the given metric.

In order to send a message, e.g., for session establishment, a peer has to discover the resource location first. dSIP uses REGISTER messages for resource discovery: a REGISTER message without a contact header implements locating target end-terminals. Therefore, peers have to calculate the target resource ID based on the target's SIP URI according to the given DHT algorithm. dSIP specifies a resource URI for identifying resources (e.g., SIP URI), such as *<sip:bob@xyz.org;resource-ID=1257ebd371>* (the resource ID represents the hashed value of the SIP URI). The peer queries its routing table to select the closest known entity (or entities) with respect to the target resource ID (specified by the closeness metric of the DHT algorithm). Then, the REGISTER message is sent to these closest known peers. dSIP specifies three possible mechanisms for the routing process: iterative (see Figure 1.3a), recursive (see Figure 1.3b) and semi-recursive. In all three cases, the final response is either a 200 OK message with the required contact information in case of success, or a 404 Not Found message if the target end-terminal cannot be located. For iterative routing, the source (sending peer) receives 302 Redirect response messages from the contacted peers that contain peer IDs that are closer to the sought resource ID according to the given metric. Then, the source iterates sending messages to more and more close peers until the response is either 200 OK or 404 Not Found. In the recursive routing process, a peer forwards messages in a proxy-like behavior to the closest known node on its own. The response is also sent to each former peer in the forwarding chain. Semi-recursive routing is almost identical as messages are also forwarded by peers on their own. In contrast to recursive routing, the response is sent directly to the requesting peer. On the basis of the contact information given in the 200 OK message in case of successful discovery, messages, e.g., for session establishment, are directly sent to the target end-terminal.

Initial registration at the network requires two successive steps: basic registration of the peer for integration into the P2P network and registration of the SIP contact information in order to be available for other SIP terminals. First,

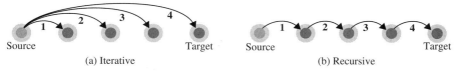

Source   Target   Source   Target

(a) Iterative   (b) Recursive

FIGURE 1.3
Routing mechanisms for dSIP.

peers send a REGISTER message with their own calculated node ID (according to the DHT algorithm) to a so-called *bootstrap peer* (see Section 1.4). Therefore, dSIP specifies a peer URI that identifies the node's actual address, such as *<sip:peer@10.0.0.34;peer-ID=efdab45629>*. The REGISTER message is forwarded to the peer closest to the new peer's GUID according to the given metric (either iterative or recursive forwarding is used). Then, the new peer exchanges its DHT state information with the closest peer in order to learn about neighbors. In the next step, the new peer has to register its SIP URI in order to be discoverable as a SIP end-terminal. Equal to the initial peer's registration, the peer calculates the resource's (SIP URI) GUID and sends a REGISTER message to the closest peer. The registration is successful if the peer receives a 200 OK message.

REGISTER messages are also used to maintain the overlay. Such messages are periodically exchanged between peers for managing DHT neighbor bindings (see Section 1.2.1). Therefore, a peer calculates the GUID of a peer it wants to connect to and sends the message to the peer with the closest known GUID. This message is forwarded until it arrives at the addressed peer; then, both peers are able to exchange maintenance information. Whenever a node joins or leaves the network, messages are exchanged between peers in order to keep the network up-to-date. Session management messages are also used for maintaining the overlay. Therefore, these messages contain information about a peer's neighbor nodes as well. Two new SIP headers are introduced in order to maintain the overlay. A *DHT-PeerID* identifies the sending peer's GUID and *DHT-Link* describes other peers that are located in the neighborhood.

The calculation of GUIDs is very expensive. Thus, calculated resource and peer GUIDs are attached to messages as part of the headers for performance reasons. However, whenever messages result in changing the overlay structure, network entities have to check if the GUIDs are hashed correctly.

For connecting standard SIP user agents to the overlay, dSIP suggests so-called *adapter peers*, which act like gateways between standard SIP and the overlay.

1.3.2 Extension of the SIP Location Service

Beside changing the SIP protocol itself as described before, another approach is an extension of the SIP location service. In contrast to a standard SIP location service, a P2P SIP location service (P2P-LOC) stores contact information in a P2P network. Therefore, the location service interface provides means for registration as well as for discovery of SIP-URI's contact information. The P2P-LOC can either be integrated into the SIP proxy server (P2P-PROXY) or directly into the user agent (P2P-UA). A hybrid approach with P2P-PROXYs and P2P-UAs is possible as well. As shown in Figure 1.4, an integration into the SIP user agent leads to reduced communication in comparison to standard SIP (see Figure 1.2); messages are saved due to user-agent-internal location of the target's contact information.

With SIPPEER, Singh and Schulzrinne developed a P2P SIP adaptor that allows for participation into a P2P network without changes in the user agent by implementing a P2P-PROXY [9, 10]. SIPPEER should run on the same host as the user agent and has to be configured as an outbound proxy. Then, all messages of the user agent are automatically sent through SIPPEER, which provides an extended location service on basis of the OpenDHT [11] infrastructure for registration and discovery (OpenDHT uses the Bamboo [12] DHT implementation as underlying infrastructure).

Schmidt et al. developed another P2P-LOC on the basis of JXTA, an open and generic P2P infrastructure [6, 13]. Their P2P-LOC can either be integrated into a P2P-PROXY (comparable to SIPPEER) or directly into the user agent (P2P-UA). The P2P-PROXY should run on the user agent's machine for allowing local communication with the user agent for reducing network traffic.

In both systems, REGISTER messages result in publishing resources in the particular P2P network. Contact information (e.g., IP address and port) is stored according the corresponding SIP URI, which acts as the key. SIP messages, which require routing, lead to a P2P network discovery process: corresponding resources with the given SIP URI as key have to be discovered. Then, this contact information is used to directly send the SIP message to the target (see Figure 1.4).

However, there are differences between both systems as well. In contrast to SIPPEER, which is based on OpenDHT, the P2P-LOC of Schmidt et al. uses JXTA. JXTA provides the ability to integrate arbitrary P2P mechanisms in order to support a variety of application scenarios. For instance, in some mobile scenarios it is not feasible to use DHTs because of the maintenance overhead [14]. Additionally, Schmidt et al. seamlessly integrated service location into their P2P-LOC concept, which is especially important for dynamic networks. Therefore, the P2P SIP proxy is able to interpret extended SIP methods [15]. The standard SIP message REGISTER with an arbitrary attachment is used for service registration. Thus, it can be combined with standard SIP registration, which leads to reduced network traffic. An OPTIONS message with a query as attachment can be used for services discovery. Both requests lead to the P2P-LOC either storing or discovering service information within the JXTA network. In contrast to the work of Schmidt et al., SIPPEER allows the use of the SIP extension as presented in Section 1.3.1 as well.

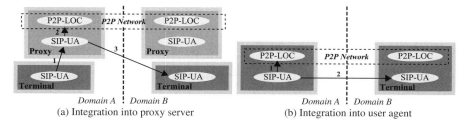

(a) Integration into proxy server (b) Integration into user agent

FIGURE 1.4

Message flow of P2P SIP session establishment with extended location service.

1.4   Issues in P2P SIP

There are several issues in P2P SIP that have to be considered. First of all, there are general issues that are common to all P2P systems. The system has to ensure that IDs are globally unique within the whole system. Otherwise, the overlay structure and consequently the message routing does not work. Another common issue is the so-called first-contact-problem targeting the bootstrapping mechanism for the overlay. In order to join the network, a new node has to know what participating peers (i.e., bootstrap peers) to contact. Recent literature provides many different ways to solve this issue. For instance, a fixed Web address can be used to provide random bootstrap peers or broadcasting mechanisms are used. Another severe problem within P2P infrastructures is security. To solve this, either the protocol ensures security or there has to be some trust relationship between the participating entities.

A severe issue in P2P SIP is identity management. The infrastructure has to ensure that identities cannot be stolen, i.e., that SIP-URIs cannot be used by unauthorized entities. Skype, for instance, solves this by introducing a central entity for identity management. However, in the case that the system should not rely on central entities, ensuring identity is very difficult. For instance, Seedorf tried to approach this issue with self-certifying SIP URIs [16].

A common problem with SIP is Network Address Translation (NAT), which is, for instance, used for locating users behind firewalls. However, there are solutions for NAT traversal, e.g., Simple traversal of UDP over NATs (STUN [17]). For P2P SIP solutions based on SIP messages only, these standard NAT traversal mechanism can be reused [5]. If P2P SIP approaches use particular P2P systems, these approaches have to ensure that these systems are capable of NAT traversal (e.g., JXTA allows this by introducing specific router peers [13]).

Performance is another issue that has to be addressed by the infrastructure. However, as only session establishment is based on a P2P discovery and multimedia communication traffic is transferred directly, the requirements to latency of the P2P infrastructure are quite low.

1.5   Field of Application of P2P SIP

P2P SIP can be used for a broad range of applications. Bryan et al. wrote an IETF Internet draft summarizing possible use cases [18].

Amongst others, P2P SIP can be used for providing a P2P VoIP service comparable to Skype. However, P2P SIP builds on a standard protocol, whereas Skype uses proprietary protocols. As Skype, a P2P SIP VoIP provider may rely on more or less central entities, for instance for accounting and identity management. Nevertheless, in the future there might also be an open global P2P VoIP network in which anybody connected to the Internet is able to join with a compatible device (this requires standard protocols as proposed by P2P SIP).

P2P SIP can also be used for providing anonymous communication. As there are no central servers within a P2P SIP infrastructure users cannot be easily monitored. Another application are ad-hoc networks, connecting a set of devices with each other spontaneously. Examples are spontaneous meeting networks, spontaneous emergency and crisis networks, or even networks inside the developing world. P2P SIP can provide a VoIP service to such environments.

With regard to fault tolerance, P2P SIP can be used for the implementation of redundant standard SIP proxy servers. Therefore, SIP providers have to provide server farms, which are able to use P2P SIP for data synchronization. P2P SIP may also be used as failover in case of server crashes.

Last but not least, P2P SIP can be an alternative for internal VoIP networks in small companies. Often, these companies do not have the administrative staff to manage a central SIP-based system.

1.6 Standardization

Currently, there is a lot of work on standardizing P2P SIP. The P2P SIP working group within the IETF developed a first Internet draft on basic concepts and terminology regarding P2P SIP [19]. The document includes high-level descriptions of relationships between network entities, a reference model and a discussion of open problems that should be addressed within the working group. The document is expected to provide the general framework for P2P SIP.

Besides, there are many efforts leading to standardization of concepts regarding P2P SIP. The Web site p2psip.org provides a current overview of these efforts.

1.7 Conclusion

To sum up, there are two general approaches to integrate P2P mechanisms into standard SIP. Either the SIP protocol itself or the location service behavior is extended. The SIP extension leads to protocol changes, which prohibits its use in environments with standard SIP entities. Adapter peers, which act as some kind of gateway between P2P SIP and standard SIP environments, are a solution to this. The location service extension to a P2P-LOC leads to P2P-PROXYs, which are compatible to standard SIP and thus can be seamlessly integrated into a standard SIP environment. The direct integration into a P2P-UA leads to changes to the user agent's code.

Currently, there are many efforts made to standardize P2P SIP. P2P SIP will have its field of application as Skype has already shown that VoIP using P2P mechanisms works, even within the Internet. The great strength of P2P SIP is the reduction of administrative efforts by eliminating central servers.

References

[1] Rosenberg, J., H. Schulzrinne, G. Camarillo, A. Johnston, J. Peterson, R. Sparks, M. Handley, and E. Schooler. 2002. SIP: Session Initiation Protocol. IETF RFC 3261, June.

[2] 3GPP. 2006. IP multimedia call control protocol based on Session Initiation Protocol (SIP) and Session Description Protocol (SDP). TS 24.229 V7.4.0, June.

[3] Skype Limited. 2008. Skype. http://www.skype.com, July 21.

[4] Baset, S., and H. Schulzrinne. 2004. An analysis of the skype peer-to-peer internet telephony protocol. Columbia University Technical Report CUCS-039-04, New York: NY, December.

[5] Bryan, D.A., B.B. Lowekamp, and C. Jennings. 2007. dSIP: A P2P approach to SIP regis- tration and resource location. IETF draft-bryan-p2psip-dsip-00, February.

[6] Schmidt, H., T. Guenkova-Luy, and F.J. Hauck. 2007. A decentral architecture for sip-based multimedia networks. In *KiVS* '07, 63–74. Springer: Informatik aktuell, February.

[7] Stoica, I., R. Morris, D. Karger, F. Kaashoek, and H. Balakrishnan. 2003. Chord: A scalable peer-to-peer lookup service for internet applications. *IEEE/ACM transactions on networking* 11(1): 17–32.

[8] Zangrilli, M., and D. Bryan. 2007. A chord-based DHT for resource lookup in P2PSIP. IETF draft-zangrilli-p2psip-dsip-dhtchord-00, February.

[9] Singh, K., and H. Schulzrinne. 2005. Peer-to-peer internet telephony using SIP. In *NOSSDAV '05*. Washington: Skamania, June.

[10] Singh, K., and H. Schulzrinne. 2006. Using an external DHT as a SIP location service. Columbia University Technical Report CUCS-007-06, New York: NY, February.

[11] Rhea, S., B. Godfrey, B. Karp, J. Kubiatowicz, S. Ratnasamy, S. Shenker, I. Stoica, and H. Yu. 2005. OpenDHT: A public DHT service and its uses. In *SIGCOMM '05*, 73–84, New York, NY, USA: ACM.

[12] Rhea, S., D. Geels, T. Roscoe, and J. Kubiatowicz. 2003. Handling churn in a DHT. Technical Report UCB//CSD-03-1299, University of California: Berkeley, December.

[13] Gong, L. 2001. Industry report: JXTA: A network programming environment. *IEEE Internet Computing* 5(3): 88–95.

[14] Eberspächer, J., R. Schollmeier, S. Zöls, and G. Kunzmann. 2004. Structured P2P networks in mobile and fixed environments. In *HET-NETs '04*. Ilkley, UK, July.

[15] Schmidt, H., T. Guenkova-Luy, and F.J. Hauck. 2006. Service location using the session initiation protocol (sip). In *ICNS '06*. IEEE Computer Society, July.

[16] Seedorf, J. 2006. Using cryptographically generated SIP-URIs to protect the integrity of content in P2P-SIP. In *3rd annual VoIP security workshop*, June.

[17] Rosenberg, J., J. Weinberger, C. Huitema, and R. Mahy. 2003. STUN— Simple traversal of User Datagram Protocol (UDP) through Network Address Translators (NATs). IETF RFC 3489, March.

[18] Bryan, D.A., E. Shim, and B.B. Lowekamp. 2007. Use cases for peer-to-peer session initiation protocol. IETF draft-bryan-p2psip-usecases-00, July.

[19] Bryan, D., P. Matthews, E. Shim, and D. Willis. 2007. Concepts and terminology for peer to peer SIP. IETF draft-ietf-p2psip-concepts-01, November.

2

SIP: Advanced Media Integration*

Marisol Hurtado
i2CAT Foundation

Andre Rios, Antoni Oller, and Jesus Alcober
Polytechnic University of Catalonia

Sebastia Sallent
i2CAT Foundation

CONTENTS

2.1 Introduction 15
2.2 Building Advanced Media Services 16
2.3 Service-Oriented Architecture (SOA) 25
2.4 MPEG-21/SIP-Based Cross-Layer 28
2.5 Service Convergence Platform of Telecom Operators 28
2.6 SIP and the Future of the Internet 31
2.7 Experimental Results 33
    References 40

2.1 Introduction

The explosive growth of Internet Protocol (IP) networks capable of providing sophisticated convergent services based on multimedia applications has increased the development of new services as an alternative to traditional communication systems. In this way, the service convergence is growing under an IP-based provisioning framework and controlled mainly by lightweight application level protocols like Session Initiation Protocol (SIP) [1].

SIP, a key technology enabling this evolution, is creating new possibilities to develop multi-modal communications. SIP increases the possibilities of developing integrated multimedia services and offers advantages such as independence from the underlying transport network, session establishment with the involvement of multiple devices and the support for mobility.

*This research was supported by the i2CAT Foundation and the MCyT (Spanish Ministry of Science and Technology) under the projects TSI2005-06092 and TSI2007-66637-C02-01, which are partially funded by FEDER.

End-users need request, find and consume advanced media services, regardless of terminal capabilities and location. SIP is an effective way to do this, not by providing services, but just primitives that can be used to implement them.

Fundamentally, SIP is a control protocol that offers additional capacities such as user location, availability, and capabilities. SIP depends on relatively intelligent endpoints that require little or no interaction with servers. Each endpoint manages its own signaling, both to the user and to other endpoints. There is a separation between signaling and media—any kind of media supported in next generation networks can be signaled, such as uncompressed high definition or 3D video application.

But the real powerful advantage of SIP is that, with the help of companion protocols, it can provide an Internet-style plethora of continuous media, value-added services, such as conferencing, instant messaging, or even streaming and gaming. This concept is named SIP-CMI (SIP-based continuous media integration) [2].

On the other hand, the non-continuous media, value-added services can be provided by Service-Oriented Architecture (SOA) [3]. Furthermore, a combination of SIP and Web services can fulfill all the needs of the current and future Internet user. Web services and SIP services are the paradigm of converged communication services [4].

Users demand personalized services that can be accessed from any terminal within any network. Therefore, service providers must deploy more advanced services with shorter time-to-market and development cycles in heterogeneous networks. SIP and SOA have characteristics that satisfy these criteria, and this is the main reason SIP has been selected by the Third Generation Partnership Project (3GPP) as a major component of IP Multimedia Subsystem (IMS).

This chapter explains the SIP capabilities of building advanced media services and the possibilities of performing advanced services in a SIP environment.

In Section 2.2, the SIP application server and the control of media servers are explained as the basis of value-added service implementation and delivery. After that, in Section 2.3, the SIP SOA paradigm is described. In Section 2.4, an important feature of SIP, flexible control and management mechanisms based in SIP, such us cross-layering, are drawn. Section 2.5 covers the foreseeable role of SIP in the Internet of the future. This chapter ends with experimental results of these concepts, developed within the scope of the i2CAT Foundation.

2.2   Building Advanced Media Services

From the architectural point of view, a SIP platform can be composed of elements grouped in two categories: clients and servers.

SIP clients can act not only as User Agent Client (UAC), but also as User Agent Server (UAS). For instance, they can be a SIP terminal, (e.g., a VoIP telephone in the case of IP telephony), and SIP gateways. These are devices that provide services such as formats translation and communication procedures between SIP terminals and non-SIP terminals, or transcoding between audio and video codecs.

Commonly, SIP servers work as proxy, redirect, and registrar. Besides these types, two other servers have special relevance to developing advanced media services: media servers and applications servers.

Media servers allow the possibility of having immediate access to any audiovisual (audio and video) content stored in their repository. When users request this type of services, they should know, only and exclusively, what service to request or with whom they want to communicate. With current paradigms, when users ask for access to streaming content, they should know which server will provide them this content. They should know the characteristics and present conditions of their network to request the exact quality that it should be able to access. Users see the network as an information transport system which should be adapted if they want to use it.

In case of continuous media services (CM), and in the special case of streaming services, SIP needs to be extended with media server control mechanisms. In this way, there are works in progress in IETF and W3C that propose an architectural framework of media servers control. On the other hand, there are other proposals focused on exploiting the cooperation between SIP and RTSP.

The latter, the applications servers, are components in charge of external applications execution, which adds intelligence to the services. Examples of advanced services in a VoIP environment are call filtering, destination blockage, call forwarding, etc. The number of services deployed in this environment has no limit and service providers have been forced to offer a high level of personalized services according to the necessities of a user or group of users. Furthermore, service providers look for a service delivery platform environment that minimizes the time-to-market before the incorporation of new services. Advanced mechanisms to implement such services can be provided by the Java community and the open source software community.

In the following, approaches of media server control and application servers are explained in more detail.

2.2.1 Media Server Control Mechanisms

A media server is a media processing network element that performs all the media processing for advanced services. It is controlled by the application server, which provides a service execution environment, application-specific logic, and all the signaling for one or more services.

Application servers and media servers work together in a client–server or master–slave relationship. The application servers provide the service logic for each specific application and the media server acts as a media processing resource for the applications.

There is a clear tendency to standardize this control, and currently there are two main approaches: SIP-based and Java-based.

```
<?xml version="1.0" encoding ="UTF-8"?>
<ccxml>
   <var name="theState" expr=" start" />
   <eventprocessor state ="theState">
     <transition state ="start" event="conn.alerting" name="evt">
        <assign name="theState" expr="connected"' />
        <accept connectionid="evt.conn.connectionid"/>
     </transition>
     <transition state="connected" event="conn.disconnected"
                    name="evt">
       <exit/>
     </transition>
   </eventprocessor>
</ccxml>
```

FIGURE 2.1
CCXML for an incoming call.

In the SIP domain, several SIP-based control markup languages extend SIP capabilities, such as VoiceXML and CCXML [5], MSCML [6], MSML [7], and MEDIACTRL [8]. In the Java-based approach, the main programming interface is the Media Server Control API (JSR 309) [9].

2.2.1.1　CCXML and VoiceXML

Call Control eXtensible Markup Language (CCXML) is a W3C Working Group XML-based standard designed to provide telephony support to VoiceXML. CCXML is designed to inform the voice browser how to handle the telephony control of the voice channel, whereas VoiceXML is designed to provide a Voice User Interface to a voice browser.

CCXML with SIP integrates the CCXML browser, dialog servers (IVR platforms), and conference bridges.

CCXML defines an XML document that describes a VoIP service in terms of its state diagram and its transitions. When an event occurs in the signaling layer, a CCXML interpreter maps the diagram transitions in call control actions. CCXML offers a higher abstraction level for programming complex services. Figure 2.1 shows an example of service implementation that automatically answers incoming calls and ends its execution when the caller hangs up.

As depicted in Figure 2.2, several entities can appear in a CCXML session:

- Connection. It represents the endpoints of a communication

- Conference. It manages the mixing of audio video streams when there are several participants in a multiconference

- Voice Dialogs. They are associated with an existing Connection or Conference and they allow executing media actions in a media server, for example, music on hold, DTMF, IVR, Early Media, etc

CCXML brings Web and XML advantages to call control and provides advanced call control features that VoiceXML lacks. It allows flexible development

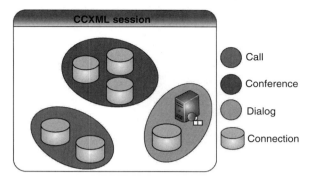

FIGURE 2.2
CCXML entities.

of applications such as multi-party conferencing, call center integration, and follow me and find me services. Its main features are:

- Basic call control features such as answer, create call, disconnect, reject, and redirect

- JCC/JAIN-based call model

- Flexible asynchronous event model

- Multi-party conferencing

- IVR/Dialog integration (VoiceXML, SALT, among others)

2.2.1.2 MSCML

The Media Server Control Markup Language (MSCML) is an IETF RFC proposal aimed at enabling enhanced conference control functions such as muting individual callers or legs in a multi-party conference call, the ability to increase or decrease the volume, and the capability of creating sub-conferences. MSCML also addresses other feature requirements for large-scale conferencing applications, such as sizing and resizing of a conference. MSCML implements a client server model to perform the media control server.

2.2.1.3 MSML

The Media Session Markup Language (MSML) is an IETF Draft proposed as an extension to SIP that aims to control media servers with dialogs through a master-slave scheme. MSML makes use of XML content in the message bodies of SIP INVITE and INFO requests, without modifying the SIP protocol in any way.

2.2.1.4 MEDIACTRL

Media Server Control (MEDIACTRL) is a new IETF working group formed to create a standard protocol for controlling media servers. The group intends

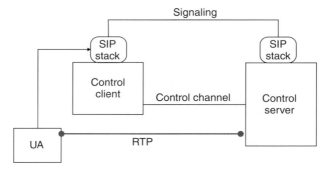

FIGURE 2.3
Illustration of a simplified view of the proposed mechanism.

to focus on the use of SIP and XML for its protocol suite. Figure 2.3 shows their proposal of a framework and protocol for application deployment where the application logic and processing are distributed.

Control channels are negotiated using SIP standard mechanisms that would be used in a similar manner to create a SIP multimedia session.

It highlights a separation of the SIP signaling traffic and the associated control channel that is established as a result of the SIP interactions.

The use of SIP for the specified mechanism provides many inherent capabilities which include:

- Service location. Use SIP Proxies or Back-to-Back User Agents (B2BUA) for discovering Control Servers

- Security mechanisms. Leverage established security mechanisms such as Transport Layer Security (TLS) and Client Authentication

- Connection maintenance. The ability to re-negotiate a connection, ensure it is active, audit parameters, etc.

- Agnostic. Generic protocol allows for easy extension

A CONTROL message is used by the Control Client to invoke control commands on a Control Server.

2.2.1.5 Media Server Control API (JSR2309)

The goal of JSR 309 is to specify an API that standardizes access to external media server resources and to provide their multimedia capabilities from services built on application servers. JSR 309 defines a mechanism to provide:

- Media network connectivity to establish media streams

- IVR functions to play/record/control multimedia contents from a file or streaming server

- A way to join/assemble IVR function to a network connection, following any topology, to compose conferences, bridges, etc.

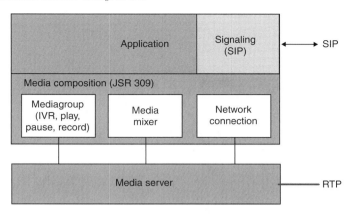

FIGURE 2.4
Media control architecture.

The JSR 309 API may be structured following three main sets of functionalities, drawn in Figure 2.4.

2.2.1.6 SIP and RTSP Synergy

SIP and RTSP [10] are both related to each other, but they are independent protocols that allow initiating and controlling of stored, live, and interactive multimedia sessions in the Internet. The former is used mainly for inviting participants into a multimedia session in order to initiate, for instance, a multi-party conference. The latter is used to control playback and recording for stored continuous media, in order to offer a video-on-demand like service. However, in both scenarios, these protocols can work separately. The improvement of the IP networks in terms of bandwidth capacity and QoS, and their expansion in terms of reachability and value addition are increasing the number and variety of deployment scenarios for streaming media applications. These scenarios impose new requirements on signaling protocols in order to offer more flexible and scalable applications. In this sense, cooperation between SIP and RTSP can be key in creating new collaborative applications. Due to that, their synergy expands the number of media control possibilities not available with only one of them.

Currently, there is no integrated solution but there are some approaches for SIP and RTSP convergence. For instance, in [11], a proxy architecture based on collaborative session transfer is proposed and, in [12], a merged scheme is described, where the re-use of RTSP as a stream control is proposed. This control is negotiated by SIP/SDP in order to keep the compatibility with existing streaming solutions. Another option is to use a gateway to translate SIP messages in RTSP and vice versa. This last scheme is described in Figure 2.5.

2.2.2 SIP Application Servers

SIP can also be used to provide more advanced and rich services. Application servers are those that provide added functionalities. SIP application servers

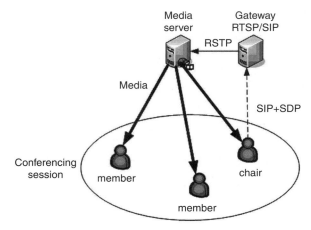

FIGURE 2.5
Internet conferencing example combining SIP and RTSP.

manage incoming requests and run associated specific applications. Services can be customized by the parameters on specific fields of the request, origin information, or other variables.

Currently, several initiatives aim at providing advanced mechanisms to implement SIP services. Initiatives within the community of Java developers and in the open source community are highlighted.

2.2.2.1 Java-Based Approach

Java for Advanced Intelligent Networks (JAIN) [13] aims at standardizing different APIs that enable the fast development of next generation Java-based communications. SIP has been covered by the JAIN initiative, producing de JAIN-SIP API. This API encapsulates a SIP stack behind some well-defined interfaces.

Given the amazing evolution of dynamic Web-services, the SIP community has also adopted one of the most widespread environments to create and deploy SIP services. It is called the Servlet programming API, specified within the Java Community Process. The next section will describe the JAIN initiative and the SIP Servlet Programming model.

The JAIN initiative has defined a set of Java technology APIs that enable the rapid development of next generation Java-based communications products and services for the Java platform.

Within the JAIN community, JAIN SIP is the part in charge of implementing interfaces for the use of the SIP protocol. The JAIN SIP specification provides functionalities from RFC 3261 (SIP) and the RFCs with SIP extensions. Some of these features are methods to format the messages, application ability to receive and send messages, analysis ability about incoming messages by accessing their fields, transport of control information generated at a meeting (RFC 2976), and instant messaging (RFC 3428). Other RFCs that complement the SIP specification are RFC 3262, 3265, 3311, 3326, and 3515.

Jain SLEE [13] is the Java standard for Service Logic Execution Environment (SLEE). SLEE is defined as runtime services with low latency, high performance and asynchronous event oriented; i.e., basic telecommunication services characteristics.

Jain/Java Call Control (JCC) [13] is a specification (JSR 21) that abstracts the creation of VoIP services from the underlying protocol. It supports different signaling protocols such as SIP, SS7, H.323 or MGCP. Its operation philosophy is based on subscription to the receipt of different events associated with the following network roles.

The SIP Servlet API is a specification of the Java community that defines a high-level API for SIP servers, enabling SIP applications to be deployed and managed based on the well-known HTTP Servlet model.

SIP Servlets achieve their highest potential when they can be deployed along with HTTP Servlets in the same Application Server (also known as Servlet Container). This environment, which combines SIP and HTTP, truly realizes the power of converged voice/data networks. The HTTP protocol represents one of the most powerful data transmission protocols used in modern networks (think of the SOAP Web-services protocol). At the same time, SIP is the protocol of choice in most modern and future voice over IP (VoIP) networks for the signaling part. Therefore an Application Server capable of combining and leveraging the power of these two APIs will be the most successful, as is shown in Figure 2.6.

Convergence of SIP and HTTP protocols into the same Application Server offers, among others, a key advantage. It is not necessary to have two different servers (HTTP and SIP), so it reduces from maintenance problems, and eases user and configuration provisioning.

Several SIP Application Servers have been developed such as WeSIP AS [14], implemented by Voztelecom company and Broadband Research Group from Universitat Politecnica de Catalunya (UPC), IBM Websphere, BEA WebLogic SIP Server or Glassfish, among others.

2.2.2.2 Developing Advanced Media Services Using Open Source Tools

Open SIP Express Router (OpenSer) [15], Asterisk [16] and RTP Proxy are the three key components to build an open source VoIP platform.

OpenSer is one of the most widely used SIP Proxies, and is usually the entry point for most SIP networks, providing authentication, registration, and other basic functionalities.

It's a VoIP server based on SIP used to build a large-scale IP telephony infrastructure. It can operate as Registering, Proxy, Redirect, etc. When operating with SIP standard, it makes interoperability with other SIP systems (from other vendors) easy. OpenSer is composed of a core system and several modules, with support of presence, authentication by AAA (e.g., RADIUS),

FIGURE 2.6
SIPServlets API.

remote calls (XML-RPC) and Call Processing Language (CPL). It also offers a Web-based interface application/server where it can monitor the state of the server and negotiate all features.

The OpenSer configuration file is more than just a typical configuration file. It combines both static settings and a dynamic programming environment. In fact, configuration file is a program that is executed for each message received by the OpenSer.

Also, OpenSer uses advanced techniques for executing programmable logic. One of them is called Call Programming Logic (CPL). CPL scripts are XML-based documents. They describe operations that the server performs upon a call setup event. CPL scripts can reside on a SIP proxy server, an application server, or intelligent agent. Figure 2.7 shows how SER works with CPL.

Asterisk is an open source IP PBX, composed of several modules that operate as a simple PBX, Gateway or Media Server. Asterisk has a General Public

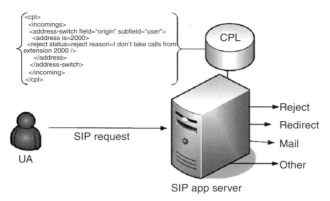

FIGURE 2.7
Call processing logic.

License (GPL) and although it was originally developed for the Linux operating system, it currently works on other operating systems.

Asterisk has many features only available in expensive PBX systems such as conferences, IVR, voice mailbox, automatic distribution of calls, etc. It's possible to add new features on its contexts collection, called dialplan, which is written in the characteristic script language of Asterisk, and to add modules written in C language or in another programming language supported by Linux. Asterisk can be installed on a traditional computational platform working as a gateway interconnecting the PSTN and IP environments.

RTP Proxy is a server that operates together with the SER and any proxy server, solving the topic of NAT Traversal with the appropriate handling of ports. One of the most used is Portaone RTP Proxy.

2.3 Service-Oriented Architecture (SOA)

Service-Oriented Architecture (SOA) is an architectural pattern for creating business processes organized as services. It provides an infrastructure that allows applications to exchange information and participate in business processes. While the SOA concept is general and is based on the idea of any type of message provision of service, it is most often used in Web environments and based on Web standards [17], such as HTTP, XML, SOAP, BPEL, WSDL, or UDDI.

SOA is characterized by the following properties:

- A logic view. It is an abstraction of services such as programs, databases, and business processes within the application. They are defined in terms of what they do. Thus, its basic components are the services that implement the business logic and the basic functionality of the system.

- Message oriented. As part of the services description, the exchanged messages between suppliers and applicants are defined. The internal structure of the service (programming language, internal processes, etc.) remains hidden at this level of abstraction.

- Description oriented. A service is described in processable metadata. The description supports the public nature of SOA. Only those public details are included in the description and are important for the use of the service. Thus, a simple Web service is characterized by four standards: XML, SOAP, WSDL, and UDDI, which work according to the basic model request / response.

- Granularity. Services tend to use a small number of operations with relatively complex messages.

- Network oriented. Services tend to be used over a network, although this is not a prerequisite.

- Platform independent. Messages are delivered in a standard format and neutral to the platform (usually XML).

SOA is associated with a client–server relationship between software modules, where services are subroutines serving customers. However, not all environments and software topologies fit this model, such as advanced multimedia services. With SOA 2.0, an event-oriented architecture is deployed, in which software modules are related to business components, and provide alerts and notifications events.

2.3.1 SOA and SIP Convergence

The goal of the SOA and SIP convergence is to offer an infrastructure to implement services with high complexity using collaborative services or an orchestration of services (e.g., in the form of a BPEL process [18]). They can be implemented in different environments, programming languages, and platforms and can be located in different service providers.

In the case of advance media services, it takes advantage of the power of SOA and SIP convergence, because of the SOA. This paradigm can be applied with Web service orchestration, transparent integration with other services and integration with continuous media services such as session and stream control, multimedia services composition, and call signaling, as shown in Figure 2.8. SIP works as an application layer signaling protocol and media control in parallel with SOAP in the data plane, as depicted in Figure 2.9.

In the specific case of the Web services, the service provider develops their application and provides a Web Service Description Language (WDSL) interface that describes the capabilities of the service. The location of the service is stored in a Universal Description Discovery Integration (UDDI) repository that manages the publication, search, and discovery of services.

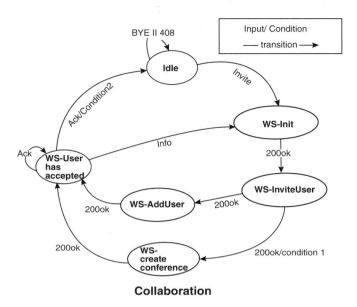

FIGURE 2.8
SIP/Web services collaboration.

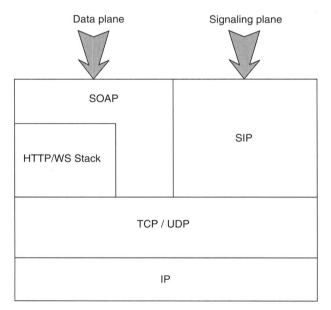

FIGURE 2.9
Data plane and control plane.

2.4   MPEG-21/SIP-Based Cross-Layer

In recent years, network connections are becoming ubiquitous and, therefore, people expect to access multimedia content anywhere and anytime. One approach to solve this requirement is to use a technique known as session mobility, which allows a user to maintain a media session even while changing terminals. For example, upon entering the office, a person may want to continue a session they initiated on a mobile device on their desktop PC. However, in order to develop this solution, it is necessary to consider that the delivery and adaptation of multimedia contents in distributed and heterogeneous environments requires flexible control and management mechanisms in terminals and in control entities inside the network. On one hand, there is MPEG-21 Digital Item Adaptation (DIA), which provides normative descriptions for supporting adaptation of multimedia content, but does not define interactions with transport and control mechanisms. On the other hand, there is SIP, which is an application-layer protocol used for establishing and tearing down multimedia sessions. It uses Session Description Protocol (SDP) which is a format for describing streaming media initialization parameters. Recently, an evolution of SDP, SDPng [19], was proposed in order to address the perceived shortcomings of SDP.

Both technologies come from different worlds, and in order to develop a solution for advanced signaling, it is important to reach interoperability between the IETF approaches on multimedia session establishment and control, and the MPEG-21 efforts for multimedia streaming and adaptation to bring advanced multimedia service provisioning and adaptation services to the customer. Some proposals that join MPEG-21 and SDPng exist [20], [21], but they are at draft level and their focus is general, without detailing implementation aspects in solutions such as session mobility. The main idea is to contribute to the XML model in which MPEG-21 and SDPng are based.

There are proposals where MPEG-21 is used to provide a common support for implementing and managing end-to-end QoS [22]. As Figure 2.10 shows, MPEG-21 is proposed as a Cross-Layer Adaptation.

2.5   Service Convergence Platform of Telecom Operators

Nowadays, there are different initiatives for the integration of telecommunications networks. They share the common objective of defining platforms that allow the development of new and complex services that can be used by end users independent of the transport network used. In this way, software developer communities will be allowed to operate the characteristics of the present

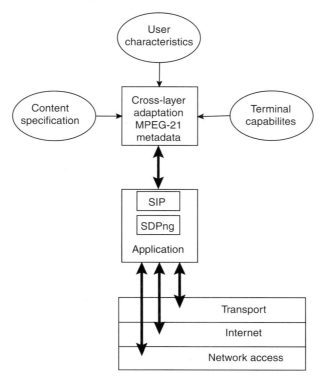

FIGURE 2.10
MPEG-21 enabled cross-layer interaction.

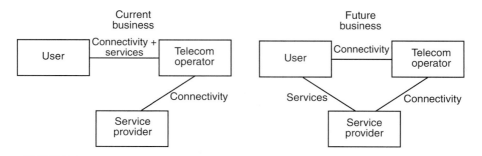

FIGURE 2.11
The telecommunication business model.

and future telecommunications networks for the benefit of telecom operators and service providers. Therefore, the goal is to define a new framework that makes it possible to offer new business models for application development and telecommunication services, creating new business models and sources of incomes for network operators, as shown in Figure 2.11. The current user demands services that cannot always be offered by the operator whose main product is connectivity. In this new model, application service providers (ASPs) can easily integrate their solutions on the network of the operator. All of them lead to a rise in traffic and in a higher number of user services. As a result, the business of telecommunications opens up to other actors.

The current trend for access to these value-added services is based on the concept of service-oriented architectures. The idea of these architectures is the definition of a framework that makes standardized access to the different available resources in the network possible. This allows for integration and interoperability among services built from different vendors, which can be built in different technologies and deployed in different networks. One of the most prominent technologies in this field is the so-called Web services.

In the case of implementing telephony services, video transfer or real-time communications, called continuous media services, the help of specific protocols is necessary, such as SIP, SDP, RTP, and RTSP. These protocols provide the specific transport and signaling needs of the media content. In this sense, within the New Generation Networks (NGN) scenario, a new set of standards known as IP Multimedia Subsystem (IMS) is gaining popularity.

IMS is an evolving architecture for providing voice, video, and other advanced multimedia services to mobile and fixed environments. It promises to have an important impact on the whole of the telecommunications industry, including network providers, applications developers, and end users. Though IMS standards come from 3GPP, they have also been adopted by 3GPP2 and TISPAN (a European organization focused on next generation fixed networks). IMS can be considered the basis for all variants of the next generation network (NGN) envisioned by the traditional telecommunication industry.

The key technology behind IMS is SIP, which was chosen as the underlying protocol in many of the important interfaces between elements in an IMS-based network. IMS specification is available from the 3GPP Website [23]. The reason for this choice is that SIP presents several advantages to building new features or services. SIP is simple because it is based on a straightforward request-response interaction model, making life simple and comprehensible for developers. It is extensible because it can set up sessions for any media type. It is also flexible, making it possible to interact with individual protocol messages without breaking anything. Finally, it is well-known because it is based on HTTP and other IETF standards, using many Web-based technologies to develop innovative SIP applications.

IMS basically comprises three layers: the Connectivity Layer, the Control Layer, and the Service Layer. Figure 2.12 shows a simplified view of the layered architecture of IMS. The connectivity layer comprises devices, such as routers or switches, for backbone and network access (multiple types of network access are allowed). The control layer comprises network control servers for managing call or session set-up, modification, and release. Several roles of SIP servers, collectively called Call Session Control Function (CSCF) are used to process SIP signaling packets in the IMS. In the security and user subscription functions, are the Authentication Center (AuC) and the HSS (Home Subscriber Server), which communicate with CSCF in order to provide security

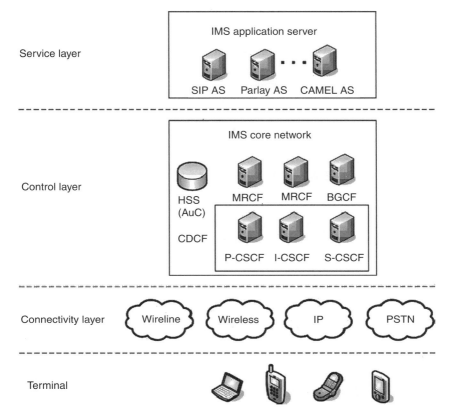

FIGURE 2.12
IMS architecture overview.

management for subscribers. The interconnections with other network operators and/or other types of networks are handled by border gateways (BGCF). Finally, the service layer comprises application and content servers to execute value-added services for the user. Generic service enablers, as defined in the IMS standard, are implemented as services in a SIP application server, which is the native server in IMS.

2.6 SIP and the Future of the Internet

There are efforts of an Internet redesign after thirty years of great success. GENI [24], FIND [25], FIRE [26] are among the major initiatives that are set to address the urgent need of defining research programs for achieving this goal. A need for new architectures and protocols is envisioned. However, it is

foreseen that SIP and its extensions accomplishes with the major parts of the
expected requirements for signaling and management planes.

The content for future Internet will be able to be adapted to the users with
respect to their preferences, terminal capabilities, and access networks. The
nature of this content will be 3D, including haptic features. Now, of the five
senses (hearing, sight, taste, touch, and smell), only the first two are used for
the Internet, the rest will appear progressively. This multimodality will be
interactive to achieve the maximum level of collaboration among users.

To accomplish this new featured content, the network should be content-
and service-centric, regardless of the location of the end user, and must transport real-time 3D multi-modal media to multi-functional devices.

SIP decouples the media from the signaling, and if the network is able to
transmit this media, SIP can signal it. The underlying idea is because SIP
uses companion protocols to perform its functions, and if, up until now, SIP
used SDP to transport media information, in the future, it will be SDPng with
embedded XML that provides this function, with an evolution of MPEG-21
performing a five-senses-content-exchange-framework, for example.

On the other hand, the capacity of networks has grown during recent years.
New generations of optical networks are enabling many different services that
until now were not accessible to users. Video engineers need to transmit video
signals, such as plugging wires locally, and users want to consume this studio-
quaility video in cultural events or health scenarios. SIP can provide the signaling framework for the transmission of these standardized digital interfaces,
such as HDSDI, SDTI, SDI, and Firewire (IEEE 1394), even if the transport of these signals is carried out by a new media network architecture (see
Figure 2.13). In this case, SIP would be able to signal these current and future
media contents, due to its decoupled nature from media. In the case of SIP
the major involvement would emerge, since IP addresses, routing, and DNS
names, among others, are closely tied to an underlying structure evolving to
a clean-state Internet [27].

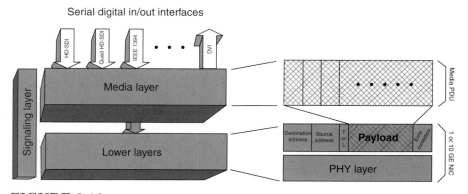

FIGURE 2.13
New media architecture and its signaling.

2.7 Experimental Results

The advent of high-speed fiber-optic networks has increased the capacity in bandwidth and the current supported high speed consolidates them as the ideal medium for high-bandwidth transmissions. Currently, the real-time transport of high quality video over IP has been successfully tested in research networks expanding the range of video formats to be used on IP networks and the improvement of new high-quality video services. Real-time transport of high-quality live video is one of the main current efforts in innovation and research of next-generation Internet organizations, such as the i2CAT Foundation. In i2CAT, a SIP-CMI testbed has been implemented as an add-on of its SIP platform (SIPCAT). The first one has provided the SIP-CMI subsystem and the second one, the SIP infrastructure.

In the following, several examples of real experiences are outlined, focusing on the concepts that have been explained previously in this chapter.

2.7.1 Example of Advanced Service: On-Line Transcoding

The goal of the platform is to deploy a system through which users can access high-quality contents from their terminal, regardless of its characteristics, by interacting with different modules in a transparent way from the users' point of view, by using Web services and SIP services.

Within this context, a content transcoding module was developed using the SIP-CMI approach. The main functionalities of the transcoding module are to provide:

- Real-time transcoding, in order to adapt to user specifications or QoS network limitations

- Full information for transcoders and media servers

- A list of available resources, such as high-quality videos or different coding schemes, among others

- Resource updating and inserting mechanisms

In order to develop this transcoding module, it has been necessary to implement different service elements and deploy the SIP-CMI platform to validate them. An adapter to be plugged in the media plane and on the control plane has been implemented, as shown in Figure 2.14.

As a proof of concept, related to the media plane, VideoLAN has been used as a media server and a VideoLAN control interface has been implemented (VideoLAN Service Control).

Related to the control plane, a driver for the VideoLAN Control Service and a transcoding service have been implemented, based on states machines. After that, they have been deployed on a SIP Application Server. As a result,

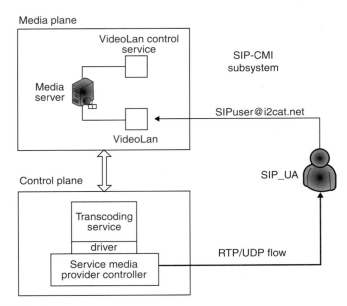

FIGURE 2.14
Transcoding service implemented on the SIP-CMI platform.

the transcoding module has become a video on demand HDV system. Current video on demand HDV systems [28] solve this control of the server by using Web applications to initiate a video stream from the transmitting unit to the receiving unit, with VideoLAN.

2.7.2   SIP Infrastructure

SIPCAT is a value-added services integrated platform based on SIP. As Figure 2.3 shows, this platform offers the main SIP components needed to develop the SIP-CMI approach. The main goal of this implementation is to offer a SIP test platform where a developer community (application service providers) and user community are testing SIP value-added services.

The SIPcat platform offers advanced communication services of voice, video, and data using several communication protocols such as H.323, PSTN, H.320, 3G domains, and SIP gateways.

All core components have been implemented using open source solutions and its core architecture has been enriched with components from commercial vendors, such as Radvision, Polycom, and Tandberg.

The components that integrate this configuration are the classic elements of the SIP framework. Proxy Server and Registrar are implemented by OpenSer. Additionally, the management services guarantee the proper operation of the platform. The management includes four fundamental aspects: location, accounting, security, and monitoring.

The location scheme is based on DNS, ENUM, LDAP, and H.350 protocols. The security includes the authorization and authentication performed by

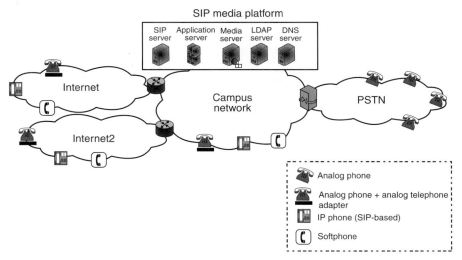

FIGURE 2.15
SIPcat scenario.

RADIUS. The monitoring is performed by automatic SNMP-based scripting and service control implemented using Nagios. The accounting is a proprietary design oriented to provide the information of resource utilization.

In SIPcat (see Figure 2.15), the dialing scheme was designed according to Internet2 testbed platforms: Videnet and SIP.edu. The H.323 dialing plan is compatible with the Global Dialing Scheme (GDS) and the SIP-URI is compatible with the e-mail contact addresses. The GDS is a new numbering plan for the global video and voice over IP network testbed, developed by ViDeNet [29].

This infrastructure is based on the above-mentioned Internet2 SIP.edu [30] experience. It promotes the convergence of voice and e-mail identities, increasing SIP-reachability, and builds community schools that are developing and deploying campus SIP services in the Internet2 community. Another similar experience is the Global University Phone System (GUPS) [31], using mainly Asterisk.

The service delivery is supported by providing different service platforms such as Web services, SIP Servlets [13], and Call Processing Language (CPL) [32].

The main element of the application and media control is the Application Server implemented through a SIP Servlets Framework. A framework has been designed for SIP-based services development. The design of the framework is based on a three-tier architectonic model.

First of all, the core component is a SIP Application Server implemented by a SIP stack based in Jain SIP [37]. Another main component is a SIP Servlets container.

The next layer is composed by the CallControl, which is responsible for the main control of the calls and the redirection to the respective service. The implementation follows the Front Controller and the Jain Java Call Control

Specification [13]. The second part is a group of components designed to facilitate the design and implementation of value-added services. They provide a
level of abstraction for the service implementation.

Over this middleware, the services are executed. A states machine-oriented
deployment is proposed in order to diminish the development of the services.

2.7.3   Example of SIP Cross-Layering: Lightpath Establishment

As explained in the cross-layering section, the session information contained in
SDP can be used by the application server to request, for example, a lightpath
establishment between the end points. This idea was deployed in Provisioning
and Monitoring of Optical Services (PROMISE) [33], a European co-operation
project under the CELTIC-EUREKA cluster, focused on supporting a range of
advanced high capacity services over an adaptable optical network infrastructure, which GMPLS protocols make possible. The main results of the project
were a control plane-enabled testbed to showcase the provisioning and maintenance of optical services in a real-world context. One of these services was
the provision of an optical path between a HD transmission unit and a HD
receiving unit, using the SIP-CMI approach. In order to establish this path,
the SIP Application Server (SMPC) had to request, using Web services, the
optical service before successfully establishing the SIP session (Figure 2.16),
and initiating the HD transmission.

2.7.4   Examples of High-End Videoconferencing

In recent years, video conferencing systems have been an important area of
telecommunications research, resulting in a large number of products that can
be used to transmit and receive video in real-time over IP networks. These
systems vary mainly in terms of quality and in the amount of bandwidth used
to transmit video over networks. In this sense, the telecom vendors and manufacturers have focused on developing low- to medium-bandwidth conferencing
systems, typically based on H.320 and H.323 standards. In addition to these
commercial products and systems, academic researchers and the university
community have also looked for a higher-quality and higher-bandwidth video
system that operates over high-speed research and educational networks (e.g.,
Internet2-Abilene Network).

One of the main key aspects of SIP signaling is the decoupling between
signaling and media. As a consequence, any kind of media can be signaled by
SIP, if SDP can support it.

Three examples are briefly explained about videoconferencing using new
media:

- Tele-immersion

- Digital video (DV)

- Uncompressed High Definition Video

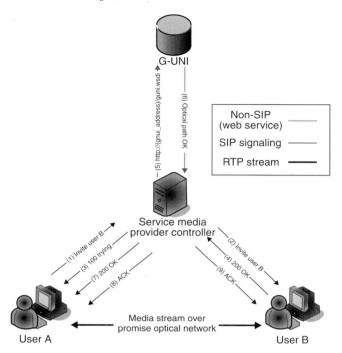

FIGURE 2.16
SIP session setup using GMPLS promise optical network.

Tele-Immersion For Applications Supporting New Interactive Services (TIFANIS) [33] was a European cooperative project under the CELTIC-EUREKA cluster, focused on designing and implementing a full tele-immersive system, with several trial scenarios. Basic SIP services to tele-immersive systems have been provided and tested by developing a Transmission Reception Subsystem (TRS) module and integrated in one of the cubicles (see Figure 2.17).

With the new generation of camcorders, a user can manage high-quality of video on its own. Editing tools are known and common among users. They only need to plug the digital interface of the camcorder into the PC and capture its signal.

Digital Video Transport System (DVTS) [34] and Ultragrid [35] are examples of software that receive these signals and transmit them directly to the network. If users can get enough bandwidth, they are ways to transmit high-quality video, and moreover, there is no compression delay. SIP can fill the gap to build a high-quality videoconferencing system (end-to-end).

DVTS is software that allows encapsulating DV format from a Firewire interface for transmission over IP networks, resulting in a high-quality DV stream that consumes roughly 30 Mbps of bandwidth.

UltraGrid is a high definition (HD) video conferencing and distribution system. It is also considered one of the first systems capable of supporting

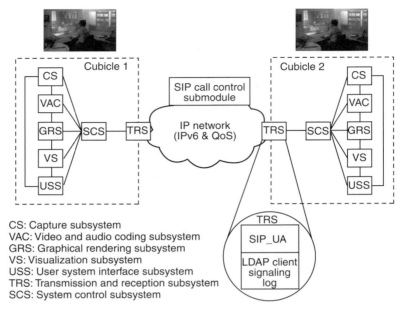

FIGURE 2.17
TRS subsystem and SIP-CMI subsystem in Tifanis project.

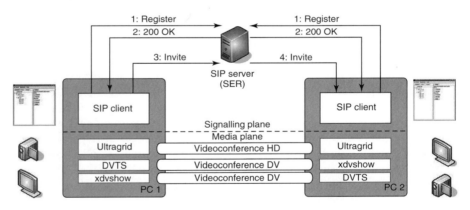

FIGURE 2.18
HD video conferencing with SIP messages.

uncompressed gigabit rate high definition video over IP. In fact, an Ultragrid node converts SMPTE 292M high definition video signals into RTP/UDP/IP packets, which can then be distributed across an IP network reaching transmission rates to 1.2 Gbps. There are few systems with these characteristics, such as iHDTV, i-Visto gateway [36].

The i2CAT Foundation research project called MACHINE has developed an open-source graphical SIP UA to establish and control high-quality

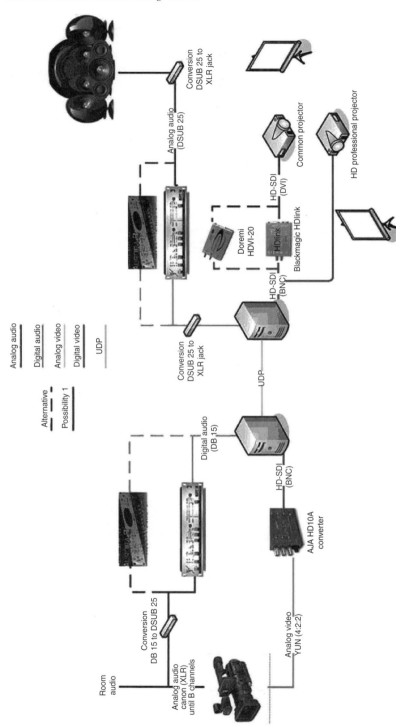

FIGURE 2.19
Ultragrid testing scenario.

videoconferencing, becoming a DVTS and UltraGrid videoconferencing system. Figure 2.18 shows the SIP message exchange to establish and terminate a basic videoconferencing session, using an OpenSer as a SIP server.

Figure 2.19 shows the current testing scenario with Ultragrid, to demonstrate that, regardless of the complexity of the media transport, SIP can signal it.

References

[1] Rosenberg, J., H. Schulzrinne, G. Camarillo, A. Johnston, J. Peterson, R. Sparks, M. Handley, and E. Schooler. 2002. SIP: Session Initiation Protocol. IETF RFC 3261. June.

[2] Hurtado, M., A. Oller, and J. Alcober. 2006. The SIP-CMI platform – An open testbed for advanced integrated continuous media services. TRIDENTCOM 2006. Barcelona, Spain, March.

[3] Hiroshi, W., and S. Junichi. 2006. A service-oriented design framework for secure network applications. In *AU: Proceedings of the 30th IEEE International Conference on Computer Software and Applications Conference* (COMPSAC). Chicago, September 17–21.

[4] Liu, F., W. Chou, L. Li, and J. Li. 2004. WSIP – Web service sip endpoint for converged multimedia/multimodal communication over IP. In *Web Services, 2004. Proceedings. IEEE International Conference*, 690–97. June 6–9.

[5] W3C. 2005. Voice Browser Call Control: CCXML Version 1.0, W3C Working Draft. http://www.w3.org/TR/ccxml/, June 29.

[6] Dyke, J.V., E. Burger, and A. Spitzer. 2006. Media Server Control Markup Language (MSCML) and Protocol. IETF RFC 4722. November.

[7] Saleem, A., Y. Xin, and G. Sharratt. 2007. Media Server Markup Language (MSML). Internet Draft (draft-saleem-msml-05). August 17.

[8] Melanchuk, T. 2008. An Architectural Framework for Media Server Control. Internet Draft (draft-ietf-mediactrl-architecture-02). February 5.

[9] JSR 309. "Media Server Control API", http://jcp.org/en/jsr/detail?id=309.

[10] Schulzrinne, H. 1998. Real Time Streaming Protocol (RTSP). IETF RFC 2326. April.

[11] Kahmann, V., J. Brandt, and L. Wolf. 2006. Collaborative streaming in heterogeneous and dynamic scenarios. ACM, Volume 49(11). Entertainment networking. November.

[12] Marjou, X., S. Whitehead, S. Ganesan, M. Montpetit, D. Ress, and D. Goodwill. 2006. "Session Description Protocol (SDP) Format for Real Time Streaming Protocol (RTSP) Streams", Internet Draft (draft-marjou-mmusic-sdp-rtsp-00), October 12.

[13] Java Community Process. 2002. http://www.jcp.org/en/jsr/detail?id=21, January 22.

[14] "WeSIP". 2006. http://www.wesip.com, July 17.

[15] "OpenSer". 2007. http://www.openser.org/, March.

[16] "Asterisk". 2004. http://www.asterisk.org/, January.

[17] Booth, D., H. Haas, F. McCabe, E. Newcommer, M. Champion, and C.F.D. Orchard. 2004. Web Service Architecture. In *W3C Recommendation* [Online]- http://www.w3.ord/TR/2004/NOTE-ws-arch-20040211/, November.

[18] Fu, X., T. Bultan, and J. Su. 2004. Analysis of interacting BPEL web services. In *Proceedings of the 13th International World Wide Web Conference (WWW'04)*. USA: ACM Press.

[19] Kutcher, Ott, Borman, "Session Description and Capability Negotiation", Internet Draft, (draft-ietf-mmusic-sdpng-08), February 20, 2005.

[20] Kassler, A., T. Guenkova-Luy, A. Schorr, H. Schmidt, F. Hauck, and I. Wolf. 2006. Network-Based Content Adaptation of Streaming Media Using MPEG-21 DIA and SDPng. *7th International Workshop on Image Analysis for Multimedia Interactive Services (WIAMIS06)*. Invited Paper to Special Session on Universal Multimedia Access (UMA), Seoul (Korea), April.

[21] Guenkova-Luy, T., A. Schorr, F. Hauck, M. Gomez, Ch. Timmerer, I. Wolf, and A. Kassler. 2006. Advanced Multimedia Management – Control Model and Content Adaptation. IASTED International Conference on Internet and Multimedia Systems and Applications (EuroIMSA 2006), Innsbruck (Austria), February.

[22] Toufik Ahmed, A., and I. Djama. 2006. Delivering Audiovisual Content with MPEG-21-Enabled Cross-Layer QoS Adaptation. *Packet Video 2006 published in J Zhejiang Univ SCIENCE A* 7(5): 784-93.

[23] 3GPP Website. 2002. http://www.3gpp.org/, October.

[24] GENI (Global Environment for Network Innovations). 2007. http://www.geni.net/, May.

[25] Fisher, D. 2007 US national science foundation and the future internet design. *SIGCOMM Comput. Commun. Rev.* 37(3): 85-87.

[26] Gavras, A., A. Karila, S. Fdida, M. May, and M. Potts. 2007. Future internet research and experimentation: The fire initiative. *SIGCOMM Comput. Commun. Rev.* 37 (3): 89-92.

[27] Balakrishnan, H., K. Lakshminarayanan, S. Ratnasamy, S. Shenker, I. Stoica, and M. Walfish. 2004. A layered naming architecture for the internet. In *SIGCOMM '04: Proceedings of the 2004 conference on Applications, technologies, architectures, and protocols for computer communications*. New York, NY, USA: ACM Press, 343–52.

[28] "VOD Services for High Definition Video Streaming over IPv6," Networked Media Lab., Gwangju Institute of Science and Technology (GIST). 2004. http://hdtv.nm.gist.ac.kr/v2/main.php, September.

[29] "ViDe // Vide Development Initiative". 2002. http://www.vide.net/, June.

[30] SIP.edu. 2003. http://mit.edu/sip/sip.edu/index.shtml, March.

[31] GUPS. 2005. http://www.aboutreef.org/gups-press.html, August.

[32] Lennox, J., X. Wu, and H. Schulzrinne. 2004. Call processing language (CPL): A language for user control of internet telephony services. IETF RFC 3880. October.

[33] Celtic Initiative. 2004. http://www.celtic-initiative.org/Projects/project-info.asp, May.

[34] DVTS, DV Stream on IEEE1394 Encapsulated into IP. 2004. http://www.sfc.wide.ad.jp/DVTS/, May.

[35] Perkins, C.S., and L. Gharai. 2002. UltraGrid: a high definition collaboratory. http://ultragrid.east.isi.edu/, November.

[36] i-Visto Internet Video Studio System for HDTV Production. 2006. http://www.i-visto.com/, April.

[37] "The Java Community Process (SM) Program – JSR 32: JAIN SIP API Specification, "http://www.jcp.org/en/jsr/detail?id=116.

3

SIP and IPv6 – Migration Considerations, Complications, and Deployment Scenarios*

Mohamed Boucadair and Yoann Noisette
France Telecom

CONTENTS

3.1   Introduction   43
3.2   Why Introducing IPv6 Could Be Problematic for
        SIP-Based Architectures   46
3.3   Business and Service Considerations Regarding
        IPv6 VoIP Migration   47
3.4   Handling Technical Issues Raised When Migrating
        SIP-Based Services to IPv6   58
3.5   Conclusions   67
        References   68

3.1   Introduction

3.1.1   General Context

Telephony over Internet Protocol (IP) has succeeded to federate a large number of researchers and engineers in both standardization and industrial fora. Among other results, several protocols, such as Session Initiation Protocol (SIP) (Rosenberg 2002) and H.323 (ITU-T H.323 1998) have been proposed and promoted for the deployment of IP-based telephony service offerings. These protocols have been introduced progressively into operational platforms. Recently, new architectures such as IP Multimedia Subsystems (IMS) (Camarillo and Garcia-Martin 2005) and TISPAN Telecommunications and Internet converged Services and Protocols for Advanced Networks (TISPAN) (TR180001 2006) have been proposed to meet service providers' requirements (mainly to implement the fixed-mobile convergence). Most current big service providers deployments are inspired from IMS and/or TISPAN architectures.

---

*Some parts of this chapter are based on "Migrating SIP-based Conversational Services to IPv6: Complications and interworking with IPv4," by M. BOUCADAIR and Y. NOISETTE which appeared in Proc. of Second International Conference on Digital Telecommunications (ICDT'07) © 2007 IEEE and "Towards Smooth Introduction of IPv6 in SIP-based Architectures," by M. BOUCADAIR and Y. NOISETTE, which appeared in Proc. Local Computer Networks (LCN'07) © 2007 IEEE.

In parallel and since the early nineties, service providers, manufacturers and standardization bodies have investigated the necessity of introducing a new IP to solve some critical problems encountered by the Internet architecture. After a large amount of standardization effort has been conducted, especially within the Internet Engineering Task Force (IETF), IPv6 has been adopted as the next generation IP connectivity protocol. Today, IPv6-related standards are judged mature enough to consider an operational deployment. Indeed, some service providers envisage adopting IPv6 as their new connectivity protocol not only for the abundance of IPv6 addresses, but also to take advantage of the routing hierarchy spirit of this new IP version (notably in order to reduce the size of routing tables) and to benefit from its auto configuration features. Furthermore, issues related to IPv6 migration strategies have been investigated and discussed exclusively in the IETF. Several working groups have thus been chartered to work on operational and migration mechanisms (e.g., v6ops Working Group). Nevertheless, this effort focused essentially on the migration of connectivity services (i.e., IP transfer capability) and not the services to be deployed over an IP infrastructure. Concretely, problems related to migrating "high level" services such as Voice over IP (VoIP) and TV broadcasting, built above an IP infrastructure composed of heterogonous nodes (i.e., IPv4 and IPv6 ones), didn't benefit from a large amount of investigations by the Internet community. The rare initiatives that have been launched were about Domain Name Service (DNS) and File Transfer Protocol (FTP) concerns. This effort dealt mainly with developing Application Level Gateways (ALGs).

From this perspective and as far as VoIP is concerned, service providers should evaluate the impact of IPv6 migration on their service offerings and associated architectures. This chapter tackles the case of telephony over IP service offerings and the impact of the introduction of IPv6 on deployed service platforms and architectures. Considering the panel of telephony over IP is diverse and rich, this chapter focuses only on SIP. Within this chapter, and if no explicit reference to IMS is quoted, a "SIP service" (or also "SIP-based service") denotes a SIP-based telephony service involving at least a SIP proxy server and a Registrar. In such a service, the proxy server and the Registrar have IPv4 or/and IPv6 connectivity. These functional entities could parse IPv4 and/or IPv6 addresses but could not use all of these IP versions to transfer IP data. A service provider offering a "SIP service" provides its customers with relevant information in order to reach SIP entities, especially the addresses of outbound and inbound proxy servers and the Registrar. These addresses could be IP addresses, or Fully Qualified Domain Name (FQDN), or the IP address of an intermediary node that will isolate the SIP service platform (such as a Session Border Controller (SBC) to ensure topology hiding of the SIP-based service architecture).

3.1.2   Beyond the State of the Art

IPv6 complications in a SIP environment (including IMS/TISPAN architectures) have not been studied deeply enough in the literature. Few papers and

Internet drafts have been edited to capture and to discuss related issues. For instance, Internet draft (Gurbani, Boulton, and Sparks 2007) proposes a set of recommendations to ease configuration of a SIP-based service that interconnects SIP user agents (UA) from both IPv4 and IPv6 worlds. Nevertheless, this effort isn't possibly viable in operational architectures since the adopted architecture is *a la* Internet and not *a la* IMS. A paper (Gurbani, Boulton, and Sparks 2007) identifies issues related to IPv6 porting that developers of SIP applications should take into account when introducing IPv6 support to SIP applications running on either dual-stack (i.e., supporting both IPv4 and IPv6) hosts or IPv4-only nodes. As far as IMS is concerned, (3GPP TR 23.981 2005) details some scenarios to interconnect IPv4 and IPv6 realms to enable heterogeneous sessions (i.e., involving participants attached to distinct realms (IPv4 and IPv6)). Unfortunately, (3GPP TR 23.981 2005) is not complete and proposed solutions are not optimal. This document is also used to undertake the effort documented in (Chen, Lin, and Pang 2005), which focuses mainly on Universal Mobile Telecommunications System (UMTS) networks. Finally, (Chen and Wu 2005) provides details about an IPv6-based SIP VoIP network experiment deployed in Taiwan. This paper does not handle issues related to interworking with IPv4 realm or how operational service offerings may be migrated to support IPv6. It focuses only on the usage of TElephone NUmber Mapping (ENUM) within a full IPv6-enabled SIP-based architecture.

As a step beyond the current state of the art, this chapter highlights issues related to migration of SIP-based conversational services to IPv6 and complications to interconnect with IPv4-enabled ones. A set of operational migration strategies are proposed associated with pertinent criteria to be taken into account by service providers in order to select a suitable migration scenario. This chapter does not recommend any particular IPv6 transition scenario (e.g., Tunnel Broker, Network Address Translation – Protocol Translation (NAT – PT), etc.) nor argue in favor of a given IP transport interconnection mechanism in order to interconnect IPv4 and IPv6 realms. Nevertheless, this chapter assumes that one possible transport interconnection mechanism is activated to allow communication between heterogeneous realms (i.e., between IPv4-enabled and IPv6-enabled networks). Additionally, this chapter introduces a novel procedure to allow SIP service platform to be aware of the type of the involved peers before forwarding session establishment requests so as to prevent from heterogeneous sessions failures and avoid invocation of adaptation functions. This procedure does not rely on "Layer 3"-related information but only on an "Application Layer"-related one. Owing to the support of this procedure, SIP proxy servers will be able to establish successful sessions between heterogeneous peers and optimize the invocation of ALGs modules. Moreover, a novel procedure called "blank SDP" is described within this chapter. This procedure aims at putting flexible means at the disposal of service providers to enforce their policies, especially to promote usage of IPv6 as the main transport protocol to convey traffic between SBC elements. More details about this new procedure are elaborated in this chapter.

3.1.3   Chapter's Structure

This chapter is structured as follows: The second section describes why introducing IPv6 into SIP-based architectures may be problematic. This section identifies a set of encountered technical issues. The third section discusses business and service concerns regarding IPv6 SIP-based architectures migration. Then, this it describes three viable migration scenarios. Furthermore, this section lists a set of pertinent criteria to be taken into account when envisaging adopting IPv6 as an IP connectivity protocol. This section also analyzes the technical issues related to each migration scenario. Technical solutions to solve identified technical hurdles are finally introduced in the fourth section.

3.2   Why Introducing IPv6 Could Be Problematic for SIP-Based Architectures

SIP and IPv6 have been envisaged at the early stages of IMS specifications. SIP is an IETF signalling protocol used for initiating, modifying, and closing multimedia sessions. SIP works jointly with Session Description Protocol (SDP) (Handley and Jacobson 1998). The latter is used to describe multimedia parameters such as IP address and port numbers that will be activated for Real-Time Transport Protocol (RTP) (Schulzrinne 1996) streams. SIP manipulates IP addresses in order to route signalling messages. These addresses can be found, for example, in the **Request URI**, **Contact**, **Via** headers or in the SDP headers such as **Originator**, **Contact** and **Media description** headers.

Since its first version, presented and documented in RFC 2547, the IPv6 address scheme is supported by SIP. SIP implementations should parse IPv6 addresses, which can even be stored as an Address of Record (AoR). Therefore, IPv6 protocol is not a problem per se; the challenge lies more in preventing session establishment failure when end users are not located in the same IP version routing plane.

The following call flow illustrates the results of a SIP session establishment request between heterogeneous user agents. For the sake of simplicity, we consider a dual-stack (DS) proxy server and two UAs "**A**" and "**B**." **A** is an IPv4-enabled UA and **B** is an IPv6-enabled one. During the registration process, **A** (respectively **B**) has registered its IPv4 (respectively IPv6) address as AoR. Let's suppose now that the IPv6-enabled UA **B** sends an **INVITE** message destined to the IPv4-enabled UA **A**. Figure 3.1 illustrates the beginning of the SIP messages exchange that occurs between all involved parties.

The SIP proxy server relays the received **INVITE** message to **A**. After receiving this message, **A** parses its SDP part and formulates an answer. This could lead to a closure of the application or simply to sending back an error message because the media line of the SDP part doesn't include any IPv4 address. As a consequence, it is impossible to set up a SIP session initiated by an IPv6-enabled user agent toward an IPv4-enabled user agent.

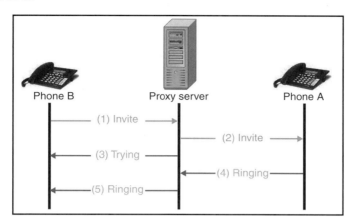

FIGURE 3.1

Example of heterogeneous call flow.

A new proposal dedicated to dual-stack user agents (i.e., owning both IPv4 and IPv6 connectivity), (Chen and Wu 2005) describes a mechanism allowing multiple alternative network addresses to be enclosed in a single SDP offer. This proposal consists of introducing a new attribute called Alternative Network Address Types (**ANAT**). This attribute allows inserting multiple media/connection lines in the same SDP offer (or SDP answer). IETF RFC 4092 (Schulzrinne 2002) defines how SIP can exploit the **ANAT** semantic by introducing a new option tag called "**sdp-anat**." This tag can be useful for SIP user agents to be aware of the capabilities of each other and then select from the supported media/network description lines the ones that are suitable for setting up the SIP communication (and also according to local preferences). A use case for illustrating the usage of this tag is a dual stack SIP user agent which can communicate either using its IPv6 or its IPv4 connectivity. This type of SIP user agent can also set a preference associated with each type of enclosed connectivity type. The proposal described in (Schulzrinne 2002) provides guidelines to ensure correct processing of this new tag by SIP user agents not supporting the **ANAT** attribute. But this procedure alone cannot lead to a successful call establishment between heterogeneous nodes. Consequently, enhancements to SIP behavior should be investigated, proposed, and promoted. This chapter proposes some insights about some new proposals.

3.3 Business and Service Considerations Regarding IPv6 VoIP Migration

In general, migration operations are always seen as risky tasks to perform by service providers. As far as migration to IPv6 of SIP-based service in concerned, and in order to ease this task, the following sections describe the overall context and list a set of pertinent indicators and parameters to be taken into account. Both service-specific and transport considerations have been mentioned. Moreover, several viable scenarios are identified and their technical hurdles highlighted.

3.3.1   Context and Objectives

The following section focuses on the context and environment in which IPv6 migration of SIP-based architectures is to be undertaken. The ultimate target is to identify migration strategies and to sketch a set of recommendations aiming at offering a full IPv6-enabled SIP-based telephony service to pure IPv6 customers. This target is a long-term objective and can be reached incrementally through one or several intermediate steps, depending on the strategies adopted by each service provider. These intermediate steps' perimeters differ from one service provider to another depending on (1) the business context, (2) the service opportunities targeted when enabling IPv6-related services, and (3) their position with regard to IPv6 (i.e., adopt an aggressive or passive position). Nevertheless, all these intermediate steps alternatives serve the same mid-term objectives listed hereafter:

- *First Objective—IPv6 SIP Support*: This means allowing IPv6-enabled customers, who should be in majority dual stack ones at early stages, to benefit from SIP-based telephony services using their IPv6 connectivity capabilities. This objective is valid when there are IPv6-enabled customers. These customers could prioritize one of their interfaces, logical or physical, and use IPv6 access for VoIP services or would like to load balance their VoIP traffic between their IP connectivity interfaces. In such a case, service providers should be able to provide service access to these users because IPv6 can be used as a criterion for selecting a VoIP service provider. It is obvious that this constraint is more valid for enterprise customers than for residential ones who in general do not care about the enabled technology, apart from those who (want to) own static IP addresses affected to specific usages such as FTP (File Transfer Protocol) servers and for whom the evolution of IP connectivity protocols could impact their IP-enabled services. When enforcing this objective, the following requirements should be met:

1- Ability of the customers to exclusively access SIP-based services via their IPv6 connectivity: This requirement assumes that customers who subscribed to IPv6 connectivity services are able to configure their IP access to use exclusively IPv6 for SIP-based service offerings. The service platform should be aware of the nature of the interface that is configured to receive VoIP traffic (i.e., SIP signalling or Real-Time Transport Protocol (RTP) streams) so as to allow successful communication (to illustrate this requirement, service discovery means should be adapted so as to provide end-users with appropriate information to access the service. Providing exclusively IPv4 service contact point address would lead to unsuccessful subscription of IPv6-enabled customers to the service.). This VoIP traffic can enclose both signalling and media streams.

2- Ability of the customers to select on a per-call basis the interface to be used to access to SIP-based service: In some scenarios where customers are multi-homed, they can apply some policies so that the decision-making process regarding the interface to be used to place a SIP call can lead to use distinguished Network interfaces for each SIP call. As a consequence, the SIP-based service should not be mapped statically to a given interface. The service platform should not use a static interface to contact a SIP user agent.

3- Ability of the customers to send signalling data and media streams using distinct interfaces: Customers should be able to send their signalling messages using a distinct interface than the one used to transmit and receive media streams. Consequently, the SIP-based service platform should have means to enforce such a requirement. For instance, it should not consider a customer runs an IPv4-enabled SIP user agent based on its lonely subscription using its IPv4 address in this context of service subscription. The context of the service subscription, especially activated socket type, should not be exploited for the placement of further SIP calls. A SIP-based service platform should be IP version-agnostic unless this interferes with the call routing logic and can lead to unsuccessful SIP sessions. In such a case, the last requirement should be relaxed and considered as optional.

- Second Objective—SIP IPv4–IPv6 interworking: The goal is to ensure SIP-based calls between all service subscribed customers and especially between those using heterogeneous SIP user agents. This objective is valid when it remains some IPv4-only customers or at least when all customers are not dual stack. This objective aims at eliminating service access' segregation based on IP connectivity considerations and not creating VoIP communities based on the type of the activated connectivity protocol. The heterogeneity of IP versions should be transparent for end users and no service degradation should be observed when interconnecting two heterogeneous SIP user agents. In the case of full dual stack deployment, this objective can be considered as obsolete and service providers could judge appropriate to exclusively restrict their service offerings to IPv6 (i.e., Only IPv6 SIP support is considered as valid goal) to speed up the migration process. Nevertheless, this information is not available for "Virtual" service providers because they have no control of the underlying infrastructure. When enforcing this objective, the following requirements should be addressed:

1- Ability of heterogeneous customers to place SIP Calls: This means that an IPv4 and IPv6-enabled SIP user agents can place successful SIP calls between them even if there is a break of realm type. The deployed solution should act as a "broker," ensuring the call is successful. Furthermore, the solution adopted to implement this feature should satisfy the following constraints:

1-1- The nature of the call should be transparent for the call participants: This means that the complexity of establishing this type of SIP session should be maintained in the service side and that the SIP user agents believe that the call is issued by a homogenous party (i.e., a customer in the same IP realm type).

The customer should not be aware that the call is traversing heterogeneous realms.

1-2- The perceived QoS of the call should not be severely altered compared to homogenous calls: This requirement expresses that the deployed solution should not have a negative impact on the end-to-end QoS observed by the call participants. In other words, the IP transport interworking mechanisms should be deployed in such a way the path of RTP flow is optimal and no significant delay is introduced due to protocol translation operations. This requirement is to be taken into account when deploying Intermediate nodes that will act as relays between IPv4 and IPv6 realms. Routing should be configured in order to select the shortest path in both realms even if this sub-optimality in each realm may not lead to a global optimal path. This requirement should be passed to control layers in order to meet service layers requirements.

1-3- End-to-end security mechanisms should be supported and not altered: In case the VoIP service offering allows end-to-end security mechanisms, this should be valid for both homogenous and heterogeneous calls. Hop-by-hop security mechanisms should also be supported. The impact of the presence of IP addresses and port numbers in SIP messages is to be carefully studied. A focus should also be put on how to ensure privacy of the call participant.

1-4- Both call direction should be supported: This means that IPv4 to IPv6 calls must be supported as well as IPv6 to IPv4 calls. The direction of the call must not have an impact on the result of call set-up. The Quality of Service (QoS) of both ways should be similar and no difference should be perceived by service end users.

2- Ability of a dual-stack user agent to place a call using its IPv4 connectivity with an IPv6-capable customer: This means that a dual-stack SIP user agent could select its IPv4 connectivity to place a call to IPv6 SIP user agents. The service platform should not impose the IP connectivity to be used by a dual-stack SIP user agent. This requirement should be relaxed for some service considerations like: traffic matrix of IPv4 clients, the success rate of IPv4/IPv6 calls, to speed the migration to IPv6, etc.

3- Ability of a dual-stack user agent to place a call using its IPv4 connectivity with an IPv4-capable customer: This means that a dual-stack SIP user agent could use its IPv4 connectivity to place a call to IPv4 SIP user agents. The service platform should not impose the IP connectivity to be used by a dual-stack SIP user agent.

4- Ability of a dual-stack user agent to place a call using its IPv6 connectivity with an IPv4-capable customer: This means that a dual stack SIP user agent could use its IPv6 connectivity to place a call to IPv4 SIP user agents. The service platform should not impose the IP connectivity to be used by a dual-stack SIP user agent. This requirement should be relaxed for some service considerations such as traffic matrix of IPv4 clients, the success rate of IPv4/IPv6 calls, to speed the migration to IPv6, etc.

5- *Ability of a dual-stack user agent to place a call using its IPv6 connectivity with an IPv6-capable customer*: This means that a dual-stack SIP user agent could select its IPv6 connectivity to place a call to IPv6 SIP user agents. The service platform should not impose the IP connectivity to be used by a given dual-stack SIP user agent.

3.3.2   Input from IPv6 Transport Migration

The introduction of IPv6 into public networks is becoming a reality. Several service providers have enabled IPv6 in their routers and launched their IPv6 migration operations. The portion of IPv6-enabled routers differs between service providers. These migration operations should be tackled in the light of the following factors:

- *Backward compatibility*: Network providers should ensure that running applications still operate and are not disturbed when activating IPv6. Dedicated ALGs should be configured and installed in this aim, for instance, for applications involving DNS and FTP. These ALGs will act as relays between the IPv4 and IPv6 realms and ensure a correct behavior of the end-to-end behavior of the involved applications.

- *Full coverage*: Network providers should exploit the current global Internet connectivity, which is worldwide. Consequently, IPv6 realms should be seen as an extension of the already existing Internet and not a parallel one. IPv4 Internet should be seen from an IPv6 customer side as a continuity of its IPv6 connectivity.

- *Reduce CAPEX*: Introducing IPv6 in the network should not be synonym of network nodes each node is backuped. This task should be achieved without duplicating the network infrastructure. The spirit of the migration should be *Evolve instead of Replace.*

- *Introduction of IPv6 should be transparent*: Customers should not observe alteration of their connectivity services and the migration should be transparent for them.

The current trade is that network providers offer dual connectivity to their customers, i.e., IPv4 and IPv6 access. IPv4 connectivity usage should be gradually decreased in favor of IPv6 one. This convergence phase toward a pure IPv6 connectivity will take several years depending on the policies adopted by service providers. For operators adopting an aggressive position with regards to the activation of IPv6, this transition phase could be small compared to passive operators. Nevertheless, the overall Internet IPv6 coverage will take at least one decade.

Based on this IPv6 transport migration, "high level" service providers can conclude that two categories of customers will co-exist during the transition phase:

- Pure IPv4 customers who don't have an IPv6 access,

- Dual-stack customers who benefit from the two access types.

As a consequence, and from a VoIP standpoint, the deployed solutions for conversational services should be designed taking into account these requirements:

- *Pure IPv4 customers must still be able to access VoIP services*: Meaning that the dimensioning of the service platform should be aware of the amount of IPv4 traffic that is still being routed by SIP proxy servers and media gateways. For this category of customers, the SIP service should not be disturbed.

- *The service provider should own the means to prioritize IPv6 traffic*: This means that service providers will activate some means in their service platform to promote the usage of IPv6 instead of IPv4 when possible.

3.3.3   Business and Strategy Considerations

The enforcement strategy of the IPv6 SIP support and the SIP IPv4-IPv6 interworking objectives depends on the type of the SIP-based offerings managed by a given VoIP service provider. For a newcomer who does not manage any existing service platform, and for whom no backward constraints are to be considered, the task is similar to the introduction of new IPv4-enabled SIP services since there is no risk regarding the impact on the stability of the already offered services at least for the enforcement of IPv6 SIP support objective. For a VoIP service provider that already offers an IPv4-enabled SIP-based telephony service, there are sensitive precautions to be taken into account before modifying the existent service platform and the risk of disturbing the service offerings that a number of customers have paid for is critical. This category of service providers should clearly identify the risks related to the evolution of the deployed service offerings and evaluate the *Quality of the Service (QoS)* that will be perceived by their customers. What follows are a set of pertinent indicators that should be taken into account in order to guide the IPv6 migration operations:

- *The celerity of introduction of IPv6 in operational networks*: Service providers should evaluate if there will be a massive migration to IPv6 or only a partial migration. They should evaluate if the coverage will be global or only reduced to a limited set of IP network providers' domains. This input will facilitate the phasing of the service migration operations and also the dimensioning of traffic matrix in terms of calls to be supported in both IPv4 and IPv6.

- *Constraints on the IP connectivity service provider to service provider interface*: Service providers should identify their needs regarding specific information to be carried at the control plane and if IP network providers should provide service providers with pertinent information for the correct behavior of their conversational service offerings. In addition, service providers should decide where to put the complexity of the migration solution(s) and examine if it is possible to maintain the service layer working independently of the type of IP connectivity protocol.

- *The risk on the stability of IPv4-enabled service offerings*: Service providers should evaluate the impact of the migration operations on the (already deployed) IPv4-enabled service offerings and that there is no interference with these services in terms of technical constraints and other critical aspects such as billing, numbering, etc.

- *Manageability of the deployed interworking solution*: Service providers should evaluate the complexity of the management operations and required maintenance tasks. The solution design should take into consideration reducing the cost of maintenance operations and simplifying engineering procedures owing to the support of automatic means dedicated to configuration and monitoring of the SIP-based VoIP service.

Based on the aforementioned indicators and in order to kick off SIP IPv6 migration operations, several options may be envisaged by service providers. These options are described below:

- *If the* new platform *will serve only IPv6-capable customers or both.* By "new platform" we mean the new changes made in the context of the IPv6 introduction operations. This could enclose new (physical) elements or new functions that should be supported by the already deployed physical entities. Service providers should clearly decide if the new platform will serve only IPv6 customers or if it is designed also for IPv4 ones. This clarification is useful to help the solution designer and could guide the dimensioning of the SIP-based VoIP service platform (including both IPv4 and IPv6 traffic matrix).

- *If the* new platform *should include means to ease the final migration to a pure IPv6 connectivity.* In other words, this aims at preparing the final transition (i.e., only IPv6 is supported by all customers) so that the same problems found in the current stage are not encountered. The shifting to a situation, where only pure IPv6-enabled customers are present, should be achieved with no major (technical) difficulties.

Specifically, service providers can opt for distinct strategies to offer IPv6-enabled SIP-based services as described in the next section.

3.3.4   Migration Scenarios

3.3.4.1   First Alternative

The spirit of this alternative is to be ready to support IPv6 in case IP networks are evolving toward an introduction of IPv6 transfer capabilities (i.e., the density of IPv6-enabled nodes becomes important and not negligible). This first migration scenario consists of exploiting the existing IPv4-enabled conversational service infrastructure(s) as it is and adding new functions where required to support IPv6-enabled customers and co-exist with IPv4 ones. The main characteristic of this alternative is to keep the core of the existing conversational services platform untouched, while IPv6 functions are enabled at the access and border segments only when required. This alternative is pertinent to be considered and studied at earlier stages of IPv6 migration. This alternative should be seen as a transition phase in order to prepare further migration operations, especially in the core segment. This migration scenario can be maintained until the number of IPv6-enabled users becomes significant and will induce significant amount of IPv6 traffic.

This alternative is viable, and even interesting to investigate if the service provider has a clear strategy for progressive deployment of IPv6. In other words, this means that the service provider has clear views and roadmaps to progressively support and activate IPv6 capabilities within its service infrastructure. The service chain can be divided into three major segments: access, core, and border segments. Within this alternative, the support of IPv6 will be mainly enabled in the extremities segments, i.e., access and border segments. This will be enabled in the access segment since it is where IP addresses are assigned and SIP user agents are provisioned with service-specific data. Also, the border segment can be impacted, since the service provider should interconnect with other service providers that can support only IPv6. In such a case, some adaptation functions should be implemented to allow successful communications between these two service providers' IP telephony domains. This latter case could seem to be fantastic and not operational but we believe that for global and universal reachability considerations this scenario is not far from reality (e.g., if China Telecom decides to enable IPv6 in its Next Generation Core Network, the interconnection with this market-sensitive network is no more than the interconnection of an IPv4-only network and an IPv6-only one). In addition, the support of IPv6 in these extremities is a first step for preparing a full migration toward an IPv6-enabled SIP-based service platform (i.e., all service chain segments). The deployment of IPv6 in the access segment is not justified nor motivated in case conversational service offerings are closely bound to the physical access of the customer and the IP connectivity services are based on IPv4. For service providers adopting this scenario (i.e., those having a clear strategy of progressive deployment of IPv6), the next transition step is conditioned by their service opportunities regarding IPv6 and the current limitations of IPv4-based deployment.

After discussing some conditions under which this alternative is viable to investigate, some details about implementation scenarios of this alternative are presented below.

Specifically, this scenario consists of maintaining IPv4-enabled SIP-based VoIP service core elements while only introducing functions to handle IPv6-capable customers when required. Therefore, the service is transparent for the end users as the service is offered by elements using the same connectivity as their SIP user agents. In other words, IPv6-enabled SIP user agents use IPv6 transport protocol to communicate with the service platform. This is notably due to the fact that IPv6-enabled SIP user agents must gain connectivity to the service (e.g., addressing and authentication), these functions being essentially provided by the access functional blocks.

When considering current service provider architectures, a technical realization of this alternative is to let the core platform pure IPv4 and handle IPv6 at the access segment, especially in the SBC. The communication between SBCs and other elements in the service platform will be based on IPv4 and communication between the SBC and the IPv6-enabled SIP user agents will be based on IPv6. Direct communications between SBCs can be either in IPv4 or IPv6, depending on the service provider policy. One concern is to provision SIP user agents with consistent configuration data leading them to connect to the appropriate SBC (all SBCs may not be dual-stack, notably in the early stages of IPv6 deployment). The provisioning of these SIP user agents can be achieved for instance via Dynamic Host Configuration Protocol for IPv6 (DHCPv6). Service providers can indeed tune their DHCP servers so as to provide appropriate information based on appropriate access control criteria.

Within this scenario, SBCs must adapt and relay messages in order to filter IPv6 data. Therefore, for dual-stack SIP user agents, SBCs must retain and associate potentially two contact address types (IPv4 and IPv6) to a single IPv4 address used to represent the SIP user agent toward the core. SBCs must be enhanced in order to allow this correlation. SBC behavior must also evolve, especially regarding operations related to SIP registration.

One challenge is that SBC should be able to identify customers' capabilities in terms of connectivity types. In other words, SBCs should have means at their disposal in order to discriminate the different SIP user agents. Another challenge is that these SBC elements should be configured so as to optimize the media routing (minimize IP version changes). A simple call can indeed be decomposed at least in three legs, where at least two adaptation functions may intervene. SBCs can limit the need for such operations when routing optimization is considered. Solutions handling the two aforementioned challenges are described in the section, "Handling Technical Issues Raised when Migrating SIP-Based Services to IPv6."

3.3.4.2 Second Alternative

This alternative is an aggressive position toward the support of IPv6, but with precautions. This alternative suggests adding to the existing IPv4-enabled SIP-based service infrastructure new elements and to study the deployment and the support of IPv6 in all aforementioned segments (i.e., access, core, and border segments). These new elements may be IPv6-enabled or dual-stack ones. This alternative is suitable in case the density of IPv6-enabled customers becomes important and judged to evolve rapidly in the (near) future. Several scenarios to implement this alternative can be envisaged. An implementation example of this alternative is provided below.

This alternative keeps untouched the existent IPv4 service platform in order not to perturb the service perceived by the already subscribed IPv4 customers (i.e., keep operational the physical equipment providing the conversational service). This option aims at exploring possible scenarios for enhancing the deployed IPv4 SIP-based VoIP platform by adding new service elements without impacting the existing service in order to offer similar service to IPv6-enabled customers. Specifically, it consists in the introduction of dual-stack elements where required in the operational architecture, in order to accompany the IPv6 deployment. This enables to isolate somehow IPv6 for the sake of IPv4 stability, while ensuring in the same time a smooth migration. Moreover, while this architecture will be field-proven, it should be possible migrating pure-IPv4 elements to become dual-stack ones, through software upgrade hopefully, thus optimizing material investment.

Unlike the previously described scenario, dual-stack elements might be introduced to the core platform, which is not likely to remain IPv4-only. However, we don't expect dual-stack core nodes to provide the service to all kind of SIP user agents. Before this happens, it is rather reasonable to consider that IPv4-only SIP user agents remain managed by IPv4-only service nodes, while IPv6-enabled SIP user agents are managed by dual-stack core service nodes in the early stages of deployment. This leaves time to field-prove the IPv6 solution before considering a full dual-stack core. In this context, it could be possible for dual-stack core nodes to handle both IPv4 and IPv6 contact addresses of dual-stack SIP user agents, providing potentially more flexibility to the service and especially for call routing optimization.

Another solution would consist in relying on the same procedures as described in the previous section at the access segment. SIP user agents would be handled by dual-stack SBCs so as to be represented by a single IP contact address toward the core, potentially IPv6 for IPv6-enabled SIP user agents (i.e., including dual-stack SIP user agents). Whatever the adopted solution, IPv4-only SIP user agents keep being managed relying on IPv4 legacy operations. This means that a distinction (i.e., Discovery of SIP user agents types) must be performed "somewhere" by the service platform. Maintaining IPv4-only nodes in the core also means that IP version changes should be carefully looked at. One purpose of this solution's design would be to know how to exploit dual-stack core nodes in order to improve call routing and limit these IP version changes.

Therefore, this scenario presents the same technical challenges as the previous one, i.e., the need for SIP user agent discrimination in order to know what treatment must be applied by SBCs and routing optimization.

3.3.4.3 Third Alternative

This third alternative is radical and proposes to change the existing infrastructure by eliminating existing IPv4-enabled SIP elements and deploy new SIP elements so as to serve both IPv4 and IPv6-enabled customers (either by replacing them physically or updating all nodes). This alternative is not viable for service providers offering IPv4-enabled SIP services.

3.3.5 Service Engineering Recommendations

The main objective of this section is to identify a non-exhaustive list of technical issues that should be handled by a solution pretending to solve SIP-based services migration to IPv6 and associated complications to interconnect with IPv4-enabled SIP-based services realm. These issues may be considered as recommendations for solution designers to take into account.

First, in order to avoid confusions, between the ":" usage to separate the IPv4 address and the port number and its usage in IPv6, SIP IPv6 elements should enclose IPv6 addresses between "**[**" and "**]**" (Schulzrinne 1996). Second, in order to avoid the usage of "hard" IP addresses inside SIP messages, UA should invoke means for discovering SIP servers (for instance, through DHCP and particularly the procedures defined in (Schulzrinne 2002) and (Schulzrinne and Volz 2003)). Therefore, the SIP user agent will retrieve an IP address in the same realm version it is attached to. SIP user agents can also be provisioned with a Fully Qualified Domain Name (FQDN) of the first contact point (e.g., SBC or Proxy Call Session Control Function (P-CSCF) in the context of IMS/TISPAN architectures). Of course, the DNS should be provisioned with both IPv4 and IPv6 entries.

For call routing optimization purposes, SIP proxy servers need to know the type of the called and the caller parties so as to determine if there is a need to alter the SDP body of the received SIP messages. SIP proxy servers should implement means to discriminate SIP user agents based on their connectivity type. This feature may optimize the invocation of adaptation functions, such as ALG or NAT – PT blocks. In addition, service providers may implement means to verify the consistency between the SIP user agent connectivity type and the content of the SDP offers (or SDP answers) carried in its SIP messages. This is a control operation that ensures the successfulness of the telephony communications.

For a dual-stack SIP user agent, it is recommended to indicate its two addresses in the SDP offers (or SDP answers). In such a case, the SIP sessions towards both IPv4 and IPv6 SIP user agent have more chance to succeed without any alteration of SDP offers. In addition, this will reduce the number of assigned addresses in the case the proxy servers check the presence of two address types and inserts the missing one. This procedure can be offered if (Camarillo and Rosenberg 2005) and (Schulzrinne 2002) are supported. Furthermore, a double **Record Route** procedure may ease the routing of SIP messages in case SIP proxy servers are dual-stack. In this case, it could be convenient to add two **Route** headers each of them indicating an IP address (IPv4 and IPv6) of the crossed server/element. Of course, this can be achieved owing to FQDN usage (instead of "hard" IP addresses).

3.3.6 Focus on Quality of Service

QoS is one of critical issues related to IP telephony services. Therefore, IPv6-enabled services should provide good QoS similar to the one experimented by IPv4-enabled ones. The introduction of IPv6 does not change the big picture of the QoS (since IPv6 does not generate bandwidth!) unless there is a need to interconnect IPv6 realms to IPv4 ones.

As far as IPv4 is concerned, several initiatives have been launched within the IETF in this track, especially those related to resource reservation associated with call setup like the framework described in (Camarillo 2002). This proposal introduces a procedure for resource reservation negotiated during the call setup phase. Solutions adopted by service providers should evaluate if this procedure applies for IPv6 and evaluate the impact of introducing IPv6 on such mechanisms.

Within a heterogeneous context, the positioning of intermediary nodes (those elements acting as relays between IPv4 and IPv6 domains) should take into consideration the end-to-end QoS as perceived by end users. The routing to a given intermediary node should be optimal so as the delivered QoS is similar to communication within a homogenous realm. This routing issue is tricky since routing is usually (except when static routes are activated) done based on shortest path, and thus the end-to-end route is the result of two sub optimal functions in two realms. These two sub optimal operations can result in a non end-to-end optimal solution. Solutions deployed by service providers should be able to redirect IP packets destined to heterogeneous realms to cross these intermediary nodes. In addition, as far as QoS is concerned and when DiffServ (Differentiated Services) are implemented, these solutions should provide techniques so as to preserve the signalling of the requested QoS and that consistent and coherent treatment is experienced by IP packets destined to remote destined located in heterogeneous domains.

3.3.7   Focus on Security

Solution adopted by service providers to serve both IPv4 and IPv6-enabled customers must support both hop-by-hop and end-to-end security modes as described in (Rosenberg 2002). In addition, these solutions may support an end-to-middle security scheme as described in (Ono and Tachimoto (RFC4189) 2005) and (Ono and Tachimoto 2005). Both Signalling and media streams must be secured. For instance, Secure Real-Time Transport Protocol (SRTP) (Baugher 2004) and Internet Protocol Security (IPSec) (Kent and Atkinson (RFC 2402) 1998) (Kent and Atkinson (RFC 2406) 1998) may be activated to secure media streams. In addition, these solutions should ensure that when interworking between heterogeneous realms, security mechanisms are not broken. Previous work has been conducted like the one introduced in (Xing and Atwood 2005), which transcribes a mechanism for guaranteeing end-to-end security between SIP entities in the IPv4 domain and others in the IPv6 realm.

3.4   Handling Technical Issues Raised When Migrating SIP-Based Services to IPv6

This section describes some proposals made by the authors of this chapter to solve some of the critical technical issues that arise when migration SIP-based conversational services to IPv6. These issues are pertinent when both IPv4 and IPv6 conversational realms co-exist in the same IP Telephony Administrative Domain (ITAD) managed by a single service provider. The first technical issue is how to discover the connectivity type of SIP user agents. This discovery operation's outcome will be used to drive the routing logic so as to optimize the invocation of interworking and adaptation functions. The second technical issue is how to tune the media traffic process between service nodes in an optimal scheme, i.e., avoid IP version change.

3.4.1 Telephony Routing Optimization: SIP User Agent's Type Discovery

The SIP service platform should be aware of the type of the involved call participants before forwarding session establishment requests. If these means are not supported, SIP proxy servers have no way to predict a session failure even if **ANAT** procedure is implemented. An alternative may be to add dual-stack elements in the core service platform and to force the signalling and media path to cross these dual-stack elements. This latter alternative is viable for big service providers but not for smaller one especially at the earlier stages of deployment of IPv6. For these reasons, we propose lightweight means to discriminate the IP connectivity types as supported by end users' machines. A SIP ALG will be invoked only if necessary. This section focuses on call routing and optimization means to avoid as possible requesting SIP ALGs and NAT-PT boxes when establishing multimedia sessions between two SIP user agents.

3.4.1.1 Use Classical SIP Behavior

The first alternative to notify the service platform about the type of the SIP user agent is to send a SIP **REGISTER** message that encloses all available IP addresses. For IPv4-only and IPv6-only SIP user agents, only one single IP address is carried in SIP **REGISTER** messages. For dual-stack SIP user agents, two IP addresses are enclosed. Upon receipt of this message, the Registrar Server stores these addresses. Two registration databases may be maintained, one for IPv4 and another one for IPv6. A second alternative is to send several **REGISTER** requests using available IP connectivity types. Thus, a dual-stack SIP user agent must send two **REGISTER** messages. The first one is sent using its IPv4 connectivity and the second one using its IPv6 one(s). Two databases are maintained by the conversational service platform.

Consequently, upon reception of an **INVITE** message, the SIP proxy server may question its registration databases to retrieve the types of the involved parties. A SIP ALG is invoked accordingly. The drawback of this procedure is that it depends on the behavior of the terminal user agents. Instead, we propose to modify the SIP protocol in order to have a standardized behavior. This behavior is described in the next sub-section.

3.4.1.2 New SIP Attribute: "atypes"

The purpose of this procedure is to put at the disposal of SIP proxy servers means to discover the connectivity types of involved SIP user agents in order to ease the call routing and to optimize the assignment of temporary addresses through SIP ALGs. This procedure uses only information related to the Application Layer and not ones related to the Transport Layer to execute its routing logic.

This procedure consists in modifying SIP messages in order to carry a new indicator to notify SIP proxy server about the supported IP interfaces (logical or physical ones) and therefore connectivity types. This new SIP indicator is called "**atypes**" and is enclosed in **CONTACT** header of SIP requests, mainly **REGISTER** and **INVITE**.

Below the modified Augmented Backus-Naur Form (ABNF) syntax of **CONTACT** header (the new attribute is written in bold):

```
Contact=("Contact"/"m") HCOLON (STAR / (contact-param *(COMMA
contact-param))) COMMA atypes
!For all attributes except «atypes», refer to (Rosenberg 2002)
atypes = "atypes" EQUAL 4/6/0
```

During the registration phase, a SIP user agent sends a **REGISTER** message with the new modified format of the **CONTACT** header, especially with the new **atypes** attribute. This tag may be positioned to **4** in order to indicate that the SIP user agent is IPv4-only (respectively **6** for IPv6-only and **0** for dual-stack).

A given SIP user agent may restrict its SIP communications to its IPv4 interface or IPv6 ones or may use all available ones. Its local preference is conveyed in the **atypes** attribute. Within this context, an IPv6 interface is judged available only if the scope of this interface is global and not local to the link.

After receiving the **REGISTER** message, the SIP Registrar Server stores, in addition to the AOR (Address Of Record) and the expire timer, the value of the **atypes** tag. This information may be refreshed by other upcoming **REGISTER** messages. Consequently, this new registration procedure replaces the ones defined in (Rosenberg 2002). Based on **atypes** tag value, the SIP Registrar Server classifies its customers to three categories: IPv4-only, IPv6-only and dual-stack ones.

The following provides an example of a **REGISTER** message **atypes** tag:

```
REGISTER sip:r.test.biz SIP/2.0
Via: SIP/2.0/UDP 192.168.25.5:5062; branch=z9hG4bK00e31d6ed
Max-Forwards: 70
Content-Length: 0
To: DS <sip:DS@test.biz>
From: DS <sip:DS@test.biz>; tag=ed3833bd768
Call-ID: 422fd431sdsd3878b7f1d8@test.biz
CSeq: 425542 REGISTER
Contact: DS<sip:DS@192.168.25.5>;expires=90, atypes=0
```

In order to route calls and to decide the need to invoke a SIP ALG or to alter SIP messages, which leads to a successful call between heterogeneous parties, a SIP proxy server may act as follows:

– A first alternative is to question the registration database maintained by the registrar server: the SIP proxy server asks the registrar server about the type of the called and the caller parties. The SIP proxy server decides to invoke ALG in case the two involved parties are from different realms, i.e., IPv4-only and IPv6-only.

– The second alternative is to examine the **atypes** tag as conveyed in the **INVITE** request and ask the registrar server about the type of the called party. In this scenario the SIP proxy server routes the call by comparing the compatibility of the two retrieved values (IP connectivity types of called and caller parties).

For SIP user agents that don't support **atypes** attribute, the SIP service platform generates an **OPTIONS** method destined to these SIP user agents. The purpose of this procedure is to check the capabilities of the SIP user agents and to drive the call optimization process. Once receiving the **OPTIONS** message, a given SIP user agent must answer with a "**200 OK**" message enclosing its network and media capabilities. Based on the content of this answer, the service platform will be able to classify its attached SIP user agents according to their network capabilities.

3.4.2 Blank SDP Procedure

The procedure described hereafter has been designed in order to improve IP versions coexistence (i.e., manage service platform IP heterogeneity) without requiring deep modifications of the involved service nodes. In this aim, this procedure presents the advantage of relying on standard SIP messages, thus allowing a simple and faster deployment into operational networks. The proposed procedure is suitable for service providers deploying intermediate nodes to hide the service core elements from their customers. The first service element visible to end users is one of these intermediate nodes. In such as case, driving the call routing process and therefore optimization the media path shall take into account the presence of three call legs: called SIP user agent until the intermediate node (denoted as originating SBC), leg between the originating SBC and the terminating SBC, and finally the leg between the terminating SBC and called SIP user agent.

This procedure is applicable to both "Alternative 1" and "Alternative 2" solutions, with different outcomes as shown below:

- Alternative 1: The procedure allows the setup of IPv6 SIP sessions between IPv6-capable SBC, while neither disturbing nor requiring modifications on IPv4-only SBCs. Notably, it ensures that as soon as an IPv4-only SBC intervenes in the exchange, this latter will occur only in IPv4. Therefore, IPv6 usage can be promoted whenever possible, keeping IPv4 service and behavior unchanged.

- Alternative 2: The procedure enables to lower IP version changes. Therefore, it provides appropriate means to detect the preferred IP version of the distant peer, thus ensuring the originating peer can adapt to the "unknown yet" destination peer to prevent IP version changes. In an environment where dual stack will be predominant, the SIP-based service can afford choosing the better IP version to avoid IP version changes all along the call path.

The principle of this procedure is to rely on the ability of SIP elements to generate and treat "blank" SDP offers, i.e., messages where the SDP part is empty. This means that the initial SDP offer is to be done by the receiving node, not the originating one. This simple operation enables the discovery of the SBC connectivity type. This procedure is likely to be implemented in early stages of IPv6 deployment and can be enforced when deploying Alternative 1-like solutions. The proposed solution may be fully adopted, by all intervening SBCs, when envisaging Alternative 2-like solutions.

3.4.2.1 Application to Alternative 1

As far as IPv4 SBCs are concerned, and owing to the activation of this procedure, IPv6 introduction is transparent. Indeed, IPv4-cabapble SBCs still perform the same operations they are used to in the already deployed IPv4-only SIP-based service. The only slight modification is that they may receive messages containing no SDP offer from other SBC peers. Anyway, they must keep answering these messages as usual (i.e., adopting the guidelines described in (Rosenberg 2002)). The procedure operations are captured in Figure 3.2.

As far as dual-stack SBCs are concerned, the implications are the following:

- When serving as originating SBC, a dual-stack SBC will never enclose a SDP offer.

- When receiving a message with IPv4 contact information only in the SDP part, the dual-stack SBC must consider the originating peer is IPv4-only. Therefore, it must process the message (and subsequent messages) as an IPv4-only SBC.

- When receiving a message with no SDP offer, the dual-stack SBC may, depending on the service provider preference, either (1) consider the originating peer is IPv6-enabled then answer with IPv6 information enclosed (2) or consider the originating peer is dual-stack and include both IPv6 and IPv4 information.

- Once an answer has been received from the destination SBC, the dual-stack SBC will treat the enclosed contact information as described in (Rosenberg 2002), and generate an answer accordingly (IPv4, respectively IPv6, if the enclosed data is IPv4, respectively IPv6). If both IP versions SDP offers are available, the originating SBC will decide which version to use and formulate a SDP answer. This SDP answer is to be enclosed in an **ACK** message sent to the terminating SBC.

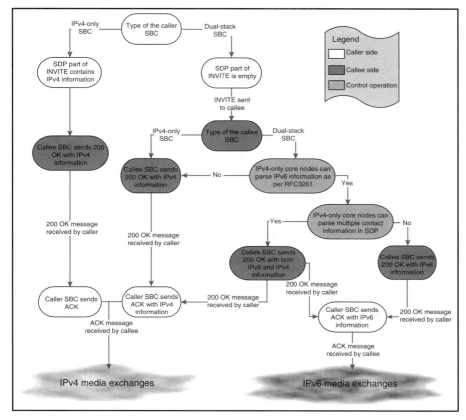

FIGURE 3.2
Overview of "blank SDP" procedure.

This procedure ensures that IPv4 exchanges will occur between SBCs when an IPv4-only SIP user agent intervenes in this communication. Owing to the discrimination techniques described above, SBCs are also capable of triggering the type of the SIP user agent, and adapt the IP version used on the extremity legs. Consequently, SBCs can adapt the call context so that most of the time it is full IPv4 or full IPv6 all along the media chain. As for the signaling plane, the content of the SDP offer doesn't need to be modified.

Figure 3.3 illustrates the SDP procedure when activated within a SIP-based architecture.

In the case the originating SBC is dual-stack and the Terminating SBC is IPv4-only, the main operations are:

– Step 1: The calling SBC receives an **INVITE** message. In its mapping table, the SBC retrieves the IPv4 IP address associated with the calling SIP user agent toward core network. The SBC uses this latter IPv4 address to forward **INVITE** message with blank SDP offer.

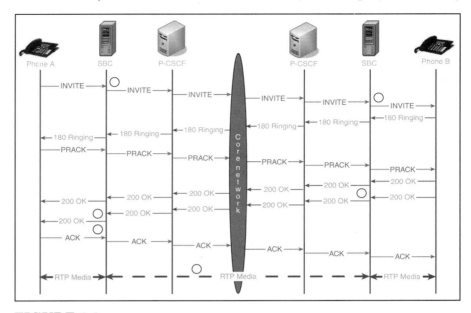

FIGURE 3.3
Example of call flows with blank SDP.

- Step 2: the terminating SBC receives the **INVITE** with no SDP offer and forwards it to the Called user agent following classical operations, i.e., including a SDP offer.

- Step 3: upon reception of **200 OK** message from the called party, terminating SBC forwards "**200 OK**" following classical operations, i.e., including an SDP offer.

- Step 4: upon reception of **200 OK** from the called SBC, originating SBC determines the distant SBC is IPv4-only thanks to the SDP offer. Then it forwards a **200 OK** with relevant IP version (IPv4 or IPv6, possibly **ANAT** respectively for IPv4-only and dual-stack SIP user agent) to calling user agent.

- Step 5: upon reception of an **ACK** message from the calling party, originating SBC forwards the **ACK** message with IPv4-only SDP offer towards the terminating SBC.

- Step 6: RTP media exchanges can then take place, relying on IPv4 except if the calling SIP user agent is dual-stack, where IPv6 could be used.

3.4 2.2   Application to Alternative 2

Within the context of Alternative 2, the connectivity type of the called SIP user agent can be taken into account to adapt the IP version used within a given call, i.e., influence the SDP offer returned by the terminating SBC. This is summarized in Figure 3.4.

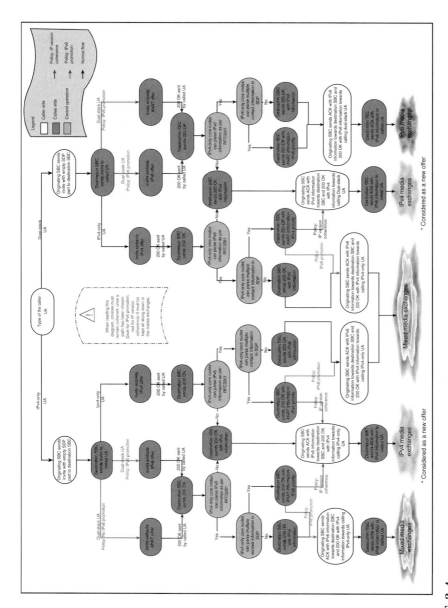

FIGURE 3.4
Blank SDP procedure applied for Alternative 2.

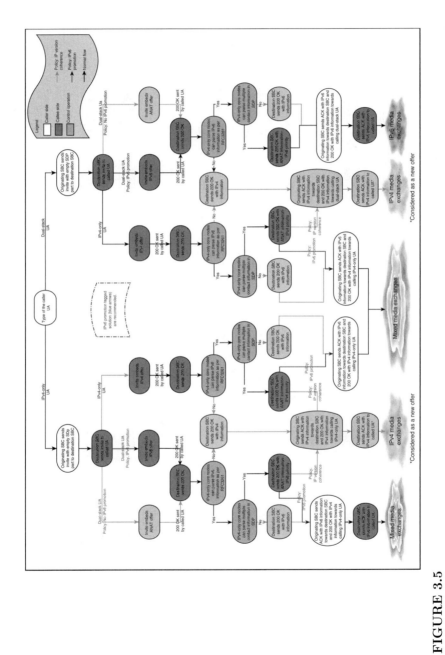

FIGURE 3.5

Blank SDP procedure applied for Alternative 2 (IPv6 Promotion policy).

Moreover, the proposed solution allows different implementations of the SIP service, depending on the service provider policy, whether it favors IPv6 promotion (as illustrated in Figure 3.5) or IP version coherence.

The main differences with Alternative 1 application are:

– The SDP offers made by SBC towards SIP user agents are more "targeted."

– When several IP connectivity type candidates are enclosed in the SDP offer made by the terminating SBC, a priority information is enclosed, depending on the IP connectivity type of the called SIP user agent, providing complementary information to the originating SBC, which is still responsible for the final decision of the IP version to use for a given call.

As far as SBCs are concerned, their implications are as follows:

– When serving as originating SBC, the SBC will never enclose contact information in the SDP part. In the same way, when serving as terminating SBC, the SBC will always receive "blank SDP" offers.

– On reception of an answer message, the SBC will treat the enclosed contact information as described in (Rosenberg 2002), and generate an answer accordingly. If both IP versions offers are available, the originating SBC will make a local decision. The decision-making process of that SBC is driven by the service provider policies; however, the terminating SBC is likely to have positioned priority information allowing to enforce IP version coherence all along the call path.

In case a service should be migrated from Alternative 1 to an Alternative 2 environment, the procedure allows for coexistence of both SBC behaviors. In addition, and as far as Alternative 1 SBCs are concerned, the enforcement of this procedure makes the migration transparent to end-users. Therefore, Alternative 1 SBCs still perform the same operations as per Alternative 1. The only slight modification is that a preference information is likely to be enclosed in the messages they receive, which contain SDP offers with several IP connectivity type candidates. In such a case, Alternative 1 SBCs may either ignore or exploit this additional information depending on their capabilities and configuration.

3.5 Conclusions

This chapter has identified a set of strategic and technical issues arising when introducing IPv6 within conversational services and describes why IPv6 may be problematic for SIP-based service platforms. This chapter proposed three scenarios for migrating SIP-based conversational services to IPv6. It also enumerated a set of engineering recommendations to ease the migration to IPv6 and interworking between IPv6 and IPv4 realms. Furthermore, this chapter introduced two new procedures for routing optimization purposes when heterogeneous SIP user agents are involved in the same SIP session.

References

3GPP TR 23.981. 2005. Interworking aspects and migration scenarios for IPv4 based Implementations (release 6).

Baugher, M. 2004. The Secure Real-time Transport Protocol (SRTP). *RFC* 3711.

Camarillo, G. 2002. Integration of resource management and session initiation protocol (SIP). *RFC* 3312.

Camarillo, G., and M.A. Garcia-Martin. 2005. *The 3G IP multimedia subsystem- merging the internet and the cellular worlds.* John Wiley, UK.

Camarillo, G., and J. Rosenberg. 2005. The Alternative Network Address Types (ANAT) Semantics for the Session Description Protocol (SDP) Grouping Framework. *RFC* 4091.

Chen, W., Y. Lin, and A. Pang. 2005. An IPv4-IPv6 translation mechanism for SIP overlay network in UMTS all-IP environment. *IEEE Journal on Selected Areas in Communications*, 23(11): 2152–60.

Chen, W., and Q. Wu. 2005. Development and deployment of IPv6-based SIP VoIP networks. *Proceedings Symposium on Applications and the Internet Workshops*, 76–9.

Gurbani, V. 2007. IPv6 Transition in the Session Initiation Protocol (SIP). Draft-ietf-sipping-v6-transition.

Gurbani, V., C. Boulton, and R. Sparks. 2007. Session Initiation Protocol (SIP) Torture Test Messages for Internet Protocol Version 6 (IPv6). Draft-ietf-sipping-ipv6-torture-tests.

Handley, M., and V. Jacobson. 1998. SDP: Session Description Protocol. *RFC* 2327.

ITU-T Recommendation H.323. 1998. SERIES H: AUDIOVISUAL AND MULTIMEDIA SYSTEMS, Infrastructure of audiovisual services – Systems and terminal equipment for audiovisual services, Packet-based multimedia communications systems.

Kent, S., and R. Atkinson. 1998. IP Authentication Header. *RFC* 2402.

Kent, S., and R. Atkinson. 1998. IP Encapsulating Security Payload (ESP). *RFC* 2406.

Ono, K., and S. Tachimoto. 2005. Requirements for End-to-Middle Security for the Session Initiation Protocol (SIP). *RFC* 4189.

Ono, K., and S. Tachimoto. 2005. End-to-middle security in the Session Initiation Protocol (SIP). Draft-ietf-sip-e2m-sec-06.

Rosenberg, J. 2002. SIP: Session Initiation Protocol. *RFC* 3261.

Schulzrinne, H. 1996. RTP: A transport protocol for Real-Time applications. *RFC* 1889.

Schulzrinne, H. 2002. Dynamic Host Configuration Protocol (DHCP-for-IPv4) Option for Session Initiation Protocol (SIP) Servers. *RFC* 3361.

Schulzrinne, H., and B. Volz. 2003. Dynamic Host Configuration Protocol (DHCPv6) Options for Session Initiation Protocol (SIP) Servers. *RFC* 3319.

TISPAN, TR180001. 2006. Telecommunications and Internet converged Services and Protocols for Advanced Networking, NGN Release 1.

Xing, J., and J.W. Atwood. 2005. SIP end-to-end security between IPv4 domain and IPv6 domain. SoutheastCon, 2005. *Proceedings IEEE 8–10, April 2005*, 501–6.

4

SIP Mobility Management

Chung-Ming Huang
National Cheng Kung University

Chao-Hsien Lee
Kaohsiung Medical University

CONTENTS

4.1 SIP-Based Host Mobility 71
4.2 SIP-Based Network Mobility 74
4.3 Protocol Analysis 81
     References 85

Session Initiation Protocol (SIP) is an application-layer signaling protocol for real-time multimedia network applications, e.g., instant messaging, Voice-over-Internet Protocol (VoIP) and video conference. The Internet Engineering Task Force (IETF) proposed and standardized SIP for session establishment, maintenance, and termination between an association of users (Rosenberg et al. 2002; Johnston et al. 2003). However, since SIP can support name mapping and redirection services, SIP is also used for handling several kinds of mobility, in which the most famous and fundamental mobility is Internet Protocol (IP) mobility (Vali et al. 2003a, 2003b; Kim et al. 2004). IP mobility concentrates on how to let nodes access the Internet using the same IP address. According to the node type, IP mobility can be divided into (1) host mobility and (2) network mobility.

4.1 SIP-Based Host Mobility

Host mobility occurs when a single node changes its original point of attachment in the Internet. In other words, as a node roams from its home network to other networks, it needs a certain mechanism to let it use the same IP address for communication. In order to solve this problem, the IETF addressed host mobility by Mobile IP (Perkins 2000, 2002; Johnson, Perkins, and Arkko 2004), which is a network-layer protocol. Based on the Mobile IP proposal, each node has its own and unique IP address called home address (HoA) in its home network. Once it leaves its home network, it acquires a new IP address

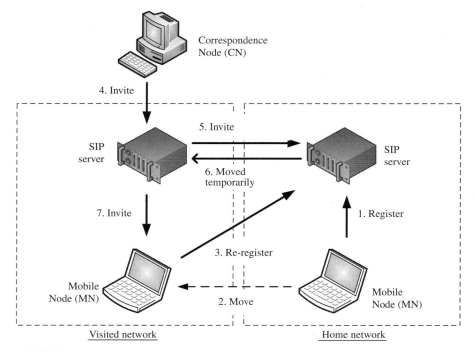

FIGURE 4.1
The pre-call operation.

called care-of-address (CoA). The node should register its new CoA to its home network. Therefore, the home network is able to tunnel those packets destined to this node to the current CoA. Although Mobile IP can handle host mobility and hide the host mobility issue from upper layers, it would induce an indirect routing path and require protocol modification of each node.

Past works (Banerjee, Basu, and Das 2003) have discussed and improved the fact that SIP has the built-in characteristics for solving the host mobility problem. According to the handoff situation, SIP supports two kinds of handoff operations which are (1) the pre-call operation and (2) the mid-call operation.

4.1.1   Pre-Call Operation

Pre-call means that a mobile node (MN) moves to a new visited network before a session is established. The pre-call operation utilizes the SIP REGISTER method to resolve the MNs' reachability. Figure 4.1 depicts the pre-call handoff operation. An MN is supposed to be located in its home network. The detailed steps are described as follows:

1. The MN must register itself with its SIP server using the SIP REGISTER method.

2. The MN roams to another network and acquires a new contact address.

3. The MN must re-register its new contact address with its corresponding SIP server using the SIP REGISTER method.

4. After the re-registration, if a correspondence node (CN) wants to establish a session with the MN, the CN uses the SIP INVITE method to invite the MN.

5. The INVITE request is routed to the SIP server located in the MN's home network.

6. The SIP server responses that the MN has moved temporarily to the re-registered contact address.

7. The INVITE request is finally re-routed to the MN.

After the pre-call handoff operation, the MN's reachability is recovered, i.e., any CN can invite and create a session with the MN. Thus, the pre-call operation guarantees that an MN can be reachable even if the MN roams the Internet.

4.1.2   Mid-Call Operation

Mid-call means that an MN moves to a new visited network during an ongoing session. The mid-call operation utilizes the SIP INVITE method to recover the original session between the MN and the CN. Figure 4.2 depicts

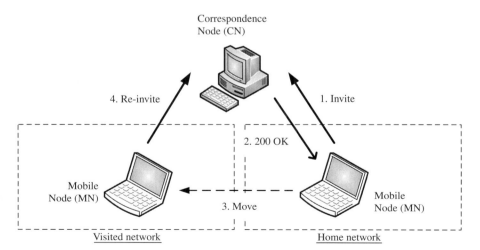

FIGURE 4.2
The mid-call operation.

the mid-call handoff operation. An MN is supposed to be located in its home network. The detailed steps are described as follows:

1. The MN sends an INVITE request to a CN for session establishment.

2. The CN accepts the INVITE request and responds with an OK message. The session between the MN and the CN is established.

3. The MN roams to another network and acquires a new contact address.

4. The MN must re-invite the corresponding CN for session recovery using the SIP INVITE method.

After the mid-call operation, the CN is able to realize the MN's movement and get the MN's new contact address. Hence, the mid-call operation guarantees that an MN can keep sessions continuous even if the MN is roaming.

In spite of the pre-call operation or the mid-call operation, SIP keeps data exchange directly between MNs and CNs. Furthermore, SIP is an application-layer protocol. Any SIP-compatible node can possess the mobile capability without any lower-layer modification. The advanced comparison between Mobile IP and SIP is presented in Section 4.3.

4.2   SIP-Based Network Mobility

Network mobility occurs when a group of nodes moves as a unit over the Internet. For example, users take some transportation carriage, e.g., a bus, a train or an aircraft. The transportation carriage can be regarded as a movable network that is made up of a set of nodes and called "mobile network." In contrast to host mobility in which each node manages its movement individually, there is at least one central point inside a mobile network for managing the movement of the entire mobile network. In order to resolve this problem, the IETF addressed network mobility by NEMO (Lach, Janneteau, and Petrescu 2003; Devarapalli et al. 2005; Ernst 2005), which is an extension of Mobile IP. According to the NEMO basic support protocol (BSP), the central point of each mobile network is called "mobile router" (MR) and operates the Mobile IP mechanism. In other words, when an MR leaves its home network and gets a new CoA, it would re-register its new CoA for keeping the correct mapping between the HoA and the CoA. Once the re-registration process is finished, a bi-directional tunnel would be established between the MR and its home network. All traffic to or from the mobile network must pass through the tunnel. Hence, the tunnel of the NEMO BSP would reinforce the sub-optimal route problem of Mobile IP (Ng et al. 2006a; Ng et al. 2006b).

A recent work (Huang and Lee 2006; Huang, Lee, and Zheng 2006) has extended SIP to support network mobility. Since the proposed SIP-based network mobility (SIP-NEMO) inherits from the SIP, it can avoid the drawbacks due to the bi-directional tunnel. In the following statements, the SIP-NEMO would be presented in terms of: (1) system components, (2) re-registration, (3) re-invitation and (4) route optimization.

4.2.1 System Components

Figure 4.3 depicts the architecture of the SIP-NEMO. In order to apply SIP into the network mobility environment, three types of the SIP Back-To-Back User Agent (B2BUA) which are: (1) SIP Home Server (SIP-HS), (2) SIP Foreign Server (SIP-FS) and (3) SIP Network Mobility Server (SIP-NMS) are defined and configured over the Internet.

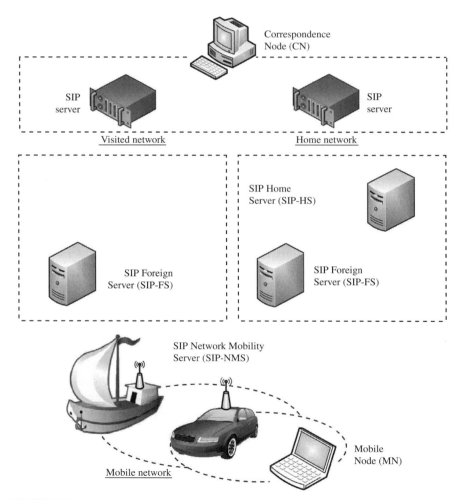

FIGURE 4.3
The SIP-NEMO architecture.

A SIP Network Mobility Server (SIP-NMS) is the application-layer gateway (ALG) of a mobile network and has the capability of routing between its attached network and its carried mobile network. In other words, the SIP-NMS is the central point of a mobile network and responsible for managing all traffic to and from the mobile network. A mobile network can consist of nested subnets, i.e., a SIP-NMS can be attached to another SIP-NMS.

For convenience, the upstream SIP-NMS is called "parent-SIPNMS" and the downstream SIP-NMS is called "sub-SIP-NMS." The SIP-NMS at the top level of nesting is called "root-SIP-NMS."

Each SIP-NMS has one corresponding SIP Home Server (SIP-HS) in its home network. A SIP-HS takes charge of recording the current location of the registered SIP-NMSs. When each SIP-NMS gets a new point of attachment, it should re-register its new contact address to the corresponding SIP-HS like the pre-call operation of the SIP-based host mobility. In addition, since a mobile network may form a complex levels of nesting, the SIP-HS can help to determine routing paths inside a nested mobile network.

A lot of SIP Foreign Servers (SIP-FSs) distributed in the Internet play the role of providing a URI-list service. When a SIP-NMS roams over the Internet, the SIP-NMS should re-invite the on-going sessions as the mid-call operation of the SIP-based host mobility. Thus, a burst of INVITE requests are produced. In order to avoid congestion over wireless links, the SIP-NMS would send an INVITE request with a URI list to a local SIP-FS. Once a SIP-FS receives this kind of INVITE requests, it would re-produce similar requests based on the embedded URI list and then forward the reproduced requests to all targeted CNs.

4.2.2   Re-Registration Operation

Like the SIP-based host mobility, if each SIP-NMS changes its location, it needs reregister its new contact address to the corresponding SIP-HS using the pre-call operation. Once a SIP-NMS updates its new contact address, the carried mobile network can recover its reachability. However, when an MN moves into a mobile network and is attached to a SIP-NMS, the MN needs to re-register its new location to the corresponding SIP server for redirecting its sessions to the newly-attached mobile network. In order to achieve this goal, the SIP-NMS would do some header translation. In other words, the SIP-NMS would replace some fields in the SIP header in order to re-route packets destined to the MN to the correct mobile network. Figure 4.4 depicts the SIP-NEMO re-registration operation. An MN is supposed to locate in its home network. The detailed steps are as follows:

1. The MN must register itself to the SIP server in its home network.

2. The MN roams into a mobile network and acquires a new contact address.

3. The MN must re-register its new contact address to the corresponding SIP server in its home network.

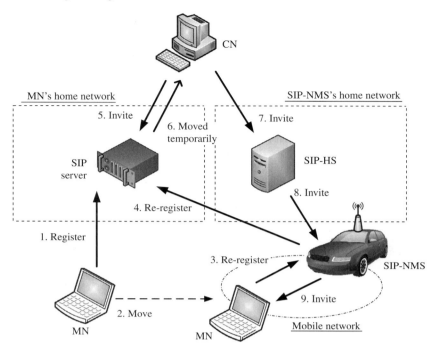

FIGURE 4.4
The SIP-NEMO re-registration operation.

4. The SIP-NMS, which is the gateway of this mobile network, would translate the REGISTER request, i.e., use the SIP-NMS's URI instead of the
SIP-MN's new contact address at the CONTACT field, and then forward
the translated request to the targeted SIP server.

5. After the re-registration, if a CN wants to create a session with the MN,
it would send an INVITE request to the MN's SIP server.

6. The SIP server would respond that the MN has moved temporarily to
the re-registered contact address.

7. Since the re-registered contact address is the SIP-NMS's URI, the CN
would re-send the INVITE request to the corresponding SIP-HS.

8. The SIP-HS redirects the INVITE request to the current contact address
of the SIP-NMS.

9. The SIP-NMS can re-route the INVITE request to the targeted MN
based on the TO field.

After the above operation, any MN that enters a mobile network can be
reachable again. One point to note is that each SIP-NMS must translate the

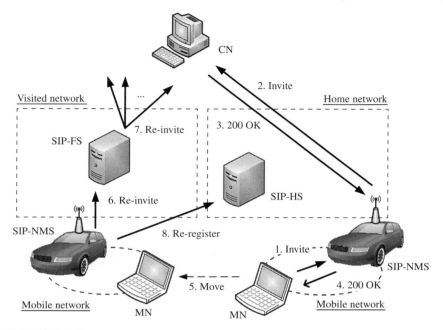

FIGURE 4.5
The SIP-NEMO re-invitation operation.

REGISTER requests sent by the registered MNs. If a SIP-NMS receives any REGISTER request that is not sent by the registered MN, e.g., sent by the MN located in different levels of nesting, it should bypass the request without any modification.

4.2.3   Re-Invitation Operation

As the aforementioned assumption of network mobility, the central point of a mobile network must handle the location change of the entire network. Although the above re-registration operation can recover the SIP-NMS's reachability, the SIP-NMS must be responsible for re-inviting all ongoing sessions like the mid-call operation of the SIP-based host mobility. In order to achieve this goal, when a session is established between an MN and a CN using the SIP INVITE method, the SIP-NMS would do some header translation in order to: (1) route packets targeted to the MN to the corresponding SIP-NMS directly and (2) log the information about this session. Figure 4.5 depicts the SIP-NEMO re-invitation operation. An MN is assumed to locate in a mobile network and wants to establish a session with a CN. The detailed steps are as follows:

1. The MN sends an INVITE request to the targeted CN for session establishment.

2. The SIP-NMS, which acts as the gateway of this mobile network, would translate the INVITE request, i.e., use the SIP-NMS's contact address instead of the SIP-MN's contact address at the CONTACT field and add the SIP-NMS's URI into the RECORD-ROUTE field, and then forward the request to the destination.

3. Once the CN accepts the INVITE request, it must respond with an OK message. Because the contact address of the INVITE request is translated to the SIP-NMS's contact address, the OK message is routed to the SIP-NMS directly.

4. The SIP-NMS can re-route the OK message to the targeted MN based on the TO field. The session between the MN and the CN is established.

5. The SIP-NMS changes its point of attachment. In other words, the entire mobile network roams from one network to another network.

6. The SIP-NMS must re-invite all on-going sessions by sending an INVITE request with a URI list to a local SIP-FS. All corresponding CNs are indicated in the embedded URI list.

7. Once a SIP-FS receives the INVITE request embedded with a URI list, it reproduces similar INVITE requests destined to different recipients according to the embedded URI list.

8. The SIP-NMS must re-register its new contact address to the corresponding SIP-HS for recovering the reachability of the entire mobile network.

After the re-invitation operation, any session established before the mobile network moves can be resumed again. One point to note is that each SIP-NMS should translate all received INVITE requests to or from the mobile network. Furthermore, each SIP-HS should also translate received INVITE requests, i.e., add the RECORD-ROUTE field. If a SIP-NMS or a SIP-HS does not translate a received INVITE request, the SIP-NMS can not actively re-invite this session after roaming. At the same time, if this mobile network is nested, the route inside the mobile network would become incomplete.

4.2.4 Route Optimization

The characteristic of the SIP-based mobility is that data can be delivered directly between MNs and CNs. Regarding the network mobility environment, the shortest routing path between MNs and CNs can be represented by $(CN) \leftrightarrow (SIP\text{-}NMS)^n \leftrightarrow (MN)$, in which n denotes the number of the nesting level. In other words, the transmission path is called "optimal" if data are delivered only via corresponding SIP-NMSs without other particular nodes, e.g., SIP-HS. Figure 4.6 depicts the route optimization of the proposed SIP-NEMO. A CN wants to communicate with an MN that locates inside a mobile

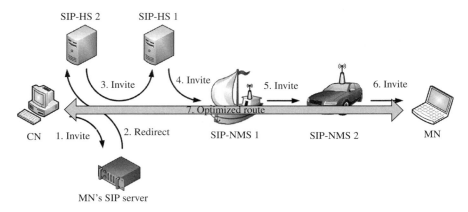

FIGURE 4.6
The SIP-NEMO route optimization.

network with two levels of nesting. The steps to achieve route optimization are described as follows:

1. A CN sends an INVITE request in which the destination is an MN's URI. According to the URI, the request is routed to the corresponding SIP server located in the MN's home network.

2. Referring to the re-registration operation, when the MN roams into a mobile network, SIP-NMS 2 would translate the REGISTER request of the MN. Thus, the MN's SIP server would redirect the INVITE request from a CN to SIP-HS 2, which is the corresponding SIP-HS of SIP-NMS 2.

3. Since SIP-NMS 2 is attached to SIP-NMS 1, SIP-NMS 1 would do the similar header translation of the re-registration operation when SIP-NMS 2 performs the pre-call operation. Hence, SIP-HS 2 would add the RECORD-ROUTE field and then re-route the INVITE request to SIP-HS 1, which is the corresponding SIP-HS of SIP-NMS 1.

4. When SIP-HS 1 receives the INVITE request and checks that the targeted destination is SIP-NMS 1, it would add the RECORD-ROUTE field and then re-route the request to the current contact address of the SIP-NMS 1.

5. Once SIP-NMS 1 receives the INVITE request and detects that the targeted node is not registered, the RECORD-ROUTE field indicates the next hop is SIP-NMS 2. SIP-NMS 1 would forward the INVITE request to SIP-NMS 2.

6. When SIP-NMS 2 receives the INVITE request and determines that the destination is a registered node, SIP-NMS 2 forwards the request to the corresponding MN.

7. Once the MN accepts the invitation, it would respond with an OK message back to the CN. The response is translated by SIP-NMS 2 and

SIP-NMS 1 in sequence. Since SIP-NMS 1 is the root-SIP-NMS, the contact address finally indicates the current location of SIP-NMS 1 when the CN receives the response.

The steps above explain how data can be exchanged directly from a CN to an MN via only corresponding SIP-NMSs using the designed header translation.

4.3   Protocol Analysis

Regarding the host mobility issue, many measurements have studied the comparison between SIP and Mobile IP (Kwon et al. 2002; Nakajima et al. 2003). In this n, we first survey the essential differences between SIP and Mobile IP. Next, we analyze the differences between the SIP-based network mobility (SIP-NEMO) and the Mobile IPv6-based network mobility (MIPv6-NEMO) in advance.

Five essential differences between SIP and Mobile IP are: (1) protocol layer, (2) involved address, (3) handoff operation, (4) data delivery, and (5) mobility support. Table 4.1 depicts the comparison summary.

4.3.1   Protocol Layer

SIP and Mobile IP belong to different protocol layers. Mobile IP is a network-layer protocol. Thus, Mobile IP can hide IP mobility from upper layers and reduce the development complexity of applications. However, Mobile IP requires changing the existing network-layer protocol stack. A node can possess

TABLE 4.1
SIP vs. Mobile IP

	Session Initiation Protocol (SIP)	Mobile IP (MIP)
Protocol Layer	Application layer	Network layer
Lower-Layer Modification	No effect	Required to Modify
Transparency	Only Users	Users and Applications
Involved Address	URI and Contact Address	HoA and CoA
Handoff Operation	Pre-call and Mid-call	Binding Update
Data Delivery	Direct Routing	MIPv4 : Triangle Routing MIPv6 : Direct Routing
Mobility Support	Host Mobility, Network Mobility, Session Mobility, Personal Mobility	Host Mobility, Network Mobility

the roaming capacity only if it add the Mobile IP module into its protocol stack. On the other hand, applying SIP to IP mobility avoids lower-layer protocol modification and keeps it compatible with the existing network environment. Any existing node can possess the roaming capacity by installing a SIP-based application.

4.3.2 Involved Address

Basically, two mechanisms match two kinds of addresses in order to let users roam with the same IP address. First, Mobile IP defines two types of IP addresses: (1) home address (HoA) and (2) care-of-address (CoA). Once an MN leaves its home network, it should reregister its new CoA. Then, the MN's home network can tunnel those packets targeted to the MN to the current CoA. On the contrary, SIP uses (1) URI and (2) contact address. The URI address is responsible for identifing a node like the HoA of Mobile IP. Nevertheless, a contact address could be another logical URI or a physical IP address. The involved addresses of SIP are more flexible since SIP can utilize any format of address as the contact address.

4.3.3 Handoff Operation

As described before, two handoff operations supported by SIP are: (1) the pre-call operation and (2) the mid-call operation. The pre-call operation is performed when a session is not established yet; the mid-call operation is performed when a session has been established. On the other hand, the handoff operation of Mobile IP is a "binding update," i.e., the process of re-registering a new CoA with the corresponding home agent (HA).

4.3.4 Data Delivery

According to the SIP proposal, data delivery between MNs and CNs is established by the INVITE method. No matter where MNs or CNs are located, they can determine each other's current locations by the CONTACT field of an INVITE request. Once the session is established successfully, data are exchanged directly between MNs and CNs. On the contrary, MNs using Mobile IPv4 must re-register their new CoAs by sending Binding Updates (BU) back to their home agents (HA). Then, the Mobile IPv4 is able to tunnel data from HA to foreign agents (FA). In order to improve performance, Mobile IPv6 not only sends the BU message to the corresponding HA but also sends BU to all communicating CNs. Therefore, data delivery using Mobile IPv6 can become direct after a short time of tunneling.

4.3.5 Mobility Support

The last difference between two protocols is that Mobile IP is capable of handling IP mobility only, including host mobility and network mobility. On

TABLE 4.2
SIP-NEMO vs. MIPv6-NEMO

	SIP-NEMO	MIPv6-NEMO
Signaling Mechanism	Off-plane	In-plane
Signaling Overhead	Improved	Light
Header overhead	No	Heavy
Involved Entities	SIP-HS, SIP-NMS, SIP-FS	HA and MR
Transmission Route	Optimal	Sub-optimal
Handoff Latency	Determined by CNs	Determined by HA
Nesting Impact	Light	Heavy

the contrary, SIP possesses a flexible addressing mechanism and provides a dynamic name mapping and redirect service. In addition to IP mobility introduced in this chapter, SIP is also able to support several other types of mobility. For example, SIP can support session mobility, which is the ability of a user to keep ongoing sessions between different devices, and personal mobility, which is the ability to identify a user with the same address across different devices.

After realizing the essential differences of these two protocols, we discuss and analyze the differences of applying SIP and Mobile IP to the network mobility problem on the perspectives of: (1) signaling overhead, (2) header overhead, (3) involved entities, (4) transmission route, (5) handoff latency, and (6) nesting impact. Table 4.2 depicts the summary between the SIP-based network mobility (SIP-NEMO) and the Mobile IPv6-based network mobility (MIPv6-NEMO).

4.3.6 Signaling Overhead

Since SIP is a control protocol for signaling, the SIP-NEMO adopts off-plane signaling, i.e., signaling messages are independent from data packets. Thus, the signaling overhead of the SIP-NEMO includes a burst of signaling messages during the re-invitation operation. According to the studies about wireless links, the quality of wireless links is easily affected by a large number of small-sized packets. In order to improve the quality, the SIP-NEMO aggregates a lot of small-size INVITE requests into a URI list. On the other hand, the MIPv6-NEMO adopts in-plane signaling, i.e., signaling messages are embedded into data packets. That is, the signaling overhead of the MIPv6-NEMO is distributed into every data packet. Besides, the MIPv6-NEMO only needs to send a BU back to the corresponding HA while roaming in the Internet. Therefore, the signaling overhead of the MIPv6-NEMO is lighter than the one of the SIP-NEMO.

4.3.7 Header Overhead

Building on the previous description, the SIP-NEMO has no IP header overhead problem because the signaling of the SIP-NEMO is off-plane. Furthermore, the SIP-NEMO utilizes the header translation to resolve the nesting issue, i.e., adding the RECORD-ROUTE fields. The size increase of a SIP header is slight and not proportional to the nesting level. On the contrary, the MIPv6-NEMO embeds signaling messages into the IP header of each data packet. All packets must add a new IP header before entering a tunnel or remove the outer IP header after leaving a tunnel. With the increase of the nesting level, the MIPv6-NEMO that employs the tunneling mechanism to resolve mobility induces the heavy IP header overhead.

4.3.8   Involved Entities

When SIP is extended to support network mobility, three back-to-back user agents (B2BUAs), i.e., (1) SIP-HS, (2) SIP-NMS, and (3) SIP-FS, are defined and involved in the SIP-NEMO. The SIP-HS and the SIP-NMS are responsible for header translation and the SIP-FS is used to providing a URI-list service. On the other hand, the MIPv6-NEMO involves two entities: (1) the mobile router (MR) and (2) corresponding HA. These two entities are in charge of maintaining a bi-directional tunnel for data exchange.

4.3.9   Transmission Route

As described before, the transmission route of the SIP-NEMO INVITE requests from a CN to an MN can be represented as $(CN) \leftrightarrow (SIPserver) \leftrightarrow (SIP\text{-}HS)^n \leftrightarrow (SIP\text{-}NMS)^n \leftrightarrow (MN)$, in which n is the level number of nesting inside a mobile network. After the session establishment, the transmission route is reduced and denoted by $(CN) \leftrightarrow (SIP\text{-}NMS)^n \leftrightarrow (MN)$. However, the MIPv6-NEMO must transmit packets through bi-directional tunnels. Thus, the transmission route is represented as $(CN) \leftrightarrow (HA)^n \leftrightarrow (MR)^n \leftrightarrow (MN)$.

4.3.10   Handoff Latency

Handoff latency is determined on how to recover the originally ongoing sessions after roaming to another point of attachment. For the SIP-NEMO, the SIP-NMS would actively re-invite all corresponding CNs. Thus, the handoff latency of the SIP-NEMO is determined on the route between the SIP-NMS and each CN. On the contrary, the MR recovers the reachability of the carried mobile network by re-register its new CoA to its HA. Then, the HA can tunnel packets to the mobile network. Hence, the handoff latency of the MIPv6-NEMO is determined on the route between the MR and its HA.

4.3.11   Nesting Impact

Based on the above analysis, some issues are affected by the nesting level, i.e., transmission route and header overhead. For the SIP-NEMO, the nesting impact is light and limited because traffic need not pass through any tunnel. Although the routing path of the invitation process is proportional to the level number of nesting, the SIP-NEMO can keep route optimization of data transmission. On the contrary, it is a great effect on MIPv6-NEMO if a mobile network has more than one nesting level. The traffic of the downstream mobile networks must pass through the bi-directional tunnel of the upstream mobile networks. Each tunneling would add one extra IP header. Therefore, the routing path and header overhead of the MIPv6-NEMO are both proportional to the nesting level.

References

Banerjee, N., K. Basu, and S. K. Das. 2003. Handoff delay analysis in SIP-based mobility management in wireless networks. *Proceedings of the International Parallel and Distributed Processing Symposium (IPDPS)*, 22–26.

Devarapalli, V., R. Wakikawa, A. Petrescu, and P. Thubert. 2005. Network Mobility BasicSupport Protocol. *IETF RFC 3963*.

Ernst, T. 2005. Network Mobility Support Goals and Requirements. *IETF DRAFT: draft-ietf-nemo-requirements-05.txt (work in progress)*.

Huang, C. M., and C. H. Lee. 2006. Signal reduction and local route optimization of SIP-based network mobility. *Proceedings of the 11th IEEE Symposium on Computers and Communications (ISCC)*, 482–87.

Huang, C. M., C. H. Lee, and J. R. Zheng. 2006. A novel SIP-based route optimization for network mobility. *IEEE Journal on Selected Areas in Communications* 24(9): 1682–91.

Johnson, D., C. Perkins, and J. Arkko. 2004. Mobility Support in IPv6. *IETF RFC* 3775.

Johnston, A., S. Donovan, R. Sparks, C. Cunningham, and K. Summers. 2003. Session Initiation Protocol (SIP) Basic Call Flow Examples. *IETF RFC 3665*.

Kim, W., M. Kim, K. Lee, C. Yu, and B. Lee. 2004. Link layer assisted aobility support using SIP for real-time multimedia communications. *Proceedings of the 2nd International Workshop on Mobility Management and Wireless Access Protocols*, 127–129.

Kwon, T. T., M. Gerla, S. Das, and S. Das. 2002. Mobility management for VoIP service: Mobile IP vs. SIP. *IEEE Wireless Communications* 9(5): 66–75.

Lach, H. Y., C. Janneteau, and A. Petrescu. 2003. Network mobility in beyond-3G systems. *IEEE Communications Magazine* 41(7): 52–57.

Nakajima, N., A. Dutta, S. Das, and H. Schulzrinne. 2003. Handoff delay analysis and measurement for SIP Based Mobility in IPv6. *Proceedings of the IEEE International Conference on Communications (ICC)* 2: 1085–1089.

Ng, C., P. Thubert, M. Watari, and F. Zhao. 2006a. Network mobility route optimization problem statement. *IETF DRAFT: draft-ietf-nemo-ro-problem-statement-03.txt (work in progress)*.

Ng, C., F. Zhao, M. Watari, and P. Thubert. 2006b. Network mobility route optimization solution space analysis. *IETF DRAFT: draft-ietf-nemo-ro-space-analysis-03.txt (work in progress)*.

Perkins, C. 2000. Mobile IP and the IETF. *ACM SIGMOBILE Mobile Computing and Communications Review* 4(3): 6–11.

———. 2002. IP Mobility Support for IPv4. *IETF RFC* 3220.

Rosenberg, J., H. Schulzrinne, G. Camarillo, A. Johnston, J. Peterson, R. Sparks, M. Han-dley, and E. Schooler. 2002. SIP: Session Initiation Protocol. *IETF RFC 3261*.

Vali, D., S. Paskalis, A. Kaloxylos, and L. Merakos. 2003a. A SIP-based method for intra-domain handoffs. *Proceedings of the IEEE Vehicular Technology Conference (VTC)* 3(8): 2068–2072.

———. 2003b. An efficient micro-mobility solution for SIP networks. *Proceedings of the IEEE Global Telecommunications Conference (GLOBECOM)* 6:3088–3092.

5

SIP Event Notification and Presence Information

Bo Zhao
The Pennsylvania State University

Chao Liu
Yahoo! Inc.

CONTENTS

5.1   Introduction 87
5.2   Session Initiation Protocol (SIP)-Specific Event Notification 88
5.3   Simple Presence Information 94
5.4   Presence Information Data Format (PIDF) 99
5.5   Event List–Notification Extension 102
5.6   Current Work 115
      References 121

5.1   Introduction

SIP, the Session Initiation Protocol [1], is a signaling protocol for Internet conferencing, telephony, presence, events notification, and instant messaging. SIP can be used in any applications where session initiation is required, which includes event subscription and notification. Presence information is a status indicator that conveys the ability and willingness of a potential communication partner. The ability to request notification of events proves useful in many types of SIP services for which cooperation between end nodes is required. The SIP specification has been extended to support a notification mechanism, by which SIP nodes can request notification from remote nodes indicating that certain events have occurred, allowing subscription to present information and convey information on the presence of users [2]. SIP nodes can request notification from remote nodes indicating that certain events have occurred and convey the presence information. Presence information conveys the ability and willingness of a user to communicate across a set of devices. A presence service is a system that accepts, stores, and distributes presence information to interested parties (called watchers). A presence protocol is a protocol for providing a presence service over the Internet or any IP (Internet Protocol) networks.

In this chapter, we first introduce the Session Initiation Protocol (SIP)-Specific Event Notification mechanism, followed by an explanation of what the presence information is. We will then see, based on Presence Information Data Format, how the notification mechanism works to convey presence information. Then we will introduce the event list notification mechanism that allows for requesting and conveying notifications and presence information for lists of resources. Finally, we will describe the current work, service, and application of notification mechanism and presence information. We can find that the protocols introduced in this chapter are broadly used in third-generation wireless networks. Because of their popularity and usefulness, these protocols are extended in many ways to make them fit the requirements of the new services and applications in recent years.

5.2   Session Initiation Protocol (SIP)-Specific Event Notification

This section discusses the event notification mechanism, which provides a SIP-specific framework for event notification. The discussion is mainly based on [2].

5.2.1   Definitions

In order to understand the operation of notification mechanism, we first define the following terminologies:

Definition 1 (Event Package) *An event package is an additional specification that defines a set of state information to be reported by a notifier to a subscriber.*

Definition 2 (Notifier) *A notifier is a user agent that generates NOTIFY requests for the purpose of notifying subscribers of the state of a resource. Notifiers typically also accept SUBSCRIBE requests to create subscriptions.*

Definition 3 (Notification) *Notification is the act of a notifier sending a NOTIFY message to a subscriber to inform the subscriber of the state of a resource.*

Definition 4 (Subscriber) *A subscriber is a user agent that receives NOTIFY requests from notifiers; these NOTIFY requests contain information about the state of a resource in which the subscriber is interested. Subscribers typically also generate SUBSCRIBE requests and send them to notifiers to create subscriptions.*

Definition 5 (Subscription) *A subscription is a set of application states associated with a dialog. This application state includes a pointer to the* *associated dialog, the event package name, and possibly an identification token. Event packages will define additional subscription state information.*

5.2.2 Operation Overview

A typical flow of messages would be:

```
        Subscriber          Notifier
|-----SUBSCRIBE---->|      Request state subscription
|<-------200--------|      Acknowledge subscription
|<------NOTIFY------|      Return current state information
|--------200------->|
|<------NOTIFY------|      Return current state information
|--------200------->|
|-----SUBSCRIBE---->|      Subscriptions are expired and
                          refreshed
|<-------200--------|
|<------NOTIFY------|      Return current state information
|--------200------->|
|----UNSUBSCRIBE--->|      Unsubscribe the subscription
|<-------200--------|
|<------NOTIFY------|      Return terminate notify
|--------200------->|
```

When a subscriber wishes to subscribe to a particular state for a resource, it forms a SUBSCRIBE message. This SUBSCRIBE request will be confirmed with a final response. A 200-class responses indicate that the subscription has been accepted, and that a NOTIFY message will be sent immediately to communicate the current resource state to the subscriber. As in SUBSCRIBE requests, NOTIFY "Event" headers will contain a single event package name for which a notification is being generated. The "Expires" header in a 200-class response to SUBSCRIBE indicates the actual duration for which the subscription will remain active. The body of the NOTIFY request provides the event package information.

When a change in the subscribed state occurs, the notifier immediately constructs and sends a NOTIFY request to inform subscribers of changes in the state to which the subscriber has a subscription. At any time before a subscription expires, the subscriber may refresh the timer on such a subscription by sending another SUBSCRIBE request. When a notifier receives a subscription refresh, the notifier updates the expiration time for the subscription and sends a NOTIFY message with current state information to the subscriber immediately. A SUBSCRIBE with an "Expires" of 0 constitutes a request to unsubscribe from an event. A subscription is destroyed when a notifier sends a NOTIFY request with a "Subscription-State" of "terminated." So a subscriber sends a subscribe request with an "Expires" of 0 in order to trigger the sending of such a "terminated" NOTIFY request.

5.2.3   A Walk-Through Example

This section shows an example of using the NOTIFY mechanism method for subscribing a state information of the resource from a subscriber user agent to a notifier user agent. The subscriber in this example is subscribing to the state information of the resource. In the message, "xxx" means some event type. "xxxxxx" means some content type. When the value of the content-length header field is "...", this means that the value should be whatever the computed length of the body is.

```
        Subscriber              Notifier
|-----SUBSCRIBE---->|    M1:Request state subscription
|<-------200--------|    M2:Acknowledge subscription
|<------NOTIFY------|    M3:Return current state information
|--------200------->|    M4:Acknowledge notification
|<------NOTIFY------|    M5:Return current state information
|--------200------->|    M6:Acknowledge notification
|-----SUBSCRIBE---->|    M7:Subscriptions are expired and
                                refreshed
|<-------200--------|    M8:Acknowledge subscription
|<------NOTIFY------|    M9:Return current state information
|--------200------->|    M10:Acknowledge notification
|----UNSUBSCRIBE--->|    M11:Unsubscribe the subscription
|<-------200--------|    M12:Acknowledge subscription
|<------NOTIFY------|    M13:Return terminate notify
|--------200------->|    M14:Acknowledge terminate notification
```

Message flow:

M1: The subscriber subscribe@handbook.com initiates a new subscription to the notifier notify@handbook.com. Expire time is 3600 sec.

```
SUBSCRIBE sip:resource@handbook.com SIP/2.0
Via: SIP/2.0/UDP example.com;branch=z9hG4bKnashds7
To: <sip:notify@handbook.com>
From: <sip:subscribe@handbook.com>;tag=12341234
Call-ID: 10275342@handbook.com
CSeq: 1 SUBSCRIBE
Max-Forwards: 70
Expires: 3600
Event: xxx
Contact: sip:user@handbook.com
Content-Length: 0
```

M2: The notifier for notify@handbook.com processes the subscription request and creates a new subscription. A 200 (OK) response is sent to confirm the subscription.

```
SIP/2.0 200 OK
Via: SIP/2.0/UDP example.com;branch=z9hG4bKnashds7;
```

```
received=168.0.2.5
To: <sip:notify@handbook.com>;tag=abcd1234
From: <sip:subscribe@handbook.com>;tag=12341234
Call-ID: 10275342@handbook.com
CSeq: 1 SUBSCRIBE
Contact: sip:server.handbook.com
Expires: 3600
Content-Length: 0
```

M3: In order to complete the process, the notifier sends the subscriber a NOTIFY with the current state of the resource.

```
NOTIFY sip:user@handbook.com SIP/2.0
Via: SIP/2.0/UDP server.handbook.com;branch=z9hG4bK8sdf2
To: <sip:subscribe@handbook.com>;tag=12341234
From: <sip:notify@handbook.com>;tag=abcd1234
Call-ID: 10275342@handbook.com
CSeq: 1 NOTIFY
Max-Forwards: 70
Event: xxx
Subscription-State: active; expires=3599
Contact: sip:server.handbook.com
Content-Type: xxxxxx
Content-Length: ...

[the state information of the resource]
```

M4: The subscriber confirms receipt of the NOTIFY request.

```
SIP/2.0 200 OK
Via: SIP/2.0/UDP server.handbook.com;branch=z9hG4bK8sdf2;
received=168.0.2.6
To: <sip:subscribe@handbook.com>;tag=12341234
From: <sip:notify@handbook.com>;tag=abcd1234
Call-ID: 10275342@handbook.com
CSeq: 1 NOTIFY
Content-Length: 0
```

M5: The notifier detects that a reportable change has been made to the resource's information, and sends a new notification to the watcher.

```
NOTIFY sip:user@handbook.com SIP/2.0
Via: SIP/2.0/UDP server.handbook.com;branch=z9hG4bK4cd42a
To: <sip:subscribe@handbook.com>;tag=12341234
From: <sip:notify@handbook.com>;tag=abcd1234
Call-ID: 10275342@handbook.com
CSeq: 2 NOTIFY
Max-Forwards: 70
```

```
Event: xxx
Subscription-State: active; expires=3400
Contact: sip:server.handbook.com
Content-Type: xxxxxx
Content-Length: ...
```

[New state information of the resource]

M6: The subscriber confirms receipt of the NOTIFY request.

```
SIP/2.0 200 OK
Via: SIP/2.0/UDP server.handbook.com;branch=z9hG4bK4cd42a;
received=168.0.2.6
To: <sip:subscribe@handbook.com>;tag=12341234
From: <sip:notify@handbook.com>;tag=abcd1234
Call-ID: 10275342@handbook.com
CSeq: 2 NOTIFY
Content-Length: 0
```

M7: The subscriber subscribe@handbook.com initiates a refresh subscription to the notifier notify@handbook.com.

```
SUBSCRIBE sip:resource@handbook.com SIP/2.0
Via: SIP/2.0/UDP example.com;branch=z9hG4bKnashds7
To: <sip:notify@handbook.com>;tag=abcd1234
From: <sip:subscribe@handbook.com>;tag=12341234
Call-ID: 10275342@handbook.com
CSeq: 2 SUBSCRIBE
Max-Forwards: 70
Expires: 3600
Event: xxx
Contact: sip:user@handbook.com
Content-Length: 0
```

M8: The notifier for notify@handbook.com processes the subscription request. A 200 (OK) response is sent to confirm the subscription.

```
SIP/2.0 200 OK
Via: SIP/2.0/UDP example.com;branch=z9hG4bKnashds7;
received=168.0.2.5
To: <sip:notify@handbook.com>;tag=abcd1234
From: <sip:subscribe@handbook.com>;tag=12341234
Call-ID: 10275342@handbook.com
CSeq: 2 SUBSCRIBE
Contact: sip:server.handbook.com
Expires: 3600
Content-Length: 0
```

M9: In order to complete the process, the notifier sends the subscriber a NOTIFY with the current event state of the resource.

```
NOTIFY sip:user@handbook.com SIP/2.0
Via: SIP/2.0/UDP server.handbook.com;branch=z9hG4bK8sdf2
To: <sip:subscribe@handbook.com>;tag=12341234
From: <sip:notify@handbook.com>;tag=abcd1234
Call-ID: 10275342@handbook.com
CSeq: 3 NOTIFY
Max-Forwards: 70
Event: xxx
Subscription-State: active; expires=3599
Contact: sip:server.handbook.com
Content-Type: xxxxxx
Content-Length: ...
```

[The state information of the resource]

M10: The subscriber confirms receipt of the NOTIFY request.

```
SIP/2.0 200 OK
Via: SIP/2.0/UDP server.handbook.com;branch=z9hG4bK8sdf2;
received=168.0.2.6
To: <sip:subscribe@handbook.com>;tag=12341234
From: <sip:notify@handbook.com>;tag=abcd1234
Call-ID: 10275342@handbook.com
CSeq: 3 NOTIFY
```

M11: The subscriber subscribe@handbook.com unsubscribe a subscription to the notifier notify@handbook.com.

```
SUBSCRIBE sip:resource@handbook.com SIP/2.0
Via: SIP/2.0/UDP example.com;branch=z9hG4bKnashds7
To: <sip:notify@handbook.com>;tag=abcd1234
From: <sip:subscribe@handbook.com>;tag=12341234
Call-ID: 10275342@handbook.com
CSeq: 3 SUBSCRIBE
Max-Forwards: 70
Expires: 0
Event: xxx
Contact: sip:user@handbook.com
Content-Length: 0
```

M12: The notifier for notify@handbook.com processes the subscription request. A 200 (OK) response is sent to confirm the subscription.

```
SIP/2.0 200 OK
Via: SIP/2.0/UDP example.com;branch=z9hG4bKnashds7;
received=168.0.2.5
To: <sip:notify@handbook.com>;tag=abcd1234
From: <sip:subscribe@handbook.com>;tag=12341234
```

```
Call-ID: 10275342@handbook.com
CSeq: 3 SUBSCRIBE
Contact: sip:server.handbook.com
Expires: 0
Content-Length: 0
```

M13: In order to complete the process, the Notifier sends the subscriber a terminated NOTIFY to destroy the subscription.

```
NOTIFY sip:user@handbook.com SIP/2.0
Via: SIP/2.0/UDP server.handbook.com;branch=z9hG4bK8sdf2
To: <sip:subscribe@handbook.com>;tag=12341234
From: <sip:notify@handbook.com>;tag=abcd1234
Call-ID: 10275342@handbook.com
CSeq: 4 NOTIFY
Max-Forwards: 70
Event: xxx
Subscription-State: terminated; expires=0
Reason:timeout
Contact: sip:server.handbook.com
Content-Type: xxxxxx
Content-Length: ...
```

```
[empty]
```

M14: The subscriber confirms receipt of the NOTIFY request.

```
SIP/2.0 200 OK
Via: SIP/2.0/UDP server.handbook.com;branch=z9hG4bK8sdf2;
received=168.0.2.6
To: <sip:subscribe@handbook.com>;tag=12341234
From: <sip:notify@handbook.com>;tag=abcd1234
Call-ID: 10275342@handbook.com
CSeq: 4 NOTIFY
Content-Length: 0
```

5.3 Simple Presence Information

5.3.1 Overview

Presence, also known as presence information, conveys the ability and willingness of a user to communicate across a set of devices [3]. It makes use of the general event notification framework and the SUBSCRIBE and NOTIFY methods mentioned before. When an entity, the subscriber, wishes to learn about presence information from some user, it creates a SUBSCRIBE request. It eventually arrives at a presence server, which generates a response to the request and sends a notify request including the presence information to the subscriber.

5.3.2 Definition

A presence service is a system that accepts, stores, and distributes presence information to interested parties called watchers [4]. The PRESENCE SERVICE has two distinct sets of "clients".One set of clients is PRESENTITIES, the other set is WATCHERS. PRESENTITIES provide PRESENCE INFORMATION to be stored and distributed. WATCHERS receive PRESENCE INFORMATION from the service.

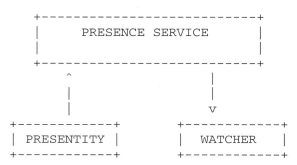

Presence server: A presence server is a physical entity of a SIP user agent that is capable of receiving SUBSCRIBE requests, responding to them, and generating notifications of changes in presence state. So in the notify mechanism, the subscriber is the watcher, and the notifier is the presence server.

5.3.3 Operations and Presence Event Package

5.3.3.1 Operations of the Presence Service

Changes to PRESENCE INFORMATION are distributed to SUBSCRIBERS via NOTIFICATIONS. Figure 5.1 through Figure 5.3 show the flow of

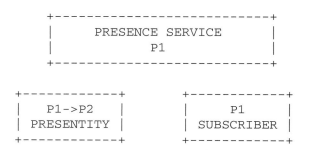

FIGURE 5.1
NOTIFICATION (Step 1).

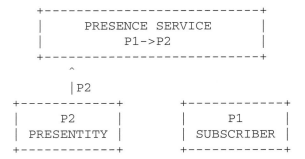

FIGURE 5.2
NOTIFICATION (Step 2).

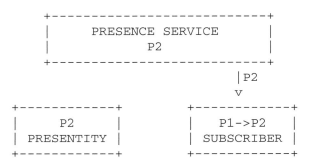

FIGURE 5.3
NOTIFICATION (Step 3).

information as a piece of PRESENCE INFORMATION is changed from P1 to P2. Initially, the subscriber and presence service have the P1 presence information of the presentity. In Figure 5.1, presentity changes its presence information from P1 to P2. Then in Figure 5.2, presentity updates the P2 presence information to presence service. In Figure 5.3, the presence server knows that the subscriber is subscribing to this presentity's presence information. So it sends a notify request to inform the subscriber and updates its presence information of the presentity.

5.3.3.2   The Presence Event Package

The presence event package is based on a presence server that is capable of accepting subscriptions, storing subscription states, and generating notifications when there are changes in presence [3]. The SIP notify mechanism mentioned before defines a SIP extension for subscribing to and receiving notifications of events. It leaves the definition of many aspects of these events to concrete extensions, known as event packages. The name of this package is "presence." This value appears in the event header field present in SUB-SCRIBE and NOTIFY requests. All subscribers and notifiers support the "application/pidf+xml" presence data format described in the next section. The subscribe request contains an accept header field. It includes "application/pidf+xml." As described in the notify mechanism, the NOTIFY message will contain bodies that describe the state of the subscribed resource.

In the presence event package, the body of the notification contains a presence document. This document describes the presence of the presentity that was subscribed to. It will be introduced in next section. The body of the NOTIFY message is sent using the type listed in the accept header field in the most recent SUBSCRIBE request, or using the type "application/pidf+xml" if no accept header field is present.

5.3.4   A Walk-Through Example

This message flow illustrates how the presence server can be responsible for sending notifications for a presentity. This flow assumes that the watcher has previously been authorized to subscribe to this resource at the server. In this flow, the presentity informs the server about the updated presence information through some non-SIP means. The [PIDF Document] and [New PIDF Document] contain the presence document that will be introduced in next section. When the value of the content-length header field is "..." this means that the value should be whatever the computed length of the body is.

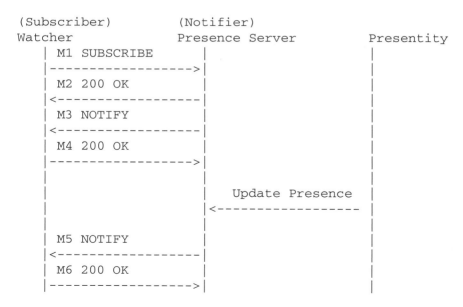

Message Details

M1: The watcher subscribe@handbook.com initiates a new subscription to the presence server notify@handbook.com. Expire time is 3600 seconds.

```
SUBSCRIBE sip:resource@handbook.com SIP/2.0
Via: SIP/2.0/UDP example.com;branch=z9hG4bKnashds7
To: <sip:notify@handbook.com>
From: <sip:subscribe@handbook.com>;tag=12341234
Call-ID: 10275342@handbook.com
CSeq: 1 SUBSCRIBE
Max-Forwards: 70
Expires: 3600
Event: presence
Accept: application/pidf+xml
Contact: sip:user@handbook.com
Content-Length: 0
```

M2: The presence server for notify@handbook.com processes the subscription request and creates a new subscription. A 200 (OK) response is sent to confirm the subscription.

```
SIP/2.0 200 OK
Via: SIP/2.0/UDP example.com;branch=z9hG4bKnashds7;
received=168.0.2.5
To: <sip:notify@handbook.com>;tag=abcd1234
From: <sip:subscribe@handbook.com>;tag=12341234
Call-ID: 10275342@handbook.com
CSeq: 1 SUBSCRIBE
Contact: sip:server.handbook.com
Expires: 3600
Content-Length: 0
```

M3: In order to complete the process, the presence server sends the subscriber a NOTIFY with the current presence information of the presentity.

```
NOTIFY sip:user@handbook.com SIP/2.0
Via: SIP/2.0/UDP server.handbook.com;branch=z9hG4bK8sdf2
To: <sip:subscribe@handbook.com>;tag=12341234
From: <sip:notify@handbook.com>;tag=abcd1234
Call-ID: 10275342@handbook.com
Event: presence
CSeq: 1 NOTIFY
Max-Forwards: 70
Subscription-State: active; expires=3599
Contact: sip:server.handbook.com
Content-Type: application/pidf+xml
Content-Length: ...

[PIDF Document]
```

M4: The watcher confirms receipt of the NOTIFY request.

```
SIP/2.0 200 OK
Via: SIP/2.0/UDP server.handbook.com;branch=z9hG4bK8sdf2;
received=168.0.2.6
To: <sip:subscribe@handbook.com>;tag=12341234
From: <sip:notify@handbook.com>;tag=abcd1234
Call-ID: 10275342@handbook.com
CSeq: 1 NOTIFY
Content-Length: 0
```

M5: The presence server detects that a reportable change has been made to the presentity's presence information and sends a new notification to the watcher.

```
NOTIFY sip:user@handbook.com SIP/2.0
Via: SIP/2.0/UDP server.handbook.com;branch=z9hG4bK4cd42a
To: <sip:subscribe@handbook.com>;tag=12341234
From: <sip:notify@handbook.com>;tag=abcd1234
Call-ID: 10275342@handbook.com
CSeq: 2 NOTIFY
Max-Forwards: 70
Event: presence
Subscription-State: active; expires=3400
Contact: sip:server.handbook.com
Content-Type: application/pidf+xml
Content-Length: ...

[New PIDF Document]
```

M6: The watcher confirms receipt of the NOTIFY request.

```
SIP/2.0 200 OK
Via: SIP/2.0/UDP server.handbook.com;branch=z9hG4bK4cd42a;
received=168.0.2.6
To: <sip:subscribe@handbook.com>;tag=12341234
From: <sip:notify@handbook.com>;tag=abcd1234
Call-ID: 10275342@handbook.com
CSeq: 2 NOTIFY
Content-Length: 0
```

5.4 Presence Information Data Format (PIDF)

5.4.1 Overview

Presence Information Data Format (PIDF), as a common presence data format for presence protocols, allows presence information to be transferred across these protocols [5]. In the notify mechanism, the PIDF is stored in the body of the notify message, such as M3 and M5 in previous message flow Figures. The PIDF encodes presence information in eXtensible Markup Language (XML). PRESENCE INFORMATION consists of an arbitrary number of elements called PRESENCE TUPLES.

5.4.2   Presence Information Contents

5.4.2.1   Content

- PRESENTITY URL: specifies the "pres" URL of the PRESENTITY.

- List of PRESENCE TUPLES

- PRESENTITY human readable comment: free text memo about the PRE-SENTITY (optional).

5.4.2.2   XML-Encoded PIDF

```
   +------------------------------------+
   | <presence>                         |
   +------------------------------------+
    | +--------------------------------+
  =>| <tuple>                          |
    | +--------------------------------+
    |  | +----------------------------+
    |  =>| <status>                   |
    |  | +----------------------------+
    |  |  | +------------------------+
    |  |  =>| <basic>                |
    |  |  | +------------------------+
    |  | +----------------------------+
    |  =>| <contact>                  |
    |  | +----------------------------+
    |  |  | +------------------------+
    |  |  =>| <priority>             |
    |  |  | +------------------------+
    |  | +----------------------------+
    |  =>| <note>                     |
    |     +----------------------------+
    |  | +----------------------------+
    |  =>| <timestamp>                |
    |     +----------------------------+
    | +--------------------------------+
  =>| <tuple>                          |
    | +--------------------------------+
```

```
 |  +-------------------------------+
=>|  <tuple>                        |
 |  +-------------------------------+
 |     . . .
```

A PIDF object is a well-formed XML document containing an encoding declaration in the XML declaration, e.g., "<?xml version='1.0' encoding= 'UTF-8'?>."

The <presence> element: The root of an "application/pidf+xml" object is a <presence> element associated with the presence information namespace. This contains any number of <tuple> elements. The <presence> element MUST have an 'entity' attribute. The value of the 'entity' attribute is the URL of the PRESENTITY. The <presence> element contains a namespace declaration to indicate the namespace on which the presence document is based.

The <tuple> element: The <tuple> element carries a PRESENCE TUPLE, consisting of a mandatory <status> element, followed by any number of OPTIONAL extension elements, followed by an OPTIONAL <contact> element, followed by any number of OPTIONAL <note> elements, followed by an OPTIONAL <timestamp> element.

The <status> element: The <status> element contains one OPTIONAL <basic> element that contains one of the following strings: "open" or "closed." The values "open" and "closed" indicate availability to receive the message.

The <contact> element: The <contact> element contains a URL of the contact address. It optionally has a 'priority' attribute, whose value means a relative priority of this contact address over the others.

The <note> element: The <note> element contains a string value, which is usually used for a human readable comment.

The <timestamp> element: The <timestamp> element contains a string indicating the date and time of the status change of this tuple.

3.2.3 Status Value Extensibility: the existing PIDF definition allows arbitrary elements to appear in the <status> element. The following example XML Schema defines an extension for <location> presence information, which can have the values of 'home,' 'office,' or 'car.'

5.4.3   Examples

Simple example:

```
<?xml version="1.0" encoding="UTF-8"?>
<presence xmlns="urn:ietf:params:xml:ns:pidf"
entity="pres:somebody@handbook.com">
     <tuple id="dsfrl3">
          <status>
               <basic>open</basic>
          </status>
```

```
        <contact priority="0.5">tel:8145557777
          </contact>
      </tuple>
</presence>
```

Example with Status Extensions

```
<?xml version="1.0" encoding="UTF-8"?>
<presence xmlns="urn:ietf:params:xml:ns:pidf"
      xmlns:im="urn:ietf:params:xml:ns:pidf:im"
      xmlns:myex="http://id.hamdbook.com/"
      entity="pres:somebody@handbook.com">
      <tuple id="dstw234">
          <status>
              <basic>open</basic>
              <im:im>idle</im:im>
              <myex:location>car</myex:location>
          </status>
          <contact
priority="0.8">im:somebody@homepage.net</contact>
          <note xml:lang="en">Don't block Please!</note>
          <timestamp>2007-10-27T17:10:10Z</timestamp>
      </tuple>
      <tuple id="dsfds2e">
          <status>
              <basic>open</basic>
          </status>
          <contact
priority="1.0">mailto:somebody@handbook.com</contact>
      </tuple>
      <note>I am going to the supermarket now</note>
</presence>
```

5.5   Event List–Notification Extension

5.5.1   Introduction

The SIP-specific event notification mechanism [6] allows a user (the subscriber) to request to be notified of changes in the state of a particular resource. This is accomplished by having the subscriber generate a SUBSCRIBE request for the resource, which is processed by a notifier that represents the resource.

In many cases, a subscriber has a list of resources they are interested in. Without some aggregating mechanism, this will require the subscriber to generate a SUBSCRIBE request for each resource about which they want information. For environments in which bandwidth is limited, such as wireless networks, subscribing to each resource individually is expensive. The primary benefit of the resource list server is to reduce the overall messaging volume to a subscriber.

To solve these problems, this event list mechanism defines an extension to notification mechanism, which allows for requesting and conveying notifications for lists of resources. The notifier for the list is called a "resource list server," or RLS.

5.5.2 Definition

Definition 6 (Back-end Subscription) *Any subscription (SIP or otherwise) that an RLS creates to learn of the state of a resource. An RLS will create back-end subscriptions to learn of the state of a resource about which the RLS is not an authority. For back-end subscriptions, RLSes act as a subscriber. [7]*

Definition 7 (List Subscription) *A subscription to a resource list. In list subscriptions, RLSes act as the notifier.*

Definition 8 (Resource List) *A list of zero or more resources that can have their individual states subscribed to with a single subscription.*

Definition 9 (RLMI) *Resource List Meta-Information. RLMI is a document that describes the state of the virtual subscriptions associated with a list subscription.*

Definition 10 (RLS) *Resource List Server. RLSes accept subscriptions to resource lists and send notifications to update subscribers of the state of the resources in a resource list.*

Definition 11 (Virtual Subscription) *Virtual subscriptions are a logical construct within an RLS that represent subscriptions to the resources in a resource list. For each list subscription it services, an RLS creates at least one virtual subscription for every resource in the resource list being subscribed to.*

5.5.3 Operation

When users wish to subscribe to the resource of a list of resources, they can use the mechanisms described in this section. The first step is creation of a resource list. This resource list is represented by a SIP URI. The list contains a set of URIs, each of which represents a resource for which the subscriber wants to receive information.

To learn the resource state of the set of elements on the list, the user sends a single SUBSCRIBE request targeted to the URI of the list. This will be routed to an RLS for that URI. The RLS acts as a notifier and accepts the subscription.

The RLS may have direct information about some or all of the resources specified by the list. If it does not, it could subscribe to any non-local resources specified by the list resource. As the state of resources in the list change, the RLS generates notifications to the list subscribers.

The list notifications contain a body of type multipart/related. The root section of the multipart/related content is an XML document that provides meta-information about each resource present in the list. The remaining sections contain the actual state information for each resource. Multipart/related and application/rlmi+xml MIME types are included in an accept header sent in a SUBSCRIBE message

Notifications contain a multipart document, the first part of which always contains meta-information about the list. Remaining parts are used to convey the actual state of the resources listed in the meta-information.

The "state" attribute of each instance of a resource in the meta-information is set according to the state of the virtual subscription. When sending information in a terminated resource instance, the RLS indicates a state of "terminated" and an appropriate reason value.

When the first SUBSCRIBE message for a particular subscription is received by a RLS, the RLS will often not know state information for all of the resources specified by the resource list. For any resource for which state information is not known, the corresponding "uri" attribute will be set appropriately, and no(instance) elements will be present for the resource.

Immediate notifications triggered as a result of subsequent SUBSCRIBE messages SHOULD include an RLMI document with full state indicated.

The XML document present in the root of the multipart/related document contains a <resource> element for some or all of the resources in the list. Each <resource> element contains a URI that uniquely identifies the resource to which that section corresponds. When a NOTIFY arrives, it can contain full or partial state (as indicated by the "fullState" attribute of the top-level <list> element). If full state is indicated, then the recipient replaces all states associated with the list with the entities in the NOTIFY body. If full state is not indicated, the recipient of the NOTIFY updates information for each identified resource. Information for any resources that are not identified in the NOTIFY are not changed, even if they were indicated in previous NOTIFY messages.

5.5.4   Resource List Attributes

In order to convey the state of multiple resources, the list extension uses the "multipart/related" mime type. The root document of the multipart/related body is a Resource List Meta-Information (RLMI) document. It is of type "application/rlmi+xml." This document contains the meta-information for the resources contained in the notification.

The <list> element present in a list notification contain three attributes. The first mandatory <list> attribute is "uri," which contains the uri that corresponds to the list. The second mandatory <list> attribute is "version," which contains a number from 0 to $2^{32} - 1$. The third mandatory attribute is "fullState." The "full-State" attribute indicates whether the NOTIFY message contains information for every resource in the list. Finally, <list> elements may contain a "cid" attribute. If present, the "cid" attribute identifies a section within the multipart/related body that contains aggregate state information for the resources contained in the list.

```
    +------------------------------------------+
    | <list>                                   |
    +------------------------------------------+
     | +--------------------------------------+
    =>|  <resource>                           |
     | +--------------------------------------+
     |    | +------------------------------+
     |   =>|  <name>                       |
     |    | +------------------------------+
     |    | +------------------------------+
     |   =>|  <instance>                   |
     |    | +------------------------------+
     | +--------------------------------------+
    =>|  <resource>                           |
     | +--------------------------------------+

     | +--------------------------------------+
    =>|  <resource>                           |
     | +--------------------------------------+
     |    ...
```

The resource list contains one <resource> element for each resource being reported in the notification. These resource elements contain attributes that identify meta-data associated with each resource. The "uri" attribute identifies the resource to which the <resource> element corresponds. Each list and resource element contains zero or more name elements. These name elements contain human readable descriptions or names for the resource list or resource. Each resource element contains zero or more instance elements. These instance elements are used to represent a single notifier for the resource. The "id" attribute contains an opaque string used to uniquely identify the instance of the resource. The "state" attribute contains the subscription state for the identified instance of the resource. This attribute contains one of the values "active," "pending," or "terminated." Finally, the "cid" attribute, which is present if the "state" attribute is "active," identifies the section within the multipart/related body that contains the actual resource state.

An example of a multipart/related document format follows.

```
<?xml version="1.0"?>
<list xmlns="urn:ietf:params:xml:ns:rlmi"
uri="sip:my_friends@lists.handbook.com" version="2"
fullState="true">
   <name language="en">Friend List</name>
   <resource uri="sip:John@handbook.com">
      <name>John Smith</name>
      <instance id="ewruwoe" state="active"
cid="wqe21ew@handbook.com"/>
   </resource>
   <resource uri="sip:Erwin@handbook.com">
      <name>Erwin Jones</name>
      <instance id="werqesdfd" state="active"
cid="kiuuoui@handbook.com"/>
   </resource>
   <resource uri="sip:Edwin@handbook.com">
      <name>Edwin</name>
      <instance id="uyiyi" state="pending"/>
   </resource>
   <resource uri="sip:Bill@handbook.com">
      <name>Bill Smith</name>
      <instance id="uyuiy" state="terminated" reason="rejected"/>
   </resource>
</list>
```

The cid attribute is the content-ID for the corresponding section in the multipart body. The cid attribute refers only to top-level parts of the multipart/related document for which the RLMI document in which it appears is the root.

We consider a multipart/related document containing three parts; we'll label these parts A, B, and C. Part A is type "application/rlmi+xml," part B is type multipart/related, and part C is type "application/pidf+xml." Part B is in turn a document containing three parts: D, E, and F. Part D is of type "application/rlmi+xml," and parts E and F are of type "application/pidf+xml." Any cid attributes in document A must refer only to parts B or C. Referring to parts D, E, or F would be illegal. Similarly, Any cid attributes in document D must refer only to parts E or F. Referring to any other parts would be illegal.

```
+------------------------------------------------+
| Top Level Document: multipart/related          |
|                                                |
| +------------------------------------------+ |
| | Part A: application/rlmi+xml             | |
| +------------------------------------------+ |
| | Part B: multipart/related                | |
| |                                          | |
```

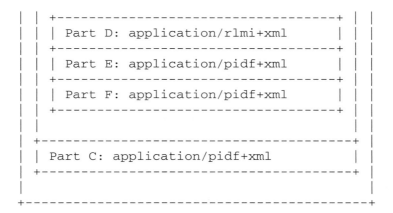

```
    |  |  +-------------------------------------+  |  |
    |  |  | Part D: application/rlmi+xml        |  |  |
    |  |  +-------------------------------------+  |  |
    |  |  | Part E: application/pidf+xml        |  |  |
    |  |  +-------------------------------------+  |  |
    |  |  | Part F: application/pidf+xml        |  |  |
    |  |  +-------------------------------------+  |  |
    |  |                                           |  |
    |  +----------------------------------------------+  |
    |  | Part C: application/pidf+xml              |  |
    |  +----------------------------------------------+  |
    |                                                    |
    +-------------------------------------------------------+
```

5.5.5 A Walk-Through Example

This section gives an example call flow. In this example, we request a subscription to a presence list. The subscriber's URI is "sip:me@chicago.handbook.com," and the name of the list resource that we are subscribing to is called "sip:me-list@chicago.handbook.com." The underlying event package is "presence."

In this example, the RLS has information to service some of the resources on the list, but must consult other servers to retrieve information for others. The implementation of the RLS in this example uses the SIP SUBSCRIBE/NOTIFY mechanism to retrieve such information. The local RLS is chicago.handbook.com.

```
     Terminal   chicago.handbook.com   new_york.handbook.com
        |                 |                            |
M1   |---SUBSCRIBE--->|                            |
M2   |<-----200-------|                            |
M3   |<----NOTIFY-----|                            |
M4   |------200------>|                            |
M5   |                |-----------SUBSCRIBE----------->|
M6   |                |<--------------200--------------|
M7   |                |<-------------NOTIFY------------|
M8   |                |--------------200-------------->|
M9   |<----NOTIFY-----|                            |
M10  |------200------>|                            |
```

M1: We initiate the subscription by sending a SUBSCRIBE message to our local RLS.

```
Terminal -> Local RLS chicago.handbook.com

SUBSCRIBE sip:me-list@chicago.handbook.com SIP/2.0
Via: SIP/2.0/TCP terminal.chicago.handbook.com;
  branch=z9hG4bKwYb6QREiCL
```

```
Max-Forwards: 70
To: <sip:me-list@Chicago.handbook.com>
From: <sip:me@chicago.handbook.com>;tag=ie4hbb8t
Call-ID: cdB34qLToC@terminal.Chicago.handbook.com
CSeq: 322723822 SUBSCRIBE
Contact: <sip:terminal.Chicago.handbook.com>
Event: presence
Expires: 7200
Supported: eventlist
Accept: application/pidf+xml
Accept: application/rlmi+xml
Accept: multipart/related
Accept: multipart/signed
Accept: application/pkcs7-mime
Content-Length: 0
```

M2: The local RLS completes the SUBSCRIBE transaction.

```
Local RLS -> Terminal
SIP/2.0 200 OK
Via: SIP/2.0/TCP terminal.chicago.handbook.com;
  branch=z9hG4bKwYb6QREiCL
To: <sip:me-list@Chicago.handbook.com>;tag=zpNctbZq
From: <sip:me@chicago.handbook.com>;tag=ie4hbb8t
Call-ID: cdB34qLToC@terminal.Chicago.handbook.com
CSeq: 322723822 SUBSCRIBE
Contact: <sip:Chicago.handbook.com>
Expires: 7200
Require: eventlist
Content-Length: 0
```

M3: the RLS sends a NOTIFY message immediately upon accepting the subscription. The NOTIFY contains an RLMI document describing the entire buddy list, as well as presence information for the users about which it already knows. Note that, since the RLS has not yet retrieved information for some of the entries on the list, those <resource> elements contain no <instance> elements.

```
Local RLS -> Terminal

NOTIFY sip:terminal.chicago.handbook.com SIP/2.0
Via: SIP/2.0/TCP chicago.handbook.com;
  branch=z9hG4bKMgRenTETmm
Max-Forwards: 70
From: <sip:me-list@chicago.handbook.com>;tag=zpNctbZq
To: <sip:me@chicago.handbook.com>;tag=ie4hbb8t
Call-ID: cdB34qLToC@terminal.chicago.handbook.com
CSeq: 997935768 NOTIFY
```

```
Contact: <sip:chicago.handbook.com>
Event: presence
Subscription-State: active;expires=7200
Require: eventlist
Content-Type: multipart/related;type="application/rlmi+xml";
  start="<nXYxAE@chicago.handbook.com>";
  boundary="50UBfW7LSCVLtggUPe5z"
Content-Length: 1560

--50UBfW7LSCVLtggUPe5z
Content-Transfer-Encoding: binary
Content-ID: <nXYxAE@chicago.handbook.com>
Content-Type: application/rlmi+xml;charset="UTF-8"

<?xml version="1.0" encoding="UTF-8"?>
<list xmlns="urn:ietf:params:xml:ns:rlmi"
      uri="sip:me-buddies@chicago.handbook.com"
      version="1" fullState="true">
  <name language="en">Buddy List at Handbook</name>
  <resource uri="sip:bob@chicago.handbook.com">
    <name>Bob Smith</name>
    <instance id="juwigmtboe" state="active"
              cid="bUZBsM@chicago.handbook.com"/>
  </resource>
  <resource uri="sip:dave@chicago.handbook.com">
    <name>Dave Jones</name>
    <instance id="hqzsuxtfyq" state="active"
              cid="ZvSvkz@chicago.handbook.com"/>
  <resource uri="sip:me-list@new_york.handbook.com">
    <name language="en">My Friends</name>
  </resource>
</list>

--50UBfW7LSCVLtggUPe5z
Content-Transfer-Encoding: binary
Content-ID: <bUZBsM@chicago.handbook.com>
Content-Type: application/pidf+xml;charset="UTF-8"

<?xml version="1.0" encoding="UTF-8"?>
<presence xmlns="urn:ietf:params:xml:ns:pidf"
    entity="sip:bob@chicago.handbook.com">
  <tuple id="sg89ae">
    <status>
      <basic>open</basic>
    </status>
    <contact priority="1.0">sip:bob@chicago.handbook.com
      </contact>
  </tuple>
</presence>
```

```
--50UBfW7LSCVLtggUPe5z
Content-Transfer-Encoding: binary
Content-ID: <ZvSvkz@chicago.handbook.com>
Content-Type: application/pidf+xml;charset="UTF-8"

<?xml version="1.0" encoding="UTF-8"?>
<presence xmlns="urn:ietf:params:xml:ns:pidf"
          entity="sip:dave@chicago.handbook.com">
  <tuple id="slie74">
    <status>
      <basic>closed</basic>
    </status>
  </tuple>
</presence>
--50UBfW7LSCVLtggUPe5z--
```

M4: The terminal completes the transaction.

```
Terminal -> Local RLS
SIP/2.0 200 OK
Via: SIP/2.0/TCP chicago.handbook.com;
  branch=z9hG4bKMgRenTETmm
From: <sip:me-list@chicago.handbook.com>;tag=zpNctbZq
To: <sip:me@chicago.handbook.com>;tag=ie4hbb8t
Call-ID: cdB34qLToC@terminal.chicago.handbook.com
CSeq: 997935768 NOTIFY
Contact: <sip:terminal.chicago.handbook.com>
Content-Length: 0
```

M5: The local RLS subscribes to the state of the non-local resource new york.handbook.com.

```
Local RLS -> RLS in new_york.handbook.com
SUBSCRIBE sip:me-list@chicago.handbook.com SIP/2.0
Via: SIP/2.0/TCP chicago.handbook.com;
  branch=z9hG4bKFSrAF8CZFL
Max-Forwards: 70
To: <sip:me-list@new_york.handbook.com>
From: <sip:me@chicago.handbook.com>;tag=a12eztNf
Call-ID: kBq5XhtZLN@chicago.handbook.com
CSeq: 980774491 SUBSCRIBE
Contact: <sip:chicago.handbook.com>
Identity: Tm90IGEgcmVhbCBzaWduYXR1cmUsIGVpdGhlci4gQ2VydGFp
          bmx5IHlvdSBoYXZlIGJldHRlcgp0aGluZ3MgdG8gYmUgZG9p
          bmcuIEhhdmUgeW91IGZpbmlzaGVkIHlvdXIgUkxTIHlldD8K
Identity-Info: https://chicago.handbook.com/
Event: presence
Expires: 3600
Supported: eventlist
Accept: application/pidf+xml
```

```
Accept: application/rlmi+xml
Accept: multipart/related
Accept: multipart/signed
Accept: application/pkcs7-mime
Content-Length: 0
```

M6: The RLS in new york.handbook.com completes the SUBSCRIBE transaction.

```
RLS in new_york.handbook.com -> Local RLS
SIP/2.0 200 OK
Via: SIP/2.0/TCP chicago.handbook.com;
  branch=z9hG4bKFSrAF8CZFL
To: <sip:me-list@new_york.handbook.com>;tag=JenZ40P3
From: <sip:me@chicago.handbook.com>;tag=a12eztNf
Call-ID: kBq5XhtZLN@chicago.handbook.com
CSeq: 980774491 SUBSCRIBE
Contact: <sip:new_york.handbook.com>
Expires: 3600
Content-Length: 0
```

M7: The NOTIFY contains an RLMI document describing the contained buddy list, as well as presence information for those users.

```
RLS in new_york.handbook.com -> Local RLS
NOTIFY sip:chicago.handbook.com SIP/2.0
Via: SIP/2.0/TCP new_york.handbook.com;
  branch=z9hG4bKmGL1nyZfQI
Max-Forwards: 70
From: <sip:me-list@new_york.handbook.com>;tag=JenZ40P3
To: <sip:me@chicago.handbook.com>;tag=a12eztNf
Call-ID: kBq5XhtZLN@chicago.handbook.com
CSeq: 294444656 NOTIFY
Contact: <sip:new_york.handbook.com>
Event: presence
Subscription-State: active;expires=3600
Require: eventlist
Content-Type: multipart/signed;
  protocol="application/pkcs7-signature";
  micalg=sha1;boundary="l3WMZaaL8NpQWGnQ4mlU"
Content-Length: 2038
--l3WMZaaL8NpQWGnQ4mlU
Content-Transfer-Encoding: binary
Content-ID: <ZPvJHL@new_york.handbook.com>
Content-Type: multipart/related;type="application/rlmi+xml";
             start="<Cvjpeo@new_york.handbook.com>";
             boundary="tuLL131DyPZX0GMr2YOo"

--tuLL131DyPZX0GMr2YOo
Content-Transfer-Encoding: binary
```

```
Content-ID: Cvjpeo@new_york.handbook.com
Content-Type: application/rlmi+xml;charset="UTF-8"

<?xml version="1.0" encoding="UTF-8"?>
<list xmlns="urn:ietf:params:xml:ns:rlmi"
            uri="sip:me-list@new_york.handbook.com" version="1"
            fullState="true">
  <name language="en">Buddy List at handbook</name>
  <resource uri="sip:joe@new_york.handbook.com">
    <name>Joe Thomas</name>
    <instance id="1" state="active"
            cid="mrEakg@new_york.handbook.com"/>
  </resource>
  <resource uri="sip:mark@new_york.handbook.com">
    <name>Mark Edwards</name>
    <instance id="1" state="active"
            cid="KKMDmv@new_york.handbook.com"/>
  </resource>
</list>

--tuLL13lDyPZX0GMr2Yoo
Content-Transfer-Encoding: binary
Content-ID: mrEakg@new_york.handbook.com
Content-Type: application/pidf+xml;charset="UTF-8"

<?xml version="1.0" encoding="UTF-8"?>
<presence xmlns="urn:ietf:params:xml:ns:pidf"
    entity="sip:joe@new_york.handbook.com">
  <tuple id="7823a4">
    <status>
      <basic>open</basic>
    </status>
    <contact priority="1.0">sip:joe@new_york.handbook.com
      </contact>
  </tuple>
</presence>

--tuLL13lDyPZX0GMr2YOo
Content-Transfer-Encoding: binary
Content-ID: KKMDmv@new_york.handbook.com
Content-Type: application/pidf+xml;charset="UTF-8"
<?xml version="1.0" encoding="UTF-8"?>
<presence xmlns="urn:ietf:params:xml:ns:pidf"
     entity="sip:mark@new_york.handbook.com">
  <tuple id="398075">
    <status>
      <basic>closed</basic>
    </status>
  </tuple>
</presence>
```

```
--tuLLl31DyPZX0GMr2YOo--

--l3WMZaaL8NpQWGnQ4mlU
Content-Transfer-Encoding: binary
Content-ID: K9LB7k@new_york.handbook.com
Content-Type: application/pkcs7-signature

[PKCS #7 signature here]

--l3WMZaaL8NpQWGnQ4mlU--
```

M8: The local RLS completes the NOTIFY transaction.

```
Local RLS -> RLS in new_york.handbook.com
SIP/2.0 200 OK
Via: SIP/2.0/TCP new_york.handbook.com;
branch=z9hG4bKmGL1nyZfQI
From: <sip:me-list@new_york.handbook.com>;tag=JenZ40P3
To: <sip:me@chicago.handbook.com>;tag=a12eztNf
Call-ID: kBq5XhtZLN@chicago.handbook.com
CSeq: 294444656 NOTIFY
Contact: <sip:chicago.handbook.com>
Content-Length: 0
```

M9: At this point, the local RLS decides it has collected enough additional information to warrant sending a new notification to the user. Although sending a full notification would be perfectly acceptable, the RLS decides to send a partial notification instead. The RLMI document contains only information for the updated resources, as indicated by setting the "fullState" parameter to "false."

```
Local RLS -> Terminal
NOTIFY sip:terminal.chicago.handbook.com SIP/2.0
Via: SIP/2.0/TCP chicago.handbook.com;
branch=z9hG4bK4EPlfSFQK1
Max-Forwards: 70
From: <sip:me-list@chicago.handbook.com>;tag=zpNctbZq
To: <sip:me@chicago.handbook.com>;tag=ie4hbb8t
Call-ID: cdB34qLToC@terminal.chicago.handbook.com
CSeq: 997935769 NOTIFY
Contact: <sip:chicago.handbook.com>
Event: presence
Subscription-State: active;expires=7200
Require: eventlist
Content-Type: multipart/related;type="application/rlmi+xml";
              start="<2BEI83@chicago.handbook.com>";
              boundary="TfZxoxgAvLqgj4wRWPDL"
Content-Length: 2862
```

```
--tuLLl3lDyPZX0GMr2Yoo
Content-Transfer-Encoding: binary
Content-ID: Cvjpeo@new_york.handbook.com
Content-Type: application/rlmi+xml;charset="UTF-8"

<?xml version="1.0" encoding="UTF-8"?>
<list xmlns="urn:ietf:params:xml:ns:rlmi"
      uri="sip:me-list@new_york.handbook.com" version="1"
      fullState="true">
  <name language="en">Buddy List at Handbook</name>
  <resource uri="sip:joe@new_york.handbook.com">
    <name>Joe Thomas</name>
    <instance id="1" state="active"
              cid="mrEakg@new_york.handbook.com"/>
  </resource>
  <resource uri="sip:mark@new_york.handbook.com">
    <name>Mark Edwards</name>
    <instance id="1" state="active"
              cid="KKMDmv@new_york.handbook.com"/>
  </resource>
</list>

--tuLLl3lDyPZX0GMr2Yoo
Content-Transfer-Encoding: binary
Content-ID: mrEakg@new_york.handbook.com
Content-Type: application/pidf+xml;charset="UTF-8"

<?xml version="1.0" encoding="UTF-8"?>
<presence xmlns="urn:ietf:params:xml:ns:pidf"
     entity="sip:joe@new_york.handbook.com">
  <tuple id="7823a4">
    <status>
      <basic>open</basic>
    </status>
    <contact priority="1.0">sip:joe@new_york.handbook.com
      </contact>
  </tuple>
</presence>

--tuLLl3lDyPZX0GMr2Yoo
Content-Transfer-Encoding: binary
Content-ID: KKMDmv@new_york.handbook.com
Content-Type: application/pidf+xml;charset="UTF-8"

<?xml version="1.0" encoding="UTF-8"?>
<presence xmlns="urn:ietf:params:xml:ns:pidf"
     entity="sip:mark@new_york.handbook.com">
  <tuple id="398075">
    <status>
      <basic>closed</basic>
```

```
    </status>
  </tuple>
</presence>

--tuLL131DyPZX0GMr2YOo--

--13WMZaaL8NpQWGnQ4mlU
Content-Transfer-Encoding: binary
Content-ID: K9LB7k@new_york.handbook.com
Content-Type: application/pkcs7-signature

[PKCS #7 signature here]

--13WMZaaL8NpQWGnQ4mlU--

--TfZxoxgAvLqgj4wRWPDL--
```

M10: The terminal completes the NOTIFY transaction.

```
Terminal -> Local RLS
SIP/2.0 200 OK
Via: SIP/2.0/TCP chicago.handbook.com;
  branch=z9hG4bK4EPlfSFQK1
From: <sip:me-list@chicago.handbook.com>;tag=zpNctbZq
To: <sip:me@chicago.handbook.com>;tag=ie4hbb8t
Call-ID: cdB34qLToC@terminal.chicago.handbook.com
CSeq: 997935769 NOTIFY
Contact: <sip:terminal.chicago.handbook.com>
Content-Length: 0
```

5.6 Current Work

5.6.1 Third Generation Wireless Networks

In this section, we introduce the SIP application service in third generation wireless networks. We do not expect readers to understand the third
generation wireless networks through the content of this section. We only
generally introduce some organization of third generation telecommunication
and its system architecture to indicate that our previous technical knowledge
is widely used in the latest technology of the industry. If readers are interested
in third generation wireless networks, they can read the reference documents.

5.6.1.1 IP Multimedia Subsystem (IMS) of 3GPP

IMS [8]: In 1998, the mobile world realized that migration to all-IP would
help to globalize the telecommunication market and the 3GPP effort was
launched. In 2000, 3GPP made SIP part of the all-IP mobile standard under

the name IMS-Internet multimedia subsystem. The IP Multimedia Subsystem (IMS) is an architectural framework for delivering IP multimedia to mobile users.

In order to achieve access independence and to maintain a smooth interoperation with wireline terminals across the Internet, the IP multimedia subsystem attempts to be conformant to IETF "Internet standards." Therefore, the interfaces specified conform as far as possible to IETF "Internet standards" for the cases where an IETF protocol has been selected, e.g., SIP.

The SIP application server may host and execute services. The SIP application server can influence and impact the SIP session on behalf of the services and it uses the ISC interface (IP Multimedia Subsystem Service Control Interface) to communicate with components of the IMS core network.

The Third Generation Partnership Project (3GPP) [8]: The 3GPP is a collaboration agreement that was established in December 1998. The collaboration agreement brings together a number of telecommunications standards bodies that are known as organizational partners.

The original scope of 3GPP was to produce globally applicable technical specifications and technical reports for a Third Generation Mobile System based on evolved GSM core networks and the radio access technologies that they support (i.e., Universal Terrestrial Radio Access (UTRA) both Frequency Division Duplex (FDD) and Time Division Duplex (TDD) modes). The scope was subsequently amended to include the maintenance and development of the Global System for Mobile communication (GSM) technical specifications and technical reports including evolved radio access technologies (e.g., General Packet Radio Service (GPRS) and Enhanced Data rates for GSM Evolution (EDGE)).

Notification: The ISC interface shall be able support subscription to event notifications between the application server and S-CSCF to allow the application server to be notified of the implicit registered public user identities, registration state and user end capabilities and characteristics in terms of SIP user agent capabilities and characteristics.

The SIP-event notification mechanism allows a SIP entity to request notification from remote nodes indicating that certain standardized events have occurred. Examples of such of events are changes in presence states, changes in registration states, changes in subscription authorization policies (see [6]) and other events that are caused by information changes in application servers or S-CSCF. It shall be possible to either fetch relevant information once or monitor changes over a defined time. It shall be possible for a user to subscribe to events related to his/her own subscription (e.g., when the user subscribes to his own registration state) or to events related to other users' subscriptions (an example is when a watcher subscribes to presence information of a presentity, see [6]).

Group Services and Presence Service in 3GPP: [6]

The presence service provides the ability for the home network to manage presence information of a user's device, service or service media even whilst roaming. A user's presence information may be obtained through input from

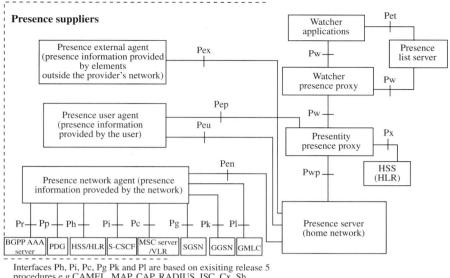

Interfaces Ph, Pi, Pc, Pg Pk and Pl are based on exisiting release 5
procedures e.g CAMEL, MAP, CAP, RADIUS, ISC, Cx, Sh
The Pr, Pp interfaces are based on existing release 6 procedures of
the 3GPP-WLAN interworking architecture.

FIGURE 5.4
Reference architecture to support a presence service.

the user, information supplied by network entities, or information supplied by
elements external to the home network. Consumers of presence information
(watchers) may be internal or external to the home network. It uses [1, 2],
and [3] see Figure 5.4 below.

The presence server shall be able to receive and manage presence information that is published by the presence user/network/external agents, and
shall be responsible for composing the presence-related information for a certain presentity from the information it receives from multiple sources into a
single presence document.

The presence list server stores grouped lists of watched presentities and enables a watcher application to subscribe to the presence of multiple presentities
using a single SUBSCRIBE transaction. The presence list server also stores
and enables the management of filters associated to presentities in the presence
list. The presence list server shall attach an associated filter to each individual
SUBSCRIBE transaction. It is based on [7]

For detail, please see the standard: [6]

5.6.1.2 OMA Presence Service

OMA was formed in June 2002 by nearly 200 companies, including the world's
leading mobile operators, device and network suppliers, information technology
companies, and content and service providers. The fact that the whole value

chain represented in OMA marks a change in the way specifications for mobile services are done. Rather than keeping the traditional approach of organizing activities around "technology silos," with different standards and specifications bodies representing different mobile technologies, working independently, OMA is aiming to consolidate into one organization all specification activities in the service enabler space. [9]

The OMA presence service is a set of system capabilities and related applications that provides both a framework for future service development and an application presence (i.e., sharing of status and reachability information) between users when using mobile networks.

3GPP and 3GPP2 have defined an aligned presence service framework in [6] and [10]. This framework has a defined presence reference architecture both in "network layer" and "application layer" meaning that 3GPP and 3GPP2 specifications [6] and [10] respectively define end-to-end presence information flows and content for the presence information document. The term "network layer" refers to the communication that is required between the presence service functional elements (e.g., presence server) and various network elements as they are defined in the network architectures of 3GPP and 3GPP2 (e.g., MSC, HLR). The term "application layer" refers to the communication that is required between the various presence service elements (e.g., presence server and presence source), as it is shown in Figure 5.4, which includes the "application layer" functional entities.

Additionally, there are presence services that exist or can be envisaged that do not leverage core network infrastructure as defined by 3GPP and 3GPP2. However, those presence services are still relevant to the mobile domain and thus supported by this OMA presence architecture.

See Figure 5.5 below.

The following Figure illustrates the OMA presence architecture.

See Figure 5.6 below.

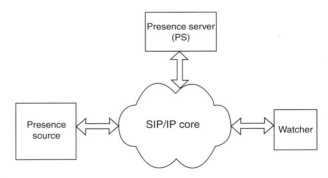

FIGURE 5.5

Abstraction of the presence reference architecture.

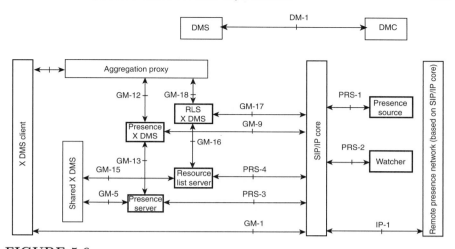

FIGURE 5.6
SIMPLE presence reference architecture.

Presence Server: The presence server (PS) supports the following: Accepts and stores presence information published to it. Distributes presence information and watcher information. The PS is able to subscribe to changes to documents stored in the shared and presence XDMS. The PS is able to fetch documents from the shared XDMS and the presence XDMS.

Presence Source: The presence source is an entity that provides presence information to a presence service [4]. The presence source can be located in the user's terminal or within a network entity.

Watcher: A watcher is an entity that requests presence information about a presentity, or watcher information about a watcher, from the presence service.

Resource List Server (RLS): The RLS is the functional entity that accepts and manages subscriptions to presence lists, which enables a watcher application to subscribe to the presence information of multiple presentities using a single subscription transaction. The RLS is able to subscribe to changes to documents stored in the shared and RLS XDMS. The RLS is able to fetch documents from the shared XDMS and the RLS XDMS.

The SIP/IP Core: The SIP/IP core is a network of servers, such as proxies and/or registrars, that perform a variety of services in support of the presence service, such as routing, authentication, compression, etc. The specific features offered by different types of SIP/IP core networks will depend on the particulars of those networks.

The presence service uses the PIDF [11] as the base format through which presence information is represented. The most basic unit of presence information in PIDF is called a tuple. The OMA presence service defines the most basic units of presence information as presence information elements, in order to avoid terminology conflicts (e.g., in OMA IMPS/Wireless Village those basic units are called attributes). The OMA presence service will define mappings of those presence information elements onto the PIDF format.

5.6.2   Current Research and Standard of Presence Features That Have Been Developed for SIP

PIDF is a very extensible format, and being based on XML allows the trivial addition of namespaces to the basic structure. The most important presence features of SIMPLE are provided by extensions to PIDF. Several of those are outlined below [12]:

(1) CIPID [5]: CIPID, Contact Information in Presence Information Data Format is an extension to PIDF. CIPID adds multiple new optional elements for presenting contact information more diversely, such as <card>, <display-name>, <homepage>, <icon>, <map> and <sound>. CIPID's elements can be easily combined into a buddy list and it does not extend the formal way of presenting the list of tuples, or devices a presentity may have listed.

(2) User Agent Capability [13]: User agent capability defines a mechanism for SIP user agents to transport their technical capabilities and characteristics to other user agents. User Agent Capabilities Extension (henceforth UACE) to PIDF introduces an extension that can convey SIP specific media feature tags. These tags contain information on what types of media in what way can a tuple handle.

(3) RPID [14]: The rich presence information data format defines a large body of properties that can be included in a presence information entry.They define logical properties such as user mood and activity, as well as physical properties such as location and environment. This information is designed with mobile devices in mind, and provides very specific information about both the device and the user.

(4) Partial Presence (PIDF-diff) [15]: Presence Information Data Format (PIDF) Extension for Partial Presence, or PIDF-diff, is an extension to PIDF. The problem with PIDF is that it demands full presence information with every single change or update in presence of an entity. This can cause congestion problems with mobile and other low-bandwidth and high latency devices. This document introduces a new MIME-Type "application/pidf-diff+xml" that enables transporting of either only the changed parts or the full PIDF-based presence information. The root element of the document distinguishes whether the partial or full PIDF document content was transported.

(5) Timed Presence [16]: Traditionally, presence information, e.g., represented as PIDF and RPID, describes the current state of the presentity. However, a watcher can better plan communications if it knows about the presentity's future plans. For example, if a watcher knows that the presentity is about to travel, it might place a phone call earlier. It is also occasionally useful to represent past information since it may be the only known presence information; it may give watchers an indication of the current status. For example, indicating that the presentity was at an off-site meeting that ended an

hour ago indicates that the presentity is likely in transit at the current time. This document defines the <timed-status> element that describes the status of a presentity that is either no longer valid or covers some future time period.

(6) Location data [17]: While not specifically part of the SIMPLE work, a related working group called GEOPRIV has developed an extension to PIDF for transporting location data. Location data is likely to become more important and pervasive in the future and it makes sense to define a way of its expression since it is closely related to the properties defined above. Also, location data can facilitate location-based services, can be gathered automatically with no effort from the user, and is a valuable bit of information in determining the situation the user is in.

References

[1] Rosenberg, J., H. Schulzrinne, G. Camarillo, A.R. Johnston, J. Peterson, R. Sparks, M. Handley, and E. Schooler. 2002. SIP: session initiation protocol. In *RFC 3261, Internet Engineering Task Force*, June.

[2] Roach, A.B. 2002. Session initiation protocol (SIP)-specific event notification. In *RFC 3265, Internet Engineering Task Force*, June.

[3] Rosenberg, J. 2004. A presence event package for the session initiation protocol (sip). In *RFC 3856, Internet Engineering Task Force*, August.

[4] Day, M., J. Rosenberg, and H. Sugano. 2000. A model for presence and instant messaging. In *RFC 2778, Internet Engineering Task Force*, February.

[5] Schulzrinne, H. 2006. CIPID: Contact information in presence information data format. In *RFC 4482. Internet Engineering Task Force*, July.

[6] 3GPP. 2005. Presence service: Architecture and functional description for release 6. version 6.9.0. *3GPP*, http://www.3gpp.org/ftp/Specs/html-info/23141.htm.

[7] Roach, A.B., B. Campbell, and J. Rosenberg. 2006. A session initiation protocol (SIP) event notification extension for resource lists. In *RFC 4662. Internet Engineering Task Force*, August.

[8] 3GPP. 2003. IP multimedia subsystem. version 6.0.0. *3GPP*, http://www.3gpp.org/Specs/html-info/23228.htm.

[9] OMA. 2004. Group management architecture. version 1.0. *OMA*, http://member.openmobilealliance.org/ftp/public_documents/PAG/Permanent_documents/OMA-PAG-GM-AD-V1_0_0-20041118-D.zip.

[10] 3GPP2. 2004. Presence service: Architecture and functional description. *3GPP2*, http://www.3gpp2.org/Public_html/Specs/X.S0027-001-0_v1.0_041004.pdf.

[11] Sugano, H., S. Fujimoto, G. Klyne, A. Bateman, W. Carr, and J. Peterson. 2004. Presence information data format (PIDF). In *RFC 3863. Internet Engineering Task Force*, August.

[12] Rajaniemi, H.-P., and K. Yanev. SIP and presence. http://www.cs.helsinki.fi/u/yanev/simplep.pdf.

[13] Rosenberg, J., H. Schulzrinne, and P. Kyzivat. 2004. Indicating user agent capabilities in the session initiation protocol (SIP). In *RFC 3840. Internet Engineering Task Force*, August.

[14] Schulzrinne, H., V. Gurbani, P. Kyzivat, and J. Rosenberg. 2006. RPID: Rich presence extensions to the presence information data format (PIDF). In *RFC 4480. Internet Engineering Task Force*, July.

[15] Lonnfors, M., E. Leppanen, H. Khartabil, and J. Urpalaninen. 2007. Presence information data format (PIDF) extension for partial presence. *draft-ietf-simple-partial-pidf-format-10. Internet Engineering Task Force*, November.

[16] Schulzrinne, H. 2006. Timed presence extensions to the presence information data format (PIDF) to indicate presence information for past and future time intervals. In *RFC 448. Internet Engineering Task Force*, July.

[17] Peterson, J. 2005. A presence-based GEOPRIV location object format. In *RFC 4119. Internet Engineering Task Force*, December.

6

Group Conference Management with SIP

Thomas C. Schmidt and Matthias Wählisch
Hamburg University of Applied Sciences

CONTENTS

6.1 Introduction 123
6.2 Group Conferencing Concepts and Technologies 132
6.3 Infrastructure-Assisted Conferences 145
6.4 Peer-Managed Conferences 147
6.5 Summary and Conclusions 152
    References 153

6.1 Introduction

Voice and video conferencing on the Internet is about to become a lightweight day-to-day application. This trend follows from high-bandwidth data connections, which are increasingly available to the public at reasonable prices. There is also some remarkable progress in video/audio compression algorithms, which reduce a media data stream considerably and reconstruct it again to a high-quality playout sequence on the receiver site. Furthermore, battery powered mobile devices rapidly gain processing and communication performance. They currently can host desktop conferencing software, thereby seamlessly using any Internet connectivity available.

Conferencing through the International Telecommunication Union telephone architecture has been known for years as a feature-rich centralized service, suffering from its static, complex, and expensive nature. Many of the issues in multi-party communication can be solved in a simpler, more generic way over the Internet, where users, groups, and devices can be addressed individually and independent of each other. Session Initialization Protocol (SIP) [1] as presented by the Internet Engineering Task Force (IETF) forms a flexible, comprehensive signaling solution and allows for a lightweight deployment of many, but not all of the well-known service features. Since media sessions are only negotiated by, but not tight to SIP control streams, calls may be arranged in a variable, adaptive fashion. This degree of freedom is of particular importance in large conferences, where media transmission and processing can easily rise beyond the capacities of single devices.

The original development of SIP was inspired by connection-oriented telephone services, whence its nature derives from a point-to-point model. It is designed as a multi-layered application protocol that interacts between components in a transactional way. Each (asynchronous) request initiates an open transaction state and requires completion by at least one response. Group communication complicates this process significantly. A newly joining member faces an entire group, which requires appropriate addressing and transactional state management. Negotiations on media parameters grow complex as common parameter intersections may have to be evaluated for many members. Extensions to perform scalable group session management are not easy to achieve, while schemes refrain from central control.

In multimedia conference scenarios, each member commonly operates as receiver and as sender on a group communication layer. In addition, real-time communication such as voice or video over Internet Protocol (IP) places severe Quality of Service (QoS) requirements: Seamless distribution services need to limit disruptions or delays to less than 100 ms. Jitter disturbances should not exceed 50 ms. Note that 100 ms is about the duration of a spoken syllable in real-time audio.

As an extensible protocol, SIP is open for the creation of new methods, header fields and protocol semantics. This opportunity is extensively used for conferencing, resulting in a large number of conceptual documents describing standard extensions, best current practices, draft proposals, etc., see [2] for a guided overview. The present contribution does not enumerate proposed features, but rather concentrates on core concepts and tries to outline conferencing solutions of common use as well as future directions of promising development.

This chapter will first illustrate the fundamental issues and SIP concepts for multiparty conversations along the line of examples and characteristic applications. An overview of core concepts and technologies for SIP-initiated group conferencing follows in Section 6.2. Point-to-point schemes and multicast solutions are covered herein, as well as mobility aspects. Section 6.3 then takes a closer look at SIP standard infrastructure components and their potential to facilitate group conferencing in an uncomplex manner. A detailed discussion of conferences solely managed by peers is the focus of the subsequent Section 6.4. Finally, a summary and conclusions will complete this report on conferencing.

6.1.1 Two Introductory Scenarios

Group communication can be of manifold nature and a variety of quite different views or scenarios have been contributed to the field. For a start at a prime perspective, two well-known and established synchronous group communication services from everyday life are considered: The three-way conference, such as Integrated Services Digital Network (ISDN) telephone systems on the one hand, and the reception of live broadcast media such as radio and television, with an optional offer of selective feedback channels, on the other. While in

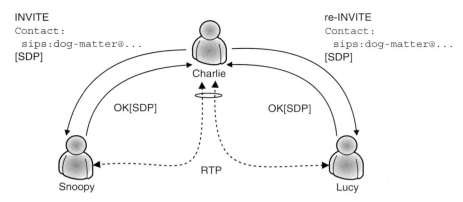

FIGURE 6.1
A three-party dialog.

the first scenario users act primarily dialog-oriented, most of them are bound to passive reception in the distribution-oriented second setup. In the following, a closer inspection will reveal more characteristic differences.

6.1.1.1 Three-Way Conference

Often a three-way conference is initiated spontaneously from a two-party session. For example, while Charlie and Lucy are on a call, Charlie decides to add Snoopy into the conversation or Snoopy rings him, being aware or unaware of the already established dialog. In either case, Charlie is in the role to act upon Snoopy joining in and thereby turning the session into a conference. As visualized in Figure 6.1, parties are in individual, point-to-point contacts and naturally manage conference negotiations and policy operations between peers. Switching the communication context from a two-party session to a three-party conference raises two additional duties at Charlie's site. The parallel calls, one with Lucy and the other with Snoopy, need a logical join to form one conference. Also, media data have to be arranged to arrive at all three participants.

The simplest solution for this end-point hosted conference would work without additional signaling, when realized within the conferencing application. Based on user reactions to calls, Charlie's end system could sort the two sessions into a virtual conference, and at the same time, start to mix media so that each correspondent receives all the information from Charlie within one stream*. Lucy and Snoopy could thus participate without logical or technical awareness of the multi-party situation.

There are, however, drawbacks of this simplistic approach. Obviously, all communication relies on the presence of Charlie—his disappearance will terminate both calls. In particular, there will be no seamless mechanism to turn

---

*Media mixing is easily achieved for voice, but does likewise work for video with combined pictures or virtual environments with merged update sets.

the conference into a call between Lucy and Snoopy. Scaling issues may arise from media mixing, which requires transcoding in the absence of a common coding scheme. Without explicit group management, no opportunity is given to negotiate on common codecs, nor are means provided to distribute mixing tasks or redirect media streams in many-party scenarios. Finally, privacy concerns are raised by a solution that allows for an undisclosed third party joining in a conversation—an explicit IETF policy holds off Internet Protocols from "wiretapping" [3].

SIP resolves these issues by explicitly defining a conference focus, which is identified by a Uniform Resource Identifier (URI). This URI represents the conference, and while it is uniquely created, additional Session Description Protocol (SDP) [4] media negotiations are foreseen [5]. The focus forms the central point of control only for the SIP conference management, which is free to define media distribution or mixing otherwise. The focus may be altered within an ongoing multi-party conversation, but this will lead to the creation of a new conference instance, distinguished by a newly defined conference URI.

In detail, SIP operations for our example will proceed as follows (see [6]). After Charlie decided to invite Snoopy into the conversation with Lucy, he generates a conference URI and issues a SIP INVITE to Snoopy using this URI in his contact field as follows:

```
INVITE sips:snoopy@dog.net SIP/2.0
Via: SIP/2.0/TLS
alpha.brown.com:5061;branch=z9hG4bcHlkapff
Max-Forwards: 70
From: Charlie <sips:charlie@brown.com>;tag=4576932
To: Snoopy <sips:snoopy@dog.net>
Call-ID: 777777@alpha.brown.com
CSeq: 1024 INVITE
Contact: <sips:dog-matter@alpha.brown.com>;isfocus
Content-Type: application/sdp
...
```

Even though the contact URI addresses Charlie directly, it is not inherently bound to a conference situation. Charlie needs to add an "isfocus" feature tag to make the multi-party situation transparent and to mark himself as the focus point. After the session has been successfully established, with SDP offer/answer negotiations aware of Lucy's media capabilities and the conference media distribution policy included [7], Charlie analogously re-invites Lucy into the conference (see Section 6.2.2):

```
INVITE sips:lucy@psychic.org SIP/2.0
Via: SIP/2.0/TLS
alpha.brown.com:5061; branch=z9hG4bKnashds
Max-Forwards: 70
From: Charlie <sips:charlie@brown.com>;tag=23431
To: Lucy <sips:lucy@psychic.org>;tag=1234567
```

```
Call-ID: 888888@alpha.brown.com
CSeq: 1024 INVITE
Contact: <sips:dog-matter@alpha.brown.com>;isfocus
Content-Type: application/sdp
...
```

Note that Lucy can explicitly react upon the group context and—if desired—decline the re-invitation.

If Snoopy wishes to contact Charlie instead, he may be aware or unaware of an already ongoing conference. In the latter case, he will just call Charlie who will reply with his conference contact information or re-invite Snoopy after their two-party dialog has been established. In the first case, when Snoopy knows about the conference including the identification of one ongoing call, he can explicitly express his will to participate using the join header field (see Section 6.2.2).

This simple, end-point hosted conferencing scheme can be extended to an arbitrary number of participants, only limited by scalability issues. Two extensions beyond the base specification have been in use, the "isfocus" tag and the join header. Any conference-unaware user agent implementing only RFC 3261 cannot act as a focus or request to join an ongoing conference, but may, nevertheless, participate in a regular client role. The presence of the 'isfocus' tag in a contact header field does not cause interoperability issues, since it will be simply ignored as an unknown header parameter.

6.1.1.2 Large-Scale Conference

Large conferencing instances usually go along with some external occasion or scheduling. This may be a program announcement via a Web page, SAP [8], or printed guides like in Internet Protocol Television (IPTV) offers or a streamed real-world conference, a meeting organized through personal communication or e-mail like in appointed conference calls, or well-known community information in contexts like gaming. These application scenarios all have in common not only large scaling requirements, but also the need for explicit policy management. In cases where feedback is foreseen and distribution does not remain unidirectional, floor control mechanisms must guide party interactions. Media distribution will scale up to millions of users, if IP multicast is used (see Section 6.2.3). For smaller numbers, powerful multipoint control units (MCUs) may satisfy the demands, or multicast implementations on the application layer will enable an optimized stream replication without infrastructure assistance.

Distribution-oriented large-scale conferences do not necessarily require SIP. Charlie, for example, can subscribe to an open media channel just by joining a multicast group. He may use SIP to inquire on multicast addresses, for authentication and authorization and for enabling feedback control. Conference control thereby may be fully decoupled from media distribution and reside on a separate entity as displayed in Figure 6.2.

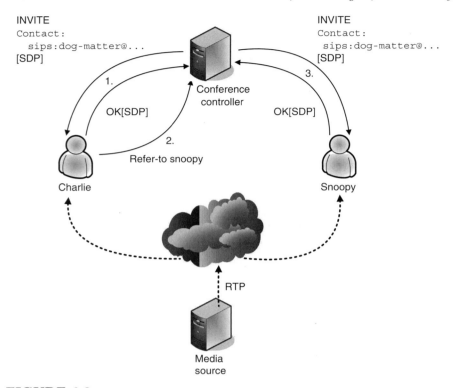

FIGURE 6.2
A multi-party distribution session.

Beyond minimal management, SIP supplies a number of optional features, which facilitate interesting interactive enrichments of a basic, reception-oriented application. Charlie could want Snoopy to participate in an ongoing conference. By subscribing to the SIP event package for conferencing states, see Section 6.2.2, he could inquire on Snoopy and discover his absence. To invite Snoopy into the conference, either for raising his awareness or for seamlessly integrating his conference-unaware user agent client, he could use the REFER method as described in Section 6.2.2. Charlie is also able to submit feedback after registering with the conference, potentially under a basic floor control mechanism [9] smoothly embedded in SIP addressing.

6.1.2 The Application Domain for SIP-Managed Group Conferencing

6.1.2.1 Audio Conferences

The primary dedication of SIP lies in a standardized signaling for transparent voice-over IP services. SIP thereby simultaneously targets campus solutions in the range of Private Branch Exchanges (PBXs) or the ITU H.323 [10] VoIP architecture, and global call handling based on Internet routing or, possibly in parts, on the public switched telephone network. For the sake of competitiveness, core voice conferencing functionality as offered by the digital telephone infrastructure or available through in-house conference control units must be part of the service spectrum of SIP. Traditional voice conference calls allow for easy media mixing, consist of limited group sizes, and SIP-initiated Internet telephony solutions can easily cope with underlying demands. Obviously, such characteristics likewise hold for simpler or supplementary group applications as chat or whiteboards. However, as a flexible signaling protocol suitable for decentralized relaying architectures, SIP may give rise to a multitude of additional services, unknown in traditional telephone networks.

6.1.2.2 Videoconferencing over IP

The idea of augmenting voice conferences by video has been around for several decades, but only the flexibility of the Internet generated a noticeable deployment. As compared to audio, video processing places significantly higher demands on end system and network transmission capabilities. The rapid evolution of networks and processors have paved the way for realistic group conferences conducted at standard personal computers, combining about a dozen visual streams of Half-Quarter Video Graphics Array (QVGA) (240 × 160 pixel @ 15–25 fps) resolution. Thus, initial centralized implementations based on H.323 are now superseded by peer-centric personal systems with SIP. Lightweight videoconferencing software today smoothly integrates desktop video of high quality with thin clients on mobile phones [11] like the daViKo system†.

In concordance with communication capabilities, video coding techniques have evolved as well. The latest standard for video coding H.264/AVC [12], although designed as a generic standard, is predestined for applications like mobile video communications. Besides enhanced compression efficiency, it also delivers network-friendly video representation for interactive (video telephony) and non-interactive applications like broadcast, streaming, storage, and video on demand. H.264/AVC provides gains in compression efficiency of up to 50% over a wide range of bit rates and video resolutions compared to previous standards. While H.264/AVC decoding software has been successfully deployed on handhelds, its high computational complexity still challenges current mobile devices when implemented as pure software encoders, however, fast hardware implementations are available. Next generation codecs like Scalable Video Coding (SVC) are already approved [13]. The main new feature, scalability, addresses schemes for delivery of video to diverse clients over heterogeneous networks, particularly in scenarios where the downstream conditions are not known in advance. The basic idea is that *one* encoded stream can serve networks with varying bandwidths or clients with different display resolutions or systems with different storage resources. This not only enables media mixers

---

†see http://www.daviko.com

or conference bridges to simultaneously serve streams of different resolution without transcoding, but is an obvious advantage in heterogeneous networks prevalent in mobile applications, as well.

6.1.3 Gaming

Multiplayer games fall into the realm of SIP conferencing services for two reasons. At first, interactive multi-party gaming is built on session-oriented group communication, which relies on standard signaling relations and largely benefits from additional presence and conference information as provided by SIP. Second, an increasing number of massive multiplayer games offers sidebar voice and chat conferencing services to enrich communication among players.

Sidebar conferences in games can just be treated like any (cascaded) SIP session. A context-aware integration of voice conferencing with multiplayer networked gaming has been presented in [14] along with a SIP-based reference architecture. The authors foresee a tight coupling of game and conference servers by SIP means, which will lead to an automated conference reconfiguration based on third party call control (see Section 6.2.1), whenever a switch in the gaming context occurs. For example, if a player changes a room in some game arena, its audio conference will then include the players in the new room and not the old ones. This solution fully complies with the SIP conferencing approach, for which reason the architectural realizations extend from a strictly centralized to a fully distributed model.

Even though today's popular games do not employ SIP standard messaging, this large software market remains open as an appropriate candidate for migrating to SIP, which in turn may facilitate more elaborate, less expensive, distributed architectures for game deployment.

6.1.4 Presence and Instant Messaging

Session dialogs and conferencing are person-oriented services that abstract from specific communication channels and from devices. The way and the instance of favorable information exchange not only depends on content and external demands, but on the individual contexts of each involved party. Compared to the Public Switched Telephone Network (PSTN), it is one of the predominant features of the Internet to rigorously preserve this abstraction of user presence over technical entities[‡]. Congruously—in taking up earlier, proprietary solutions—SIP fills this paradigm by providing presence information about parties. Depending on present state, participants may choose to communicate by short messages, e-mail, voice and video, interact within dedicated collaborative environments or not at all.

Presence information not only resolves the physical location of a person or an application, but may indicate the ability and willingness of a user to communicate. Rich presence information conveys the personal and contextual

---

[‡]Even though this is merely a reformulation of the layered design principle, it is worth noting that at this occasion we benefit from a long-term resistance to layer violations.

circumstances of a contact, and a caller *a priori* may judge whether to pursue a contact while the callee is busy at the airport and in a bad mood. This may be particularly useful in larger or extensive conferences as an unpretentious way of feedback from silent parties.

Presence indicators follow the SIP event state model, which is outlined in Section 6.2.2. They enrich a community model of "buddies," whose presence is concurrently cultivated and silently enriched by instant message (IM) exchange. SIP enables instant messaging in two ways: Short textual news may be submitted in pager mode via the MESSAGE method [15] aside from session establishment. For more extensive IM exchange, the session mode negotiates media exchange via SDP, where transport in IM sessions is facilitated by the Message Session Relay Protocol (MSRP) [16]. Messages can be exchanged individually or in conferences, realized as client–server or peer-to-peer communication. In particular, IM may be integrated in a voice conferencing infrastructure as a complementary communication format or an orthogonal channel. The latter allows a phoning party to exchange messages with a third person within the same user agent.

IM is now part of a large number of tremendously popular personal applications. The technical agreement by AOL, IBM, and Microsoft on presence and IM interoperability, SIP for IM and Presence Leveraging Extensions (SIMPLE), has tied this success to SIP.

6.1.5 SIP and the IMS

An alternate approach to SIP conferencing is taken with the ITU-T next generation network (NGN) architecture. Using Internet technologies, 3G mobile operators and wire phone companies are in the process of building next generation telephone services based on the IP Multimedia Subsystem (IMS) or its wireline emulation TISPAN. SIP has been selected as the session signaling protocol as RTP was formerly chosen in H.323 for media transport. However, this provider-centric, network-controlled and application-aware network architecture takes a completely different perspective.

The IMS accounts for VoIP/VCoIP, presence, messaging, gaming, etc., services, which as physical server instances, reside on the application server layer. A likewise centralized server instance at this layer will enable conference management. For integration purposes and ease in billing, all these session-oriented services reside on a common session layer, which aggregates states from SIP message exchanges with application servers. As of 3GPP release 6, Multimedia Broadcast and Multicast Services (MBMS) are foreseen. Anchored at a regional Gateway GPRS Support Node (GGSN), group communication may thus be used to enhance conference data distribution. This strongly debated, complex architecture provides SIP-initiated group conferencing services within a provider domain. The future will uncover the validity of the IMS business model and reveal whether people are willing to pay for dedicated conferencing services offered by operators or just use applications available from independent sources.

6.2 Group Conferencing Concepts and Technologies

Conferencing enriches the semantic of basic group communication by providing presence or session logic to the application. Conversely, it widens the meaning of dialog-oriented conversation by augmenting signaling relationships with the multi-party paradigm. Examining both aspects separately, concepts and technologies are around for serving all needs. Group communication is enabled by multicast session management by SIP, which has a base standard [1] that already defines a minimal interplay with IP-layer multicast. However, to arrive at conferencing solutions of full functionality and comfort, further efforts of integration are needed. This section provides an overview of the main service models and their technological primitives on the SIP and distribution layer within the framework of group conference management.

6.2.1 Common Service Models for Conferencing

Group conferencing must be considered a generic term for a wide variety of meanings in different contexts. Generally, it should be understood as an instance or realization of a multi-party conversation and may consist of one or several SIP sessions, each of which combines SIP dialogs as required for communication between participants. Such multi-party conversation space may include human and non-human users, e.g., tone-generating robots or recording monitors, which may be active and noticeable or hidden. Conferencing may be essentially realized by the three following service models of multi-party communication [5, 17].

6.2.1.1 Loose Coupling

Within conferences following a loosely coupled model, no signaling relationship is maintained between conference participants. This considerably lightweight approach is exceptionally scalable, easy to implement, but does not provide any SDP offer/answer dialog. Also, no standard way of authentication, policy execution, or conference coordination is provided. There is no single instance of control nor a conference server or focus. Instead, participation may be gradually learned by RTCP [18] control streams. Its realization requires an any-source multicast distribution layer for media streams. A conference can be entered by simply joining a multicast group, while addresses and media information may be pre-shared by non-SIP means. Alternatively, a SIP party may be invited by any single conference member or issue its INVITE to an ASM group established for SIP signaling, as described in Section 6.2.3.

6.2.1.2 Tight Coupling

Tightly coupled conferences follow a simple, matured communication model and, as of today, provide the most elaborate services to the users. They rely on a central managing instance, granting a signaling relationship to all participants through a globally routable conference URI (GRUU) [19]. Such a conference controller acts as a focus, which governs the conference by applying policy rules and can provide a variety of convenience functions. It may additionally perform media mixing and redistribution, in which case its functionality is equivalent to a conference bridge or multipoint control unit (MCU) known from PSTN and H.323 [10]. To avoid scalability limits inherent to the ITU architecture, media distribution in a tightly coupled conference is not bound to centralized mixing. The general architecture as defined in [5] is drawn in Figure 6.3. The focus, assisted by policy and notification service functions, forms the center of a star topology for SIP session management. Notification and policy service may be part of the conference server or distributed and coupled by non-SIP means. RTP [18] media streams flow in a serverless way, e.g., via multicast, in the upper part of the graph, while a cascade of mixers in the lower part distributes media as instructed by the focus.

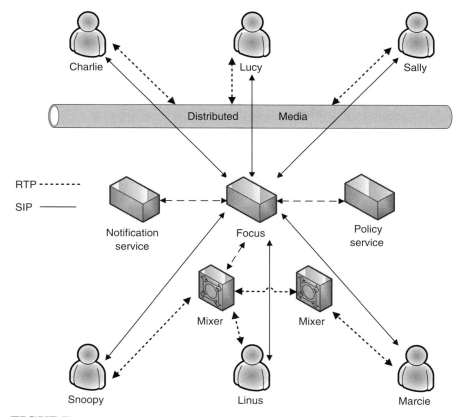

FIGURE 6.3
General architecture of a tightly coupled conference.

The central conference focus is responsible for maintaining dialogs with all participants and to ensure that everyone is served with all the media for the conference. Therefore, the conference functions minimally offered are authenticated and authorized session establishments, including media negotiations. In the presence of media mixers, it uses its session states and the media policy to appropriately configure one or several mixers. For this purpose, the focus interacts with the mixer either through a local interface or, in a distributed scenario, via INVITE/re-INVITE sequences allowing for third party call control (3pcc) [20]. It thereby is fully enabled to (re-)organize media streams at its convenience, e.g., to optimize media distribution according to network and device capabilities. It may follow an efficient strategy of balancing (cascaded) mixers. Additional convenience functions offered by the conference focus most likely are a third party invite (dial-out) for conference-unaware user agents, removal of third parties from the conference, and notification services for conference event states. Extended services may be equally included, such as floor control, conference announcements and recordings, Network Address Translation (NAT) traversal assistance, user directory, integration and billing functions, etc.

6.2.1.3   Full Distribution

The fully distributed multi-party model is built upon peer-to-peer signaling relationships between conference members in the sense that:

- Participants have the means to contact each other on an individual basis and negotiate SDP; likewise they may inquire on their presence;

- There is no central point of control nor individual instance of a focus;

- Some distributed algorithm is cooperatively performed to achieve appropriate conference coordination and media control.

Distributed peer-to-peer management allows for infrastructureless, instantaneous conference establishment and operation and, if properly designed, may encounter faster and error-resilient dialog handling. In contrast to tight coupling, definitions for this conference model are not widely matured, but remain an active area of research and future protocol development.

Most simplistically, such a model can be realized in a pairwise unicast or full mesh relationship between n parties. However, due to its n^2 signaling complexity, this approach remains suitable only for small conferences. Improvements are subject to specialized mesh optimizations. Scalability up to very large groups can be achieved by the use of multicast packet distribution, which – in contrast to the loosely coupled model – can follow a source specific model on the IP layer for both session management and media distribution; see Section 6.2.3 for a detailed discussion. User agent clients may also make use of a general multicast service resident on the application layer. Subject to its algorithmic realization, distribution efficiency of application layer multicast commonly increases with a growing number of participants and promotes scalability.

Complementary approaches to distributed conferencing are opened up by structured peer-to-peer technologies. Distributed hash tables (DHTs) offer an efficient, fully distributed persistence layer and can facilitate a decentralized realization of some or all functions of a conference focus. Such a distributed conference coordination scheme raises global routability and NAT traversal issues as major problems to solve. Fundamental work for this purpose is currently chartered in the IETF p2psip working group, complemented by research activities addressed within the Internet Research Task Force (IRTF) p2prg research group. A detailed discussion on peer-to-peer SIP technologies is given in Chapter 9. Beyond signaling, the forwarding capabilities of DHTs give rise to enhanced application layer multicast solutions, which will be reconsidered in more detail in Section 6.2.3.

6.2.2 SIP Extensions Used in Conferencing

SIP syntax and semantics require extensions whenever the support of feature-rich conferencing scenarios is desired. Following SIP design principles, invoker-oriented primitives are the common basis to enable a dial-in and dial-out call control for user agents. This simultaneously holds for ad-hoc and scheduled conferences. However, the main design guidelines of SIP extensions keep changes to a minimum and should guarantee seamless backward compatibility with conference-unaware user agents [21]. The subsequently outlined extensions reflect these principles.

6.2.2.1 Join Header

The join header [22] expresses the request of the caller to participate in an existing dialog. It is solely used in invitations and designed as the generic operation for a third party to initiate a conference. The user agent receiving a join normally needs to create a new conference URI, which is then handed to the joining party via a re-INVITE. Conference creation is needless if the dialog to be joined is already part of a conference. The existing dialog:

```
From: Charlie <sips:charlie@brown.com>;tag=7654321
To: Lucy <sips:lucy@psychic.org>;tag=1234567
Call-ID: 333333@alpha.brown.com
```

is identified in the join header by its call-ID, the to-tag and the from-tag:

```
INVITE sips:charlie@brown.com SIP/2.0
From: Snoopy <sips:snoopy@dog.net>;tag=95148
To: Charlie <sips:charlie@brown.com>
...
Call-ID: 2772@beta.dog.net
...
Join: 333333@alpha.brown.com; to-tag=1234567;
from-tag=7654321
```

Note that even though the joining party need not know about an ongoing conference nor a conference URI, it must acquire call information of the existing remote session. These IDs may be learned from non-SIP means or via a SIP event package [23].

6.2.2.2   Re-INVITE

The SIP re-INVITE operation [1] is not an independent method, but an iterated INVITE within an ongoing session, i.e., with unaltered call-ID. It allows for a change of SIP session characteristics and always triggers SDP media negotiations anew. Thus, in conferencing situations it can be used to adapt control- and media-session parameters at the same time. Commonly, conference URIs and focus points or multicast groups used for signaling are distributed via re-invitation. Similarly, modified media types and distribution parameters like the definition of new mixers or network and multicast specific parameters are negotiated along the lines. Any failure of a re-INVITE will lead to session continuation with the previously established parameter set. Existing conferences therefore cannot be disturbed or destroyed by inappropriate re-INVITE requests.

6.2.2.3   REFER Method

The SIP REFER method [24] allows a user agent to request another user agent to access a resource it refers to. The URI of the referred resource is given within the refer-to argument:

```
REFER sips:lucy@psychic.org SIP/2.0
Via: SIP/2.0/TLS
     alpha.brown.com:5061; branch=z9hG4bKnashds
Max-Forwards: 70
From: Charlie <sips:charlie@brown.com>;tag=23431
To: Lucy <sips:lucy@psychic.org>;tag=1234567
Call-ID: 787878@alpha.brown.com
CSeq: 9380 REFER
Refer-To: <sips:hypnotic-talks@circles.com>
Content-Length: 0
```

A REFER request can be sent either inside or outside an existing dialog and also provides mechanisms to notify the originator of the outcome of his referenced request. Initially created for call transfer, it is used in conferencing to add third parties. Any client may send a REFER request to a partner, asking him to send an INVITE to an established conference URI. Equally, the initiator may send a REFER to the conference focus asking it to invite the partner. The latter way provides the benefit of allowing a client to add a conference-unaware user agent that does not support the REFER method. Analogously, a client may ask the conference focus via REFER to terminate conference membership for a third party.

REFER requests may be cascaded in the following way. A participant, who wishes that a focus or another participant refers a third party into the conference by sending a REFER method, may express this by adding an escaped refer-to header field within its refer-to argument:

```
REFER sips:hypnotic-talks@circles.com SIP/2.0
[...]
Refer-To: <sips:lucy@psychic.org;method=REFER?Refer-to=sips:
hypnotic-talks\% 40circles.com>
```

Furthermore, the refer-to URI argument is not limited to SIP, but may point to any valid Internet resource. In referring to non-SIP resource URIs, user agent clients are entitled to exchange input to collateral applications, e.g., as part of collaborative tasks or environments.

6.2.2.4 INVITE-Contained URI Lists

All mechanisms for ad-hoc conference management described so far define incremental operations for adding or joining single users into a multi-party session. This procedure may be time-consuming for larger conferences and delay conference establishment in an alienating fashion. Current work proposes an extension of SIP operations to include URI lists [25] treated like e-mail, where individual delivery can be handled in parallel according to address lists. These lists are encoded in XML, denoted by a "recipient-list-invite" SIP option-tag, and appended to the regular conference invitation messages. A user agent client in a typical scenario sends an INVITE including a recipient list to a conference server, which then will simultaneously invite all the list members. Note that the conference server does not process the initial INVITE as a nested transaction, but will acknowledge positively, whenever the conference was created. To inquire about actual members, the client needs to actively search user lists provided by the conference state event package described in the following section.

6.2.2.5 Conferencing Event States

An established conference incorporates a potentially large number of member-dependent states, exceeding those of a regular dialog. To support notification for tightly coupled conferences, an event package for conference states [26] has been defined. Its focus lies in providing membership information, but notifications about additional conference components are foreseen as well. Typically, these data are provided by the conference focus, which may learn states while performing its regular conference management tasks. Note that this model allows for cascading conferences, expressed by the sidebar conference elements. An overview of the conference state information is given in Table 6.1.

Following the general SIP event notification model, a user agent client will receive conference event state changes from the state agent via the NOTIFY method after having issued a SUBSCRIBE with reference to the package name "conference." The corresponding call flow is displayed in Figure 6.4.

TABLE 6.1

The Conference State Event Groups

Conference-info	Identifies this event set by a conference URI and a version.
Conference-description	Describes the content of the conference by textual meta-data like subject, keywords, additional (service) URIs and available media.
Host-info	Contains information about the host of the conference.
Conference-state	Expresses the conference state (active, locked) and current user count.
Users	This large container block enumerates individual users. A user record consists of a display name, user URI, role and a comprehensive characterization of the employed devices.
sidebars-by-ref	A reference pointer to a sidebar conference, given by its conference URI.
sidebars-by-val	Full representation of a sidebar conference, optionally including all conference state events.

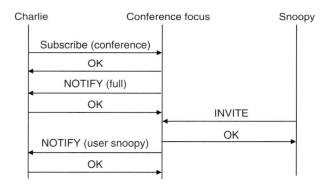

FIGURE 6.4

Call flow for SIP conference event state subscription and notification.

Event notification is performed incrementally on changes except for the initial NOTIFY response to subscription. Due to the expected amount of data in large conferences, a user agent client can receive partial notification about those compound, volatile elements assembled in container groups.

6.2.3   SIP with Multicast

Conferencing requires some basic group communication mechanisms for distributing media data to all parties and for mutual signaling, if applicable. The Internet most prominently offers IP layer multicast for this purpose, which has been designed about 20 years ago following the model of any source multicast

(ASM). The advantages of multicast over unicast-based multi-party communication must be seen in its scalability and lightweight interface at the client side. Deployment of IP multicast has been hesitant over the past decade, although it has emerged in recent times with the remarkable spread of IPTV offers. In the meanwhile, the infrastructure-friendly source specific multicast (SSM) model has been developed, as well as infrastructure-agnostic multicast on the application layer with promising performance potentials. Keeping deployment complexity in mind, it is desirable for any multicast conferencing solution to restrict group communication for signaling and media data to *only one* of the multicast models.

Due to its bandwidth requirements, *transmission of media* data takes the largest advantage of multicast. RTP streams transparently conform to group communication. While SDP session descriptions convey sufficient information for a user agent client to join a multicast group by providing multicast destination addresses within connection type fields on a per session and per media basis [4], SDP offer/answer negotiations [7] only partially fulfill multicast requirements. Supporting only a uniform view of the multicast session among all participants, RFC 3264 requires an SDP answer to match a multicast offer in address, port, and directionality. Thus, the roles of multicast sender and receiver remain in disguise and neither multicast send/receive capabilities nor SSM specifics may be arranged. Current work in the IETF sipping working group addresses this issue.

Multicast communication may be desired for *SIP signaling* whenever a previously configured group of receivers is to be contacted or in distributed scenarios with severe scalability constraints. The basic interaction with IP multicast in SIP is defined by the maddr parameter in the via header field. All SIP users of the community dog.net listening to the multicast address 239.12.11.10 could thus be contacted using the URI:

```
sip:*@dog.net;maddr=239.12.11.10;ttl=3
```

Syntax specification of RFC 3261 suffers from the nit of allowing only IPv4 address arguments with the ttl field serving as scoping limiter, while the semantic is bound to ASM use.

6.2.3.1 Any-Source Multicast

Deering's early host group model [27], which enables single packet submission to an unrestricted group of delocalized receivers, is the most general form of group communication. It is also known as any-source multicast (ASM), since any sender is entitled to issue data to a group of potentially unknown receivers. Receivers simply join to a group address, more specifically a $(*, G)$ channel, and the multicast routing will provide a transparent delivery service. ASM is fundamental to rendezvous processes and service discovery—SIP takes advantage for registrar access, accordingly. A user agent client may send register messages to the well-known "ALL SIP Servers" URI **sip:sip.mcast.net**, mapped to the IPv4 address **224.0.1.75**, without being aware of any individual registrar.

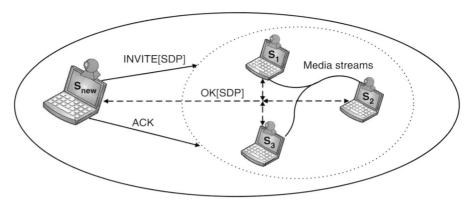

FIGURE 6.5
SIP call initiation based on ASM – a callee negotiates with a previously defined multicast group.

For multicast conference management, SIP defines only a limited "discovery-like" service, delivering a request to a group of homogeneous servers. A client wishing to initiate or join into a multi-party conference sends its INVITE request to a multicast group by employing the maddr attribute in the SIP VIA header. Group members subsequently indicate their presence by responding to the same group (see Figure 6.5). The transactional nature of SIP dialogs is preserved in the sense that the inviting party interprets the first arriving OK as the regular completion, while additional messages are treated as irrelevant iterates. Note that multicast responses require the requestor to subscribe to the same group, and prevent routing of multicast requests via SIP proxy servers. Thus, by pure SIP means, a caller cannot issue an INVITE to a remotely located user group, e.g., route to ***@dog.net**, without having established a global multicast connectivity.[§] Suitable for large, loosely coupled and mutually unknown parties, this simple scheme only operates through the use of any source multicast and does not allow for dynamic SDP negotiations.

ASM media distribution is easily achieved for homogeneous conference members with uniformly matching session descriptions. Such media parameter settings may be preappointed out of band or learned from a SAP online announcement. However, SAP is rarely used. Once established, the ASM distribution tree serves as a transparent connectivity layer, delivering all media streams to all parties.

6.2.3.2 Source-Specific Multicast

Source-specific multicast (SSM) [28,29], recently released as an initial standard, is considered a promising improvement of group distribution techniques. In contrast to ASM, optimal (S, G) multicast source-rooted trees are constructed immediately from source-specific, i.e., (S, G) subscriptions at the client side,

---

[§]Such a scenario could only be realized by tunneling the multicast SIP request into the destination domain.

without utilizing network flooding or rendezvous points. ASM and SSM can be distinguished from a partitioning of the multicast address space.¶ It is widely believed that simpler and more selective mechanisms for group distribution in SSM will globally disseminate to many users of multicast infrastructure and services. While any source multicast accounts for easy common addressing, SSM presupposes source-specific subscriptions. Hence, its use requires a dedicated distribution of source addresses for newly joining session members, which otherwise remain unnoticeable in previously established SSM groups. SIP session initiation in conferencing scenarios could facilitate this requirement [30].

The media parameters in tightly coupled or fully distributed conferences are mutually known to all peers. Despite the limitations of the SDP offer/answer model in the multicast context discussed above, each party will be aware of all source addresses of the correspondents and thus be enabled to issue source-specific joins. Without protocol extensions, user agent clients can thus participate in an SSM-based media session by simply performing IGMP/MLD [31] (S, G) joins for all peers. A simpler model well-suited for medium-size conferences has been introduced in the conferencing framework [5]: Each participant sends its media stream to a central media point or dedicated user agent using unicast. This central point will then redistribute the media using a source-specific multicast address. Whenever a new party joins the conference, the focus will perform the necessary third party call control to assure media reception. This scenario, which may lead to triangular traffic detours, is illustrated in Figure 6.6.

Up until now, SIP signaling based on SSM has not been defined. Simple possible extensions for SIP over SSM will be discussed in Section 6.4.

6.2.3.3   Application Layer Multicast

Exploiting the novel approach of structured peer-to-peer routing, a collection of group communication services has been developed, with the aim of seamless depoyability as application layer or overlay multicast. Among the most popular approaches are multicast on CAN [32], Bayeux [33], as derived from Tapestry, and Scribe [34] or SplitStream [35], which inherit their distributed indexing from Pastry. Approaches to multicast distribution in the overlay essentially branch in two algorithmic directions. In the first case, distributed hash tables (DHTs) are used to generate a structured sub-overlay of group members, which is then flooded with multicast packets. This mechanism underlies multicast on CAN. In the other, a distribution tree is erected within the full overlay, to be used as a shared- or source-specific tree. The latter schemes are used in Scribe and SplitStream, where a rendezvous node is chosen from group key ownership, or in Bayeux.

The performance of DHT–based multicast has been thoroughly studied in [36] with the comparative focus on tree–based versus flooding approaches

---

¶In IPv4 the 232/8 range and the prefix FF3x::/32 in IPv6 are designated as source-specific multicast destination addresses.

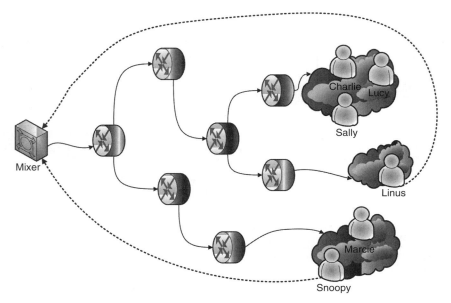

FIGURE 6.6
Media distribution in a centralized source-specific multicast model.

built onto CAN and Pastry. The separate construction of mini-overlays per group as needed for a selective flooding incurred significant overhead. In addition, flooding was found to be outperformed by forwarding along trees, where a shared group tree combined with proximity-aware routing as in SCRIBE could minimize the overlay delay penalty down to a factor of *two*. For the sake of completeness, we mention that application-layer multicast concepts concurrently exist for unstructured peer-to-peer approaches. They operate at lower algorithmic complexity, but show significantly higher efforts in coordinative signaling and thereby admit performance measures far below native multicast.

From a conceptual point of view, a user agent client can transparently take advantage of its multicast access on the application layer by submitting media data or SIP requests to the DHT. Overlay routing will just transport native packets in an application layer tunnel to the corresponding group of receivers. However, standardization work on peer-to-peer SIP has just begun [37,38] and currently concentrates on distributed user location and NAT traversal. A significant amount of prestandard work is required until an efficient, globally scalable, and robust multicast distribution layer becomes available for wider use. The IRTF p2p and SAM research groups are dedicated to corresponding work.

6.2.4 Mobile Group Members

Mobile end systems today escort people from puberty to old age. Standard systems such as mobile phones or PDAs, but also specialized gadgets, i.e., gaming stations or dedicated professional equipment, must be increasingly considered as people's primary access devices to information, communication, and entertainment. At the same time, most of public and private spaces have turned into areas of ubiquitous network operations. Wireless Internet connectivity is offered by a 3GPP or IEEE 802.11 WLAN infrastructure, but emerging 802.16 environments and mobile ad hoc network (MANET) mesh networks have started to complement the established access networks.

Voice calls dominate the mobile market, gradually complemented by selective video services. Mobile telephone operators are on the spot of migrating to entirely IP-based fourth-generation networks in accordance with 3GPP release eight standard track. SIP has been selected to play a dominant role in both: provider-centric architectures based on the IP Multimedia Subsystem (IMS) and provider-independent end-to-end session signaling, the latter being particularly obstructed by NAT and port barriers in the mobile regime. As SIP/IP 2-way dialogs are expected to dominate the mobile world soon, demands for conferencing are likewise foreseeable. Mobile group conferences meet multipoint transmission capabilities in virtually all wireless technologies, e.g., in 802.11 and 802.16, in DVB-H, as well as in 3GPP, which defines the Multimedia Broadcast and Multicast Services (MBMS) in release 6. Issues of providing seamless communication services under mobility-related handovers remain though on Internet Protocol layers. They will require an explicit handover management whenever provider or IP subnet addresses change between attachments. Schemes for achieving a seamless mobility management in conferencing vary depending on the communication model, i.e., the use of unicast or multicast.

6.2.4.1 Unicast-Based Mobility

A tightly coupled conference scenario is likely to operate in unicast mode. This does not only hold for SIP signaling relations, but may also apply for media distribution in a central mixing or conference bridge scheme. On handover, any mobile member changing within a given IP subnet may just continue operation on upper layers, any disturbance being bound to disconnection times on the network access layer. In cases where movement occurs between providers or internal subnets, e.g., triggered by a vertical handover between different access technologies, the mobile conference member changes its IP address and needs to procure conference session persistence through additional means. Session continuation can be accomplished by SIP mobility management as described in Chapter 4 and Chapter 17. Alternatively, transparent handover management may be operated on the transport layer, provided (mobile) SCTP is in use. The most general approach to network layer mobility resides in IPv4/v6 protocol extensions, which have been the focus of IETF work over recent years (see [39] for an excellent discussion).

In conferencing situations, special care is needed whenever the mobile node performs internal relaying, as its disconnection will cause data delay or damage for downstream members. Such threat of service degradation is obvious for a peer-managed focus or mixer in tightly coupled conferences, but may result from routing and forwarding functions in distributed settings or application layer multicast, as well.

6.2.4.2   Mobile Multicast

Multicast mobility management must be assured for mobile user agents whenever native multicast routing serves for media or signaling distribution. IP multicasting is of particular importance to mobile environments, where users commonly share frequency bands of limited capacities. In general, multicast routing dynamically adapts to session topologies, which may then change under mobility. However, depending on the topology and the protocol in use, routing convergence may be far too slow to support seamless handovers.

In multicast routing, the roles of senders and receivers need to be distinguished. Any listener subscribed to a group while in motion, requires delivery branches to pursue its new location; any mobile source requests the entire delivery tree to adapt to its changing positions. Operations should facilitate seamless data flows compliant to real-time requirements and at the same time ensure routing convergence without compromising network functionality, see [40] for a detailed discussion. In a conference, multicast routing will always be exposed to listener mobility, while source movement in some schemes may be hidden to the network layer, e.g., by a central static mixer.

Like in the unicast case, SIP may assist multicast mobility on the application layer, even though the undertaking is more complex. Subsequent to handover, a listening user agent client can initiate unicast data forwarding by third party call control until multicast routing has converged. A moving source may either rely on static entities for packet redistribution or take advantage of application layer tunneling between conference members. Note that a change of its source address will trigger a user agent to issue SIP/SDP updates within immediate signaling relations. Thereafter, address information needed for SSM specific rejoins are in place.

In concordance with mobile IP, a natural and efficient way to manage multicast mobility will reside on the network layer. Three approaches to mobile multicast are commonly around:

Bi-directional tunneling guides the mobile node to tunnel all multicast data via its home agent and thereby hides its mobility.

Remote subscription forces the mobile node to re-initiate multicast distribution subsequent to handover by submitting an IGMP join or MLD listener report within the subnet it newly attached to.

Agent-based solutions attempt to balance between the previous two mechanisms. Static agents typically act as local tunneling proxies, allowing for some inter-agent handover, while the mobile node moves away.

Work on standardizing seamless solutions has just begun; the requirements are being discussed in the MobOpts research group. For further details and a brief survey on current solutions, we refer to [40].

6.3 Infrastructure-Assisted Conferences

Conventional deployments of SIP architectures are accomplished with the help of infrastructural entities, typically a SIP proxy server and registrar, a gateway, and additional convenience functions. These components typically reside on network nodes that are well connected and in unrestricted, permanent availability. A manifest strategy to advance service offers lies in augmenting these nodes with lightweight side primitives such as NAT traversal assistance or client mobility management. Likewise, a conference control function could reside on a standard SIP server instance. However, media mixing and relaying for large conferences cannot be considered lightweight, but out of range. Multicast distribution can assist in serverless delivery of media streams as discussed in Section 6.2.3, and open the floodgates to scalable conferencing at no additonal cost.

Unfortunately, IP multicast is not uniformly available, even though all major router vendors and operating systems offer a wide variety of implementations to support it. While many (walled) domains or enterprise networks operate multicast, group service rollout has been largely limited in public inter-domain scenarios. Application layer multicast based on structured peer-to-peer systems offer group communication in an infrastructure-agnostic fashion. DHT are reasonably efficient and scale over a wide range of group sizes. However, they do not allow for layer two interactions, thus do not facilitate unrestricted scaling in shared end system domains, and experience severe performance degradation when terminal mobility is introduced [41]. These drawbacks may be mitigated by hybrid approaches, where overlay multicast routing only takes place among selected nodes that are particularly stable and form a virtual infrastructure. The servers of a SIP infrastructure are apt candidates for this, as will be outlined in the following sample architecture.

6.3.1 A Hybrid Architecture for Transparent Group Communication

This section introduces a hybrid architecture for SIP proxy-assisted group conferencing, designed to enable global multicast peering at the ISP or enterprise level, and at the same time, sustain end system transparency. The basic concept preserves multicast routing and lower layer packet transmission within domains, while bridging the inter-domain gap with the help of a structured overlay network to overcome deployment problems. Its focus originates from a customer network or an ISP domain, where multicast services are locally deployed. Multicast service exchange is then implemented like unicast peering by a gateway function, resident on a SIP proxy. The primary call routing function of a SIP server is thereby augmented with a multicast overlay for conferencing and media distribution. It interconnects the local multicast routing with the distributed peering on the structured overlay.

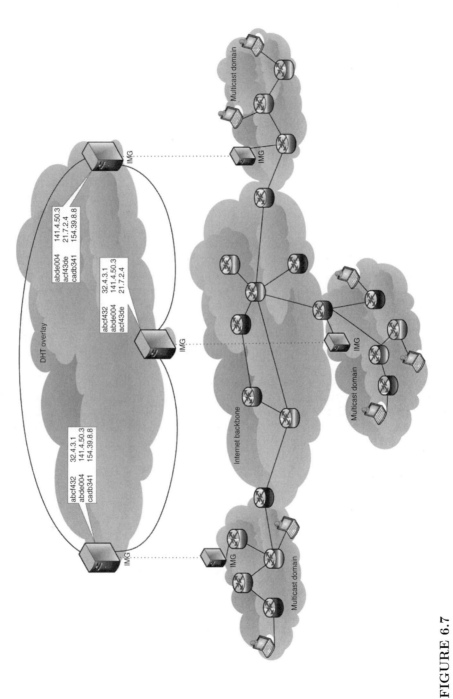

FIGURE 6.7

A hybrid group communication architecture with SIP inter-domain multicast gateways.

The new function, Inter-domain Multicast Gateway (IMG), acts as a gateway between the overlay it is a member of, and the multicast routing at the intra-domain underlay that it resides in, see Figure 6.7. Such a gateway will participate in multicast traffic originating from its residential network, which it will forward into the overlay according to the distributed multicast receiver domains of this group. It will also advertise group membership and receive data according to any subscription from its domain. On the overlay, the IMGs will jointly operate a distributed hash table, hosting an application layer multicast. Depending on specific requirements, any well-known ALM may be executed. The bidirectional shared tree approach introduced in [42] gives rise to fully transparent, mobility-agnostic, and proximity-aware multicast and broadcast overlay services.

Like a SIP proxy, the IMG function may be positioned anywhere within the multicast domain, and provides a protocol interface to the locally deployed multicast routing. Activation of inter-domain multicast gateway services requires only a small amount of selected information for bootstrapping, i.e., an arbitrary contact member of the structured overlay, authentication, and authorization credentials, if applicable. Typically, such initialization is performed at SIP domain peering. The IMG further on remains under the administrative control of the local network operator, who may restrict admission, scoping, and QoS characteristics of the group traffic flowing in and out of the intra-domain. Aside from general multicast peering policies, a service provider is thus enabled to implement firewall-type of packet filters at, or co-located with, these multicast gateways. It thereby can easily shape the inter-domain multicast layer according to conferencing application or community specific particularities—without control of the underlying network infrastructure.

This architecture allows for flexibility in several additional ways. A domain operator is enabled to connect to several multicast overlays in parallel, may choose to replicate IMGs for load balancing or redundancy purposes or may transparently take advantage of the fail-safe unicast peering realized by multi-homed network connectivity. Replication operations will be seamlessly empowered by the self-organization capabilities of the DHT overlay. Active coordination between gateway peers may be achieved by SIP or non-SIP means.

6.4 Peer-Managed Conferences

An increasing number of applications aim for simple, flexible, and cost-efficient ad-hoc conferencing functions, which scale appropriately, but avoid any infrastructure assistance. Such solutions require group session management and media distribution at peers. Commonly implemented as pure software on standard personal devices, user agent peers are exposed to severe restrictions in real-world deployments: Often they are located behind NATs and firewalls with network capacities confined to asymmetric DSL or wireless links. Multicast routing may be unsupported or only available in parts of the network facilities spanned by the conference. Clients may run on handhelds or other truly mobile devices that admit processing resources too scarce to serve for mixing and redistribution purposes. Peers may join or leave a conference in an unpredictable manner, advising other members not to rely on its relaying service. Nevertheless, real-time constraints apply to data processing and packet forwarding, whenever voice, video, or interactive elements are the purpose of the conference.

Capacity constraints and resilience to node failures require peer-managed ad-hoc conferences to organize in a distributed multi-party model. As a key component, the heterogenity of clients must be accounted for. Ranges of scalability, however, may vary. From an application point of view, many unmoderated systems are designed for only about one dozen participants, which in particular holds for dialog-oriented video conferences. Other applications, equally built of lightweight peers, may foresee media streams to reach large numbers of receivers. Two examples of SIP initiation, illustrating either side of the coin, are detailed in the following.

6.4.1   A Simple, Distributed Point-to-Point Model

Peer-to-peer conferencing systems for moderate membership face the grand challenge of remaining robust with respect to the infrastructure. The role a user agent is able to attain in a distributed scenario needs to be adaptively determined according to the constraints of its device and current network attachment. In a simplified scenario, clients may be divided into two groups, distinguished by their ability or inability to act as a focus. A focus must hold a GRUU and have access to necessary processing and network resources. This elementary adaptation scheme can be based on individual decisions of user agents and gives rise to a hybrid architecture of super peers, chosen from potential focus nodes, and remaining leaf nodes. Leaf nodes attach to super peers in subordinate position, whereas potential focus nodes may be assigned to be super peers or leaves. Super peers provide global connectivity among each other and NAT traversal assistance to leaves‖, while leaf nodes experience super peers in different roles: A leaf nodes sees its next hop super peer as the conference focus, while the remote super peers act as proxies on the path to any leave located behind them.** This set-up corresponds to the well-known architecture of Gnutella 0.6 and successive hybrid unstructured peer-to-peer systems, see [43]. Despite its architectural analogy, a routing layer for real-time group applications should follow a different design.

‖Super peers are globally addressable nodes with packet relaying function. TURN will be the natural unilateral self-address fixing (UNSAF) protocol to use.

**This architecture relies on the presence of at least one globally addressable, sufficiently powerful peer. In scenarios, where this is likely to fail, a common practice of vendors or communities is to permanently deploy a "silent" relay-peer at some unrestricted place.

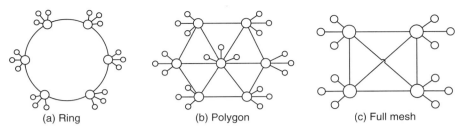

FIGURE 6.8
Peer-to-peer routing topologies on the overlay.

From the perspective of the conference, super peers form a distributed focus. To keep distribution transparent to leaves, each super peer needs to provide full conferencing service functions, e.g., synchronized policing and event notification, and most likely assistance in media mixing and redistribution. Focus nodes consequently require signaling relationships among each other, established on top of an application layer routing of sufficient performance for a simultaneous distribution of media streams. The design of a routing will admit critical impact on scalability, application performance, as well as forwarding and maintanance load of the super peers. The three characteristic topologies for routing between N super peers as displayed in Figure 6.8 explore the problem space: On one extreme, routing on a ring will minimize neighbor states and the forwarding load of each peer, but requires $\mathcal{O}(N)$ hops and thus induces large, varying delays. A full mesh, on the other extreme, places the burden of $N-1$ neighbor states to be fed in replicated forwarding, but guarantees a rigid three-hop forwarding limit and minimal delays. A polygonal mesh of dimension d keeps replication load constant (but dependent on d), while its corresponding path lengths grow as $\mathcal{O}(\sqrt[d]{N})$. Forwarding on a polygonal mesh will require routing intelligence, which is neither needed on a ring or in a full mesh topology. As routing paths in conferencing scenarios are equivalent to the signaling relationships, mesh robustness or scheme redundancy is equivalent to the number of neighbor states at each peer. Alternative routing topologies may be deployed on the basis of DHTs, admitting logarithmic scaling in all measures on the price of a higher algorithmic complexity.

Refocussing on the problem of moderately sized peer-to-peer conferences of simple and robust nature, a favorable routing scheme is easily identified. The full mesh topology outperforms alternative schemes in forwarding efficiency and robustness, while scaling well up to a hundred nodes, provided a significant fraction of unrestricted, high-performance super peers is available. In addition, this scenario will be bound to low complexity, since no routing intelligence beyond standard SIP logic of next-hop proxying is required. Full mesh topologies are thus considered here as the favorable approach to mid-size multi-party conversations.

To explore the corresponding conference scenario in detail, consider an ad-hoc join. A client submits an INVITE to any party. It thereby needs to

indicate its potential roles in some way.[††] The callee may be a conference focus
or leaf node. In the first case, it will be aware of the overall leaf node distribution from the conference event states and will transfer the newly joining
party to the least-occupied super peer by a REFER, e.g.,

```
REFER sips:lucy@psychic.org SIP/2.0
...
CSeq: 9380 REFER
Refer-To: <sips:hypnotic-talks@vain-focus.circles.com>
Content-Length: 0
```

In the second situation, the contacted leaf node will issue a re-INVITE to
attach the new conference member to its focus, which in turn may refer the
caller to another focus for the sake of load balancing. Having indicated its
ability to serve as a super peer, the newly arrived party may be selected to
join the group of focus nodes. This decision is taken by its current super peer
and realized via a third party invite issued to the group of all established focus
nodes. The elected super peer will thereby establish point-to-point signaling
relationships with all correspondents, leading to an immediate formation of
the full mesh conference topology. Focus election and leaf node distribution
may be conducted in an individual or collective way, following an eager or
lazy strategy. Its implementation most likely will depend on the overall environment and the conference persistence of super peers.

Note that media negotiations have been part of the initial arrival steps for
each party. Media distribution will naturally follow the paths of the established routing topology, where super peers can act as two-sided mixers: They
may combine media streams arriving from their attached leaf nodes before
peering them within the focus mesh, but may as well mix media arriving from
neighboring super peers for a lean transmission to leaves.

6.4.2 Scalable, Peer-Centric Conferencing Based on SSM

In the presence of source-specific multicast at the network layer (see Section 6.2.3), peer-to-peer conferences can scale to very large numbers. Com-
plying with the fully distributed ad-hoc paradigm, the following scenario is
considered throughout this section.

Some party will initiate a conference by contacting one or several peers via
unicast addresses as resolved from a SIP URI. Following an initial contact, signaling will then be turned to scalable multicast group communication. Later
on new parties will join the conference by either calling or being called by
an existing member. Such group conference initiation scheme is currently not
covered by a SIP standard, nor is the employment of source-specific multicast
for group signaling. In order to enable SSM, all dialogs must carefully provision addresses of newly arriving senders to all current group members, which

---

[††]A corresponding client protocol extension has not been specified yet, cf. [3].

need to adapt source specific subscriptions appropriately; see [30] for further details.

Central to the approach introduced here is the concept of a fully delocalized focus. Conference signaling and management is entirely delegated to the peers participating in the multicast group. This scheme may be equally applied to an undistinguished group of equal participants or to super peers in a hybrid topology as outlined in the previous section. Hybrid architectures may account for SSM accessibility and shield multicast or conference unaware user agents.

In detail, protocol operations of SIP initiated SSM proceed as follows: A caller, wishing to participate in a previously established unicast dialog, will initiate a regular INVITE request to some selected member. Eventually, after the call setup has completed, either party will decide to transfer the established session to group communication. Heading for SSM, it will create a conference URI with a locator consisting of an SSM group address G, e.g.,

```
sip:dog-matter@232.116.16.6.
```

Like in unicast conferences, the initiator will submit a re-INVITE announcing its conference URI, i.e., the desired multicast address, in the CONTACT field of the SIP header:

```
INVITE sip:lucy@psychic.org SIP/2.0
...
CSeq: 1024 INVITE
Contact: <sip:dog-matter@232.116.16.6>;isfocus
Content-Type: application/sdp
...
```

In parallel, the SIP protocol stack will submit a multicast source-specific JOIN to its underlying IGMP/MLD stack, thereby subscribing to the group and the source address of the corresponding peer, both learned from the previous SIP message exchange. Any peer will identify the multicast address in the contact field, and proceed along the protocol semantic for SSM SIP.

This two-step procedure purposefully decouples application layer session establishment and underlying multicast routing operations. Temporal progress in IP layer multicast routing and SIP transactional timers thereby remain independent for the sake of robustly layered protocol operations. Appropriate media session descriptions for source-specific multicast distribution of media streams may or may not be submitted along the re-INVITE request.

In multiparty environments, the straightforward generalization for switching a previously established unicast conference into SSM group communication is shown in Figure 6.9. If the callee decides to accept the call from S_{new}, it will forward the INVITE to its partner, thereby initiating unicast sessions among the three. Thereafter, the callee will turn the conference signaling to multicast by submitting the corresponding re-INVITE procedure.

If a new source S_{new} contacts an established SSM group conference, it will do so analogously by inviting some member S. If S decides to accept the caller, it will redistribute its INVITE to the SSM group and acknowledge the

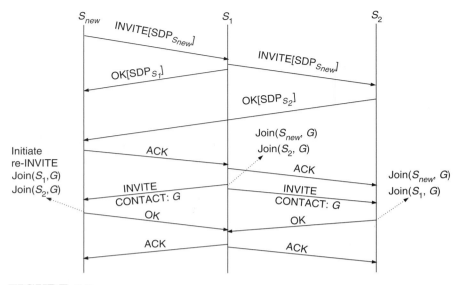

FIGURE 6.9
Call flow of switching from unicast to SSM within re-INVITE.

initial call by placing the group conference URI in the CONTACT header field. As displayed in Figure 6.10, all group members will immediately add S_{new} to their source-specific multicast filters. S_{new} subsequently will learn about all group members from (unicast) OK messages as needed for its own multicast subscriptions. Note that call redistribution will remain a point-to-point user transaction of S_{new} at the SIP layer, but a transmission from source S at the network layer, and therefore compliant with previous SSM group establishment.

Proceeding along this incremental way, a callee will never be required to redistribute messages to more than one party or group. This scheme thus remains fully scalable and fairly transparent to group sizes. Multicast initiation of media sessions may be led correspondingly.

6.5   Summary and Conclusions

The concepts and mechanisms in this chapter describe how group conferences can be realized with SIP. They apply equally to voice and video, messaging and chat, gaming and collaborative environments, as well as to many other application areas of this lively and growing field.

Starting from elementary operations to initiate a multi-party conversation, an overview of basic SIP concepts and service primitives is presented for

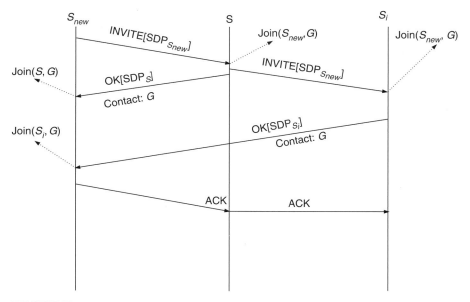

FIGURE 6.10
Call flow of extending established group sessions to a new party.

central and distributed conference management using unicast or multicast. Special focus is donated to the emerging field of mobile user agents, which are expected to soon populate daily usage and require seamless integration. Presently available and developing mobility schemes promise to assist roaming conference members at real-time compliant handovers in an end-to-end fashion.

SIP offers widespread potential for supporting group formation and communication, while the field of optimizing multi-peer sessions is under active current investigation. Comprising peer-to-peer technologies, as well as application layer and source-specific multicast, this outline explores the rich solution space of future directions in infrastructure-assisted or purely peer-managed conferences. Lightweight solutions adaptive to heterogeneous environments and varying scenarios are discussed. It is the hope of the authors to stimulate an active development of new applications that admit rich functionality and substantial deployment.

References

[1] Jonathan Rosenberg, Henning Schulzrinne, Gonzalo Camarillo, Alan Johnston, Jon Peterson, Robert Sparks, Mark Handley, and Eve Schooler. 2002. SIP: Session Initiation Protocol. *RFC 3261*, IETF, June.

[2] Jonathan Rosenberg. 2007. A Hitchhiker's Guide to the Session Initiation Protocol (SIP). Internet draft—work in progress 04, *IETF*, November.

[3] IAB and IESG. 2000. IETF Policy on Wiretapping. *RFC 2804, Internet Engineering Task Force*, May.

[4] Mark Handley, Van Jacobson, and Colin Perkins. 2006. SDP: Session Description Protocol. *RFC 4566, IETF*, July.

[5] Jonathan Rosenberg. 2006. A framework for conferencing with the Session Initiation Protocol (SIP). *RFC 4353, IETF*, February.

[6] Alan Johnston, ed. 2007. Session initiation protocol service examples. Internet draft—work in progress 13, *IETF*, July.

[7] Jonathan Rosenberg, and Henning Schulzrinne. 2002. An offer/answer model with the Session Description Protocol (SDP). *RFC 3264, IETF*, June.

[8] Mark Handley, Colin Perkins, and Edmund Whelan. 2000. Sessionannouncement protocol. *RFC 2974, IETF*, October.

[9] Gonzalo Camarillo, Jörg Ott, and Keith Drage. 2006 The Binary Floor Control Protocol (BFCP). *RFC 4582, IETF*, November.

[10] ITU-T Recommendation H.323. Infrastructure of audio-visual services— Systems and terminal equipment for audio-visual services: Packet-based multimedia communications systems. Technical report, ITU, 2000. Draft Version 4.

[11] Hans L. Cycon, Thomas C. Schmidt, Gabriel Hege, Matthias Wählisch, and Mark Palkow. 2008. An optimized H.264-based video conferencing software for mobile devices. In Antonio Navarro, eds., *ISCE2008— The 12th IEEE International Symposium on Consumer Electronics*, Piscataway, NJ, USA: IEEE, IEEE Press.

[12] ITU-T Recommendation H.264 & ISO/IEC 14496-10 AVC. Advanced video coding for generic audiovisual services. Technical report, ITU, 2005. Draft Version 3.

[13] Heiko Schwarz, Detlev Marpe, and Thomas Wiegand. 2007. Overview of the scalable video coding extension of the H.264/AVC Standard. *IEEE Transactions on Circuits and Systems for Video Technology* 17(9): 1103–20.

[14] Aameek Singh, and Arup Acharya. 2005. Multiplayer networked gaming with the session initiation protocol. *Computer Networks* 49(1): 38–51.

[15] Ben Campbell, Jonathan Rosenberg, Henning Schulzrinne, and Christian Huitema, and David Gurle. 2002. Session Initiation Protocol (SIP) Extension for instant messaging. *RFC 3428, IETF*, December.

[16] Ben Campbell, Rohan Mahy, and Cullen Jennings (Eds.). 2007. The Message Session Relay Protocol (MSRP). *RFC 4975, IETF*, September.

[17] Rohan Mahy, Robert Sparks, Jonathan Rosenberg, Dan Petrie, and Alan Johnston. 2007. A call control and multi-party framwork for the session initiation protocol (SIP). Internet draft—work in progress 9, *IETF*, November.

[18] Henning Schulzrinne, Stephen Casner, Ron Frederick, and Van Jacobson. 2003. RTP: A Transport Protocol for Real-Time Applications. *RFC 3550, IETF*, July.

[19] Jonathan Rosenberg. 2007. Obtaining and using globally routable User Agent (UA) URIs (GRUU) in the Session Initiation Protocol (SIP). Internet draft—work in progress 15, *IETF*, October.

[20] Jonathan Rosenberg, Jon Peterson, Henning Schulzrinne, and Gonzalo Camarillo. 2005. Best current practices for third party call control (3pcc) in the Session Initiation Protocol (SIP). *RFC 3725, IETF*, April.

[21] Alan Johnston, and Orit Levin. Session Initiation Protocol (SIP) Call Control—Conferencing for user agents. 2006. *RFC 4579, IETF*, August.

[22] Rohan Mahy, and Dan Petrie. 2004. The Session Initiation Protocol (SIP) "Join" Header. *RFC 3911, IETF*, October.

[23] Jonathan Rosenberg, Henning Schulzrinne, and Rohan Mahy. 2005. An INVITE-Initiated Dialog Event Package for the Session Initiation Protocol (SIP). *RFC 4235, IETF*, November.

[24] Robert Sparks. 2003. The Session Initiation Protocol (SIP) refer method. *RFC 3515, IETF*, April.

[25] Gonzalo Camarillo, and Alan Johnston. 2007. Conference establishment using request-contained lists in the Session Initiation Protocol (SIP). Internet draft—work in progress 01, *IETF*, November.

[26] Jonathan Rosenberg, Henning Schulzrinne, and Orit Levin. 2006. A Session Initiation Protocol (SIP) Event Package for Conference State. *RFC 4575, IETF*, August.

[27] Stephen E. Deering. 1989. Host extensions for IP Multicasting. *RFC 1112, IETF*, August.

[28] Supratik Bhattacharyya. 2003. An overview of Source-Specific Multicast (SSM). *RFC 3569, IETF*, July.

[29] H. Holbrook, and B. Cain. 2006. Source-specific multicast for IP. *RFC 4607, IETF*, August.

[30] Thomas C. Schmidt, Matthias Wählisch, Hans L. Cycon, and Mark Palkow. 2007. Scalable mobile multimedia group conferencing based

on SIP initiated SSM. In *Proceedings of 4th European Conference on Universal Multiservice Networks—ECUMN'2007*, 200–209, Washington, DC, USA: SEE/EUREL/IEEE, IEEE Computer Society Press.

[31] Hugh Holbrook, Brad Cain, and Brian Haberman. 2006. Using internet group management protocol version 3 (IGMPv3) and Multicast Listener Discovery Protocol Version 2 (MLDv2) for Source-Specific Multicast. *RFC 4604, IETF*, August.

[32] Sylvia Ratnasamy, Mark Handley, Richard M. Karp, and Scott Shenker. 2001. Application-Level Multicast Using Content-Addressable Networks. In Jon Crowcroft and Markus Hofmann, eds., *Networked Group Communication, Third International COST264 Workshop, NGC 2001, London, UK, November 7–9, 2001, Proceedings*, 2233 of *LNCS*, 14–29. London, UK: Springer-Verlag.

[33] Shelley Q. Zhuang, Ben Y. Zhao, Anthony D. Joseph, Randy H. Katz, and John D. Kubiatowicz. 2001. Bayeux: An architecture for scalable and fault-tolerant wide-area data dissemination. In *Proceedings of the 11th International Workshop on Network and Operating System Support for Digital Audio and Video (NOSSDAV 2001)*, 11–20, June.

[34] Miguel Castro, Peter Druschel, Anne-Marie Kermarrec, and Antony Rowstron. 2002. SCRIBE: A large-scale and decentralized application-level multicast infrastructure. *IEEE Journal on Selected Areas in Communications* 20(8): 100–10.

[35] Miguel Castro, Peter Druschel, Anne-Marie Kermarrec, Animesh Nandi, Antony I. T. Rowstron, and Atul Singh. 2003. Splitstream: High-bandwidth content distribution in cooperative environments. In M. Frans Kaashoek and Ion Stoica, eds., *Peer-to-Peer Systems II. Second International Workshop, IPTPS 2003 Berkeley, CA, USA, February 21–22, 2003 Revised Papers*, 2735 of LNCS.292–303. Berlin Heidelberg: Springer-Verlag.

[36] Miguel Castro, Michael Jones, Anne-Marie Kermarrec, Antony Rowstron, Marvin Theimer, Helen Wang, and Alec Wolman. 2003. An evaluation of scalable application–level multicast built using peer-to-peeroverlays. In *IEEE Infocom 2003*, 2, 1510–20. Piscataway, NJ, USA: IEEE Press.

[37] David Bryan, Salman Baset, Marcin Matuszewski, and Henry Sinnreich. 2007. P2PSIP protocol framework and requirements. Internet draft—work in progress 00, IETF, July.

[38] David Bryan, Philip Matthews, Eunsoo Shim, and Dean Willis. 2007. Concepts and terminology for peer to peer SIP. Internet draft—work in progress 01, *IETF*, November.

[39] Rajeev Koodli, and Charles Perkins. 2007. *Mobile Inter-Networking with IPv6*. Hoboken, NJ: Wiley-Interscience.

[40] Thomas C. Schmidt, Matthias Wählisch, and Godred Fairhurst. 2008. Multicast mobility in MIPv6: Problem statement and brief survey. IRTF internet draft—work in progress 04, MobOpts, July.

[41] Anargyros Garyfalos, and Kevin Almeroth. 2005. A flexible overlay architecture for mobile IPv6 multicast. *IEEE Journal on Selected Areas in Communications* 23(11): 2194–205.

[42] Matthias Wählisch, and Thomas C. Schmidt. 2007. Between underlay and Overlay: On deployable, efficient, mobility-agnostic group communication services. *Internet Research* 17(5): 519–34.

[43] Ralf Steinmetz, and Klaus Wehrle, (eds.). 2005 *Peer–to–Peer Systems and Applications*, volume 3485 of *LNCS*. Berlin Heidelberg: Springer-Verlag.

7

Home Networking System with SIP

Yukikazu Nakamoto and Naoko Kuri
University of Hyogo

CONTENTS

7.1 Introduction 159
7.2 UPnP 160
7.3 SIP 162
7.4 Redesign Issues 165
7.5 Redesigning UPnP Middleware 165
7.6 Middleware Implementation 168
7.7 Conclusion 169
    References 170

7.1 Introduction

Several protocols, middleware, and interoperable specifications have been proposed for home networking systems to allow devices on a network to utilize service functionalities provided by other devices or appliances. These include Jini [1], Universal Plug and Play (UpnP) [2], the Open Server Gateway Initiative (OSGi) [3], and Digital Living Network Alliance (DLNA) [4]. For these purposes, the protocols and middleware provide the following functionalities: allowing a device to join a home network and make its services available to other devices on the network, search for an expected service on the network, and exchange device information for utilizing a discovered service. Utilization of the discovered service includes control of the device and accessing its services, together with provision for the controlled device to communicate notification of an event to a control point.

On the other hand, in the field of communication, the popularity of SIP (Session Initiation Protocol) [5] continues to increase. SIP is an application-layer control protocol that can establish, modify, and terminate multimedia sessions (conferences) such as Internet telephony calls [5]. In session establishment, the functions required include registering a device to join the session, discovering appropriate devices, exchanging device capabilities for streamed communication, and controlling the communication stream. The home network protocols and middleware can be implemented by the SIP functionalities because of the similarities in the functionalities required in the home network system and SIP.

This paper presents a redesign of some parts of home network middleware with SIP, preserving the UPnP API specifications. There are advantages if SIP is used to implement home networking protocols, in particular UPnP.

- Seamless connectivity between devices both inside and outside the home: Suppose that a control device is controlling appliances at home and the user would like to control appliances outside the home in a similar way. Implementing the home network with SIP results in seamless connectivity without extra protocol conversions and gateways, allowing such a scenario to be easily realized.

- Integration of VoIP (Voice-over-Internet Protocol) applications and home network applications: In the near future, VoIP appliances and home appliances will co-exist at home. For example, a video conversation might be recorded using a VoIP video telephone and the video conversation watched at a later time. In this situation, it would be useful for a user to be able to control the video conversation in the same way as ordinary video streams arriving via TV or DVD are controlled.

There are several reports of research in which home network systems have been implemented with SIP. Bushmitch et al. extends the OSGi framework for handling SIP device facilities to support interoperability with mobile devices [6]. A novel architecture for the mobile OSGi architecture based on SIP functionalities is described In [7]. In [8], a multimedia service framework is proposed, which defines communication protocols to adjust A/V rendering among a media server, a home server, and A/V devices. Our work complements the research detailed in these papers.

This paper is organized as follows. In Section 7.2 and Section 7.3 we briefly present the UPnP and the related SIP specifications, respectively. We overview issues in redesigning the UPnP functionalities with the SIP in Section 7.4. In Section 7.5, we describe how the discovery in UPnP may be realized in the SIP functionalities. Section 7.6 briefs its implementation.

7.2 UPnP

We describe the functionalities of UPnP based on [2, 9]. A UPnP network consists of communicating devices. Some devices, known as control points, can access and control other devices, which may offer any number of services. A device may be root or may contain other devices embedded within it. A root or embedded device provides services to a user and other devices. For short, both are called "devices" in the paper. A control point and a device are depicted in Figure 7.1. A service is modeled by state variables and actions

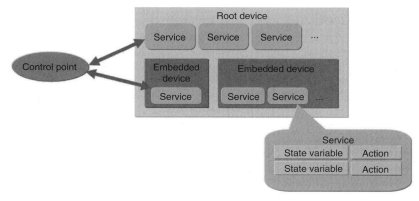

FIGURE 7.1

Structure element in UPnP. (From Nakamoto, Y. and N. Kuri, *Proc of 6th IEEE Int Conf Comp Info Tech*, 2006, p. 99. With permission.)

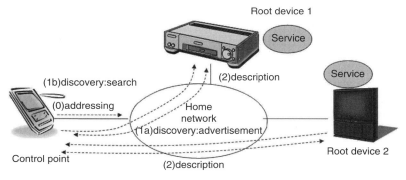

FIGURE 7.2

Addressing, discovery and description in UPnP. (From Nakamoto, Y. and N. Kuri, *Proc of 6th IEEE Int Conf Comp Info Tech*, 2006, p. 99. With permission.)

that may result in a change of state variable values. UPnP defines protocols and conventions for the following operations by which a control point accesses and controls devices (see Figure 7.2).

(0) addressing: When a device joins the home network, it obtains a unique network address.

(1) discovery: Discovery defines how a device announces its presence on the network and how control points discover it. When a device joins the network, it advertises its participation (advertisement). The device multicasts specific device unique identifier (uid), types of the device and services to devices in the home network. A device can be discovered by control points and learn about the device's capabilities by retrieving a device description.

A control point, once added to the network, may search for particular devices or services (search). A search may be specified in terms of a specific device uid, a particular service type, or a particular device type. Discovery messages are sent to devices through an IP (Internet Protocol) multicast server in the home network. Hypertext Transfer Protocol/Multicast UDP (HTTP/MU), Simple Service Discovery Protocol (SSDP), and General Event Notification Architecture (GENA) are used in the discovery phase. The message formats are described in Section 7.5.

(2) description: For a control point to learn more about the device and its capabilities or to interact with a device, the device summarizes its services and capabilities and sends them in a description to the control point. The device description includes a list of actions the service responds to and a list of variables that model the state of the services.

(3) control: Given knowledge of a device and its services, the control point can ask these services to invoke actions.

(4) eventing: The device can send an event to control points when values of state variables in a device change. A device may subscribe to receive event notification.

(5) presentation: A device should provide an HTML-based administrative interface that allows a user to directly control the device and view the device status.

7.3 SIP

In this section, we briefly describe functionalities of SIP. SIP is a signaling, presence, and instant message protocol to establish, modify, and finish multimedia session in one-to-one or among multiple participants in the application layer. SIP also provides the delivery of presence and instant messages. It is defined in RFC3261 [5].

In SIP, the following clients and servers are defined:

User agent: A user agent is a SIP-enabled end device. The user agent provides communication services to a user. A user agent client (UAC) is a functional module issuing a request and a user agent server (UAS) is a functional module responding to a request. In VoIP, a UAC is a calling party and a UAS is a called target party.

SIP server: A SIP server supports communication establishment between a UAC and a UAS. Three types of a SIP server exist. A proxy server transfers a SIP request, which a UAC issues to a UAS and other servers. Multiple-hop transfer is also available. Moreover, the server provides authentication, authorization, network access control, and routing. A redirect server receives a request and sends a current address of the target agent. A registration server accepts a REGISTER message described below and makes contact information available as described below.

The following are SIP messages to use during call establishment in SIP. The others are omitted.

REGISTER message: The REGISTER message informs a SIP server of the current contact URIs of a user agent. The SIP server transfers a SIP request to a contact URI of the user agent. Multiple contact URIs can be specified to redirect and fork a SIP message so that a SIP message is routed to the specified contact URI. In a contact header, each contact URI is separated by a comma. In the following case, a user registers a SIP address, a mail address, and a phone number as its contact address.

An expires field specifies the duration during which the registered user agent is valid. If a value expires field is zero, the registered device is unregistered from the home network.

```
REGISTER sip:userA@ai.u-hyogo.ac.jp SIP/2.0
To:userA<sip:userA@ai.u-hyogo.ac.jp; user=phone>
From:userA<sip:userA@ai.u-hyogo.ac.jp>
Expires:7200
Call-ID: 12-34-56-78@ai.u-hyogo.ac.jp
CSeq: 1 REGISTER
Contact: <sip:userA@ai.u-hyogo.ac.jp>,
         <mailto:userA@ai.u-hyogo.ac.jp>,
         tel:+81-12-XXX-YYYY@phone.com; user=phone>
Content-Type: application/sdp
Content-Length: 0
```

INVITE message: The INVITE message establishes multimedia sessions between a UAC and a UAS. An example of the INVITE message is as follows: A UAC issues an INVITE message to establish a session with a UAS. The From and To header in the INVITE message contain an address of a UAC and a UAS in a URI form, respectively. Note that the SIP message header can be extended and the extended messages are sent to the To: address without processing the headers. The following example shows a calling request from user A to user B with an audio device. Information about a communication media is described in the content of the INVITE message using Session Description Protocol (SDP).

```
INVITE sip:123@ai.u-hyogo.ac.jp user=phone SIP/2.0
Max=Forward:60
To:userB<sip:123@ai.u-hyogo.ac.jp;user=phone>
From:userA<sip:userA@ai.u-hyogo.ac.jp>
Call-ID: 12-34-56-78@ai.u-hyogo.ac.jp
CSeq: 1 INVITE
Subject: Lunch?
```

```
Contact: <sip:userA@ai.u-hyogo.ac.jp>
Content-Type: application/sdp
Content-Length: 123
```

```
m=audio 5060 udp sdp
```

SDP [10] describes general information about multimedia session established using SIP. The general form of SDP is:

```
x=parameter1  ...  parameterN .
```

In this example, the m field means that a caller would like to communicate with an audio media by User Datagram Protocol (UDP) at port 5060.

In SDP, an option field a is defined. It can be used to extend SDP to provide more information about the media.

Response message: When a SIP server and a UAS accept a SIP message such as REGISTER and INVITE, the server and the UAS reply to the SIP message. The responses 2xx, 4xx, and 5xx mean a success (OK), a client error, and a server error, respectively.

A UAS registers its own information using REGISTER message (Figure 7.3(1)). The information includes multiple contact addresses where the UAC can reach the UAS. When a UAC would like to communicate with a UAS, the UAC issues an INVITE message to the SIP server (Figure 7.3(2)). The SIP server searches for the UAS locations and routes the INVITE message using the contact URI (Figure 7.3(3)). If the UAS accepts the INVITE message, it replies a 2xx OK message to the server (Figure 7.3(4)). The server relays the OK message to the UAC (Figure 7.3(5)). After the channel is established, the communication starts between the UAC and the UAS (Figure 7.3(6)).

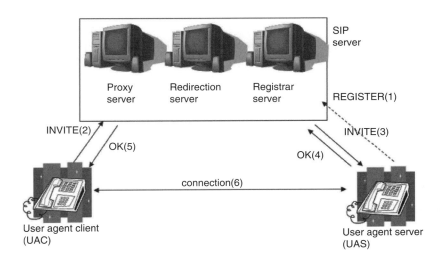

FIGURE 7.3

SIP call establishment. (From Nakamoto, Y. and N. Kuri, *Proc of 6th IEEE Int Conf Comp Info Tech*, 2006, p. 99. With permission.)

7.4 Redesign Issues

A control point and a device in UPnP are regarded as a UAC and a UAS in SIP. Therefore, straight mapping a UPnP advertisement request message to a SIP REGISTER message and a UPnP search message to an INVITE message is available. However several issues remain. A method to discover interesting devices and services is different between UPnP and SIP. In searching for an interesting device and service in UPnP, multiple search requests can be issued using HTTP/MU. Each request contains not only the specific device and service, but an interested type of defines and services. In SIP, a UAC specifies the single search target in an INVITE message, not the type of devices.

To solve the above problem, first, we utilize multiple contact URIs in the contact field in a REGISTER message. In a contact field, search target candidates including the particular device uid, types of the device, and services are specified and registered to the register server. Second, when an INVITE message is issued to search for interesting devices and services, we extend the a field in the SDP to add such search target information to the INVITE message. This extension is reasonable since SDP specifies information for connecting to a target.

Allowing multiple search targets in the contact headers can provide *abstract* home services. For example, if a control point requires an abstract TV program recording service, then actual devices such as a DVD or PC in the contact headers can be registered as suitable concrete providers of the abstract service. A SIP registry server can then select a recording device based on the contact information.

7.5 Redesigning UPnP Middleware

In this section, we describe how discovery functionalities in UPnP may be implemented with SIP functionalities, while preserving UPnP API specifications. In the following, we show a mapping between UPnP messages and SIP messages*.

7.5.1 Discovery: Advertisement

In the discovery advertisement phase of UPnP, a device multicasts a NOTIFY message as shown below, thus advertising to control points its participation in the network. NT (Notification Type) headers, defined in GENA, specify the device and its services. Multiple NOTIFY messages are sent to advertise

---

*In the messages, a typesetting face and an italics face show keywords and actual parameters for a UPnP and SIP message, respectively. A bold face in a SIP message shows the same parameters as ones in the corresponding UPnP message.

service types, device types, and a device identifier, which are described in the NT header, to other devices in the network. The information in the NT header is in the URI form and used in the discovery search phase in Section 7.5.2. The CACHE-CONTROL header specifies how long the device advertisement will continue. The USN (Unique Service Name), header is a unique identifier defined by SSDP.

```
NOTIFY * HTTP/1.1
HOST: address and port for multicast
CACHE-CONTROL: max-age = seconds until advertisement
Expires
LOCATION: URL for UPnP description for root device
NT: search target
NTS: ssdp:alive
SERVER: OS/version UPnP/1.0 product/version
USN: advertisement UUID
```

A discovery advertisement in UPnP corresponds to registration in SIP. When a device joins a network, it sends a REGISTER message to a SIP registry server. In the context of UPnP, a REGISTER message contains the following headers. Note that a text in bold in the SIP message matches a corresponding field with identical text in the UPnP message. The from and to headers show the unique identifier of the registering, or advertised, device. In the contact headers, the search targets of NT headers in the multiple NOTIFY message can be found because the contact header in a single REGISTER message shows an actual address that can be searched for in SIP.

```
REGISTER SIP URI of registry server
To: advertisement UUID
From: advertisement UUID
Expire: seconds until advertisement expires
Call-ID: call identifier
CSeq: sequence number
NTS: ssdp:alive
Contact:
search target1; host = OS/version
UPnP/1.0 product/version; location = URL for
UPnP description for root device,
search target2; host = OS/version
UPnP/1.0 product/version; location = URL for
UPnP description for root device
```

When a device leaves the home network, the following ssdp:byebye message is sent to devices in the UPnP network.

```
NOTIFY * HTTP/1.1
HOST: address and port for multicast
NT: search target
```

```
NTS: ssdp:byebye
USN: advertisement UUID
```

The corresponding operation in SIP is cancellation of registration. To cancel its registration, a device sends the following REGISTER message with an Expire header zero to the registry server.

REGISTER *SIP URI of registry server*
To: **advertisement UUID**
From: **advertisement UUID**
Expire: **0**
Call-ID: *call identifier*
CSeq: *sequence number*
NTS: ssdp:byebye
Contact: **search target**

7.5.2 Discovery: Search

In the discovery advertisement phase of UPnP, a control point multicasts a message as shown below, allowing the control point to search for a device of interest. Note that the ST header here corresponds to the NT header in the discovery advertisement message. When a device receives an M-SEARCH message with its own uid as the search target, it responds with an OK message to the control point.

```
M-SEARCH * HTTP/1.1
```
HOST: *address and port for multicast*
MAN: ssdp:discover
MX: *seconds to delay response*
ST: *search target*

In SIP, a device of interest is searched for with an INVITE message. When a VoIP client requires a certain person, an INVITE message is sent to a registry server and an actual connection destination is searched for using the contact data. In this redesign, an INVITE message with search target data is used to search for a device of interest. To add this search target information, we need to extend the a field in the SDP. An attribute with a prefix X-upnp is an extension attribute for UPnP. For example, the X-upnp: st attributed specifies search targets in the M-SEARCH UPnP message.

```
INVITE advertisement UUID
```
To: **SIP URL of registry server**
From: *requesting URI*
Call-ID: *call identifier*
CSeq: *sequence number*
NTS: *ssdp:byebye*

```
m=application 5060 udp sdp
```
a=X-upnp:st **search target;** X-upnp:mx = **seconds to delay response**

When a device receives an INVITE message with its own UID as the search target, the device responds to the control point with an HTTP OK message in the following format. Note that an OK message contains information about the discovered device such as the URL of the device description, and software versions.

```
HTTP/1.1 200 OK
CACHE-CONTROL: max-age = seconds until advertisement expires
DATE: when response was generated
EXT:
LOCATION: URL for UPnP description for root device
SERVER: OS/version UPnP/1.0 product/version
ST: search target
USN: advertisement UUID
```

A SIP OK message is sent when an actual connection with a destination device has been established and this destination device is able to communicate with the requesting device. We also extend the x SDP field to add information about the discovered device as follows:

```
SIP/2.0 OK 200
To: requesting URI
From: advertisement UUID
Call-ID: call identifier
CSeq: sequence number
Date: when response was generated
Expire: seconds until advertisement expires
Location: URL for UPnP description for root device
Server: OS/version UPnP/1.0 product/version
St: search target
Usn: advertisement UUID
```

7.6   Middleware Implementation

We have chosen CyberLink for Java [11] to provide the APIs and libraries of UPnP. We also use JAIN-SIP, which provides the SIP functionalities [12]. Why we use these programs is because both are open source software and therefore the API specifications and Java source codes for their implementations are readily available.

We have built home network middleware based on SIP called Siphnos [13], as shown in Figure 7.4. On the top of the protocol stack, we implemented class libraries with the same API specification as in Cyberlink for Java. The UPnP classes are inherited from the HTTP classes in Cyberlink for Java. Instead of modifying the HTTP classes with SIP to localize modified program parts and reduce the modification, we write *adapter* program modules that convert a UPnP message to the corresponding SIP message and vice versa.

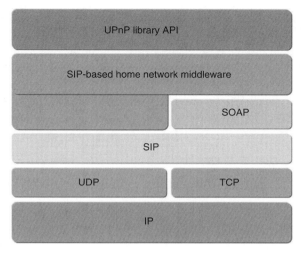

FIGURE 7.4
Protocol stack for the redesigned SIP-based home network. (From Nakamoto, Y. and N. Kuri, *Proc of 6th IEEE Int Conf Comp Info Tech*, 2006, p. 99. With permission.)

TABLE 7.1
Line Numbers in the Redesign

Module name	# of line
RegisterClient	239
RegisterServer	291
SearchClient	295
SearchServer	361
All java files in CyberLink for Java	15279

Source: From Nakamoto, Y. and N. Kuri, *Proc of 6th IEEE Int Conf Comp Info Tech*, 2006, p. 99. With permission.

The adapter modules are shown in Table 7.1. Since the total line number of all the class files is approximately 15,000 lines and the adapter modules can be implemented in the smaller program size, our intention is successful.

7.7 Conclusion

In this paper, we present a redesign of some aspects of the service functionalities within a home network system using SIP. Implementing UPnP using SIP leads to seamless connectivity between devices inside and outside the home and integration of VoIP with home network applications. In future work, we plan to use SIP to implement the other functionalities in UPnP.

One big effect of implementing this UPnP middleware in SIP is that a server is required to maintain relationships between a device and its services. In the UPnP world, however, no such server is necessarily required: relationships are transmitted upon request by an IP multicast server using the discovery protocol.

In the redesign with SIP, a single server failure in a home networks brings out a single point failure. To avoid the single server failure, we have an idea that all participants in the network have a SIP server. In this idea, every device has a name server, no need of specific name servers, and it tolerates the single point failure by the server failure, which is realized in UPnP.

References

[1] Sun Microsystems. 2003. *Jini Specifications Archive – v2.0.*

[2] UPnP Forum. 2000. *UPnP Device Architecture.*

[3] OSGi Alliance. 2003. *OSGi Service Platform Release 3.* IOS Press. San Ranon CA.

[4] Digital Living Network Alliance. 2004. *DLNA Home Networked Device Interoperability Guidelines v1.0,* June.

[5] Rosenberg, J., H. Schulzrinne, G. Camarillo, A. Johnston, J. Peterson, R. Sparks, M. Handley, and E. Schooler. 2002. *SIP: Session Initiation Protocol.* IETF.

[6] Dennis Bushmitch, Wanrong Lin, Andrzej Bieszczad, Alan Kaplan, Vasillis Papageorgiu, and Algirdas Pakstas. 2004. A SIP-based device communication service for OSGi framework. In *Proceedings IEEE Consumer Communications and Networking Conference 2004.*

[7] Alan Brown, Mario Kollberg, and Dennis Bushmitch. 2006. A SIP-based OSGi device communication service for mobile personal area networks. In *Proceedings IEEE Consumer Communications and Networking Conference 2006.*

[8] Jongwoo Sung, Daeyoung Kim, Hyungjoo Song, Junghyun Kim, Seong Yong Lim, and Jin Soo Choi. 2006. UPnP based intelligent multimedia service architecture for digital home network. In *Proceedings of Fourth IEEE Workshop on Software Technologies for Future Embedded and Ubiquitous Systems, and the Second International Workshop on Collaborative Computing, Integration, and Assurance.*

[9] Michael Jeronimo, and Jack Weast. 2003. *UPnP Design by Example.* Intel Press.

[10] Handley, M., and V. Jacobson. 1998. *SDP: Session Description Protocol.* IETF.

[11] CyberLink for Java Project. 2003. Cyberlink for java. http://sourceforge. net/ projects/cgupnpjava/, December.

[12] National Institute of Standards and Technology. 2003. Jain-sip. https:// jain-sip.dev.java.net/, December.

[13] Nakamoto, Y. and N. Kuri. 2006. Siphnos—Redesigning a home networking system with SIP, in *Proc. of 6th IEEE International Conference on Computer and Information Technology,* p. 99.

8

Compliance and Interoperability Testing of SIP Devices

Doris Bao, Luca De Vito, Sergio Rapuano, and Laura Tomaciello
University of Sannio

CONTENTS

8.1	Protocol Testing	173
8.2	Protocol Testing Methods	176
8.3	SIP Testing Tools	183
8.4	Case Study: Validation of the INVITE Client Transaction	190
8.5	Conclusions and Future Directions	196
	References	197

The Session Initiation Protocol (SIP) standards [1–5] define a common interface for SIP products made by different manufacturers. If all products of a network are compliant to the same standards, they can interoperate each other and a well-defined set of features can be provided to the users. However, SIP standards can lead to different implementations. SIP products, which are independently developed by different vendors, need compliance and interoperability tests to obtain a problem-free integration. While developing a SIP network device, an essential phase is, therefore, the verification of the compliance to the standards and the interoperability with other implementations of the same reference standards.

This chapter is organized in four main parts. In the first part, the definitions and characteristics of the compliance and interoperability testing of network protocols are presented. Then, the methods to perform such testing on SIP entities, which are commonly used in the industry and the research, are reported. Moreover, a summary of commercially available instruments and tools that provide assistance in the analysis, implementation, and robustness verification of the SIP protocol is presented. Finally, a SIP validation case study is examined in order to clarify the investigated methods and concepts.

8.1 Protocol Testing

Protocol testing is used to benchmark performance of the SIP device with respect to standard features and to the other products existing in the market.

This is particularly true in the information and communications technology industry, in which the compliance of products and the interoperability among products of different manufacturers is nonetheless an essential requirement. However, the creation and conduct of tests, on the one hand can be expensive and not cover all the situations and scenarios specified by the standards, on the other hand, they can reduce the costs to fix a problem in a pre-deployment setting [6].

8.1.1　Compliance Testing

In order to verify the correct implementation of a product and to ensure its working in accordance with the specifications, compliance testing is executed. It is able to increase confidence that a product complies with its specifications, and to reduce risk of malfunctioning when the product is put into place. The compliance testing process can identify product bugs and can verify that the eliminated bugs are fixed successfully and within the applicable standards [7].

In order to execute a compliance testing process for each characteristic of the standard specifications, as shown in Figure 8.1, a test purpose (TP) is written. Next, each TP is revised to detail (1) what is sent to the system under test (SUT), (2) what is expected from the SUT, (3) what the SUT must do in order to pass the test case, and (4) how the SUT can fail the test case. In this way an abstract test case (ATC) is defined and the resulting collection of ATCs forms an abstract test suite (ATS).

The implemented ATS becomes an executable test suite (ETS). This last permits the user to select and run one or more test cases from the ATS and save the result in a test report (TR) with a pass, fail, or inconclusive verdict. If the verdict is fail, the product designer can fix the problems found and rerun the ETS [8].

8.1.2　Interoperability Testing

Interoperability testing answers to the question: will this new product or network element work with another particular product, or with a class of products built according to the same standard? Interoperability testing aims at

FIGURE 8.1
Compliance testing.

giving information about the capability of a product to work in an Internet Protocol (IP) network according to the standard.

Interoperability testing is the next logical step after compliance testing. While compliance testing verifies that system A complies with specification X and that system B also complies with specification X, interoperability testing evaluates a successful communication between systems A and B [7]. In order to perform interoperability testing, all possible system interoperations between the end users, versus the rest of the system, have to be tested (assuming that access to the system is possible only through end user behaviors) [7].

The major steps of the interoperability testing process are illustrated in Figure 8.2:

1. By analyzing the specification, test purposes are defined and an ATS is written.

2. From this ATS, an ETS is implemented.

3. The ETS is executed on a protocol analyzer against two or more SUTs.

4. A TR is generated after execution of ETS.

Interoperability testing is like compliance testing: (1) the ATCs and ATSs of both test types appear similar, (2) the processes of implementing the ATS and running the ETS are the same, (3) the results are listed in TR and the verdicts are pass, fail, or inconclusive. The main difference between them is that interoperability testing verifies several SUTs at the same time [8].

8.1.3   Compliance Versus Interoperability Testing

"Compliance" is the ability to operate in the way defined by a standard. Two systems that comply with a standard are likely to have a high degree of interoperability but complete interoperability cannot be guaranteed [9].

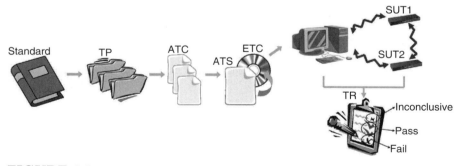

FIGURE 8.2
Interoperability testing.

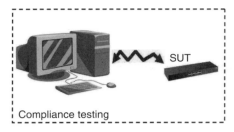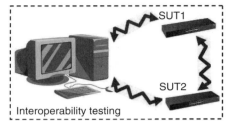

FIGURE 8.3
Compliance versus interoperability differences in practice.

Even after the appropriate standards organization's committees formalize standards, those standards might lack sufficient definition to ensure compatibility, and vendors might interpret guidelines differently. As a result, two products can comply fully with the same set of standards but fail to interoperate (Figure 8.3) [10].

Compliance testing is able to show that a particular implementation complies with all of the protocol requirements specified in the associated base standard. However, it is difficult for such testing to be able to prove that the implementation will interoperate with similar implementations in other products. On the other hand, interoperability testing can clearly demonstrate that two implementations will cooperate to provide the specified end-to-end functions but cannot easily prove that either of them complies with the detailed requirements of the protocol specification.

The compliance testing tends to verify basic protocol features, while interoperability testing is designed to replicate real-life scenarios.

Thus, interoperability testing is not better than compliance testing or vice versa. The approaches are complementary and not mutually exclusive. Complex technologies, the cost of getting things wrong, and a renewed interest in branding and certification programs encourage a combined approach.

Compliance and interoperability are both important and useful approaches to the testing of standardized protocol implementations although it is unlikely that one will ever fully replace the other.

8.2   Protocol Testing Methods

Protocol testing is carried out with the help of a test sequence generated from a protocol specification. Communication protocol specifications can be represented in the form of a finite state machine (FSM). Then the common approach in protocol testing is to cover the whole FSM and verify each state and transition. This method is suitable for protocols that have a short FSM. In other cases, knowledge of the protocol's weaknesses and vulnerable areas allows definition of more appropriate ATS, by analyzing only the key features and aspects of the protocol.

8.2.1 State Machine Approach

Since a protocol can be modeled as a state machine, to solve the problem of protocol testing many solutions based on a mathematical approach to select test cases, were proposed [11,12]. The goal is to obtain a minimum set of test sequences, constituting an ATS, that cover all machine states or transitions to perform a thorough test of an implementation under test (IUT). Various ATS development methods have been designed for the case that the reference specification is given in the form of an FSM. The best-known methods [12] are called transition tour [13], w-method [14], distinguishing sequence method [15], and unique-input-output (UIO) method [16]. These methods are applicable to software and hardware systems as long as the system behavior can be characterized as a reactive system that provides output responses depending on the received inputs and the dynamically evolving system state. In FSM-based testing [17], both the specification and the implementation can be modeled as an FSM. If the behavior of the implementation FSM is different from the specified behavior, the implementation contains a fault. The types of implementation faults are: (1) output faults (the output of a transition is wrong), and (2) transfer faults (the next state of a transition is wrong). Each test derivation method mentioned above provides the following fault coverage guaranteed: if the implementation can be modeled by an FSM with at most m states (where m is larger or equal to n, the number of states of the specification), then an ATS can be derived by the method (for this given m) and the implementation will only pass this ATS if and only if it complies with the specification (that is, it does not contain any output nor transfer faults) [17]. The ATS derived from each of the above methods will detect any output error of the implementation, that is, if the implementation follows the FSM specification except for the output produced for certain state transitions. Nevertheless, the w-method and distinguishing sequence method will find all such errors if the number of states of the implementation remains within certain boundaries. The discussion of the fault coverage of the test methods is, therefore, based on the fault model of FSM "output" and "transfer" faults.

8.2.1.1 Example: SIP REGISTER Session State Machine

The specifications for the UA REGISTER session originate from the high-level state machine shown in Figure 8.4. Following, for example, the transition tour method, a test sequence (called a transition-tour sequence) can be generated by simply applying random inputs to the UA REGISTER FSM until the machine has traversed every transition at least once. A transition-tour sequence for the FMS in Figure 8.4 is reported below:

1 2 3 4 1 2 1 2 3 4 3

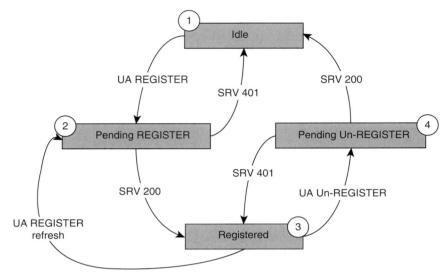

FIGURE 8.4
SIP REGISTER session state machine.

In looking at the test sequence, note that it can be divided into four parts, each one identifying a test:

- 1 2 3 : Registration without authentication: A UA in an idle state sends a SIP message, REGISTER, without authentication credentials, receives a 200 OK SIP answer from the registrar server and moves to the registered state.

- 3 4 1 : Un-registration without authentication: A UA in the registered state sends an un-REGISTER message, obtains a 200 OK SIP answer and moves to the idle state.

- 1 2 1 2 3 : Registration with authentication: A UA in an idle state sends a SIP REGISTER message without authentication credentials, receives an error answer, 401 "Unauthorized", and replies with a REGISTER message with an authentication field, receives a 200 OK SIP answer and moves to the registered state.

- 3 4 3 : Un-registration with authentication: A UA in the registered state sends an un-REGISTER message, obtains an error answer, 401 "Unauthorized", replies with an un-REGISTER message with authentication, receives a 200 OK SIP answer and moves to the idle state.

8.2.2 SIP Testing

In the place of the FSM approach, some alternative approaches [18] for the selection of test cases for the compliance tests of SIP could be obtained by using test cases provided by either (1) standard test specifications, or (2) interoperability events:

- European Telecommunications Standards Institute Test Specifications (ETSI TS 102 027 [19–21]) are designed to concentrate on areas critical to interoperability, including testing an implementation's reaction to an erroneous behavior. The goal is compliance testing for interoperability. This should not be confused with interoperability testing, which is a useful but a different activity. In many cases, development of the test specifications gives valuable feedback to the base standards. The test specification designers make a special effort to ensure that their experiences are fed back to the relevant ETSI technical committee. However, development of the base standard and test specifications must be done in a coordinated and timely manner for this input to be effective. A focused set of compliance tests can provide an excellent framework for subsequent interoperability testing.

- Interoperability events for SIP standards and product validation are open to all companies, organizations, and working and study groups implementing SIP entities. The customers of interoperability events are:

  - Operators, vendors or equipment manufacturers who (1) are about to place their products on the market, (2) want to be sure of the interoperability of their products, (3) feel comfortable with the technologies but still want to improve their know-how;

  - Standardization bodies (ETSI, Internet Engineering Task Force (IETF), etc.) or any forum or interest group that (1) develop SIP standard or specifications, (2) want to check the coherence of the SIP specifications implemented, (3) need to check progress in using their specification, and (4) want to let their members get useful feedback for quickly and efficiently adapting or improving their specifications.

Interoperability events refine both the protocol and its implementations, determine the source of incompatibilities, and, if the specification is at fault, prepare a "fix" of common bugs for the draft revision. A result of these events is an informational document that gives examples of SIP test messages designed to exercise and "torture" a SIP implementation. [22].

8.2.2.1 ETSI TS 102 027 Technical Specification for SIP IETF RFC 3261

ETSI has over ten years' experience producing compliance test specifications for key telecommunication technologies among which is Voice-over-Internet Protocol (VoIP). These technical specifications (TSs) are used by manufacturers as a basis for internal development testing. ETSI test specifications are usually developed by a specialist task force, which is a group of experts recruited from the ETSI membership and managed by the Centre for Testing and Interoperability. ETSI TS 102 027 provides TS for testing the compliance of various SIP entities to IETF RFC 3261. In order to evaluate compliance of a particular protocol implementation, a protocol implementation conformance statement (PICS) [19], a test suite structure (TSS) and a TP are defined [20]. The PICS for the SIP implementation is in compliance with the relevant requirements specified in IETF RFC 3261. The PICS contains a table for each SIP entity: User sgent, registration server, proxy application server or redirect server. These tables define the services, the service procedures and the related messages (requests and responses) correlated with mandatory (M or m), optional (O or o), not applicable (N/A or n/a) or prohibited (X or x) parameters and capabilities. An example related to a user agent entity and in particular to the REGISTER message request parameters is reported in Figure 8.5. The TSS is a test classification based on SIP entities and their role in the main functionalities: (1) registration, (2) call control, (3) querying for capabilities, and (4) messaging (Figure 8.6).

Areas of compliance investigated by TSS are: (1) capability and valid behavior to verify that mandatory capabilities and "normal" procedure handling are supported, (2) syntactically invalid behavior to verify that the IUT is able to properly react to receiving a syntactically incorrect message or stimulus, (3) invalid and inopportune behavior to verify that the IUT is able to react properly when a syntactically correct, but unexpected, message is received, and (4) timer expiry to verify the IUT reacts properly.

In each main functionality and for each entity, a TP is identified by four keywords:

- The TP Identifier (TPId) gives a unique identifier to each test purpose.

- "Status" specifies, when a test purpose is optional, recommended, and mandatory, according to the IETF RFC 3261 [1] PICS document.

- "Ref" outlines the references in IETF RFC 3261 [1] used to create the test purpose.

- "Purpose" describes the objective of the test.

Some examples are reported, for the case of registration in Figure 8.7, call control in Figure 8.8, and for registrant and originating endpoint, respectively.

8.2.3 SIP Interoperability Test Events

SIP interoperability test events (SIPITs) [23], are weeklong events where people bring their SIP implementations to ensure that they work together. The SIPITs are open to anyone with a working SIP implementation verifying compliance reaction of SIP entities to syntactically incorrect or correct but unexpected messages. During the events, problematic messages were noted and released as an IETF-draft (IETF RFC 4475) [22].

Prerequisite: A.2/1 and A.1/1			
Item	Parameters name	Reference	Status sending
1	Request line		
1.1	Method	RFC 3261, section 7.1	M (see note 1)
1.2	Request-URI	RFC 3261, section 7.1	M
1.3	SIP-Version	RFC 3261, section 7.1	M (see note 2)
2	Headers		
2.1	Accept		O
2.2	Accept-encoding	RFC 3261, section 20	O
2.3	Accept-language	RFC 3261, section 20	O
2.4	Allow	RFC 3261, section 20	O
2.5	Authorization	RFC 3261, section 20	Oa6.01
2.6	Call-ID	RFC 3261, section 20	M
2.7	Call-Info	RFC 3261, section 20	O
2.8	Contact	RFC 3261, section 20	Ca6.01
2.9	Content-disposition	RFC 3261, section 20	O
2.10	Content-encoding	RFC 3261, section 20	O
2.11	Content-language	RFC 3261, section 20	O
2.12	Content-length	RFC 3261, section 20	Ca6.04
2.13	Content-type	RFC 3261, section 20	Ca6.02
2.14	CSEq	RFC 3261, section 20	M
2.15	Date	RFC 3261, section 20	O
2.16	Expires	RFC 3261, section 20	Ca6.03
2.17	From	RFC 3261, section 20	M
2.18	Max-forwards	RFC 3261, section 20	M
2.19	MIME-version	RFC 3261, section 20	O
2.20	Organization	RFC 3261, section 20	O
2.21	Proxy-authorization	RFC 3261, section 20	Oa6.01
2.22	Proxy-require	RFC 3261, section 20	O
2.23	Require	RFC 3261, section 20	O
2.24	Route	RFC 3261, section 20	O
2.25	Server	RFC 3261, section 20	O
2.26	Supported	RFC 3261, section 20	O
2.27	Timestamp	RFC 3261, section 20	O
2.28	To	RFC 3261, section 20	M
2.29	User-agent	RFC 3261, section 20	O
2.30	Via	RFC 3261, section 20	M
2.31	Warning	RFC 3261, section 20	O
3	Body	RFC 3261, section 7.4	O
Ca6.01 if (A.3/2 OR A.3/4) then M else O.			
Ca6.02 if (A.6/3) then M else O.			
Ca6.03 if (contact/A.6/2.8 = *) then M else O.			
Ca6.04 if (A.12/2 OR A.6/3) then M else O.			
Oa6.01 if (A.8/1) then one at least shall be supported.			
Note 1: Set to "REGISTER" value in this case.			
Note 1: To be conform to RFC 3261 shall be set to "SIP/2.0"			
Comments:			

FIGURE 8.5

User agent: REGISTER request parameters.

Test suite	Main functionalities	Role	Functionalities subgroups		Test group
SIP	Registration	Registrant			V
		Registrar			V-I-O
	Call control	Originating endpoint	Call establishment		V-timers
			Call release		V-I-timers
			Session modification		V
		Terminating endpoint	Call establishment		V-I-timers
			Call release		V-I-timers
			Session modification		V-I
		Proxy	Message processing	Request	V-I
				Response	V
			Transaction	Client	V-timers
				Server	V-timers
		Redirect server	Call establishment		V
			Call release		V
	Querying for capabilities	Originating endpoint	Asking for capabilities		
		Terminating endpoint	Responding for capabilities		
		Proxy	Responding for capabilities		
	Messaging	Registrant			V-I
		Registrar			V-I
		Originating endpoint			V-I
		Terminating endpoint			V-I
		Proxy			V-I
		Redirect server			V-I

FIGURE 8.6

TSS for SIP where test group indicates: Valid behavior (V), invalid behavior (I), inopportune behavior (O).

The IETF RFC 4475 informational document tries to focus on areas that have caused interoperability problems or that have particularly unfavorable characteristics if they are handled improperly. It gives examples of SIP test messages designed to exercise and "torture" a SIP implementation. This document contains test messages based on the current SIP version (2.0) as defined in IETF RFC 3261. These test messages are organized into several sections and are presented in the text format using a set of markup conventions to avoid ambiguity and meet Internet-draft layout requirements. Some sections stress only a SIP parser, and other ones stress both the parser and the application above it (as application-layer semantics). Some messages are valid, and some are not.

Some examples of parser tests (syntax) valid and invalid messages and application-layer Semantics are reported in the Figure 8.9, Figure 8.10, and Figure 8.11, respectively. Parsers must carefully consider edge conditions and malicious input as part of their design. The first example (Figure 8.9) represents an unusual reason phrase in a 200 OK SIP response. A parser must accept this message because it does not clash with IETF RFC 3261.

Figure 8.10 shows an OPTION message with a start line and CSeq method mismatch: this request has mismatching values for the method in the start line and the CSeq header field. Any element receiving this request will respond with a 400 Bad Request.

TP for registration	
Registrant	
Group selection	Registration being listed as an option, the test purpose is applicable if the SUT is declared as supporting periodic registration and can behave as User Agent.
Status	PICS: A.1/1 and A.2/1
Valid behavior	
TPId	SIP_RG_RT_V_001
Status	Mandatory
Ref	RFC 3261 [1] section 10.2.
Purpose	Ensure that the IUT, in order to be registered, sends a REGISTER request to its registrar, without user name in the Request-URI and with a SIP-URI as request-URI.

FIGURE 8.7
Registrant: Test purpose for registration.

The last example (Figure 8.11), is an INVITE message that exercises an implementation's parser and application-layer logic. This INVITE request contains a body of unknown type. It is syntactically valid: a parser must not fail when receiving it, a proxy receiving this request would process it just as it would any other INVITE, and an endpoint receiving this request would reject it with a 415 Unsupported Media Type error.

8.3 SIP Testing Tools

In order to validate protocol implementations, various test tools are used today by the VoIP developers' community. Moreover, those tools are also used by service providers or technical marketing organizations during product interoperability testing or in order to verify protocol implementation compliance statements. This paragraph analyzes some commercially available tools for SIP testing and classifies them on the basis of their main features, in: (1) protocol analyzer tools, (2) simulation tools, and (3) robustness tools.

TP for call control	
Originating endpoint	
Group selection	IUT is an User Agent
Status	PICS: A.1/1
Call establishment	
Group selection	IUT can behave as a User Agent client to establish a call.
Status	PICS: A.16/1.1
Valid behavior	
TPId	SIP_CC_OE_CE_V_001
Status	Mandatory
Ref	RFC 3261 [1] section 8.1.1.
Purpose	Ensure that the IUT, to establish a call sends an INVITE request including at least To, From, CSeq, Call-ID, Max-Forwards, Contact and Via headers.

FIGURE 8.8

Originating Endpoint: Test purpose for call establishment.

```
Message details : unreason

<allOneLine>
SIP/2.0 200 = 2**3 * 5**2 <her>DOBDDOBE20D181D182
DOBE20DOB4DOB5DOB2D18FDOBDDOBED181D182DOBE20DOB4
DOB5DOB2D18FD182D18C202D20DOBFD180DOBED181D182DO
BEDOB5</hex>
</allOneLine>
Via: SIP/2.0/UDP 192.0.2.198;branch=z9hG4dK1324923
Call-ID: unreason.1234ksdfak3j2erwedfsASdf
CSeq: 35 INVITE
From: sip:user@example.com;tag=11141343
To: sip:user@example.edu;tag=2229
     .   .   .
```

FIGURE 8.9

Unusual reason phrase in 200 OK response. This particular response contains unreserved and non-ASCII UTF-8 characters but this response is well-formed.

```
Message details : mismatch01

OPTION sip: user@example.com SIP/2.0
To: sip:j.user@example.com
From: sip:caller@example.net; tag=34525
Max-Forward: 6
Call-ID: mismatch01.dj0234sxdf13
CSeq: 8 INVITE
Via: SIP/2.0/UDPhost.example.com;branch=z9hG4bK1324923
```

FIGURE 8.10

Start line and CSeq method mismatch.

```
Message details : invut

INVITE sip:user@example.com SIP/2.0
Contact: <sip:caller@host5.example.net>
To: sip:j.user@example.com
From: sip:caller@example.net;tag=8392034
Max-Forwards: 70
Call-ID: invut.OhaOisndaksdjadsfij34n23d
CSeq: 235448 INVITE
 Via: SIP/2.0/UDP somehost.example.com;branch=z9hG4bKkdjuw
Content-Type: application/unknownformat
Content-Length: 40

<audio>
<pcmu port="443"/>
</audio>
```

FIGURE 8.11

Unknown content-type.

8.3.1 Protocol Analyzer Tools

A number of tools exist to perform real-time network packet capture. Some of the key features include real-time capture at wire speed, viewing and browsing the capture data in details, filtering of captured packets based on specific protocols or target network addresses, exporting of capture traces in various formats to be able to share the finding, printing capture in hexadecimal form, text format with frame number, etc. They offer protocol decoders, which allow seeing the message details in a verbose format. The key is to find tools with decoders for protocols like H.323, SIP, media gateway control protocol (MGCP), real time protocol (RTP), real time control protocol (RTCP), etc.

8.3.1.1 Wireshark

Wireshark [24] is an open source network protocol analyzer for Unix and Windows. It allows examining data from a live network or from a capture file on disk. It is possible to interactively browse the capture data, viewing summary and detail information for each packet. Wireshark distribution provides decoders for H.323 (H.225 RAS, H.225.0, H.245), SIP/SDP, MGCP, RTP, RTCP, etc.

Figure 8.12 shows the analysis of the unknown content type SIP "torture" test. After the capture session, Wireshark decodes the SIP communication and displays the results in the main program window. The protocol analyzer allows stepping through each captured SIP packet, showing the details of both the packet header fields and the data portion of the packet.

8.3.1.2 LinkBit Online Protocol Analyzer/Decoder

LinkBit [25] is an online protocol analyzer and decoder for a wide range of protocols, including H.323, SIP, integrated services digital network (ISDN),

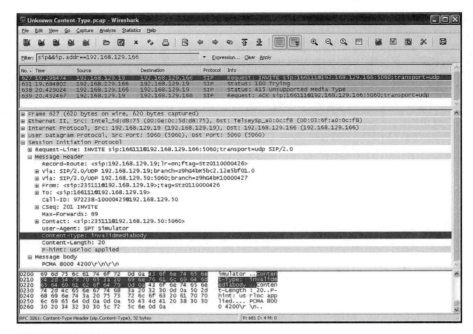

FIGURE 8.12
Wireshark analysis: Unknown content-type test.

V5.2, MGCP, and H.248. It just copies and pastes a hex dump and gets something readable.

8.3.1.3 SIP Analyzer

Distributed SIP analyzer [26] is a Web-based analyzer that can (1) capture SIP messages in different subnets (Local Area Network (LAN)) and draw SIP CALL Flow, (2) calculate talking time, call setup time, and (3) compare SIP versions, log SIP sessions. It uses MySQL database to store messages and uses PHP to show the SIP messages. It is an open source freeware.

8.3.1.4 GL Communication

GL Communications [27] provides an extensive SIP-based call emulation, SIP call analysis, SIP call trace, and call monitoring. Fully compliant to current SIP standards and latest codecs, this suite of products provides a comprehensive set of tools for network testing, monitoring, analysis, and evaluation.

8.3.1.5 Empirix

Empirix [28] Hammer Call Analyzer is a tool for VoIP environments. It is VoIP aware, which enables users to quickly visualize signaling and voice quality problems in VoIP networks. H.323, H.248, MGCP, SIP, T.38, RTP are the main protocols supported.

8.3.2 Simulation Tools

Simulation tools are used to run a representation of the code in a controlled environment. The advantages of the simulation are that the tests are cheap, quick to assemble, and easy to control and to reproduce. The disadvantages are that simulation tests may differ from real implementations.

8.3.2.1 Sipp

Sipp [29] is a GNU Operating System General Public License (GNU GPL) open source software test tool/traffic generator for SIP. It includes a few basic SipStone user agent scenarios (UAC and UAS) and establishes and releases multiple calls with the INVITE and BYE methods. It can also read custom XML scenario files describing from very simple to complex call flows. It features the dynamic display of statistics about running tests (call rate, round trip delay, and message statistics), periodic CSV statistics dumps, TCP and UDP over multiple sockets or multiplexed with retransmission management and dynamically adjustable call rates.

8.3.2.2 ProLab SIP Test Manager

ProLab SIP Test Manager by RADVISION [30] is an advanced testing system capable of performing essential tests including performance, load, stress, interoperability, and media and protocol compliance. The ProLab SIP Test Manager is a highly scalable, feature-rich testing system that is suitable for use in different stages of the product development cycle.

8.3.2.3 SIP Express Router

SIP Express Router (SER) [31] is a high-performance, configurable, free SIP (IETF RFC3261) server. It can act as registrar, proxy, or redirect server. SER features an application-server interface, presence support, SMS gateway, SIM-PLE2Jabber gateway, RADIUS/syslog accounting and authorization, server status monitoring, and FCP security.

A Web-based user provisioning, serweb, is also available.

8.3.2.4 Sipbomber

Sipbomber [32] is a tool for SIP developers intended for testing SIP implementation against IETF RFC 3261. The current version can check only server implementations (proxies, user agent servers, redirect servers, and registrars). This program is distributed under terms of GPL.

8.3.2.5 Compliance Engine

As part of the requirements for attending the SIPIT interoperability testing events, participants are expected to have already run a set of published test messages. The SIP Center's SIP Messenger tool [33] gives developers the facility to do this.

SIP Messenger is a Java software that allows the operator to send SIP test messages from text files over UDP to the SIP implementation and, optionally, listen for responses. The messages can be sent using a command line utility (messenger), suitable for invocation by automated scripting, or via a GUI (MessengerGui). Developers can use this software to construct their own SIP messages that can be pushed onto SIP servers or user agents (possibly in conjunction with the SIP Center's own SIP resources—the SIP network server and UA). This tool is especially useful for stress testing products with scenarios that are otherwise difficult to reproduce. This software has been made available by Avaya Inc., founder of the SIP Center [34].

8.3.2.6   SIPSAK

SIPSAK [35] is a small command line tool for developers and administrators of SIP applications. It can be used for some simple tests on SIP applications and devices.

8.3.2.7   Spirent Protocol Tester

The Spirent Protocol Tester (SPT) [36] is an advanced integrated network diagnostic and test platform. It is specifically designed to address VoIP signaling protocol test needs of equipment and service developers with signaling protocol testing requirements. SPT meets the challenges of VoIP network testing and the demands of high levels of automation, reusability, and scalability. SPT provides the development or test engineer with a rich set of signaling protocols, such as SIP. It provides a cost-effective, scalable, and evolving test automation solution for network equipment manufactures and service providers. Figure 8.13 shows the SPT test logic program (TLP) that is a protocol-specific call flow design diagram. In this case, the TLP is used to simulate a network device using SIP, to generate a call flow (originating endpoint). The TLP consists of a set of states and transitions between states. Users can create and/or edit a TLP with a finite set of states and transition links between the states to define the SIP test logic. The TLP editor allows users to display and modify the characteristics of any selected state (etc., sending SIP messages) in order to realize a specific SIP test.

8.3.3   Robustness Tools

Robustness is defined as the degree to which a system operates correctly in the presence of exceptional inputs or stressful environmental conditions [37]. This general class of tools (often referred to as protocol "fuzzers") includes tests designed to flush out problems and vulnerabilities with lower-level SIP stacks and parser mechanisms. The need for SIP-oriented robustness validation is important now, but will be even more important as the proliferation of SIP-based communications grows and malicious VoIP denial of service attacks become more prevalent.

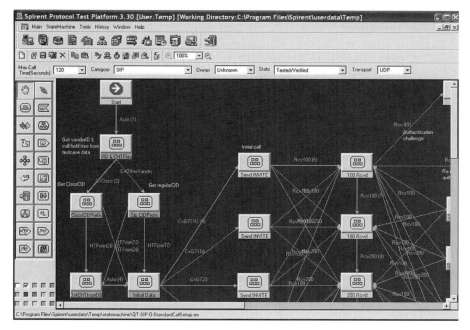

FIGURE 8.13
SPT TLP: emulation of an originating endpoint.

Using these tools it is possible to set up a special test that attempts to launch call sessions toward the SUT using SIP messages that have been "fuzzed" (intentionally corrupted) with the purpose of exposing weaknesses in the lower-level parsing and SIP message handling layer. The SUT is monitored for the appropriate response to each test case [38].

8.3.3.1 PROTOS Test-Suite c07-Sip

The publicly available University of Oulu PROTOS robustness tests [39] contain 4,527 test cases in 54 test groups. Varying SIP header fields in INVITE messages are tested. The PROTOS c07-sip test suite uncovered numerous security problems in numerous SIP products that were tested by the developers. This test suite is freely usable, as it is released under the GNU public license.

8.3.3.2 Codenomicon SIP Test Tool

The Codenomicon SIP test tool [40] is the commercial extension of the above work PROTOS test-suite, generated using the Codenomicon test generator. Currently, it contains over 20,000 test cases in over 2,000 test groups and tests all SIP header fields and all SDP fields. It contains a test documentation system and a GUI to assist in the test execution and fault elimination.

8.4　Case Study: Validation of the INVITE Client Transaction

This paragraph aims at providing a practical example about the validation phase of a SIP device to verify the compliance and interoperability of a protocol implementation.

The object of the example is the INVITE client transaction, which is the operation executed when the user starts a call. This example is not intended to provide an exhaustive testing of all facets of the SIP-originating endpoint operations: the discussion concerns the INVITE client transaction and intends to investigate just this specific feature.

8.4.1　The Invite Client Transaction

An invitation occurs when a SIP endpoint (user A) "invites" another SIP endpoint (user B) to join in a call. During this process, user A formulates an INVITE message requesting that user B joins a particular conference or establishes a two-party conversation. This INVITE request may be forwarded by proxies, and it eventually arrives to the called endpoint (user B), whichcan potentially accept the invitation. If user B wants to join the call, it sends a confirmatory response (SIP 2xx). Otherwise, it sends a failure response (SIP 4xx), or a 3xx, 5xx, or 6xx SIP response, depending on the reason of the rejection. Before sending a final response, the called endpoint can also send provisional responses that have a status code of type 1xx to advise the calling party of progress in contacting the called user.

The INVITE client transaction for an originating endpoint starts with the INVITE request sent by the endpoint and terminates with an ACK sent by the endpoint at the reception of a 2xx-6xx response.

8.4.2　Set Up of the Test Environment

In order to test a SIP-originating endpoint, the first step is to replicate a VoIP production network environment in a test laboratory. Technological advances have made it possible to use network emulators, traffic generators, and analyzers to recreate realistic conditions for testing a VoIP device. However, to perform a SIP signaling compliance test, when there are no specific network emulators' requirements, it is necessary to generate SIP signaling. It could be enough to build a SIP network installing different SIP proxy servers, such as the open source Asterisk and the SIP express router (SER). Each SIP server had to support a number of phones and be able to send calls to the other SIP servers. Basic call flow scenarios like registration, registration with authentication and SIP to SIP calls with and without authentication, and more complex situations like the use of supplementary services (call waiting, call forward, etc.) should be possible.

The above-described environment can be used to perform conformance testing, interoperability testing, or robustness testing. If the objective of the analysis is compliance or interoperability testing, one can conduct the analysis by following either an approach based on a state machine, as shown in Section 8.2.1, or one based on standard test specifications, as shown in Section 8.2.2.

In this case, it is sufficient to use one of the protocol analyzers that support SIP, like those described in Section 8.3.1.

In order to test compliance, the behavior of the IUT should be compared with the standard specifications. For interoperability test purposes, the test should instead be executed on two endpoints and one should verify that the communication behaves as specified by the standard.

Usually, an approach based on the state machine is more flexible because it allows writing a custom state machine that could better represent the IUT behavior. Instead, a standard test specification approach is less flexible, but it is easier to use because the tests are provided directly by the specifications and they should not be derived from a state machine.

Finally, in order to test the robustness of an IUT, it is necessary to conduct a stress test. This can be done by following standard test specification, such as the IETF RFC 4475, described in Section 8.2.3.3. In this case, it is necessary to use either one of the simulation tools presented in Section 8.3.2 or one of the robustness tools of Section 8.3.3.

In the following subsections, the ways to proceed with SIP testing are presented in the practical case of an invitation process. In Section 8.4.3, the method for interoperability or compliance by means of a state machine approach is used. In Section 8.4.4 a method based on standard test specifications is shown on the same example. Finally, in Section 8.4.5 the way to conduct some robustness tests on the same case is presented.

8.4.3   Testing by Means of a State Machine Approach

Figure 8.14 shows the state machine for the INVITE client transaction.

The state machine is entered when the endpoint sends an INVITE request. When it happens, the machine moves to the calling state. Then, if a 1xx response arrives, it moves to the proceeding state. Instead, if a final 2xx response arrives, it moves to the terminated state. If a response of types 3xx, 4xx, 5xx, or 6xx arrives in the calling or proceeding states, the machine moves to the completed state and waits for the expiration of a timer to go to the terminated state.

From the state machine of Figure 8.14, according to the *transition tour* method, the following sequence can be extracted, which covers all states and transitions:

1 1 2 2 3 3 4 − 1 3 4 − 1 4 − 1 2 4 − 1 2 3 4

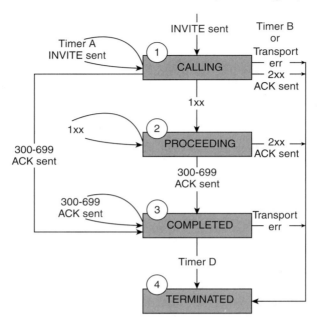

FIGURE 8.14
INVITE client transaction state machine for a SIP UAC starting an invitation process.

This sequence can be reproduced by running the following tests:

1. – Corresponding to the subsequence 1 1 2 2 3 3 4 –
   The SIP endpoint sends an INVITE request;
   timer A fires and a new INVITE is sent;
   a 1xx message is received;
   a new 1xx message is received;
   a response with code between 300 and 699 is received, and the ACK is sent;
   timer D fires.

2. – Corresponding to the subsequence 1 3 4 –
   The SIP endpoint sends an INVITE request;
   a response with code between 300 and 699 is received, and the ACK is sent;
   timer D fires.

3. – Corresponding to the subsequence 1 4 –
   The SIP endpoint sends an INVITE request;
   a 2xx response is received and an ACK is sent.

4. – Corresponding to the subsequence 1 2 4 –
   The SIP endpoint sends an INVITE request;
   a 1xx response is received
   a 2xx response is received and an ACK is sent.

In order to test for the compliance of the invitation process of the SIP IUT, it is sufficient to reproduce the steps described above, acting on the endpoint.

In addition, it is necessary that a called endpoint is configured to act as specified by the test. This can be easily done by simulating the called endpoint using one of the tools described in Section 8.3.2. How to configure the tool is not described here, because it depends heavily on the tool that has been selected.

As an example, if the IUT is a SIP phone, one should hook the phone off, compose a phone number that is configured in the network and wait for the dialing timeout (this will cause the INVITE to be sent). If it is running the test number 1, the called endpoint will not answer, causing timer A to expire. Then the IUT should send another INVITE message. At this point, the called endpoint will respond with two subsequent 1xx messages. Then, the called endpoint will send a message with a response code between 300 and 699, and the IUT should send an ACK.

In the case described, in order to check the compliance of the IUT, one should check if it correctly sends the INVITE and ACK messages, and if it correctly closes the SIP transaction when timer D expires.

The same procedure can be applied for all the other tests derived from the state machine.

8.4.4 Testing Using Standard Test Specifications

Validating the SIP UA client that begins an invitation process means: (1) in the first instance, to verify the INVITE message features according to RFC 3261 specifications; (2) subsequently, to check that the IUT follows the state machine transition according to the response received and the current state, and generates an ACK response in conformity with the RFC 3261; (3) finally, in case of no answer, to confirm that the timer and message retransmission processes act as defined in the standard.

ETSI test specifications deal in a complete treatment of these tests. The following subsections give some examples that clarify the way to validate the INVITE client transaction.

8.4.4.1 INVITE Features

ETSI TS 102 027 lists the "valid behavior TPs," with the aim of checking the mandatory requirements of the INVITE, generated by the SUT.

Figure 8.15 shows a well-formed INVITE request. A protocol analyzer, in this case Wireshark, allows checking that the INVITE message passes the ETSI TPs formulated according to the RFC 3261:

- the INVITE includes at least to, from, CSeq, call-ID, max-forwards, contact and via headers;

- the request-URI is set to the same URI value of the to header; the from header contains a TAG parameter;

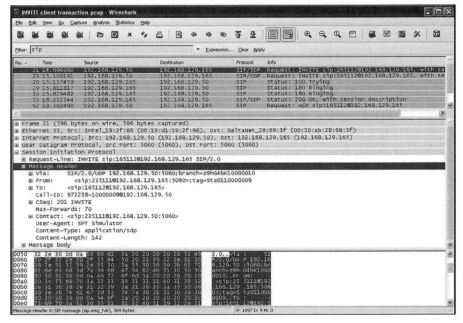

FIGURE 8.15
INVITE client transaction: analysis with Wireshark.

  – the CSeq header method matches "INVITE" and the via header has a
    protocol name set to SIP, a protocol version set to 2.0 and a branch
    parameter set to a value beginning with "z9hG4bK".

8.4.4.2   State Machine Transitions and ACK Response

Some ETSI TPs want to check the state transition at the receipt of a particular
response to INVITE.

Following the same Wireshark analysis detailed in Figure 8.15, if the IUT
is in the calling state, on receipt of a trying (100 trying) response, enters in
the proceeding state. Then, on receipt of one or more ringing (180 ringing)
responses the IUT remains in the proceeding state.

Consequently, another TP has the purpose to ensure that, on receipt of a
success (200 OK) response, the IUT sends an ACK request and enters in the
terminated state.

Some ETSI TPs have as purpose to ensure IUT behavior at the receipt of
300–699 messages. For example, the IUT when an INVITE client transaction
is in the calling state, on receipt of a multiple choices (300 multiple choices)
response must send an ACK request with the same Call-ID, From headers
and request-URI as in the original INVITE request and the same tag in the to
header as in the multiple choices (300 multiple choices) response. Figure 8.16
shows the Wireshark analysis of these specifications.

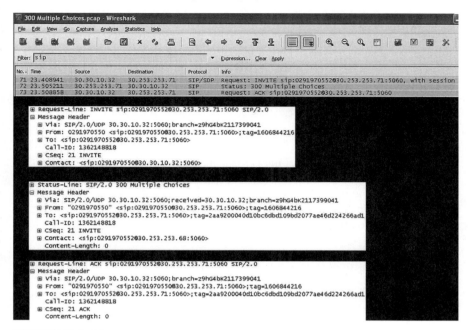

FIGURE 8.16

Wireshark analysis: SUT in the CALLING state receives a "300 Multiple Choices" message, the generated ACK must respect the test specifications.

8.4.4.3 Timers Test

Some ETSI TPs aimat verifying the correct management of timers in retransmissions of messages or state transitions when there are no responses.

For example, if an unreliable transport (UDP) is used, it should be verified that the IUT, when an INVITE client transaction is in the calling state, it repeats its INVITE request on the timeout condition of timer A, which has been initially set to a value of T1.

8.4.5 Robustness Tests

A torture test is an evaluation of the SIP IUT ability to withstand extreme conditions. The following treatment reports some example, derived from RFC 4475, useful to stress the considered INVITE client transaction.

For example the 200 success response to the INVITE sent by the IUT could be a 200 message with "unusual reason phrase" as reported previously in Figure 8.9 or with an "empty reason phrase". In fact, a 200 response containing no reason phrase is a well-formed response. A parser must accept this message. Only the space character after the reason code is required. If it were not present, this message could be rejected as invalid (a liberal receiver would accept it anyway). A simulation tool can be used to modify the reason phase of the 200 response. The obtained message, inserted in a test case simulating a UAS, allows the simulation tool to respond to the INVITE sent by the IUT and test its behavior when a 200 response with an unusual or empty reason phase is received.

Other stress messages can contain invalid scalar field headers. For example, in the INVITE client transaction case, the responses 200 or 300-699 to the INVITE could have a negative content-length. The IUT receiving these messages should respond with an error. When a message like this appeared over UDP, the remainder of the datagram can simply be discarded. If a request like this arrives over TCP, the framing error is not recoverable, and the connection should be closed. The same behavior is appropriate for messages that arrive without a numeric value in the content-length header field.

Scalar Fields with overlarge values could be used in 200 or 300-699 responses to the INVITE to stress the IUT. Scalar header field values outside their legal range are for example:

- The CSeq sequence number $>2^{31}$.

- The Max-Forwards value >255.

- The Expires value $>2^{31}$.

- The Contact expires parameter value $>2^{31}$.

The IUT receiving this request should respond with a 400 bad request due to the CSeq error. If only the max-forwards field were in error, the element could choose to process the request as if the field were absent. If only the expiry values were in error, the element could treat them as if they contained the default values for expiration (3600 in this case).

As a last torture test example, responses with missing required header fields can be considered. The IUT receiving a 200 or 300-699 containing no call-ID, from, or to header fields must not break because of the missing information. Ideally, it should respond with a 400 bad request error.

8.5 Conclusions and Future Directions

The investigation conducted on SIP testing methods and tools detects some common issues for SIP attendants. The aim of interoperability is addressed on the one hand from the production of technical standards and reports, on the other hand from the development of standardized test specifications: conformance test specifications and interoperability test specifications. SIP testing standardization is focusing on the creation of effective test suites following well-defined and transparent procedures in which all interested parties can participate. Standardization bodies are directing their efforts toward the development and application of testing methodologies, architecture, techniques, and languages designed according to the SIP features that can guarantee a certain degree of conformance and interoperability.

From a commercial point of view, production houses are addressing their attention toward the creation of SIP testing instruments and tools that are easy to use, sometimes embedded in the system, and that can allow users with a high configurability and programmability in order to emulate every particular situation.

References

[1] Rosenberg, J., H. Schulzrinne, G. Camarillo, A. Johnston, J. Peterson, R. Sparks, R. Handley, M. Schooler, and E. Schooler. 2002. SIP: Session Initiation Protocol. *RFC 3261*, 1–269.

[2] Rosenberg, J., and H. Schulzrinne. 2002. Reliability of provisional responses in the Session Initiation Protocol (SIP). *RFC 3262*, 1–14.

[3] Rosenberg, J., and H. Schulzrinne. 2002. The Session Initiation Protocol (SIP) UPDATE Method. *RFC 3264*, 1–25.

[4] Roach, A.B. 2002. Session Initiation Protocol (SIP)-specific event notification. *RFC 3265*, 1–38.

[5] Sparks, R. 2003. The Session Initiation Protocol (SIP) refer method. *RFC 3515*, 1–23.

[6] Dawson, K. 2007. Q and A: The importance of testing your technology. *New customer management insight* 34. http://www.callcentermagazine. com/showArticle.jhtml?articleID=197005862.

[7] Tretmans, J. 2001. An Overview of OSI Conformance Testing. Formal Methods & Tools group University of Twente. Translated and adapted from: Tretmans, J., and Vande Lagemaat, J. Conformance Testes. In *Handbook Telematica*, vol. II, 4400p., pp.1–19. Samson 2001.

[8] Coombs, C.F., and C.A. Coombs. 1998. *Communications Network Test & Measurement Handbook*. McGraw-Hill Professional, New York.

[9] Griffeth, N., R. Hao, D. Lee, and R.K. Sinha. 2000. Interoperability Testing of VoIP Systems. *Global Telecommunications Conference*, GLOBE-COM, San Francisco, CA, USA, 3; 1565–70.

[10] CETIS-http://assessment.cetis.ac.uk/FAQs/FAQs/Basics/compliance% 20v%20conformance. Accessed December 2007.

[11] Ural, H., and B. Yang. 1991. A test sequence selection method for protocol testing. *IEEE Trans. on Commun.* 39(4): 514–23.

[12] Sidhu, D.P., and T.-K. Leung. 1989. Formal methods for protocol testing: A detailed study. *IEEE Transactions on Software Engineering* 15(4): 413–26.

[13] Naito, S., and M. Tsunoyama. 1981. Fault detection for sequential machines by transition tours. In *Proceedings of the 11th Annual International Symposium on Fault-Tolerant Computing*, Portland, Maine, USA, (FTCS-11), pp. 238–43.

[14] Chow, T.S. 1978. Testing Software Design Modelled by Finite-State Machines. *IEEE Trans. Software Engineering*, SE-4(3): 178–87.

[15] Gonenc, G. 1970. A method for the design of fault detection experiments. *IEEE Trans. Computer* C-19(6): 551–8.

[16] Sabnani, K.K., and A.T. Dahbura. 1988. A protocol testing procedure. *Computer Networks and ISDN Systems* 15(4): 285–97.

[17] El-Fakih, K., N. Yevtushenko, and Gv.Bochmann. 2004. FSM-based incremental conformance testing methods. *Transactions on Software Engineering* 30(7): 425–36.

[18] Qiu, P.Q., O. Monkewich, and R.L. Probert. 2004. SIP Vulnerabilities testing in session establishment & user registration. *ICETE*, (2), 223–9.

[19] ETSI TS 102 027-3 V3.1.1 (2004–11): Methods for testing and specification (MTS); Conformance Test Specification for SIP (IETF RFC 3261); Conformance Test Specification for SIP (IETF RFC 3261); Part 3: Abstract Test Suite (ATS) and partial Protocol Implementation eXtra Information for Testing (PIXIT) proforma specification, 1–40.

[20] ETSI TS 102 027-2 V3.1.1 (2004–11) Technical Specification Methods for Testing and Specification (MTS); Conformance Test Specification for SIP (IETF RFC 3261); Part 2: Test Suite Structure and Test Purposes (TSS&TP), 1–120.

[21] ETSI TS 102 027-1 V3.1.1 (2004–11): Technical Specification Methods for Testing and Specification (MTS); Conformance Test Specification for SIP IETF RFC 3261; Part 1: Protocol Implementation Conformance Statement (PICS) proforma, 1–82.

[22] Sparks, R., A. Hawrylyshen, A. Johnston, J. Rosenberg, and H. Schulzrinne. 2006. Session Initiation Protocol (SIP) torture test messages. *RFC 4475*, 1–53.

[23] SIP Interoperability Test Events, http://www.cs.columbia.edu/sip/sipit/. Accessed December 2007.

[24] www.wireshark.org.

[25] www.linkbit.com.

[26] http://ant.comm.ccu.edu.tw/sip.

[27] www.sipcenter.com/sip.nsf/html/Sponsors+GL+Communications.

[28] www.empirix.com/products-services/v-hca.asp.

[29] http://sipp.sourceforge.net.

[30] www.sipcenter.com/sip.nsf/html/Sponsors+Radvision.

[31] www.iptel.org/ser.

[32] www.metalinkltd.com/downloads.php.

[33] www.sipcenter.com/sip.nsf/html/Compliance+Engine.

[34] www.avaya.com.

[35] www.sipsak.org.

[36] http://www.spirentcom.com/analysis/technology.cfm?WS=325&
SS=209&wt=2.

[37] Std 610.12-1990, IEEE Standard Glossary of Software Engineering
Terminology. Accessed December 2007.

[38] www.tmcnet.com.

[39] www.ee.oulu.fi/research/ouspg/protos/testing/c07/sip.

[40] www.codenomicon.com.

Part II

TECHNOLOGIES

9

P2P SIP: Network Architecture and Resource Location Strategy

Guiran Chang, Chuan Zhu, and Wei Ning
Northeastern University

CONTENTS

9.1 Introduction 203
9.2 Peer-to-Peer Network Technologies 204
9.3 Early Development of P2P SIP Network Architecture 210
9.4 P2P SIP Standardization 215
9.5 Protocol Development and Implementation 218
9.6 Concluding Remarks 222
    References 223

9.1 Introduction

Session Initiation Protocol (SIP) has been widely used in Internet telephony and has been chosen as the protocol for Internet Protocol (IP)-based multimedia call control for third generation (3G) wireless networks. As defined in RFC 3261 [1], SIP is a client server-based control protocol above the transport layer for creating, modifying, and terminating sessions between two or more participants. However, SIP-based telephony can be viewed as an application of peer-to-peer (P2P) architecture where the user agents (UAs) form a self-organizing P2P overlay network to locate and communicate with each other. In consideration of the characteristics of P2P, it is expected to be a perfect method to solve the scalability, robustness, and fault tolerance problems in traditional SIP networks. Skype is a free P2P application based on the Kazaa architecture for Internet telephony and instant messaging [2,3]. The protocol is proprietary and the system has centralized elements for login authentication [2]. Researchers have proposed several pure P2P architectures for SIP-based IP telephony systems [4,5]. The P2P SIP working group has been formed within the Internet Engineering Task Force (IETF) for adapting P2P with features suitable for SIP.

9.2   Peer-to-Peer Network Technologies

An overlay network is a computer network built on top of another network. Nodes in the overlay are connected by virtual or logical links, each of which corresponds to a path in the underlying network [6]. An example overlay network built on top of another IP computer network is shown in Figure 9.1. The overlay network resides at the edge of IP network and is ignorant of the underlying IP network. The nodes of the overlay network use only the IP addresses from the underlying network for discovery and routing on the application layer.

The term "peer-to-peer" (P2P) refers to a class of systems and applications that employ distributed resources to perform a function in a decentralized manner. A P2P network relies primarily on the computing power and bandwidth of the participants in the network rather than concentrating it in a number of servers. Because it exists on top of a physical layer network, a P2P network is often called a "P2P overlay" or "P2P overlay network." A typical client–server network is shown in Figure 9.2a, while a P2P network is illustrated by Figure 9.2b. In comparison with the client–server network, P2P networks are inherently scalable and reliable because there is no single point of failure. Peer-to-peer systems operate without centralized organization or control and, in the purest form, have no concept of servers. All participants are peers and communicate in a distributed, potentially untrusted environment to achieve a certain objective [6,7].

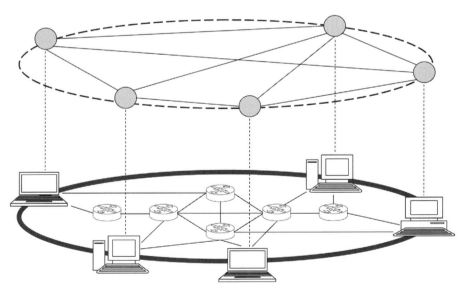

FIGURE 9.1
Example of an overlay network.

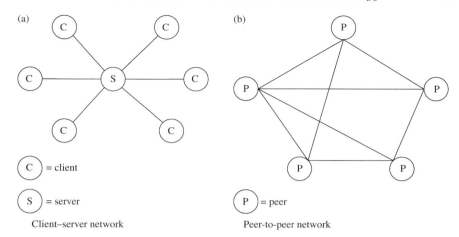

FIGURE 9.2

Client–server vs. peer-to-peer network.

The core operation in most peer-to-peer systems is efficient location of data items. The resource location strategy of a P2P network can be classified into four types: centralized unstructured directory model, pure decentralized unstructured flooded requests model, hybrid unstructured model, and decentralized structured distributed hash table (DHT) Model [8]. Currently, DHT-based structured P2P systems are the main technology used in P2P SIP researches, and the most popular algorithms include Chord, Pastry, and Bamboo, which can be easily implemented and be used to solve the problems faced by P2P SIP.

9.2.1 Chord

Chord [9,10] is a ring-based DHT with an identifier space of size N. A Chord node with identifier u has a pointer to the first node following it clockwise on the identifier space $(Succ(u))$ and the first node preceding it $(Pred(u))$. The nodes form a doubly linked list. A node keeps $M = \log_2(N)$ pointers called fingers. The finger points to the edge on the Chord, which will be used for quick resource location algorithm. The set of fingers of a node u is:

$$F_u = \{(u, succ(u + 2^{i-1}))\} \quad 1 \le i \le M, \tag{9.1}$$

here the arithmetic is modulo N.

With this choice of edges, a node perceives the circular identifier space as if it starts from its identifier. Then the edges are chosen in such a way that the space can be partitioned into two halves, one of the halves can be partitioned into two quarters, and so on.

As an example, let us look at the network in Figure 9.3a with an identifier space $N = 16$. Each node has:

$$M = \log_2(N) = 4 \text{ edges}. \tag{9.2}$$

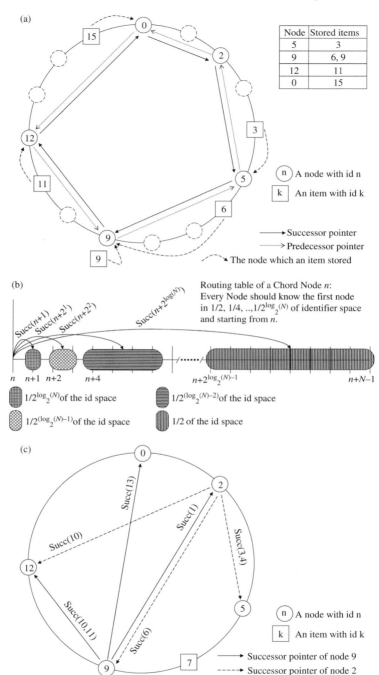

FIGURE 9.3

(a) A chord network with $N = 16$ populated with 6 nodes and 5 items. (b) The general policy for Chord's routing tables. (c) Example of inserting a new item.

The network contains nodes with identifiers 0, 2, 5, 9, 12. The policy for constructing routing tables is shown in Figure 9.3b. As pointed out earlier, Node n chooses its pointers by assuming that the identifier space starts from its identifier. It chooses to have the pointers to the successors of the identifiers:

$$n + 2^0, \quad n + 2^1, \quad n + 2^2 \quad \text{and} \quad n + 2^3.$$

In this way, the last pointer $n + 2^3$ divides the space into two halves. The one before it, $n + 2^2$, divides the first half into two quarters, and so on. When no node exists at the desired position, its successor will be taken instead.

As shown in Figure 9.3a, an item is stored at the first node that follows clockwise on the identifier space. If items with identifiers 3, 6, 9, 11, and 15 are to be stored in this network, then $\{3\}$ will be stored at 5; $\{6,9\}$ at 9; $\{11\}$ at 12; and $\{15\}$ at 0.

The lookup process follows the result of the way the identifier space is partitioned. Both the insertion and querying of items depend on how to find the successor of an id. As illustrated in Figure 9.3c, if node 12 wants to insert a new item with an identifier of 7, the lookup is forwarded to node 5, which is the closest preceding finger to the identifier 7 from the point of view of 12. Similarly, node 5 will lookup its set of fingers to find the closest successor of 7. Node 5 finds that 7 is between itself and its successor 9, and therefore, returns 9 as an answer to the query through the reverse path. Upon getting the answer, the application layer of node 12 should contact that of node 9 and ask for the storage of some value under the key 7. Any node looking for the key 7 can act similarly. And in no more than M hops, a node will discover the node at which 7 is stored. Under normal conditions, a lookup takes:

$$O(\log_2(N)) \text{ hops.} \tag{9.3}$$

To join the network, a node n performs a lookup for its own identifier through some first contact in the network, called bootstrap node, and inserts itself in the ring between its successor s and the predecessor of s. The routing table of n is initialized by copying the routing table of s, or letting s lookup each required edge of n. The subset of nodes can automatically adjust their tables to reflect the presence of n through periodically checking the routing table and looking up the value of each edge. The last task is to transfer part of the items stored at s. Items with an identifier less than or equal to n will be transferred to n by the application layers of n and s.

Graceful leave procedures are transferring all items to the successor first and then informing the predecessor and successor. To solve the negative effects of ungraceful failures, Chord let each node keep a list of the $\log_2(N)$ nodes that follow it on the circle. With this list, if a node detects that its successor is dead, it replaces it with the next entry in its successor list, and all the items stored at a certain node are also replicated on the nodes in the successor list.

9.2.2   Pastry and Bamboo

The overlay graph design of Pastry [10,11] aims not only at achieving logarithmic diameter with a logarithmic node state, but also tries to target the issue of locality. As a result of obtaining the node identifiers by hashing IP numbers/public keys, nodes with adjacent node identifiers may be farther apart geographically.

Pastry also assumes a circular identifier space and each node has a list containing $L/2$ successors and $L/2$ predecessors known as the leaf set. A node also keeps track of M nodes that are close according to some metric other than the identifier space, such as network delay. This set is known as the neighborhood set and is not used for routing but for maintaining locality properties. Another node state is the main routing table. It contains $|\log_{2^b}(N)|$ rows and $2^b - 1$ columns. L, M and b are system parameters.

Node identifiers are represented as string of digits of base 2^b. In the first row, the routing table of a node contains node identifiers that have a distinct first digit. A node needs to know $2^b - 1$ nodes for each possible digit except its own. In the second row, the routing table of a node with identifier n contains $2^b - 1$ nodes that share the first digit with n but differ in the second digit. The third row contains nodes that share the first and second digit of n but differ in the third and so on. Figure 9.4 illustrates how the identifier space is partitioned with this scheme. For each of the constraints about the node identifiers contained in a routing table, there exist many satisfying nodes. The node with the lowest network delay or the best according to some other criteria is included in the routing table.

An item in Pastry is stored at the node that is numerically closest to its id. Such a node will have the longest matching prefix. To locate the closest node to an identifier x, a node n checks first if x falls within the range of node identifiers covered by its leaf set. If so, it is forwarded to such a node. Otherwise, the lookup is forwarded to the node in the interval that x belongs to, that is to a node that shares more digits than the shared prefix between n and x. If no such node is found in the routing table of n, the lookup is forwarded to the numerically closest node to x. With the matching of one digit of the sought identifier in each hop, after $\log_{2^b}(N)$ hops a lookup is resolved.

When a node n joins an existing Pastry network, the following steps are needed. Step 1 contacts its bootstrap node g (also called gateway node). Step 2 uses g to find a closer gateway g'. Step 3 performs a lookup for its own identifier through g' to figure out the numerically closest node s to n, and gets the routing path P. Step 4 uses the leaf set of s as its own for initialization. Step 5 takes the i-th row from the i-th node on the path from t to s, which is a latency probe defined in [12], and uses those rows in initializing its routing table. Step 6 informs every node in its neighborhood set, leaf set, and routing table of its presence. The cost is about

$$3 \times 2^b \log_{2^b}(N). \qquad (9.4)$$

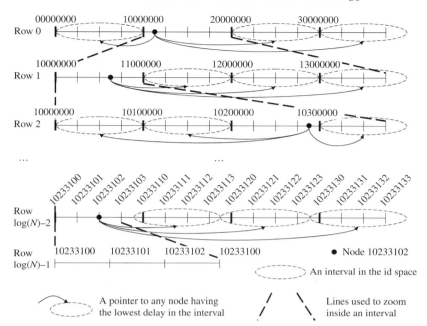

FIGURE 9.4

Illustration of how the Pastry node 10233102 chooses its routing edges in an identifier space of size $N = 2^{128}$ and encoding base $2^b = 4$.

Node departures are detected as failures and repaired in a routing table by asking a node in the same row of the failed node for its entry on the failed position.

The overlay network and routing algorithm of Bamboo is identical to Pastry. The difference lies in how Bamboo maintains the network as nodes join and leave the network and the network conditions vary. A new Bamboo node performs only step 1, 3, and 4 of the Pastry join algorithm described above, using g for $g\prime$ in step 3. It adds the nodes in P to its routing table, but sends no latency probes. After this quick bootstrapping, it considers itself joined. Its routing table is then filled and optimized through a continuous process that stops only when it leaves the network [13].

The key features of bamboo are as follows: First, Bamboo nodes acquire and maintain neighbor state by several congestion-aware, continuous optimization processes. This process is divided into separate concerns, first producing correctness through maintenance of the leaf sets, then producing efficiency by filling the routing table, and finally tuning routing table entries for proximity. In all cases, the rate at which these messages are sent is capped and is further scaled back in response to congestion or load. Second, Bamboo nodes actively probe their neighbors to compute good timeout values, and by using recursive routing, they avoid sending messages to nodes for which they do not have such timeouts [13].

9.3   Early Development of P2P SIP Network Architecture

The early work in combining SIP with P2P includes the SOSIMPLE project at the College of William and Mary. It replaces the SIP user registration and lookup by an existing P2P protocol [4]. The other is a pure P2P architecture for SIP-based IP telephony system developed at Columbia University. The system uses SIP messaging to implement P2P algorithm [5].

9.3.1   The SOSIMPLE Approach

SOSIMPLE [4] uses P2P to decentralize a SIP/SIMPLE communications system. In SOSIMPLE, there are no central servers and nodes communicate directly with one another. Peers connect to a few other peers in the overlay network and use these nodes as the destination for most SIP messages. Nodes act both as UAs and as proxies and replace the functionality of SIP registrars and proxies. Each node is responsible for those roles for some portion of the overlay.

In SOSIMPLE, nodes are organized in a DHT-based on Chord or any other P2P location strategy. Each node is assigned a node-ID created by hashing (using Secure Hash Algorithm 1 (SHA-1)) the real IP address and appending a port. The same algorithms are used by Chord to maintain the SOSIMPLE overlay network, and SIP messages are used to implement the DHT operations. The implementation of Chord used in SOSIMPLE saves data not only at the node for which the data hashes, but also at the following k nodes. Accordingly, after hashing their username to a Chord node, the UA sends a SIP REGISTER message to the node and its k successors, which means that the resources, including users' register information, are stored on the node responsible for a particular resource as well as k successor nodes.

SOSIMPLE nodes maintain a number of finger table entries for routing and prefetch entries for contacts in the buddy list of the user as well as caching previously contacted nodes for future use. All messages for maintaining the DHT, registering users, locating resources, and establishing sessions are SIP messages. As SOSIMPLE needs P2P functionality, a number of new SIP headers are defined.

SOSIMPLE uses the SIP REGISTER message to pass the overlay information between nodes. In SOSIMPLE, *User Registration* refers to the traditional usage for registering users, while *Node Registration* refers to REGISTER for DHT operations, such as entering, leaving, or maintaining the overlay.

9.3.1.1   Node-Level Operations

When a new node tries to join SOSIMPLE, it uses its IP address to calculate a node-ID. Once the new node has joined the overlay, it will be responsible for storing information associated with the portion of the overlay mapped to that

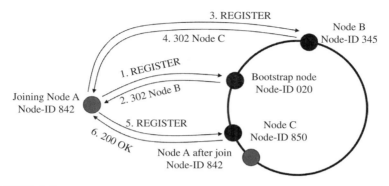

FIGURE 9.5

An example of a new node joining SOSIMPLE.

calculated node-ID. The joining node must locate the node currently responsible for that region, insert itself into the overlay, and transfer the information for the region [4].

Figure 9.5 illustrates how a new node joins SOSIMPLE. The joining node A calculates its node-ID, 842 in this example, and sends it in a REGISTER to the bootstrap node, which has node-ID 020 (1). In most cases, the bootstrap node is not the node currently responsible for this region. So the bootstrap responds with information about the nodes it knows nearest to where node A will be placed in the overlay, in this case node B, with Node-ID 345. This information is passed back in the new headers, extended by SOSIMPLE, within a SIP 302 moved temporarily response (2). Then node A will repeat the process, using this nearer node as the new bootstrap node (3–4), until it reaches node C, which is currently responsible for maintaining the appropriate section of the overlay. Node C responds with a SIP 200 OK response indicating information about nearby neighbors in the headers (5–6), thus allowing the joining node to insert itself into the overlay.

9.3.1.2 User Operations

User registration and resource location are decentralized in SOSIMPLE. In order to register, or to contact another user, the node responsible for storing the information of that user must be located. An illustration about the resource location process is shown in Figure 9.6 [4].

In this example, Alice will locate Bob, and the naming convention is followed. First, the node of Alice hashes the user name Bob to produce a resource-ID. After getting the resource-ID, the node of Alice searches the Node-ID nearest to the resource-ID of Bob in its finger table and gets node A. Then, the node of Alice sends a message to node A (1). As node A is not responsible for that resource-ID, it sends a SIP 302 moved temporarily reply. The reply includes the closest node, Node B, found in its finger table in the headers (2). The node of Alice will keep on trying the node received from the replies (3, 4, 5, 6) until it finds the node responsible for the resource-ID of Bob.

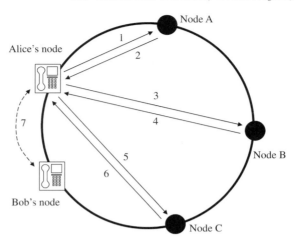

FIGURE 9.6
Alice locates user Bob and establishes a communications session.

If Alice is registering herself as a user within the overlay, the messages Alice's UA sent (1, 3, 5) are of type REGISTER. Node C appends a mapping key/value item of Alice's username and her IP address into its registration table and replies to the node of Alice with a 200 OK message (6). Alice is then registered with the system.

If Alice is trying to call Bob, the messages Alice's UA sent (1, 3, 5) are either SIP INVITE messages to establish a VoIP call or SIP MESSAGE messages containing IM data. Node C looks in its registration table to see if Bob's registration is present. If it is not, a 404 Not Found is returned (6), indicating the user Bob is not in the system. Otherwise, node C sends a 302 moved temporarily reply, providing a contact directly for Bob (6). Finally, the call between Alice and Bob can be established directly between the nodes using conventional SIP mechanisms without involving the overlay (7).

9.3.2   A P2P Architecture for SIP-Based IP Telephony System

The architecture for a P2P SIP-based IP telephony system is shown in Figure 9.7 [5]. Some of the nodes with high capacity (bandwidth, CPU, memory) and availability (uptime, public IP address) are made super-nodes and form the DHT, whereas other nodes attached to one or more super-nodes without being part of the DHT are made ordinary nodes.

Figure 9.8 shows the block diagram of components in the proposed P2P SIP node. When the node starts up and the user signs in with his identifier, the discover module is activated to initiate Network Address Translation (NAT) and firewall detection, peer discovery, and SIP registration. Multicast SIP registration, cached peer addresses from last boot cycle, and pre-configured bootstrap addresses may be used to discover an initial set of nodes. The user

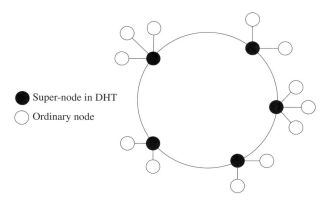

FIGURE 9.7
P2P network architecture for SIP-based IP telephony system.

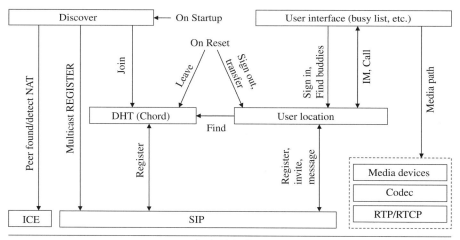

FIGURE 9.8
Block diagram of a P2P-SIP node.

interface module keeps track of the "friends list" of the user and invokes the user location module to locate these friends. User location is obtained using the SIP module or the DHT module. The DHT module maintains the peer information and performs DHT operations.

In this system, SIP is used as the underlying protocol for locating users or nodes, joining the DHT, registering users, call setup, and instant messaging. Once a user is located, the call setup or instant messages can be sent directly via the SIP module to the user's phone. SIP REGISTER refresh and OPTIONS messages are used to detect node failure. When a super-node shuts down or fails, the registrations are transferred to other super-nodes in the DHT. Other SIP functions, such as third-party-call control and call-transfer, can be implemented similarly. The media path (media device, codec and transport) is independent of the P2P-SIP operation. Some DHTs (e.g., Controller Area Network (CAN)) may allow parallel search to multiple peers. In this case, the super-node may act as a back-to-back user agent (B2BUA) and propagate the SIP message to the neighboring peers.

In a real deployment, it may be useful to allow multiple P2P-SIP networks to be interconnected. This hybrid architecture allows both the P2P-SIP networks and the server-based SIP infrastructure to coexist. There are two approaches to implementing this design: cross register all the users of one network with all the other networks, or locate the user in the other network during call setup. The first method is suitable for a small number of known P2P SIP networks, where each P2P SIP network is represented by a domain name. This is the same as a server-based SIP network where the domain name resolves to one or more bootstrap nodes in that network [14]. The second one can be implemented using a global naming service such as Domain Name System (DNS), or a hierarchy of P2P SIP networks. P2P SIP is used instead of DNS to resolve the domain name. As an example, individual large organizations can have a local P2P SIP network that is connected to the global P2P SIP network as shown in Figure 9.9. The local domain-specific DHT has representative server nodes that are also reachable in the global DHT, for example, key *private.com* maps to nodes *W* and *V* in the global DHT. Any node in the domain-specific DHT can reach the global DHT, and any node in the global DHT can reach the domain-specific DHT via the representative server nodes in the domain.

This hybrid architecture allows the user to register with his provider's SIP server as well as the P2P SIP network. Call setup is sent to the SIP destination, if resolved via DNS, as well as to the P2P SIP network.

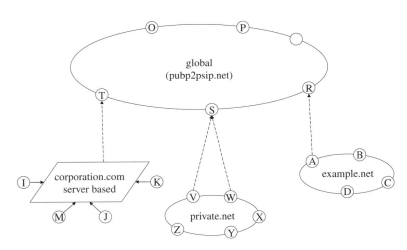

FIGURE 9.9
Example of hybrid systems.

9.3.3   Summary

The peers in SOSIMPLE replace the functionality of SIP registrars and proxies, each node being responsible for those roles for some portion of the overlay. SOSIMPLE meets the requirements for a practical P2P VoIP/IM system and keeps most of the advantages of a proxy-based system. Although a number new SIP headers are defined, SOSIMPLE offers compatibility in terms of reuse of existing SIP clients as well as the ability to interface to established SIP systems.

Different from SOSIMPLE, the hybrid architecture proposed in [5] is achieved by implementing the P2P algorithm using SIP messaging. It provides reliability and scalability inherent in P2P systems, in addition to interoperability with existing SIP infrastructure. The peer nodes form the P2P network using standard SIP messages with no change in message semantics.

9.4   P2P SIP Standardization

More and more companies and institutes focus their research work on P2P SIP, but there are no standard specifications or criterions yet. Some Internet drafts have been proposed to IETF.

The concepts and terminology for use of SIP in a peer-to-peer environment are defined in [6]. The draft describes the functional relationships between the network elements from a high-level view. An architecture for P2P SIP systems is specified in [15] in which a separate and independent peer-to-peer overlay layer provides a distributed resource placement and search service for SIP. A hierarchical P2P SIP architecture is outlined in [16], in which heterogeneous overlays are inter-connected via a decentralized manner.

9.4.1   Concepts and High-Level Description

Some basic concepts and high-level description are given in [6]. A P2P SIP overlay is defined as a collection of nodes organized in a peer-to-peer fashion for the purpose of enabling real-time communication using SIP. The nodes in the overlay, which are defined as P2P SIP peers, collectively provide a distributed mechanism for both mapping names to network locations and transporting SIP messages between any two nodes in the overlay.

The peers in the overlay collectively run a distributed database algorithm. This distributed database algorithm allows a copy of a data item stored on more than one peer and for it to be retrieved in an efficient manner. One of the most useful functions of the distributed database is to store the information required to provide the mapping between AoRs and Contact URIs for the distributed location function.

Besides offering storage and transport services, some individual peers are also expected to offer other services. These services, such as STUN [21] and voicemail services, may be required to allow the overlay to form and operate, or may enhance the basic P2P SIP functionality.

Peers in an overlay need to speak some protocol between themselves to maintain the overlay and to store and retrieve data. Currently, this protocol is called the P2PSIP peer protocol. Since P2P SIP is about peer-to-peer networks for real-time communication, it is expected that most peers will be coupled with SIP entities.

9.4.2   An Architecture for P2P SIP

The P2P SIP architecture defined in [15] has a separate layer, independent of SIP that provides the P2P overlay functions and SIP is an application using the services of the layer. A peer in the P2P SIP system is comprised of two layers: the P2P overlay layer and the SIP layer. The P2P overlay layer handles all the P2P overlay functions, but does not interpret semantic meanings of the resources placed or looked up by the higher layers, including SIP.

From the perspective of SIP, the P2P overlay layer mainly provides a location service for various SIP resources. DHT-based structured P2P networks support efficient search mechanisms when the resource names are precisely known and meet the requirements of P2P SIP well. Therefore, DHT is proposed to be used for the P2P overlay layer.

P2P overlay networks can be comprised of peers with different amount of physical resources operating in various network environments. Thus, peers are classified into two types: super nodes (SNs) and ordinary nodes (ONs). For DHT-based P2P overlay networks, super nodes participate in building and maintaining the DHT, whereas ordinary nodes do not participate in DHT. Super nodes serve search requests by routing the search traffic in the DHT or generating replies to search requests. An ordinary node relies on a super node to discover resources from other peers. In this sense, super nodes form a distributed storage and search service network.

Since only the super nodes form a DHT-based P2P overlay, there is a hierarchical structure in network: the core consisting of super nodes forming a DHT with a periphery of ordinary nodes. Each ordinary node is associated with one or more super nodes. An ordinary node uses the storage and lookup services of the overlay network core through the associated super nodes. On the other hand, SIP has its own hierarchical structure. But in general, the hierarchy of peers in the P2P overlay layer is independent of the hierarchical structure of SIP.

A UA with a P2P overlay layer is referred to as a P2P UA, and similarly, a proxy or a registrar with a P2P overlay layer is referred to as a P2P proxy or P2P registrar respectively. Super nodes are good candidates for P2P proxies, but not all super nodes have to become a P2P proxy.

There are four types of entities in the P2P overlay network. The bootstrap server and login server in a P2P overlay network facilitate the formation of the P2P overlay network. When other means become available for their functions, they can be omitted and thus are optional. Once a node knows one or multiple nodes participating in a P2P overlay network, it can start the process to join the network. The details of the entities and the related operations are similar to those described in Section 9.3.1.

9.4.3 A Hierarchical P2P-SIP Architecture

In the hierarchical P2P SIP architecture outlined in [16], P2P overlays are used by SIP as the underlying peer location protocol. Heterogeneous overlays are inter-connected via a decentralized manner. Each overlay will elect one or more powerful peers to be P2P proxies, which logically construct an upper level overlay and forward messages among heterogeneous overlays. There are also several stateful peers in an overlay to inherit the controllable feature of the stateful SIP proxies.

The proposed hierarchical P2P-SIP structure is illustrated in Figure 9.10. Various P2P overlays (such as Pastry, CAN) are respectively used as the underlying route discovery protocol. Thus the communication overhead will be significantly reduced, since relaying nodes need not maintain dialog states.

To connect peers from heterogeneous overlays, an upper level overlay, the global P2P overlay, is formed, which interconnects the local overlays. That is, each overlay elects one or more powerful peers to be the gateway-like nodes that route messages among heterogeneous overlays. The elected gateway-like node is defined as the P2P proxy. In Figure 9.10, node n joins both the local Pastry overlay and global overlay to perform the gateway-like function. P2P proxies are good candidates for such kinds of stateful nodes.

The benefits of introducing such a hierarchical architecture are that peers in heterogeneous overlays can be interconnected and the balance between low

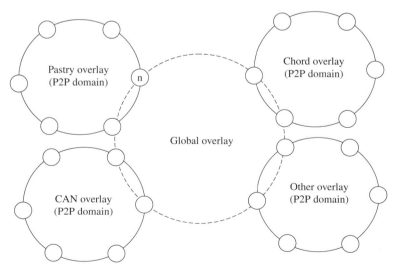

FIGURE 9.10
Hierarchical P2P-SIP structure.

signaling overhead and easy management can be achieved. More details about the architecture can be found in Section 9.3.2 and Reference [16].

9.5 Protocol Development and Implementation

Reference [17] outlines the motivation, requirements, and architectural design for a distributed Session Initiation Protocol (dSIP). dSIP is a P2P-based approach for SIP registration and resource discovery using distributed hash tables maintained with SIP messages. Reference [18] outlines the motivation, requirements, and architectural design for a extensible and lightweight distributed REsource LOcation And Discovery (RELOAD) protocol.

9.5.1 Recursive and Iterative Routing

The routing scheme used by a P2P system is very important to both its performance and resistance against attacks. Several mechanisms have been discussed. In each case, the initial message is sent from the requester to the peer in the routing table most likely to route correctly, as defined by the DHT algorithm in use. Subsequently, that peer may provide further routing using one of three mechanisms [19].

In iterative routing, the querying node contacts each intermediate node directly. If the contacted peer is not responsible for the target ID, then the contacted peer issues a 302 redirect response pointing the search peer toward the best match the contacted peer has for the target ID. The searching peer then contacts the peer to which it has been redirected and the process iterates until the responsible peer is located [17].

In recursive routing, a lookup request is forwarded from node to node until it reaches its destination. If the contacted peer is not responsible for the target ID, it will forward the query to the nearest peer to the target that it knows, and the process repeats until the target is reached. The response unwinds and follows the same path on the message return. Because dSIP uses SIP messages for transport, SIP's proxy behavior is used to enable recursive routing. Semi-recursive is the same as recursive routing on the outbound leg, but the reply "shortcuts" and is directly sent back to the requester. When discussing these techniques, we often just refer to iterative and recursive, because of the similarity between recursive and semi-recursive routing. Various mechanisms may be used within the same overlay and even within the same search [17].

9.5.2 The dSIP Protocol

dSIP is a specific proposal for the P2P SIP peer protocol proposed in the drafts discussed in Section 9.4, which uses SIP messages as the syntax for encoding the protocol. The function of the P2PSIP peer protocol is to provide for mechanisms to maintain the overlay, as well as to store and retrieve information, and to route messages when needed. dSIP's syntax is SIP with a number of newly defined headers, however, no new methods are added to SIP in dSIP. Since conventional SIP is used for this, there is no need for a P2PSIP client protocol, and therefore, dSIP defines no such protocol [17].

dSIP is modular, allowing for the use of multiple DHTs, including those defined later. DHTs can be negotiated among the peers in much the same way as codecs or features are negotiated in conventional SIP. For compatibility, support for one basic DHT algorithm, Chord, is required. Additional DHTs can be added and supported. These will be discussed in the next subsection. The details of the Chord algorithm for dSIP are described in [20] and an alternate DHT algorithm for dSIP based on Bamboo is provided in [20].

9.5.2.1 Peer Functions and Behavior in dSIP

dSIP peers provide many functions. A participating peer must be an active member of the overlay and must provide some SIP "server-like" behaviors as well. The code that implements the additional server-like and DHT behavior can be located in several places in the network. The simplest is an endpoint peer. These peers provide the basic functionality of any SIP endpoint, but additionally implement some more operations described in [17] to enable self-organization and provide SIP server-like functionality.

The behavior can also be located in an adapter peer, which allows one or more non-P2P aware SIP UAs to interact with the P2P overlay network. The adapters perform the additional self-organizing and SIP server-like behavior on behalf of the UAs they support. In this case, only the adapter peer is a peer in the overlay, the UAs are not. The adaptors speak the P2P SIP peer protocol, while the UAs speak conventional SIP. All interaction with the P2P overlay is carried out by the adapter peer.

9.5.2.2 P2P Overlay Structure

As described in Section 9.4, the P2P overlay in dSIP consists of peers, which collectively serve as a directory service for locating resources. Peers are organized using a supported DHT P2P structure. dSIP allows for pluggable DHT algorithms, the exact form of which is defined in the DHT algorithm definition.

Each peer is assigned a peer-ID, and each resource that is stored in the overlay is assigned a resource-ID. These values must map to the same name space. dSIP provides for various algorithms to be used to produce these values, but all members of the overlay must use the same algorithm. In the Chord DHT implementation, SHA-1 is used to produce 160 bit values for both the peer-ID and resource-ID.

The peer-ID assigned to each peer determines the peer's location in the DHT and the range of resource-IDs for which it will be responsible. The resource-IDs map to the same space as the peer-IDs. In the case of users, the unique keyword is the userid and the resource is the registration.

A resource with resource-ID k will be stored by the peer with peer-ID closest to the resource-ID, as defined by the DHT algorithm used.

Since each DHT is defined and functions differently, the table of other peers that the DHT maintains and uses to route requests (neighbors) are generically referred to as a routing table. dSIP defines the syntax for the headers used to exchange these entries, but leaves the exact form of the data each DHT stores in the table as a decision for the implementation. Peers may additionally maintain a list of peers to which they maintain connections for purposes other than routing.

When locating a resource with a particular resource-ID, the peer will send the request to the routing table entry with the peer-ID closest to the desired resource-ID. The peer receiving the request is assumed to know of a peer with a peer-ID closer to the resource-ID, and responds by suggesting or forwarding the message to this peer.

9.5.2.3 Use of SIP Messages in dSIP

dSIP uses SIP messages to implement the P2PSIP peer protocol. The motivation has been to preserve the semantics of conventional SIP messages. Messages are being exchanged for two purposes. The first class of messages is used to maintain the DHT. The second type of messages is that used for registering users, inviting other users to a session, etc.

The messages used to manipulate the DHT are SIP REGISTER messages. They are used to add, remove, and query bindings. REGISTER is used both for the bindings of hosts as neighbors (entries in the routing table) in DHT maintenance operations as well as the bindings of resource names to locations that are commonly maintained by SIP registrars. These operations occur within the overlay, rather than on the conventional server.

9.5.2.4 Other Features of dSIP

dSIP does not specify which of the three mechanisms will be used for routing: Iterative, recursive, or semi- recursive. In general, the messages can be routed using any of these mechanisms. Various mechanisms may be used within the same overlay and even within the same search. dSIP provides mechanisms that are used for a number of operations, including peer registration, resource registration, session establishment, and DHT maintenance.

9.5.3 The RELOAD Protocol

RELOAD is an extensible and lightweight distributed REsource LOcation And Discovery protocol. RELOAD is one possible implementation of the P2P SIP protocols. RELOAD uses binary messages, derived as much as possible from the STUN protocol [21], as the underlying protocol. In this architecture, no P2P SIP client protocol is needed, unmodified SIP is used for access by non-peers. The protocol considers NAT traversal and fragmentation, supports storage of information other than registrations, allows for multiple DHT and hash algorithms, and provides hooks for multiple security schemes.

As described earlier, dSIP transported all messages, including those used to maintain the DHT, using SIP. However, concerns were raised about the "misuse" of SIP messages and the heaviness of the SIP as the P2P peer protocol. For these concerns, the RELOAD draft presents a binary protocol for the P2P SIP peer protocol. The protocol provides the framework for exchanging P2P messages, and is designed to be extended by pluggable DHT algorithms and security solutions.

The benefit of moving away from SIP as the encoding for the P2P SIP peer protocol is the ability to design a new protocol wholly focused on the requirements of P2P SIP. The major advantages of the protocol include the following aspects.

A fixed-length header is defined that can be used for routing the message through the overlay without the contents being parsed (or even visible) by intermediate peers. The header includes no information about specific IP addresses but only includes source and destination peer-IDs as well as some minor option and version flags. Clearly separating the header components necessary for routing from the message contents simplifies processing and increases security. The content of the message is in STUN-based type-length-value (TLV) format, extending a syntax for which many implementers will already have a parser. The flexible contents allow great flexibility in supporting a variety of DHT and security algorithms using the same base peer protocol.

The protocol supports User Datagram Protocol (UDP), Transmission Control Protocol (TCP), Transport Security Layer (TLS), or Stream Control Transmission Protocol (SCTP) for the underlying transport protocol. New transports can be supported easily. Explicit support for fragmentation is provided, which is required when using UDP. UDP must be supported because TCP may not always be available in the environments this protocol will be deployed in.

The protocol supports one-to-many registrations. Fully distributing all of the states required to duplicate the functionality of central servers requires information about different types of resources for the same user: primary contact, offline voicemail, security certificate, etc. Furthermore, obtaining a list of the resources available for a particular user requires one-to-many registrations stored in a single location, rather than distributing each type of resource by a separate key. Each key to value mapping has an individual expiration time. Thus, a single key can have both long-lived and transient registrations. Each key to value mapping is individually signed by the registering entity. The protocol allows multiple entities to register under the same key, thus allowing voicemail or other data to be left for another user. Because each message lives as a separate registration, there are no race conditions associated with editing the same file, and no need for elaborate synchronization between peers. Each resource registration contains parameters that can be used to specify the type of mapping.

9.5.4   The Resource Lookup Algorithms

M. Zangrilli and D. Bryan have described how a structured peer-to-peer algorithm can be used for resource lookup by a P2P SIP peer protocol. Reference [20] describes how to integrate a DHT based on Chord with dSIP while reference [22] describes how to integrate a DHT based on Bamboo with dSIP. The dSIP draft is extended to provide possible implementations of pluggable DHT algorithms.

The Chord algorithm has been adapted to use SIP messages, as specified by dSIP, to communicate between peers in the overlay. Messages are routed by taking advantage of a key property of the Chord finger tables. A peer has more detailed, fine-grained information about peers near it than farther away, but it knows at least a few more distant peers. When locating a particular ID (either resource-ID or peer-ID), the peer will send the request to the finger table entry with the peer-ID closest to the desired ID. Because the peer receiving the request has many neighbors with similar peer-IDs, it will presumably know of a peer with a Peer-ID closer to the ID, and suggests this peer in response. The request is then resent to this closer peer. The process is repeated until the peer responsible for the ID is located, which can then determine if it is storing the information [20].

Bamboo uses prefix routing. Messages are routed so that each hop will result in a peer that either shares a larger prefix with the ID being searched for or shares the same length prefix as the previous hop's peer-ID, but the new peer's peer-ID is numerically closer to the ID than the previous hop's peer-ID. To route a message to a particular ID, the peer first looks to see if any of the peers in its leaf set are responsible for the ID. If the ID lies within the leaf set, the message is routed to the appropriate leaf set peer. If the ID is not within the leaf set, the searching peer computes the length l of the longest matching prefix between the search ID and its own peer-ID. The peer then looks in its Bamboo routing table at row l. If there is an entry in that row corresponding to the l length prefix of the ID being searched for, then the message is forwarded on to that peer. If that Bamboo routing table entry is empty, the message is routed to the peer in its leaf set with the peer-ID that is numerically closest to the ID being searched for [22].

9.6   Concluding Remarks

The goal of P2P SIP is to allow true P2P networking between SIP peers by using SIP as the signaling protocol. The concept behind P2P SIP is to leverage the distributed and failure-tolerant nature of P2P by building an overlay network above SIP to provide P2P services for a network of SIP peers. This removes the need for centralized server components and allows true P2P networking between the SIP nodes. Several tentative Internet-drafts have been made for P2P SIP, describing the overall architecture, the environment of the system, the architectural design of the overlay network, and the resource location strategy.

As pointed out by D. Bryan and B. Lowekamp [23], P2P SIP will not be a replacement for SIP. It is an enhancement and companion to SIP, enabling SIP to be used in scenarios where it might not have otherwise been easily deployed. P2P SIP is available now from a number of vendors, but is at the beginning stage of the adoption. Various implementations are still proprietary, but the standardization work of the IETF will improve the situation. P2P SIP is becoming a powerful tool in various applications, such as in VoIP for small-enterprise systems, disconnected and ad hoc deployments, and global, decentralized deployments. It is also a strong contender for a protocol in IPTV and between consumer electronics devices in the home and mobile devices [23,24].

References

[1] Rosenberg, J., H. Schulzrinne, G. Camarillo, A. Johnston, J. Peterson, R. Sparks, M. Handley, and E. Schooler. 2002. SIP: Session Initiation Protocol, *RFC 3261, Internet Engineering Task Force*, June.

[2] Baset, S. and H. Schulzrinne. 2006. An analysis of the skype peer-to-peer internet telephony protocol. In *Proceedings INFOCOM 2006 Conference on Computer Communications*, Barcelona, Spain, April 23–29.

[3] Liang, J., R. Kumar, and K.W. Ross. 2005. The KaZaA overlay: A measurement study. *Computer Networks*, 49: 6.

[4] Bryan, D., B. Lowekamp, and C. Jennings. 2005. SOSIMPLE: Towards a serverless, standards-based, P2P communication system. In *International Workshop on Advanced Architectures and Algorithms for Internet Delivery and Applications (AAA-IDEA)*, Orlando, FL, June 15.

[5] Singh, K., and H. Schulzrinne. 2005. Peer-to-peer internet telephony using SIP. In *International Workshop on Network and Operating System Support for Digital Audio and Video (NOSSDAV)*, Stevenson, Washington, USA, June 13–14.

[6] Bryan, D., P. Matthews, E. Shim, and D. Willis. 2007. Concepts and terminology for peer to peer SIP. (work in progress), IETF draft, draft-ietf-p2psip-concepts-00, June.

[7] Singh, K., and H. Schulzrinne. 2004. Peer-to-peer internet telephony using SIP. Technical Report CUCS-044-04, Department of Computer Science, Columbia University, New York, NY, October.

[8] De Boever, J. 2007. Peer-to-peer networks as a distribution and publishing model. In *Proceedings ELPUB2007 Conference on Electronic Publishing*, Vienna, Austria, June.

[9] Stoica, I., R. Morris, D. Liben-Nowell, D.R. Karger, M.F. Kaashoek, F. Dabek, and H. Balakrishnan. 2003. Chord: A scalable peer-to-peer lookup service for internet applications. *IEEE Transactions on Networking* 11: 17–32.

[10] El-Ansary, S., and S. Haridi. 2006. An overview of structured overlay networks. In: *Handbook on Theoretical and Algorithmic Aspects of Sensor, Ad Hoc Wireless and Peer-to-Peer Networks.* J. Wu ed. Boca Raton, FL: Auerbach Publications.

[11] Rowstron, A., and P. Druschel. 2001. Pastry: Scalable, distributed object location and routing for large-scale peer-to-peer systems. In *IFIP/ACM International Conference on Distributed Systems Platforms (Middleware)*, Heidelberg, Germany, November 12–16.

[12] Castro, M., P. Druschel, Y.C. Hu, and A. Rowstron. 2002. Exploiting network proximity in peer-to-peer overlay networks. Technical Report MSR-TR-2002-82, Microsoft.

[13] Rhea, S., D. Geels, T. Roscoe, and J. Kubiaatowicz. 2003. Handling churn in a DHT. Technical Report UCB/CSD-03-1299, University of California, Berkeley, December.

[14] Rosenberg, J., and H. Schulzrinne. 2002. Session Initiation Protocol (SIP): Locating SIP Servers. *RFC 3263, Internet Engineering Task Force*, June.

[15] Shim, E., S. Narayanan, and G. Daley. 2006. An architecture for peer-to-peer session initiation protocol (P2P SIP). (work in progress), IETF draft, draft-shim-sipping-p2p-arch-00, February.

[16] Shi, J., Y. Ji, H. Zhang, and Y. Li. 2006. A hierarchical P2P-SIP architecture. (work in progress), IETF draft, draft-shi-p2psip-hier-arch-00, August.

[17] Bryan, D., B. Lowekamp, and C. Jennings. 2007. dSIP: A P2P approach to SIP registration and resource location. (work in progress), IETF draft, draft-bryan-p2psip-dsip-00, February.

[18] Bryan, D., M. Zangrilli, and B. Lowekamp. 2007. REsource LOcation and Discovery (RELOAD). (work in progress), IETF draft, draft-bryan-p2psip-reload-01, July.

[19] Bryan, D., M. Zangrilli, and B. Lowekamp. 2006. Challenges of DHT design for a public communications system. Technical Report WM-CS-2006-03, Computer Science Department, William and Mary University, Williamsburg, VA, June.

[20] Zangrilli, M., and D. Bryan. 2007. A chord-based DHT for resource lookup in P2PSIP. (work in progress), Internet draft draft-zangrilli-p2psip-dsip-dhtchord-00, February.

[21] Rosenberg, J., J. Winberger, C. Huitema, and R. Mahy. 2003. STUN: Simple Traversal of User Datagram Protocol (UDP). *RFC 3489, Internet Engineering Task Force*, March.

[22] Zangrilli, M., and D. Bryan. 2007. A bamboo-based DHT for resource lookup in P2PSIP. (work in progress), Internet draft draft-zangrilli-p2psip-dsip-dhtbamboo-00, February.

[23] Bryan, D., and B. Lowekamp. 2007. Decentralized SIP: peer-to-peer SIP. *ACM Queue Magazine* 5(2): 34–41.

[24] Harjula, E., J. Ala-Kurikka, D. Howie, and M. Ylianttila. 2006. Analysis of peer-to-peer SIP in a distributed mobile middleware system. In *IEEE Global Telecommunications Conference (Globecom06)*, San Francisco, CA, USA, November 27–December 1.

10

SIP Mobility Technologies and the Use of Persistent Identifiers to Improve Inter-Domain Mobility and Security

Henry Jerez
Corporation for National Research Initiatives

Joud Khoury and Chaouki Abdallah
The University of New Mexico

CONTENTS

10.1 Introduction 228
10.2 SIP Inter-Domain Mobility 229
10.3 Host Identification and Mobility 233
10.4 Implementation 240
10.5 Future Work 247
10.6 Conclusion 248
     Acknowledgements 249
     References 249

The original implementation of the Internet assumed a set of static nodes talking to each other. This approach sets clear boundaries between different network topologies and allows mobility only through patches and adaptations that make true seamless mobility very difficult. This is due to the use of IP addresses that contain geographic semantics and are inherently not mobile and domain names that are intrinsically linked to these static servers. In recent years, the total number of mobile devices has outpaced that of static ones and supporting mobility in Internet Protocol (IP) networks has become a a crucial step toward satisfying the nomadic communication paradigms on the current Internet. Session Initiation Protocol (SIP) presents one approach toward supporting IP mobility. Additionally, SIP is increasingly gaining in popularity as the next generation multimedia signaling and session establishment protocol. It is anticipated that the SIP infrastructure will be extensively deployed all over the Internet and it has already been accepted as the signaling protocol of preference for many multimedia frameworks. In this paper, we explore the use of persistent identifiers to provide a secure and efficient approach to inter-domain SIP mobility. We will discuss the issues associated with SIP session mobility and the different approaches used to provide persistent identification that expedites mobility; some of which are explicitly applied to SIP. We will later illustrate SIP mobility and security through the implementation of an identification framework to application-level SIP addressing by introducing a level of indirection on top of the traditional SIP architecture. We refer to our approach as the Handle SIP (H-SIP). H-SIP leverages the current SIP architecture abstracting any domain binding from users. Our approach to mobility is both secure and user-controlled and has shown through our inter-domain authentication and call routing experiments to be as scalable as current solutions for static networks and more scalable and reliable in the presence of mobility.

10.1 Introduction

SIP [1] and H.323 [2] are among the most widely adopted protocols for IP telephony. While SIP and H.323 have different architectural components, the ability of these two protocols to coexist, and the simpler implementation and open collaboration of SIP makes it the signaling protocol of choice for advanced multimedia communications signaling. This fact is evidenced by its acceptance into the Third Generation Partnership Project (3GPP) as a signaling protocol for establishing real-time multimedia sessions. The protocol is continuously gaining in popularity and deployment and has been adopted by service providers like Verizon and Sprint to provide IP telephony, instant messaging, and other data services. It is anticipated that SIP will be widely deployed by operators and enterprises, thus populating the Internet with SIP infrastructural components. The widespread deployment and adoption of SIP has resulted in the need for an inter-domain mobility scheme for SIP.

SIP [1] is a signaling and control protocol for handling multimedia sessions, allowing for the establishment and termination of media streams between two or more participants. SIP works in concert with other multimedia protocols due to the independence of the protocol from the underlying transport mechanisms and session types. SIP architecture allows for its deployment as a centralized system, a distributed system, or a combination of both.

SIP architecture is also proposed as an efficient candidate that can be reused to provide personal, terminal, and session mobility [3–6] with a readily available infrastructure. This avoids the redundancy introduced by simultaneous deployment with Mobile IP [7]. The successful reuse of SIP to simultaneously support both multimedia communications and mobility leverages the issues emanating from SIP users *roaming** across multiple SIP domains. These issues are highlighted through a brief overview of SIP functionality.

SIP handles user location through the use of a proxy/location server† that accepts user registration requests and updates the respective user location in a

---

*Throughout this paper, roaming is defined as the SIP inter-domain roaming, i.e., the migration of a user between different SIP domains.

†We will use the terms SIP proxy, SIP server, SIP registrar interchangeably.

location repository. The protocol inherently implements location independence through the use of the uniform resource identifiers (URI) [8], which directly offer personal mobility. A URI acts as a location-independent identifier abstracting the actual physical location of a user with respect to the system. So, SIP allows for personal mobility whether through the use of a proxy that sets up the session between the calling parties or through the use of redirection servers. However, the protocol defines a user only within the domain boundaries of the service provider. A user must associate with a specific proxy server that handles user authentication as well as initial traffic routing. The proxy maintains a unique account for the user, who in turn, is expected to coordinate with that same proxy irrespective of his location. This requirement translates into unnecessary loads on the SIP server and on a particular domain. Additionally, it complicates the coordination of *roaming* users who must communicate with a central proxy server while roaming. Despite the possible presence of firewalls and other network restrictions on the foreign domain, roaming users are required to use the central home server instead of using the available local servers. Consequently, while URIs solve the location binding issue, they introduce the domain binding issue. Inefficient traffic routing is a direct consequence of such binding. Besides, the URI identification translates into users needing to be aware of each others' current domain associations. It also brings up the complexity of satisfying calls when initiated from regular keypad terminals.

In fact, inter-domain mobility can be addressed through the use of an abstraction framework based on a unique and persistent identification mechanism. This provides an approach that can enhance personal and terminal mobility [4] in current SIP architectures. As to session mobility, the readily available approaches like mid-call mobility [5] or enhancements to that [9] may be used. The framework we propose, which we refer to as the H-SIP, can seamlessly fit into the current SIP architecture, allowing SIP users to transparently roam across different SIP domains. H-SIP can be gradually deployed and can coexist with the traditional SIP infrastructure. User location and association is abstracted through the use of globally unique and persistent identifiers called *handles*, which are part of the Handle System [10–13]. The Handle System is a distributed system extensively used as an indirection layer for the management of persistent Identifiers. Using the Handle System as an intermediate layer on top of multiple distributed SIP implementations allows us to implement seamless multi-domain authentication and call routing.

10.2 SIP Inter-Domain Mobility

10.2.1 Sessions and Mobility

To clarify the SIP inter-domain mobility problem, we will present a simple example. Recall that SIP defines a user as an entity that associates with a

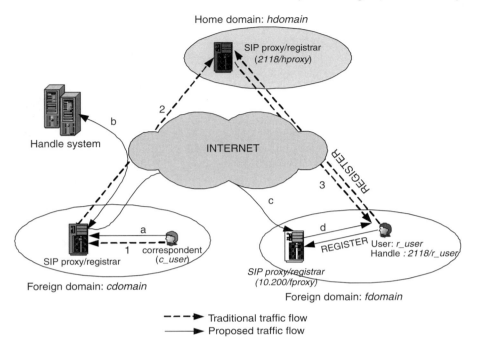

FIGURE 10.1
A reference inter-domain roaming scenario.

particular domain. Figure 10.1 depicts a simple scenario of a roaming user *r_user* who has a valid association with his home domain *hdomain* but is currently present in a foreign domain *fdomain*. SIP signaling traffic originating from (REGISTER) or terminating at (arrows 1,2,3: arbitrary SIP user trying to INVITE the roaming user) *r_user* must inefficiently pass through his home proxy server. Figure 10.1 identifies this traffic as traditional traffic flow.

There are several ways in which roaming issues can be addressed, depending on whether the SIP architecture is roaming-unaware or modified to become roaming-aware. We study these issues and we present our approach by showing a typical flow for INVITE and REGISTER requests. We also compare the different approaches and illustrate the different scenarios in Figure 10.2. A more elaborate description of these scenarios is presented in Section 10.4.

- The first scenario shows how SIP naturally handles a call flow for a roaming user. A data flow is presented in Figure 10.2a. In this case, no roaming logic is injected into the system (system is roaming-unaware). All requests to/from the roaming user must go through the central home proxy server. The home proxy thus treats both roaming and non-roaming users equally and portrays a roaming user as merely a home domain user registering with a foreign contact address. Clearly, if the user is present in another country, his traffic would still have to go

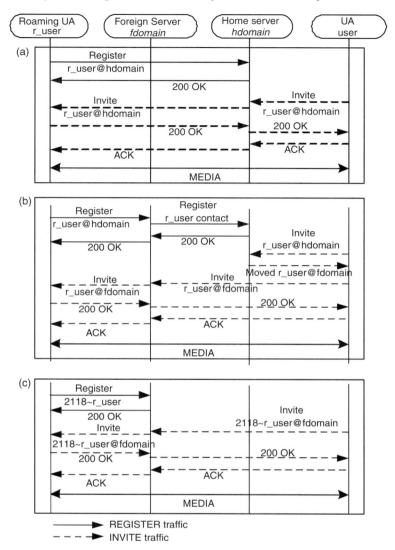

FIGURE 10.2
SIP traffic flow (a) With no roaming logic (b) With traditional roaming logic (c) With proposed roaming logic.

through his central home proxy (triangle routing) as depicted in Figure 10.2a, despite the availability of a local proxy server in the foreign domain (foreign server). This results into significant delays that are not accepted for time-sensitive applications. Even with SIP mobility management (SIPMM) [5,6] support (personal, terminal and session mobility) enabled, the same scenario occurs.

SIP mobility allows a user to roam between subnets and domains maintaining accessibility and session continuation using pre-call and mid-call mobility signaling. With pre-call signaling, the mobile user will re-REGISTER with the home proxy anytime his IP address changes. With mid-call signaling, the mobile user will negotiate an address change with the correspondent user while the session is in progress using re-INVITE messages. Mid-call mobility assumes a session is already in progress between the calling parties. Inefficient pre-call traffic routing and service centralization are obvious limitations that users roaming in these traditional and mobile SIP environments have to suffer from. This is the same case also for Mobile IP with location registers (MIP-LR) [14,15], whereas here, the SIP proxy servers are replaced with location registers. We argue that our proposed approach to roaming and inter-domain mobility in general, can significantly enhance the SIP personal and terminal mobility performance. Additionally, since our approach addresses SIP personal and terminal mobility, we can improve the pre-call portion of any SIP session mobility scheme while other features like mid-call mobility can remain unchanged. For mid-call mobility, current proposals like MIP-LR, SIPMM, or a combination of these two [16]) can be used. These approaches implement mid-call mobility by sending binding updates directly to correspondent nodes without going through home agents. Mobile IP (MIP) [7], however, uses home agents to forward traffic, which creates triangular routing issues. An enhanced version of MIP is MIPv6 [17] that avoids triangular routing and implements route optimization. As to the simultaneous mobility issue, discussed lately in [18], it is left for a future paper to offer a secure framework for simultaneous mobility in the context of H-SIP.

- A second scenario is that of a SIP roaming-aware approach such as the one proposed by double user agent servers [19] that mimics the roaming solution employed in telecommunication environments. In other words, a user who is roaming outside his home domain, registers with a foreign server. The latter consults the user's home server for redirection, authentication, and billing, and proceeds to process the user's transactions. Correspondent users trying to communicate with the roaming user will have to go through his home proxy server, which in turn, redirects them to the foreign proxy where the user is currently located. Hence, significant signaling overhead results primarily due to the nature of the SIP URI. The URI is composed of a domain part, like in *r_user@hdomain*, thus forcing the calls directed to this user to go through the *hdomain* proxy server first. The data flow for this scenario is presented in Figure 10.2b. We argue that this approach is inefficient as it introduces unnecessary overhead and load on the original server.

In the two scenarios above, the use of URIs to identify users and the inherent dependence of the URI on a particular domain, complicates message routing. One solution is to abstract the actual identifier eliminating per-call coordination to minimize the signaling traffic in highly mobile environments.

10.3   Host Identification and Mobility

Decoupling the host identity from the actual location has been the topic of extensive research. Some proposals address this issue by respecting the end-to-end argument [20–22], while others rely on an overlay [23–25]. All of these approaches are trying to address the same problem: The IP address performed well as a location identifier due to inherently embedded topological location information. When mobility is introduced, the IP can no longer be used for host identity since the host is now moving around. Thus, the IP address loses all meaning of identity reference and degenerates into a pure routing identifier. Decoupling the identity from the attachment point is currently solved by inserting a level of indirection on top of the network layer that manages the abstraction of host identities. This is essential for reasons of mobility, multi-homing, etc. In our case, handles are used to identify digital objects in contrast to using host identifiers as in the case of HIP. In this section, we will compare our solution to host identity protocol (HIP) [20].

There are several proposals that address the current need for an additional naming system on the Internet as a result of the limitations of Domain Name System (DNS) and IP namespaces. In this section, we focus on three proposals that describe global naming architectures, namely the work by Balakrishnan, Shenker and Walfish [26,27], TRIAD [28] and the Globe Project [296]. These proposals are of interest because of their close resemblance to the Handle System [11], which we consider the most accurate persistent identifier solution to entity identification.

The naming system proposed in [26] uses a very similar approach to that of the Handle System. Both proposals share the concepts of a global name service, identifier persistence, and independence of identifier from attributes, delegation, high-level identifiers, distributed administration, etc. However, in contrast to the purely flat semantic-free names used in layered naming [26] SFR [27] and the Globe Project [29], the Handle System allows, but does not require, hierarchy in the resolution mechanism and administration. This allows for delegation to enable local resolution and disconnected usage. The handle itself, although not semantic by design, may be used to convey other naming mechanisms such as DNS addresses through mapping and/or direct inclusion. This is not possible with a system that is dependent on hashes for the identifiers themselves. Another level of indirection (for mapping unfriendly names to friendly names) on top of a flat resolution infrastructure introduces unnecessary overhead. The separation between the semantic-free referencing infrastructure and the human-friendly names advertised in SFR [27] and [26] can be done within the same system. This is the approach that the Handle System takes. A particular meaningful handle can refer to another non-meaningful handle within the same system. Hence, multiple competing third parties can manage the meaningful handle space.

The basic distribution of the handle namespace [12] into prefixes allows for distributed secure delegation through the consolidation of structured prefixes and the segmentation of the namespace to accommodate commercial vendors offering dedicated resolution infrastructure. With support from the National Science Foundation (NSF), the Corporation for National Research Initiatives (CNRI) demonstrated the mapping of DNS names to handles and vice-versa. We also demonstrated this approach in enabling SIP call completion, regular Web service resolution, and e-mail services. The existence of multiple prefixes plus the currently deployed architecture of Handle-DNS and DNS-Handle proxies enables us to provide a gradual migration environment from the current Internet in order to support existing applications and to enable new ones. Another difference is that the current Handle System service model involves a global registry, and may involve local servers at sites related to specific naming authorities. This maintains some desirable features inherited from DNS like name uniqueness and fate sharing (local names can still be resolved if a site is disconnected from the global system; this design property is directly related to the underlying area of influence aggregation and fate sharing), small common infrastructure, and distributed financing.

The proposed architecture shares with Translating Relaying Internet Architecture Integrating Active Directories (TRIAD) [28] the notions of coupling of name identification and routing, as well as the participation of backbone nodes in the resolution. However, the technical details of the two implementations are completely different. We propose an implementation in the context of Digital Objects (DOs), Areas of Influence (AoIs) and Green Networks. The Handle System is a currently implemented and widely used identification system. The system is mature and satisfies most of the design concepts that references [26,27] put forth. So far, handles have been used to identify digital objects in repositories where each digital object has its own handle, allowing long-term control and flexibility in managing digital repositories. CNRI has demonstrated the application of handles to network devices and services [30], users [31], and mobile agents. An example of a system that uses handles is DSpace. Long-time users of the handle system include the Library of Congress and Defense Technical Information Center (DTIC's) Defense Virtual Library (DVL) sponsored by Defense Advanced Research Projects Agency (DARPA). The two largest users of the Handle System are the International Digital Object Identifier Foundation (IDF) and its largest registration agency, Cross-Ref. The system has proven to be extremely scalable. The largest individual Handle System implementation to-date is deployed at the Los Alamos National Laboratory, which is intended to support more than a half-billion identifiers, while providing internal resolution services to one of the largest archival collections in the United States.

There are also approaches such as Electronic Numbering (ENUM) [24], which use a translation between SIP URIs and e.164 numbers into IP addresses. Unfortunately, the IP information is stored within the DNS and is therefore restricted to the centralized administration, similar to dynamic DNS. These approaches require a third-party update and modification of the DNS entry in order to update this information. This operation is generally performed by a user with administrative rights over the DNS domain. Due to the low level of security implemented in the DNS protocol per se, attacks on ENUM and DNS are very easy to implement with approaches such as the corruption of DNS caches. The handle system has an encrypted storage and query mechanism that allows every connection to be encrypted and every response to be challenged. Furthermore, *handle* administration is completely distributed, allowing for a very secure infrastructure.

In the next sections we will explore the concepts associated with SIP mobility while contrasting the current practices with the use of persistent identifiers.

10.3.1 H-SIP: Abstraction Layer

As mentioned before, our proposed approach uses *handles* as globally unique identifiers to locate and identify SIP architectural elements. This abstraction allows the system to route calls independent of user location and domain association. We refer to the modified SIP framework as the Handle-SIP or H-SIP. Note that we have also exploited this abstraction approach at the level of network devices and services in [30,32].

Briefly, the Handle System [10–13] is intended to be a means of universal basic access to registered digital objects [33]. It provides a distributed, secure, and global name service for administration and resolution of *handles* over the Internet. A *handle* is a persistent name that can be associated with a set of attributes. Some of these attributes describe location, permissions, administrators, and state. The fact that *handles* are defined independently of any of the attributes or public keys of the underlying objects makes them persistent identifiers [34]. These identifiers are managed and resolved using a secure global name service that guarantees the association of the identifier with its respective attributes over distributed communication.

Security is a crucial property of the Handle System. The system acts as a certification authority, assuring that attributes of the name/reference are securely transferred between the communicating ends. Hence, the Handle System allows for secure name resolution and administration in a distributed fashion, making it highly scalable and suitable to operate in mobile environments. In our approach, elements of the SIP architecture, SIP users and proxy servers, are identified with *handles* abstracting any domain binding. Users will identify each other, as well as the SIP servers they associate with using *handles* instead of URIs and domain names, respectively. In Figure 10.1, the roaming user *r_user* will have his own *handle 2118/r_user* with the necessary administrative privileges over the *handle*. Additionally, the home proxy server has a *handle*‡ *2118/hproxy*, and the foreign proxy server has a *handle 10.200/fproxy*. Note that a *handle* has the form *prefix/suffix*. The prefix represents the naming authority (NA) while the suffix represents a unique

---

‡Please note that a direct mapping between domains and *handles* exists and is enabled by a *handle*/DNS proxy approach.

Handle	Field type:index	Value
2118/r_user	HS_ADMIN:100	rwr:0.NA/2118:300
	HS_ADMIN:101	rwr:2118/r_user:200
	HS_VLIST:200	rwr:2118/r_user:300
		rwr:2118/hproxy:300
		rwr:10.200/fproxy:300
	SIP_URL:250	sip:2118.r_user@x.y.z.w
	SIP_PWD:251	password
	HS_PUBKEY:300	00BE0034........

FIGURE 10.3
Sample user *handle* structure.

local name under the NA namespace [12], thus rendering the *handle* globally unique. A possible realization of the *handle 2118/r_user* inside the Handle System is depicted in Figure 10.3. The *handle* has several fields. The HS ADMIN and HS_VLIST fields determine the administrators of the *handle* who are the naming authority (0.NA/2118): the *handle* itself (*2118/r_user*) and the two proxy servers in the HS_VLIST field. Any of these administrators has the privilege to modify the fields inside the *handle* provided the administrator succeeds to authenticate with the Handle System using his private key.

10.3.2   Authentication and Registration

Currently, the most common authentication mechanism employed by SIP is the digest authentication [35] used by HTTP. When a user associates with a domain proxy server, he obtains an account on that server with a username and password, which he uses to authenticate himself to the server if asked. The digest authentication depicted here is domain-dependant, i.e., the user's credentials are valid for a particular domain. Briefly, digest authentication proceeds as follows:

1. The user sends a REGISTER request to a SIP proxy/registrar server.

2. The server replies with a 401 unauthorized response message challenging the user to authenticate himself for the requested service (realm) through a user and password prompt.

3. The user sends back a message digest of his credentials, which include his username, password, etc.

4. The same message digest is computed internally using the server's internal user information and compared to the one sent by the user.

5. Authentication is granted if the two digests match.

6. The user registers with the SIP proxy/registrar server.

In our approach, we still use digest authentication for the SIP users due to its wide support by current SIP servers and user agents, although a better authentication mechanism can be designed that would leverage the inherent security that *handles* expose.

Access to the authentication information is controlled inside the Handle System by the users. Recall that each user owns and administers his own *handle*. As part of this process, the user specifies in the HS_VLIST field the set of *handles* that have administrative rights over his *handle*. Among these *handles*, the user should include *handles* of any SIP proxy server that he wishes to register with, which could be any foreign server(s) that he trusts.

Two approaches can be exploited to implement the logic needed by the current SIP architecture for supporting *handle* authentication and registration. The first is to modify the actual SIP servers by extending their functionality through a server plug-in. This approach requires no changes to the current user agent devices whether hardphones or softphones. The devices will adapt seamlessly to the system. Alternatively, a second approach is to modify the user agent devices instead, which is a more cumbersome task that would require software upgrades for all existing user agents.

This paper implements the first approach, which deals with extending the functionality of the proxy/registrar servers. We present the proposed solution in light of the reference example of Figure 10.1. In Figure 10.3, the roaming user *2118/r_user* has granted both SIP proxy servers *2118/hproxy* and *10.200/fproxy* administrative rights over his *handle*. Note that the VLIST could refer to another *handle* containing a list of globally trusted servers. For the roaming user *r_user* present in the foreign domain fdomain, the authentication/registration process with the foreign proxy server *10.200/fproxy*, depicted in Figure 10.2c and Figure 10.1 (proposed traffic flow, arrows a, b, c, d), proceeds as follows:

1. *r_user*, after including the *handle 10.200/fproxy* in his *handle* HS_VLIST field, sends a REGISTER request to *fproxy*.

2. *fproxy* challenges *r_user* to authenticate himself.

3. *r_user* uses the same digest authentication with handle *2118˜r_user* as username and the value of the SIP_PWD field as password. He gets these values from his handle handle as shown in Figure 10.3.

4. *fproxy* uses the Handle protocol [13] to resolve the *handle 2118/r_user* into the SIP_PWD field. The server then computes a message digest over the obtained credentials.

5. Authentication is granted if the two digests match.

6. After authenticating *2118/r_user*, the foreign proxy fproxy proceeds to create an internal account for *r_user* to be able to use the SIP *services*

on fproxy. The internal user account will have a username identical to the *handle* of the registering user with the """ replaced by ".", i.e., *2118.r_user*, in this case.

7. Registration of the user follows. This requires that fproxy modifies the *handle 2118/r_user* updating the field SIP URL to point to the internal account, *2118.r_user@x.y.z.w* in this case, as shown in Figure 10.3. This means that *r_user* is currently associated with *fproxy*.

Obviously, our modified authentication algorithm is domain-independent. In other words, the user's credentials are valid for all realms, provided the correct administrative privileges are set in the Handle System. This property is essential, as it allows a particular authenticated SIP message to traverse multiple domains instead of requiring re-authentication for each domain on the path of the message. Since all communication between the proxy and the Handle System is secure [13], the proxy can be reasonably certain that the roaming user is indeed who he claims to be by validating his credentials against the secure *handle*. Internally, the proxy server monitors the user accounts created and removes an account (also updating the *handle*) due to unregister requests or account expiration. A sample *handle* for the foreign proxy is shown in Figure 10.4.

Devices, whether hardphones and softphones, are treated similarly. This depends on the ability of the device owner to present the SIP proxy with a username (this could be the *handle*) and password for authentication.

With this approach, a user no longer needs to register with a home proxy server as was required by pre-call mobility [5]. After registering with the foreign server, the user's handle-to-URI mapping remains fresh, allowing correspondent users to reach him simply by addressing his *handle*, as we will show in Section 10.3.3.

10.3.3 Routing

After abstracting any domain binding from users and allowing seamless authentication and registration with local proxy servers, the next step is to permit the user to initiate and receive calls by addressing a particular *handle* with

Handle	Field type:index	Value
10.200/fproxy	HS_ADMIN:100	rwr:0.NA/10.200:300
	HS_ADMIN:101	rwr:10.200/fproxy:300
	INET_HOST:240	x.y.z.w
	HS_PUBKEY:300	00BE00445........

FIGURE 10.4
Sample proxy *handle* structure.

no explicit reference to domain bindings (URIs). In this sense, a SIP user can INVITE any other SIP user provided he knows the latter's *handle*. From the perspective of a user, all other users seem to belong to one local domain and abstraction is complete.

It is extremely important to minimize the call setup time for the time-sensitive and interactive applications enabled by SIP. To explain how this is achieved, we will go through the steps where an arbitrary SIP user *c_user* (caller) tries to INVITE the roaming user *r_user* (callee) using the latter's *handle 2118/r_user* as shown in Figure 10.1. The call routing process, presented in Figure 10.2c, proceeds as follows:

1. Caller *c_user* sends an INVITE request to *r_user*. The invite request reaches the caller's SIP proxy/registrar containing the following header fields:

   ```
   INVITE sip:2118 ~ r_user@somedomain SIP/2.0
   To:<sip:2118 ~ r_user@somedomain>.........
   ```

   Note: In this message, the domain *somedomain* is irrelevant to our approach. We are only concerned with the *handle* part of the request-URI. To distinguish between *handle* and non-*handle* requests, we resort to the "/" character§ in the host name.

2. The proxy checks if the *handle 2118~r_user* is a locally registered user. If not, the server resolves the *handle* into the SIP URL field, which is *2118.r_user@x.y.z.w* in this case, as shown in Figure 10.3.

3. The server then rewrites the target URI of the message to the resolved URI.

4. From this point on, the natural SIP call flow is leveraged and the traditional SIP architecture [1] is utilized for efficient call routing. Note that other proxy servers on the call path treat the request as a normal request, i.e., no *handle* resolution is required.

Again, with our approach, correspondent users trying to communicate with the mobile user need not go through a home proxy for session setup or redirection. This renders the call route more efficient, eliminating unnecessary overhead and significant round-trip times.

One last point worth mentioning is the ability of a user to register with multiple servers from different devices simultaneously using the same *handle*. In our implementation, the SIP URL field of a particular *handle* can contain a list of bindings (URIs) to enable this attractive property. Exploiting this property is left for future papers.

---

§Since Internet hostnames can not contain the "/" character [13] ascii $(0 \times 2F)$ (essential character in the *handle* Namespace [35]), we replaced it with the "~" ascii $(0 \times 7E)$ character in the examples above for implementation purposes. We also allow the "#" ascii (0×23) character for compatibility with hard IP phones.

10.3.4   User-Controlled Mobility

Our mobility scheme is user-controlled in the sense that the user is responsible for the administration of his SIP identifier now that the latter is domain-independent. Consequently, routing the user's calls through a local server simply requires the user to add the server's persistent identifier (*handle*¶) to her persistent identifier's admin list. This indicates that the user trusts the local server to route her calls.

The other aspect of user control is the distributed service model that potentially eliminates the need for service level agreements (SLAs) between domains. Instead, the local domain proxy will directly challenge the user and grant her trust provided sufficient credentials exist within the user's persistent identifier. These credentials can include financial information as well as trust information that the actual server can validate before allowing the user to route traffic through.

10.4   Implementation

10.4.1   Testbed

We have implemented the functionality described in this paper as an extension to two open source SIP servers, the JAIN-SIP proxy [36] and the SIP express router (SER) [37]. The JAIN-SIP proxy server is an open source JAVA-based SIP proxy built on top of the JAIN-SIP-1.1 API. SER is an open-source, configurable SIP server that is widely deployed in the research community. We have implemented a JAVA-based H-SIP API that can be easily called from both proxy servers to expose the H-SIP interface operations. The operations mainly enable inter-domain authentication, registration, and call routing using *handles.* All the results depicted hereafter are based on the JAIN-SIP proxy.

Our testbed is a realization of the framework depicted in Figure 10.1. We are running three modified SIP servers on three separate domains:

1. *ece.unm.edu* located at the University of New Mexico, Albuquerque, New Mexico [Server IP: 129.24.24.106]

2. *cnri.reston.va.us* located in Reston, Virginia [Server IP: 132.151.9.104]

3. *istec.org* located in Panama [Server IP: 168.77.202.59]

A roaming user is allowed to move across the domains while establishing connectivity within each domain using the respective local server. The three servers are running Fedora Core 4 kernels and are identical in terms of workload and processing speed (AMD Athlon 1.1 GHz processors).

¶We have used the terms persistent identifier and *handle* interchangeably throughout this paper since our current implementation of the persistent identifier is the *handle.*

To use the framework, users are expected to be able to manage their own *handles*. The Handle System provides a free administration tool [10] for this purpose. This tool is currently implemented in JAVA.

Currently, users wishing to associate with a proxy server are required to specify the latter's IP address or domain name. Since our approach exploits *handles* instead of domain names, we have implemented a specialized gateway that translates between DNS and *handle* protocols. The gateway is responsible for protocol translation, specifically, *handle* to DNS. We refer to this gateway as the Handle-DNS proxy (HDP). HDP is a modified DNS server that communicates using the BIND protocol and implements extra functionality, allowing it to associate canonical names and aliases inside its particular naming zone with *handles*. HDP will, therefore, resolve canonical names inside its naming zone using the Handle System and will, in addition, allow any common DNS server to resolve DNS entries in the format: <handle>. [DNS proxy domain] to the actual value of the SIP HOST attribute of that particular *handle*. Details about this gateway, which can be extended to become a plug-in for current DNS servers, are presented in [32].

The ease of deployment of our framework and success of the conducted experiments encourage us to pursue this work.

10.4.2   Experimental Model

In this section, we will compare the performance of roaming and non-roaming environments in terms of registration and call establishment delays. The non-roaming case represents traditional SIP signaling between a user and his home SIP server irrespective of roaming, while the former case represents our proposed approach. Both cases were tested under the same conditions regarding UA/SIP processor speed and workloads. As UA, we used the Java SIP Communicator [38]. We also tested with the Cisco 7940/7960 IP phones.

10.4.2.1   Registration

Figure 10.5 is a magnified image of Figure 10.2 that focuses on the registration process according to the testbed. We show a comparison of the registration process for roaming (H-SIP) and non-roaming (no H-SIP) environments. *r_user* is located in *fdomain, ece.unm.edu*. For the no H-SIP case, we tested with *hdomain* being either *cnri.reston.va.us* (Reston, Virginia) or *istec.org* (Panama). Again, registration here assumes the basic digest authentication with the proxy server. In our model, the UA refreshes its registration with the proxy/registrar continuously (we used a registration TTL of one minute).

If we estimate the server's average processing time of the REGISTER request including digest authentication by a, then from Figure 10.5a, the average registration time t_A as seen by *r_user* is given by,

$$t_A = \frac{1}{n} \sum_{1 \leq i \leq n} t_i = t_1 \tag{10.1a}$$

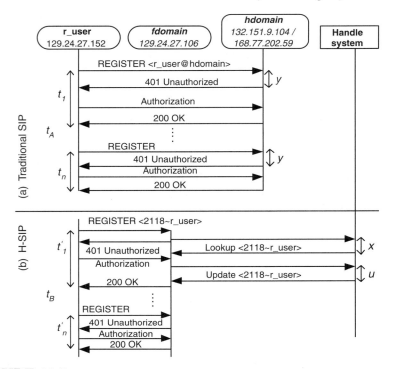

FIGURE 10.5
REGISTER message flow A. With no roaming logic C. With proposed roaming logic H-SIP.

since,

$$t_1 = t_2 = \cdots = t_n \approx \alpha + 2y \tag{10.1b}$$

where t_i is time consumed by the ith REGISTER, and y is the average round-trip communication delay between r_user and the registering SIP server. Obviously, t_A includes a relatively significant communication delay as a result of the presumably large geographical separation between the roaming user and the home SIP server.

Now, if we consider H-SIP, then from Figure 10.5c, the average registration time tc as seen by r_user is given by,

$$t_C = \frac{1}{n} \sum_{1 \le i \le n} t_i' \tag{10.2a}$$

and,

$$t_1' \approx \alpha + x + u \tag{10.2b}$$

$$t_2' = \cdots = t_n' \approx \alpha \tag{10.2c}$$

where t_1' is time to perform the ith REGISTER, x is the average *handle* resolution delay and u is the average *handle* update delay. First, we note that the round-trip delay y is negligible in this case due to the existence of *r_user* and the SIP server on the same local network. Besides, note here that the server will issue one *handle* resolution and one update for the first REGISTER request only. Additionally, the first *handle* resolution is always cached internally on the server. Unless *r_user* moves to another domain or unregisters, no *handle* resolution/update is required. This means that subsequent REGISTER requests will read the cached value and thus, for $i > 1, x = u = 0$. This also means that x and u can be discarded for n sufficiently large.

Here, t_A and t_C are measured by *r-user* as the average time between sending the REGISTER request and receiving the 200 OK response. In Figure 10.5, *r_user* performs a DNS lookup or a Handle-DNS lookup in scenarios A and C, respectively. These lookups are excluded from the performance metrics, i.e., t_1 and t_2 do not include the DNS lookups for the SIP servers. However, we closely examine and compare the two lookup times and we show the performance results in Section 10.4.3.

10.4.2.2   Call Establishment

Figure 10.6 is a magnified image of Figure 10.2 that focuses on the call establishment process according to the test-bed, where *c_user* will try to establish a call with the roaming user *r_user*. For this section, *cdomain* is *ece.unrn.edu* [New Mexico]. The *hdomain* is *istec.org* [Panama] and the *fdomain* is *cnri.reston.va.us* [Virginia]. For brevity, the 100 TRYING messages are excluded from Figure 10.6. We focus on the INVITE message flow and the servers require no authentication. We denote by $\text{rt}(_{m,n})$ the round-trip

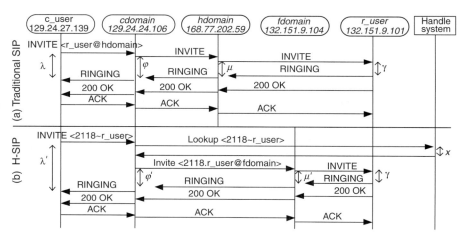

FIGURE 10.6
INVITE message flow A. With no roaming logic C. With proposed roaming logic H-SIP.

communication delay between nodes m and n. For example, in Figure 10.6, $rt_{(cdomain,hdomain)}$ is the round-trip delay between the *cdomain* SIP server and that of *hdomain*. If we estimate the server's average processing time of the INVITE request by $f3$, and the UA's average processing time of the INVITE request by 7, then again from Figure 10.6a,

$$\lambda \approx rt_{(c\_user,cdomain)} + \beta + \phi \qquad (10.3a)$$

$$\phi \approx rt_{(cdomain,hdomain)} + \beta + \mu \qquad (10.3b)$$

$$\mu \approx rt_{(hdomain,r\_user)} + \gamma \qquad (10.3c)$$

and therefore,

$$\lambda \approx rt_{(c\_user,cdomain)} + rt_{(cdomain,hdomain)} + rt_{(hdomain,r\_user)} + 2\beta + \gamma \qquad (10.3d)$$

where λ is the average call establishment time (the INVITE/RINGING round-trip) as seen by *c_user*, ϕ is the average INVITE/RINGING round-trip time between the *cdomain* and *hdomain* SIP servers and u is the average INVITE/RINGING round-trip time between the *hdomain* SIP server and *r_user*.

Now, if we consider H-SIP, then from Figure 10.6c,

$$\lambda' \approx rt_{(c\_user,cdomain)} + \beta + x + \phi' \qquad (10.4a)$$

$$\phi' \approx rt_{(cdomain,fdomain)} + \beta + \mu' \qquad (10.4b)$$

$$\mu' \approx rt_{(fdomain,r\_user)} + \gamma \qquad (10.4c)$$

and therefore,

$$\lambda' \approx rt_{(c\_user,cdomain)} + rt_{(cdomain,hdomain)} + rt_{(fdomain,r\_user)} + 2\beta + x + \gamma \qquad (10.4d)$$

where λ' is again the average call establishment time as seen by *c_user*, ϕ' is the average INVITE/RINGNG round-trip time between the *cdomain* and *fdomain* SIP servers, μ' is the average INVITE/RINGING round-trip communication time between the *fdomain* SIP server and *r_user*, and x is the average *handle* resolution delay. Note here that each INVITE request will require the server to issue one fresh *handle* resolution.

Both λ and λ' are measured by *c_user* as the time between sending the INVITE request and receiving the RINGING response.

10.4.3 Performance Measurements

The measurements of the registration times, call establishment times and communication delays in this section were all averaged from 10,000 samples dispersed over a 10-day period, i.e., $n = 1,000$ samples a day.

10.4.3.1 Registration

Examining Equation 10.1a and Equation 10.2a, we deduce that $t_c \leq t_A$ for sufficiently large n. In the general case of a roaming user, the round-trip delay

TABLE 10.1

Average Registration Delays with and without H-SIP

		Registration Delays [ms]			
		X, u	Y	t_A	t_C
No	hdomain: virginia	NA	86	209	NA
H-SIP	hdomain: panana	NA	116	275	NA
H-SIP		84, 260	0	NA	40

Note: As shown in Figure 10.5; $n = 1000$, transport = UDP.

$2y$ is expensive. Our real-time measurements are listed in Table 10.1. Clearly, our measurements show that Equation 10.1a and Equation 10.2a hold for an $a \sim 39$ ms. We see here that t_A is approximately $5t_c$ or $7t_c$ for the Virginia and Panama cases respectively. Obviously, the H-SIP approach outperforms the traditional approach as long as y is significant, which is often the case for roaming subscribers. The value of a directly depends on the implementation of the SIP server, which is the JAIN-SIP Proxy server [36] in this case. Besides, x is random and it directly depends on the location of the local handle server (LHS) [11] storing the particular *handle*.

10.4.3.2   Call Establishment

In comparing the call establishment time given by Equation 10.3d and Equation 10.4d, we are interested in computing the performance enhancement or degradation $\Delta\lambda = \lambda - \lambda'$. We consider the following variables to verify our model:

$$\tau_1 = rt_{(hdomain, r\_user)} - rt_{(fdomain, r\_user)} \gg 0 \qquad (10.5a)$$

$$\tau_2 = rt_{(cdomain, hdomain)} - rt_{(cdomain, fdomain)} \qquad (10.5b)$$

$$\sigma = \tau_1 + \tau_2 > 0 \qquad (10.5c)$$

where σ is the difference in the cumulative round-trip delay between the no H-SIP case and the H-SIP case, respectively. Equation 10.5a is true in general due to the presumably large geographical separation between the roaming user and his home server versus using a local server. We also argue that Equation 10.5c holds in general based on our model. i.e., the cumulative round-trip delays for *c_user* to reach *r_user* is smaller in the H-SIP scenario. However, τ_2 in 10.5b can be positive or negative since it directly depends on the *cdomain-hdomain* separation versus that of *cdomain-fdomain*. For our particular setup, $\tau_2 > 0$.

It follows from Equation 1.3 and Equation 1.4 that,

$$\Delta\mu = \mu - \mu' \approx \tau_1 \qquad (10.6a)$$

$$\Delta\phi = \phi - \phi' \approx \tau_2 + \Delta\mu \qquad (10.6b)$$

$$\Delta\lambda \approx \Delta\phi - x \qquad (10.6c)$$

TABLE 10.2
Comparison of Average Call Setup Delays* and Average
Round-Trip Communication Delays

Call Setup Delays [ms]			Round-trip Delays [ms]		
$\Delta\lambda$	$\Delta\phi$	$\Delta\mu$	τ_1	τ_2	x
88	168	106	111	47	85

*Per Figure 10.6
Note: transport = UDP.

Notice here that,

$$\Delta\lambda \approx \tau_1 + \tau_2 - x = \sigma - x \qquad (10.6d)$$

Consequently, H-SIP outperforms the traditional SIP approach, in general, as long as $\sigma > x$. Our conducted real-time measurements are listed in Table 10.2. In Table 10.2, we verify the validity of our model by separately comparing the real-time call setup measurements to the round-trip delays, thus asserting Equations 10.6.

10.4.3.3 DNS vs. Handle-DNS Resolution

Obviously, the only performance degradation introduced by our approach is the *handle* resolution overhead. This overhead has been extensively measured by CNRI and its partners and is illustrated in Figure 10.7. The current performance of the system that oscillates from 3 to 10 ms makes it comparable to the bind implementation of the DNS protocol. Additionally, due to its intrinsic fully distributed administration and resolution, it avoids the pitfalls

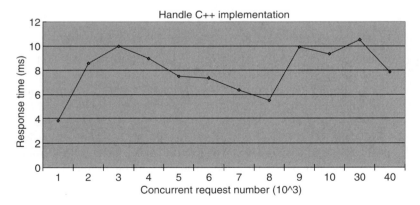

FIGURE 10.7
Handle implementation performance measurement as of August 2005. Acquired through the courtesy of Mr. Sam Sun and CN-NIC. (Courtesy of Mr. Sam Sun and CN-NIC. With permission.)

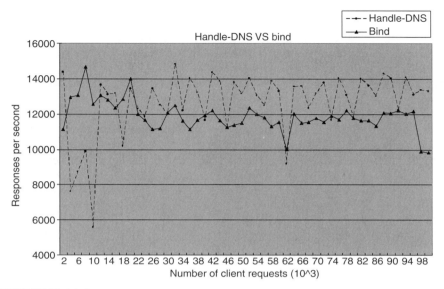

FIGURE 10.8

Handle load performance measurement as of August 2005. Acquired through the courtesy of Mr. Sam Sun and CN-NIC. (Courtesy of Mr. Sam Sun and CN-NIC. With permission.)

that plague the current DNS implementation where DNS resolution times can extend to over 100 milliseconds [39].

Overall robustness of the handle-DNS implementation has also been extensively tested along with load assessment. Results, as shown in Figure 10.8, show that the Handle System can efficiently replace the DNS system.

On top of addressing the domain resolution itself, our approach minimizes the signaling traffic needed by a roaming user to join the SIP infrastructure and be ready to initiate calls. The user is efficiently utilizing the services of a local server with no need for per-call coordination with his home server. If the home server is located in another continent per se, the round trip times for registration/redirect messages from the roaming user to home proxy/correspondent user, respectively, become significant. The proposed approach optimizes the association, authentication, and call routing times for the roaming user by guaranteeing that the mobile user will always be addressed through the closest available server in his vicinity.

10.5 Future Work

This paper used globally unique *handles* to identify SIP users. It is also possible to use a set of aliases of such *handles* that can in turn translate into *handles.* Such a service would be administered by yet another service provider using the *handle* protocol to administer and coordinate the general use of aliases in the system. This may be done using registrylike features for particular *handles*, and allows the provider to service users that opt into this service. Since *handles* can also point to other *handles*, certain intermediate *handles* may be provided to systems that choose to run private *handle* servers in a way similar to a Private Branch eXchange (PBX). The advantage of having a distributed resolution infrastructure that is also domain-independent, translates into users being able to run smaller SIP servers that communicate with the Handle System to expedite routing. You could even envision a cellular-like behavior for SIP systems in which users are able to use the resources of many smaller SIP servers along a user's roaming path. All these ideas need to be investigated as well as possible ways to expedite routing and *handle* resolution are possible future research paths.

We intend to leverage the scalable resolution, security, and administration services of the Handle System and use it to replace the DNS system within the SIP protocol. According to RFC 3263, the main reason SIP needs to use DNS is to enable the originator domain proxy to locate the SIP proxy in the destination domain (IP, port and transport protocol). The other need for DNS in SIP is for the terminating proxy to identify a backup for the originating proxy in the case the latter fails. Replacing the DNS within SIP requires carefully examining all the DNS resolutions performed by UA clients and proxy servers according to RFC 3263 and formalizing the fields to be resolved within the *handles*. Besides, we envision the future Handle System to be completely decentralized and to be based on concepts like distributed hash tables, where a particular identifier is not necessarily required to be located under a domain hierarchy. Finally, part of our current research is to focus on implementing a structured peer-to-peer form of the Handle System to expedite lookups and resolutions, and eliminate single points of failure.

10.6 Conclusion

In this paper, we outlined the use of an indirection architecture based on the Handle System to address SIP inter-domain mobility. Our approach not only enables roaming controlled by the users rather than organizations, but also provides a faster implementation than traditional approaches currently deployed. Through our work, users are able to dynamically enable their own mobility and benefit from the advantages of a secure distributed persistent identifier network. By disassociating users from DNS domains, while still providing the means to interact with traditional SIP systems, we provide a scalable interchangeable enhancement to the SIP infrastructure.

Acknowledgements

We would like to thank CNRI for their support as well as the Ibero American Science and Technology Consortium (ISTEC) and the University of New Mexico that provided the VoIP infrastructure for the test bed of this work.

The work presented in this report is partially funded by the NSF under the Future Internet Design (FIND) Grant CNS-0627067-0626380.

References

[1] Rosenberg, J., H. Schulzrinne et al. 2002. RFC 3261: Session initiation protocol. June.

[2] H.323 : Packet-based multimedia communications systems. http://www.itu.int/rec/T-REC-H.323-200307-I/en. Accessed July 25, 2008.

[3] Ashutosh Dutta, Faramak Vakil, Jyh cheng Chen, Miriam Tauil, Shinichi Baba, Nobuyasu Nakajima, and Henning Schulzrinne. 2001. Application layer mobility management scheme for wireless internet. In IEEE 36 wireless'01 San Francisco, CA. May.

[4] Pandya, R. 1995. Emerging mobile and personal communication systems. *IEEE Communications Magazine* 33: 44–52.

[5] Henning Schulzrinne and Elin Wedlund. 2000. Application-layer mobility using sip. SIGMO-BILE Mob. *Computer Communication Review* 4(3): 47–57.

[6] Elin Wedlund, and Henning Schulzrinne. 1999. Mobility support using sip. In *WOWMOM '99: Proceedings of the 2nd ACM International Workshop on Wireless Mobile Multimedia*, New York, NY, USA: ACM Press, 76–82.

[7] Charles E. Perkins. 2002. RFC 3220: IP mobility support for ipv4. January. http://www.rfc-editor.org/rfc/rfc3220.txt. Accessed July 25, 2008.

[8] Berners-Lee, T., R. Fielding, and L. Masinter. 1998. RFC 2396: Uniform resource identifiers (URI): Generic syntax.

[9] Nilanjan Banerjee, Sajal K. Das, and Arup Acharya. 2005. SIP-based mobility architecture for next generation wireless networks. In *PerCom* 181–90. IEEE Computer Society. Kauai, HI. March 8–12.

[10] The *handle* system. http://www.handle.net. Accessed 07/25/08.

[11] Sun, S., L. Lannom, and B. Boesch. 2003. Handle system overview. *RFC 3650*. November.

[12] Sun, S., L. Lannom, and B. Boesch. 2003. Handle system namespace and service definition. *RFC 3651*. November.

[13] Sun, S., S. Reilly, L. Lannom, and J. Petrone. 2003. Handle system protocol (ver2.1) specification. *RFC 3652*. November.

[14] Ravi Jain, Thomas Raleigh, Charles Graff, Michael Bereschinsky, and Mitesh Patel. 1998. Mobile internet access and qos guarantees using mobile ip and rsvp with location registers 3. 1690–95. In *ICC International Conference on Communications*. Atlanta, GA. June 7–11.

[15] Ravi Jain, Thomas Raleigh, Danny Yang, Li-Fung Chang, Charles Graff, Michael Bereschinsky, and Mitesh Patel. 1999. Enhancing survivability of mobile internet access using mobile IP with location registers. In *INFOCOM*, 3–11.

[16] Wong, K.D., A. Dutta, J. Burns, R. Jain, K. Young, and H. Schulzrinne. 2003. A multilay-ered mobility management scheme for auto-configured wireless ip networks. *Wireless Communications* 10(5): 62–69.

[17] Johnson, D., C. Perkins, and J. Arkko. 2004. RFC 3775: Mobility support in ipv6. June.

[18] Wong, K., Ashutosh Dutta, Henning Schulzrinne, and Ken Young. 2006. Simultaneous mobility: Analytical framework, theorems, and solutions. *Wireless Communication and Mobile Computing*. Pg. 623–42, 7(5): 2007.

[19] Chen Hongtao, Yang Fangchun, and Xu Peng. 2005. Analysis on sip mobility of double user agent servers. In *Communications and Information Technology* ISCIT, IEEE. October, 1: 87–90.

[20] Moskowitz, R., P. Nikander, and P. Jokela. 2006. Host identity protocol architecture. *RFC 4423*. May.

[21] Nikander, P., J. Arkko, and B. Ohlman. 2004. Host identity indirection infrastructure (hi3). In *The Second Swedish National Computer Networking Workshop*. Karlstad, Sweeden. November 23, 24.

[22] Alex C. Snoeren, and Hari Balakrishnan. 2000. An end-to-end approach to host mobility. In *Sixth Annual ACM/IEEE International Conference on Mobile Computing and Networking*. Boston, MA. August.

[23] Eriksson, Faloutsos, and Krishnamurthy. 2003. Peernet: Pushing peer-to-peer down the stack. In *International Workshop on Peer-to-Peer Systems (IPTPS), LNCS*. Berkeley, CA.

[24] Falstrom, P. Technical report. ENUM. http://www.enum.org. Accessed July 2008.

[25] Ion Stoica, Daniel Adkins, Shelley Zhuang, Scott Shenker, and Sonesh Surana. 2004. Internet indirection infrastructure. *IEEE/ACM Transactions on Networking.* 12(2): 205–18.

[26] Hari Balakrishnan, Karthik Lakshminarayanan, Sylvia Ratnasamy, Scott Shenker, Ion Stoica, and Michael Walfish. 2004. A layered naming architecture for the internet. In *Proceedings of ACM SIGCOMM 2004,* 343–52. Portland, Oregon, US: ACM. August.

[27] Michael Walfish, and Hari Balakrishnan. 2004. Untangling the web from DNS. In *NSDI.* 225–38. USENIX. Otawa, OT, Canada.

[28] Mark Gritter, and David R. Cheriton. 2001. An architecture for content routing support in the internet. In *USITS'01: Proceedings of the 3rd conference on USENIX Symposium on Internet Technologies and Systems,* 4–4, Berkeley, CA, USA. USENIX Association.

[29] Ballintijn, G., M. van Steen, and A.S. Tanenbaum. 2001. Scalable human-friendly resource names. *Internet Computing* 5(5): 20–27.

[30] Khoury, J., H. Jerez, N. Nehme, and C. Abdallah. An application of the mobile transient network architecture: Ip mobility and inter-operability. 2006. pre-print available at http://hdl.*handle*.net/2118/jk_transapp_08. Accessed July 25, 2008.

[31] Joud Khoury, Henry Jerez, and Chaouki Abdallah. 2007. Efficient user controlled inter-domain sip mobility: Authentication, registration, and call routing. Technical Report EECE-TR-07-010, University of New Mexico, May. [online]: https://repository.unm.edu/dspace/brtstraum/ 1928/3059/SIP_roamming_TR.pdf. Accessed July 25, 2008.

[32] Henry Jerez, Joud Khoury, and Chaouki Abdallah. A mobile transient internet architecture, 2006. Arxiv. http://arxiv.org/abs/cs.NI/06/0087. Accessed July 25, 2008.

[33] Robert Kahn and Robert Wilensky. 1995. A framework for distributed digital object services. Internet Whitepaper http://www.cnri.reston. va.us/k-w.html. Accessed July 25, 2008.

[34] Sun, S., 2001. Establishing persistent identity using the *handle* system. *Tenth International World Wide Web Conference.* Hong Kong. May 1–5.

[35] Franks, J., P. Hallam-Baker, J. Hostetler, S. Lawrence, P. Leach, A. Luotonen, and L. Stewart. 1999. HTTP authentication: Basic and digest access authentication. *RFC 2617.* http://www.fags.org/rfc/ rfc2617.html. Accessed July 25, 2008.

[36] Jain-sip proxy (built on jain-sip 1.1 api). https://jain-sip-presence-proxy. dev.java.net/. Accessed July 25, 2008.

[37] Rebahi, Y., D. Sisalem, J. Kuthan, A. Pelinescu-Onicicul, B. Iancu, J. Janak, and D. Mierla. The sip express router, an open source sip platform. http://www.iptel.org/ser. Accessed July 25, 2008.

[38] Sip-communicator 1.0. https://sip-communicator.dev.java.net/. Accessed July 25, 2008.

[39] Huitema, C., and S. Weerahandi. 2000. Internet measurements: The rising tide and the dns snag. In *Proceedings of the 13th ITC Specialist Seminar on IP Traffic Measurement Modeling and Management*, IPseminar, Monterrey, CA, USA, ITC. September 18–20.

11

SIP and Vertical Handoffs in Heterogeneous Wireless Networks

Jie Zhang
The University of British Columbia

F. Richard Yu
Carleton University

Xiaolei Wang
The University of British Columbia

Henry C. B. Chan
The Hong Kong Polytechnic University

Victor C. M. Leung
The University of British Columbia

CONTENTS

11.1 Introduction 253
11.2 Vertical Handoff Management 257
11.3 Seamless Vertical Handoff Support Using SIP 262
11.4 Efficient Transport of SIP Traffic over SCTP 267
11.5 Conclusions 272
    References 272

11.1 Introduction

With the widespread deployment of Internet services, it is expected that people will demand to access Internet service "anytime and anywhere." Third generation (3G) cellular mobile communication systems, which can provide very wide coverage, are contributing to the realization of this goal. However, the 384 kbps data transmission rate provided by current 3G systems is not sufficient to support many emerging multimedia services. Moreover, due to the use of licensed spectrum, the data services provided by 3G systems are very expensive. Contrarily, some other wireless access technologies such as wireless local area networks (WLAN) can provide high-speed Internet access in small areas like cafes or airports at a much lower cost because they operate on unlicensed spectrum. These access technologies can be a good complement to 3G mobile communication systems. Due to advances in signal processing technologies and microelectronics, a contemporary mobile device can be equipped with multiple radio interfaces that enable it to be connected to multiple networks. As a result, it is anticipated that the next generation (NG) wireless communication systems will integrate different wireless access technologies together and provide user "Always Best Connected" (ABC) services [1] via the heterogeneous wireless access networks.

The 3rd Generation Partnership Project (3GPP) has released several specifications for 3G wireless network development. Apart from the circuit-switched (CS) and packet-switched (PS) domains, 3GPP has defined a third signalling domain named IP Multimedia Subsystem (IMS) in release 5 to provide some administration services such as quality of service (QoS)-provisioning and charging [2]. IMS can provide access-independent services via standard access interfaces. As shown in Figure 11.1, it is foreseeable that, in next generation (NG) wireless systems, heterogeneous wireless access networks are interconnected by the IP core network for data transport and by the IMS for signalling exchanges. The signalling protocol operating in IMS is a text-format protocol named Session Initiation Protocol (SIP) [3] that was initially developed by the Internet Engineering Task Force (IETF) to invite users to join multimedia sessions or conferences over the Internet. Currently, the functionality of SIP has been extended to enable modification and termination of

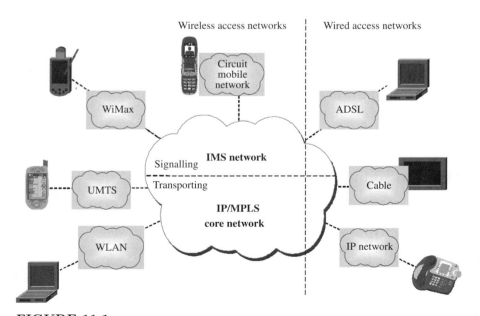

FIGURE 11.1
A next generation network structure.

multimedia sessions as well as delivery of instant messages and handling of subscriptions to events. Based on SIP, IMS supports QoS provisioning, and service charging, as well as service integration.

Due to the popularity of WLAN technologies, 3GPP has also released specifications on the integration of the Universal Mobile Telecommunication System (UMTS) and WLAN [4]. Two approaches called loose coupling and tight coupling [5] are specified. In tight coupling, WLAN access routers connect to the UMTS core network like other radio access networks (RAN) and data from WLAN are delivered to the external networks via the UMTS core network. In this case, the WLAN serves as an alternate RAN and the roaming between these two systems is similar to the inter-RAN roaming. In a loose coupling architecture, WLAN delivers data traffic to the Internet directly, while some signalling messages are transmitted to the IMS of UMTS. In a tight coupling architecture, both UMTS and WLAN belong to the same operator in order to allow the WLAN data to cross the UMTS core networks. The Serving GPRS Support Node (SGSN) and Gateway GPRS Support Node (GGSN) should also be configured to accommodate the heavy traffic load from the WLAN. On the contrary, the loose coupling architecture allows for independent traffic management of both UMTS and WLAN, which is more applicable in situations where different operators cooperate to provide service. Therefore, the loose coupling architecture is generally regarded as a mainstream integration architecture [5]. However, loose coupling of UMTS and WLAN also incurs some technical challenges. For example, although both networks are operated independently, an integrated authentication and accounting system is necessary to support roaming of users. In addition, macro-mobility management is required and results in the possibility of longer delays and more packet losses when inter-system mobility occurs. A hybrid-coupling approach is also proposed in [6], where best-effort traffic is directly delivered to the Internet while real-time traffic streams go through UMTS. This approach is subject to the same limitations as the tight coupling architecture.

In NG heterogeneous wireless networks, the mobile node (MN) initiates handoffs when it roams between different RANs. To differentiate the *horizontal handoff* that happens in homogeneous networks, a handoff taking place between heterogeneous access networks is referred as a *vertical handoff*. Vertical handoffs require the heterogeneous wireless networks to redirect in a timely manner ongoing connections to RANs employing different wireless technologies. Horizontal handoff management is normally implemented at the link layer because it involves RANs employing homogeneous wireless technologies, which belong to the same administrative domain. However, the same function cannot be applied to vertical handoff between heterogeneous RANs that are usually administrative-separated. Therefore, vertical handoff management should operate at a higher layer in order to provide a uniform management capability. Several procedures should also be executed during a vertical handoff, such as re-authentication in the new RAN, updating the addresses in the home network and transferring the context information to the new domain. The MN should also inform the corresponding node (CN) that communicates with the MN during the handoff in order to keep the connection going with minimal disruption. All these actions will incur a heavier signalling cost and more QoS degradation than horizontal handoffs. For example, some inflight packets may be lost during the handoff or the connection may even be terminated if the handoff takes too long. Moreover, vertical handoffs in NG heterogeneous wireless networks are not necessarily compulsory. For example, vertical handoffs from UMTS to WLAN (*downward handoff*) are normally executed to achieve better QoS or lower access costs rather than for the purpose of maintaining the ongoing connections, while vertical handoffs from WLAN to UMTS (*upward handoff*) usually happen when the MNs move out of the WLAN coverage areas and are therefore compulsory for the purpose of maintaining the ongoing connections. Horizontal handoffs are usually triggered by comparing the radio signal strength (RSS) or signal-to-interference ratio (SIR) from different RANs. However, this method cannot be directly applied to vertical handoffs because it is difficult to directly compare the RSS/SIR of heterogeneous RANs.

Several schemes have been proposed for vertical handoff management. Their operation scope varies from network layer to application layer. Generally, network layer solutions are based on Mobile IP (MIP) developed by the IETF [7], which requires some mobility-supporting agents to be incorporated in the various networks involved, while the transport layer solutions and the SIP-based application layer solutions are agnostic to the underlying network layer and thus easier to be adopted in the current IP-based network environment. Moreover, the timely transmissions of signalling messages is crucial for realizing seamless handoffs; however, this is a challenge since network environments during handoff periods may be unstable. Currently, SIP messages are delivered using User Datagram Protocol (UDP), or Transmission Control Protocol (TCP) in cases where the message size is large. While UDP provides an unreliable best-effort data delivery, TCP provides a reliable data delivery service without regard to latency. These two extremes in data delivery methods are not flexible enough for reliable and timely delivery of SIP signalling messages over unsteady network environments. To address this challenge, a transport layer protocol known as Stream Control Transmission Protocol (SCTP) [8] that was recently standardized by the IETF can be an attractive candidate due to its partial-reliability option for data delivery. In addition, the multihoming feature of SCTP also enables it to support vertical handoffs over the transport layer.

The objective of this chapter is to discuss the functionalities of SIP in supporting vertical handoffs. Section 11.2 explores various vertical handoff management schemes, and explains why those proposals based on transport layer and SIP-based application layer protocols are superior over other protocols. Section 11.3 discusses the seamless vertical handoff support using SIP, and presents a SIP-based soft-handoff (S-SIP) scheme. Section 11.4 demonstrates how the performance of SIP can be optimized by running the protocol over SCTP. Section 11.5 concludes the chapter.

11.2   Vertical Handoff Management

The loose coupling architecture for heterogeneous network interworking is generally considered an attractive solution for NG heterogeneous wireless networks. This architecture requires solutions for vertical handoffs to be implemented at the network layer or above. Solutions that operate in the same layers usually have some common characteristics. In this section, we briefly introduce the main solutions proposed for the network, transport, and application layers; their common characteristics, and discuss the pros and cons of these protocols.

11.2.1   Network Layer Solutions

In IP networks, if a terminal possesses a permanent IP address, then this address not only gives a unique identity for the terminal, but it also identifies the home domain of the terminal, which enables packets to be routed to the terminal from any corresponding node (CN) that knows the terminals IP address. In cases that a terminal is a MN that has roamed to another IP network outside of its home network, its IP address is no longer sufficient for routing packets from CNs, as the home domain identified by its IP address is different from the network domain that the MN is currently attached to. To solve this problem, the IETF developed MIP to facilitate packet forwarding to roaming MNs [7].

To identify the network domain that a roaming MN is currently visiting, it is assumed that the MN acquires a local address at the visited domain, known as the Care-of Address (CoA). A binding between the MN's home address and CoA is then established, so that the MN's identity continues to be specified by its home address, but its current location is specified by its CoA for routing purposes.

To support the registration of this binding and the routing of packets to the MN using this binding, MIP requires the deployment of two types of mobility support agents in IP domains associated with a MN: a home agent (HA) at the MN's home domain, and a foreign agent (FA) in each foreign network domain that the MN is visiting. In the visited domain, the FA relays the binding registration from the MN to the HA at the MN's home domain. The HA authenticates the MN, acknowledges the positive authentication to the FA, and stores the binding in a cache so that incoming packets addressed to the MN can be forwarded to the MN's current CoA by encapsulating them with the CoA as the destination address and tunnelling the packets to the FA. The FA in turn caches the binding so that packets arriving from the HA via the tunnel can be forwarded to the MN. Thus, packets sent by a CN to the MN undergo triangular routing via the MN's HA, but outgoing packets from the MN to a CN are sent directly to the CN. Usually, the address of the FA also serves as the CoA of the visiting MN, although the MN is allowed to serve as its own FA if it is able to independently obtain a CoA via other means, such as using Dynamic Host Configuration Protocol (DHCP) or IPv6 address auto-configuration. To avoid stale information, binding registrations and binding caches are soft states that need to be renewed before the timers for the respective information expire and the information is deleted.

To reduce the number of registration messages, MIP has been extended with paging functionality to reduce the signalling cost for the address update [9]. Furthermore, triangle routing leads to potential problems in long latency and high resource usage for routing of incoming packets to the MN. These problems are solved in [10], which allows the CN to cache the binding of the MN and directly tunnel packets to the MN using the MN's current CoA. The CNs cache is updated by the HA whenever the HA receives an updated registration from the MN. Note that due to security considerations, the CN only trusts the notification messages sent by the HA, which has an authenticated fixed IP address. Another problem is that a handoff from one FA to another is not seamless, as the MN is unable to receive any packets until the handoff completes a successful registration of the new CoA at the new FA. Packets forwarded to the MN via the FA during this period are either lost or deferred. To address this problem, authors of [10] also propose to set up temporary connections between the FAs of the previous network (oFA) and the new network (nFA). Once an MN registers in the new network, it will tell its new CoA to the oFA before sending out the registration message, and the oFA will forward all incoming packets to the nFA until the handoff is completed. Since the oFA is usually closer to nFA than the HA and CN, the packet delivery interruption period will be greatly shortened. To further shorten the interruption period, [11] also proposes to predict potential handoffs by tracking the MN's locations. Whenever the MN is going to move out of the coverage area of the WLAN it is currently associated with, it will initiate the handoff in advance so that the handoff completes before the WLAN signals become undetectable.

In MIP, an MN should register to the HA every time its CoA is changed. If the MN moves frequently between different subnets of a foreign network, it will send registration messages frequently back to the HA, which will generate heavy signalling traffic and cause long handoff delays if the HA is far away from the MN. A number of schemes have been proposed to manage such "intra-domain" handoffs by localizing the address updates within the foreign network. According to their modes of operation, these solutions can be classified into tunnel-based management [12,13] and routing-based management [14,15] schemes. In tunnel-based schemes, a number of registration agents are distributed hierarchically in the foreign network and data packets are tunnelled between the agents at different levels. Whenever the MN roams between two subnets in the visited domain, it will register its new CoA with the nearest common agents instead of its HA, which needs to be informed only about the network domain that the MN is visiting. In routing-based schemes, the MN will keep the same CoA address in the visited domain, which modifies the routing tables of its routers to follow the current location of the MN.

Generally, network layer solutions can provide universal mobility management over IP networks. However, this approach has two major drawbacks. First, it requires establishment of mobility support agents in all network domains that support host mobility. Second, for route optimization, it also requires CNs to implement the route optimization scheme and support tunnelling at the IP layer. The design of MIP assumes that each MN is assigned a static IP address by its home network, which is globally known by all CNs that wish to communicate with it. In IPv4 networks, due to shortage of IP addresses, MNs are most likely to be assigned temporary IP addresses via DHCP. Thus, the design premise of MIP tends not to be met in practice and the usefulness of MIP is therefore questionable.

11.2.2   Transport Layer Solutions

Transport layer solutions follow the end-to-end principle [16] that makes the Internet such a useful and flexible network: anything that can be done in the end systems should be done there and the network layer should concentrate on efficient routing of packets between the end systems. Since the transport layer is the lowest end-to-end layer in the IP stack, it is a natural candidate for vertical handoff support. Moreover, in the transport layer solutions, no third party other than the endpoints participates in vertical handoff and no modification or addition of network components is required.

Authors in [17] proposed a new set of options for migration of TCP connections from one IP address to another to support mobility. In this protocol, the MN and CN determine a unique token number to identify the TCP connection when the connection is set up, calculated according to the addresses/ports numbers of the peer nodes. The MN will notify the CN of its new IP address/port number along with the token number so that the CN can update the information in the corresponding connection. Similarly, authors of [18] and [19] propose to add a mapping layer between the IP layer and the TCP layer. All address update messages will be intercepted by the mapping layer and the data packets will be replaced with the initial connection information before they are submitted to the upper layer. It also considers the situation that two peer nodes execute handoffs at the same time. During such dual roaming situations, the MN at either end cannot deliver the address update messages to its counterpart since the destination node has also changed its IP address. To overcome this problem, it is proposed in [18,19] to set up a third party subscription/notification server. When the MN roams into a new network, it will send the new address to both the peer node and the notification server. If two MNs cannot find each other after a handoff, they can find the peer node's new address from the notification server so that the connection can be maintained. The main drawback of these approaches is that TCP has been globally deployed so that it is practically infeasible to change it in every end system.

In recent years, SCTP [8] has emerged as a standard transport protocol for the Internet that can be used in place of both TCP and UDP, and is poised to be widely deployed as the preferred transport protocol for the next generation Internet. The multi-homing, multi-stream and partially-reliable data delivery features of SCTP are especially attractive for applications that have stringent delay and high reliability requirements. By extending SCTP with

Dynamic Address Reconfiguration [36], the resulting protocol called mobile SCTP (MSCTP) [20] becomes a good choice for supporting node mobility by taking advantage of the multi-homing capability of SCTP. The use of MSCTP to manage vertical handoffs between WLANs and cellular networks was proposed in [21], which considers MNs that are configured with multiple wireless network interfaces and therefore, can acquire multiple IP addresses in association with an SCTP connection in situations where vertical handoffs can potentially take place. The vertical handoff operations can be completed simply by having the MN handshaking with the CN to reconfigure the secondary address associated with the newly preferred network as the primary address, and the current primary address as the secondary address. Using MSCTP to manage vertical handoffs has many advantages, including simpler network architecture, improved throughput and delay performance, and ease of adapting flow/congestion control parameters to the new network during and after vertical handoffs [21].

In [22], an improvement to MSCTP called Sending-buffer Multicast-Aided Retransmission with Fast Retransmission (SMART-FRX) is proposed. This scheme consists of two sub-schemes that perform different functionalities to enhance the MSCTP performance during WLAN to cellular forced vertical handoffs, namely: (1) The Sending-buffer Multicast-Aided Retransmission (SMART) sub-scheme that forces MSCTP to immediately enter into slow-start over the cellular link at the beginning of the vertical handoff period, by multicasting the data buffered in the primary WLAN link on both the cellular and WLAN links; (2) The Fast Retransmission (FRX) sub-scheme that enables MSCTP to recover error losses on the cellular link by retransmitting over the same link, thus avoiding potentially long delays of error-loss retransmissions over the possibly unreachable WLAN link. A new analytical model for SCTP is also proposed in [22] that takes into account the dynamic changes of the congestion window, the round trip time, the slow-start and congestion avoidance processes, and other factors that may affect the SCTP performance during vertical handoff in heterogeneous wireless networks.

In comparison with network layer solutions, transport layer solutions provide end-to-end mobility management without requiring establishment of any new operation entities in networks. However, the TCP-based mobility protocols require the modification of the protocol stacks in all end systems, including mobile nodes and fixed nodes, which is quite difficult to realize. On the other hand, as a new and flexible protocol having both TCP and UDP functionalities, SCTP is envisioned to be widely supported in end systems in the future. So the mobility management capability provided by MSCTP should be an attractive solution for mobility management at the transport layer.

11.2.3   Application Layer Solutions

As a signalling protocol in the application layer, SIP has been widely adopted by IETF and 3GPP to help establish sessions in packet-switched networks [2]. Each session may support communications of multiple data streams among

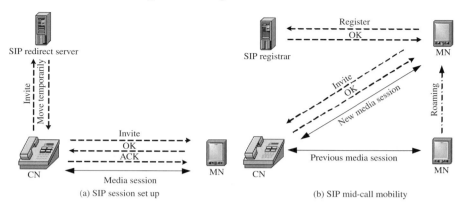

FIGURE 11.2

SIP signaling for session set up and mobility support.

multiple participants. SIP also affords mobility support by allowing a new connection to be set up in the middle of a session. It is generally envisioned that the SIP will become the main signalling protocol for NG wireless networks.

In SIP, the MN is identified with a unique logical SIP address in the format of an e-mail address, which is also called a "SIP URI"; e.g., node.name@ip. domain. When the MN roams into a foreign domain, it obtains a new contact address with the domain name of the foreign domain. The MN also registers the new contact address with the registration server in the home domain through the SIP proxy. Figure 11.2a shows the procedure of setting up a new session from a CN to MN B. In this case, the CN sends an "INVITE" message via its SIP proxy to the home network of the MN. The address of the home network can be figured out from the public SIP address of MN B. This request is captured by the SIP proxy in the MN's home network. After checking the registration server, the proxy in the MN's home network determines that MN B is currently visiting a foreign network. Then, the SIP proxy returns a "moved temporarily" message to the CN with the current contact address of MN B attached. The CN's SIP proxy re-sends the "INVITE" message MN B directly using the new contact address. The MN returns an "OK" message and negotiates session parameters with the CN. The CN completes the negotiation by sending an "ACK" message to confirm that it accepts the modification. When the negotiation is done, the session has been directly set up between the CN and MN B. As shown in Figure 11.2b, if MN B roams to a new domain in the middle of the session, it will notify the CN of its new contact address and the updated session description via another "INVITE" message. The CN then re-establishes the session with MN B at the new location.

In contrast with the mobility solutions at other layers, SIP can be easily deployed in the network without requiring the modification of network entities or end-user protocol stacks. Moreover, SIP can be easily extended or enhanced due to its operation at the application layer and the use of text-based signalling messages. Having been accepted as a standard protocol in both the Internet and public telecommunication networks, SIP is considered as an attractive candidate to support mobility in NG networks.

Despite the advantages discussed above, SIP-based handoff schemes can suffer from long handoff delays due to its operation at the highest layer. The resulting handoff delays have been shown to be unacceptable for supporting real-time multimedia services [23]; consequently many schemes have been proposed to deal with this problem. In hierarchical mobile SIP (HMSIP) [24], a global network proxy is set up for each domain to manage the roaming inside the domain. When an MN roams between different sub-nets of the domain, its registration messages are sent to the global network proxy instead of its home proxy. The global network proxy will check all incoming packets and redirect them to the current address of the MN. In this way, the signalling messages for intra-domain roaming are localized in the foreign network, and the handoff delay will thereby be greatly shortened.

HMSIP only reduces delays of handoffs between sub-nets of the same domain, which can cooperate easily due to homogeneous underlying technologies. However, handoffs between different domains are more difficult to address for the following reasons: First, it is difficult to localize signalling messages of inter-domain handoffs since it is commercially impractical to set up intermediate nodes between heterogeneous networks. Second, the above scheme reduces the handoff delay by shortening the registration period. However, inter-domain handoffs involve some additional procedures such as "authenticating to the new domain." The handoff delays could remain unacceptable even if the registration period is very short.

Several schemes have also been proposed to reduce the delays of inter-domain handoffs. In [25], the MN establishes security associations (SAs) with all neighboring domains whenever a session is set up so that neighboring networks can obtain the MN's authentication information in advance. When the MN roams into one of the neighboring networks, the authentication can be processed locally so that the handoff delay can be reduced. A fast handoff method is proposed in [26]. In this scheme, the MN reports the new contact address to both the CN and the proxy in the old domain when it registers in the new domain. The proxy in the old domain then sets up a temporary session between the old proxy and the MN to forward the received packets; this session is terminated when the MN can reliably receive packets directly from the CN.

11.3   Seamless Vertical Handoff Support Using SIP

According to the above discussions, SIP-based application layer solutions for handoff support have some overwhelming advantages over other protocols. However, the potentially long handoff delays also prevent them from becoming well-rounded mobility management solutions. Although several SIP-based protocols have been proposed to reduce handoff delays in different situations, all of them have certain limitations. HMSIP [24] is designed to reduce intra-domain handoff delays only. Although the methods in [25] and [26] are able to shorten the handoff delay in inter-domain roaming, they both require modifications to SIP operations in the terminals and networks. The method in [25] requires modifications to the existing authentication systems of all domains and may also incur heavy signalling costs to distribute the SA messages. The method in [26] requires all base stations (BSs) in the networks to be equipped with the Back-to-Back User Agents (B2BUA), which may not be preferred by some operators. Furthermore, none of these protocols can reduce handoff delays to the very small values required for seamless handoffs.

In this section, we introduce a SIP-based seamless-handoff scheme (S-SIP) [27] to support seamless inter-domain roaming. Basically, this scheme employs a "make-before-break" handoff procedure to provide seamless handoff management. It does not require any modifications to or addition of network entities, so that it easy to implement. Moreover, we also propose a suitable handoff triggering method that takes advantage of current location tracking technologies. Simulations results demonstrate that the proposed architecture performs efficiently.

11.3.1   S-SIP Handoff Scheme

The mechanism of S-SIP utilizes the existing multi-party conference capability of SIP to set up a new handoff connection before breaking the old connection. Here we take the same network environment discussed above as an example. As shown in Figure 11.3, the MN and CN first set up a session according to the normal SIP handshake procedure and they know the SIP address and contact address of each other. During the communication session, the MN detects that it is going to move out of the current network area or it finds a better wireless network, so it decides to handoff to another network. It first authenticates with the new network and acquires IP/SIP addresses through the required protocols. Afterwards, the MN will send an "INVITE" message with a special "JOIN" header to the CN via the new interface. Here the "JOIN" header contains all the information relevant to the ongoing call, such as the call ID and tags of the previous connection [28]. Such information enables the CN to determine that this new connection wants to join the related ongoing connection (note that the CN may have multiple ongoing connections, especially if it is a server). After some negotiation procedures, another connection is established between the MN and the CN via the new contact address. Note that these two connections follow different paths. The above procedure is normally employed to set up a conference session such that after receiving the "JOIN" message, the CN will send out packets or also forward received packets to both addresses. Since the MN's original SIP URI (note that it is different from the contact address) is also involved in all SIP messages, the CN will know that the new participant is logically the same as the previous peer node because they have the same SIP address. Therefore, the CN will only send packets to both interfaces but it will not send packets received from one interface to the other. At this time, the MN and the CN are communicating through

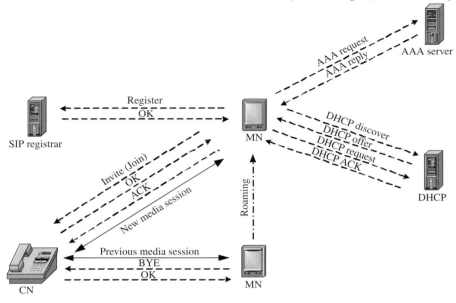

FIGURE 11.3
S-SIP signaling for mobility support.

two connections concurrently, and duplicate packets received by the MN are discarded. After updating the home registration server with the new contact address, the MN will send out a "BYE" message to the CN to terminate the connection via the old interface.

11.3.2 Handoff Period

The handoff period is the period between the handoff initiation and the handoff completion. For hard handoffs, the connections involved will be interrupted during such periods, whereas data packets continue to reach the end nodes for soft handoffs. Let $D_{handoff}$ be the handoff delay and $D_{A,B}$ be the delay of messages transmitted between nodes A and B. We have:

$$D_{handoff} = 5D_{MN,CN} + 2D_{MN,AAA} + 4D_{MN,DHCP} \qquad (11.1)$$

Generally, there are four types of delays: queuing delay, processing delay, transmission delay, and propagation delay. The propagation delay over a wireless access network is very small and can be neglected. Basically, the delays incurred by inter-domain messages involve: (a) processing and queuing delays in sending terminals, (b) transmission delays over wireless channels, (c) processing and queuing delays in the BSs or access points (APs), (d) delays in the Internet, and (e) queuing delays in receiving terminals. Usually, the processing delay depends on the workload and the serving rate of the node/server.

The queuing delay is related to the load of the terminal/server and can be determined based on a queuing model. The transmission delay over a wireless channel is determined by the size of the transmitted message as well as the bandwidth (data rate) of the wireless channel. The delay over the Internet is mostly caused by congestion over the Internet. For intra-domain messages, the delay on the Internet should be replaced by the transmission delay between the BS and the local server, which can be assumed to be negligible. Therefore, we can get:

$$D_{MN,CN} = P_{MN} + Q_{MN} + \frac{L_{SIP}}{B_{wireless}} + P_{BS} + Q_{BS} + \Delta I + Q_{CN} \qquad (11.2)$$

$$D_{MN,AAA/DHCP} = P_{MN} + Q_{MN} + \frac{L_{AAA/DHCP}}{B_{wireless}} + P_{BS} + Q_{BS}$$
$$+ P_{AAA/DHCP} + Q_{AAA/DHCP} \qquad (11.3)$$

where $B_{wireless}$ is the bandwidth of the wireless links and P_i and Q_i respectively denote the processing delay and queuing delay in node i. L_{SIP}, L_{AAA} and L_{DHCP} indicate the length of SIP messages, authentication messages, and DHCP application messages, respectively. Similarly, we can also get the value of $D_{CN,MN}$ and $D_{AAA/DHCP,MN}$. Substituting (11.2) and (11.3) into (11.1), we have:

$$D_{Handoff} = \frac{5L_{SIP} + 2L_{AAA} + 4L_{DHCP}}{B_{wireless}} + 11P_{BS} + 6P_{MN} + 2P_{CN}$$
$$+ P_{AAA} + 2P_{DHCP} + 5\Delta I + 11Q_{MN} + 11Q_{BS} + 5Q_{CN}$$
$$+ 2Q_{AAA} + 4Q_{DHCP} \qquad (11.4)$$

11.3.3   Handoff Initiation Based on User Mobility

An active MN should attach to the network offering the best service. However, when it moves out of the radio coverage of the current domain, it should switch to a less desirable access network to maintain communications connectivity. Ideally, a handoff should complete right after the MN moves out of the current domain, which means that the best handoff initiation time is $D_{handoff}$ before it moves out of current network coverage area. The remaining dwelling period of the MN in the current network can be predicted by two factors: the coverage area C of the network and mobility pattern of the MN. We define C as the area where the RSS of the network is strong enough to be detected by the MN's radio. It may not be exactly circular due to refraction and reflection of radio signals around/by obstacles such as buildings. With the advancement of radio tracking technologies, it is very likely that a future-generation MN will be equipped with a global positioning system (GPS) [29] receiver that enables the MN to determine its location with high accuracy. Therefore the network coverage area can be easily determined by having MNs report their positions to the network whenever the RSS at the MN falls below a pre-defined threshold. As the radio propagation environment varies over time with such factors as weather conditions and human activities, determination of network coverage area should be an ongoing process.

The movement of an MN can be predicted based on past location information. Assume that the MN tracks its location every ΔT time and the past h samples of the recorded positions are $(L_{x,k}, L_{y,k})$, where $k \in [0, h-1]$ and $L_{x,k}$ and $L_{y,k}$ is the position of the MN in the X and Y directions at the k-th time unit. As the handoff only lasts for several seconds, we can consider the movement trajectory of the MN during this period as a straight line. Denoting the movement direction and velocity of the MN as θ and v, we can get $L_x = L_{x,0} + vt \cos \theta$ and $L_y = L_{y,0} + vt \sin \theta$. We can estimate the value of v and θ by linear regression as:

$$v = \frac{\sum_{k=0}^{k=h-2} w_k \sqrt{(L_{x,k} - L_{x,k+1})^2 + (L_{y,k} - L_{y,k+1})^2}}{\sum_{k=0}^{k=h-2} w_k} \tag{11.5}$$

$$\theta = argtan((\phi' W \phi)^{-1}(\phi' W Y)) \tag{11.6}$$

Here w_k is the weight on the past k-th time unit and $w_{k-1} > w_k > 0$. $\phi = (L_{x,0} - L_{x,1}, L_{x,1,...}, L_{x,h-2} - L_{x,h-1})'$, $Y = (L_{y,0} - L_{y,1}, L_{y,1,...}, L_{y,h-2} - L_{y,h-1})'$ and $W = diag(w_0, \ldots, w_{h-2})$.

As the positions are sensed at regular intervals, the MN should initiate the handoff when it is leaving the access network within $\Delta T + D_{handoff}$ time units. Since the prediction may not always be correct, if the MN leaves the access network earlier than the estimated time, packets will be lost before the handoff is complete. If it leaves later than the estimated time, the overall service will be unnecessarily degraded. The balance depends on the traffic type and user's preference. Consequently, we suggest adding an adjustable factor in the expression of the remaining time threshold (i.e., it becomes $\mu(\Delta T + D_{handoff})$, where the value of μ is chosen by the user).

11.3.4   Simulations Results and Discussions

Figure 11.4 shows the delay performance for two vertical handoffs managed by S-SIP and SIP individually. Here the Y axis shows the sequence numbers of packets received at the time given by the X axis. In this Figure, the MN leaves domain A at time 4 s. The bandwidth of domains A and B are 1.5 Mbps and 1 Mbps, respectively. From the Figure, we can see that the general SIP-based hard-handoff scheme causes a handoff delay of around 1.2 s, where the old connection breaks before the new connection is set up when the MN attaches to domain B after it loses the signal from domain A. The handoff delay of the general SIP scheme exceeds 1 s, which is unacceptable to real-time multimedia traffic. Comparatively, it is evident that the proposed S-SIP scheme supports seamless roaming between domains. At time 2.65 s, the MN predicts a future movement out of the current domain. So it attaches to domain B, obtains a new domain/IP address and joins the session with the new address at about time 3.85 s. Once the MN receives the packets through the new

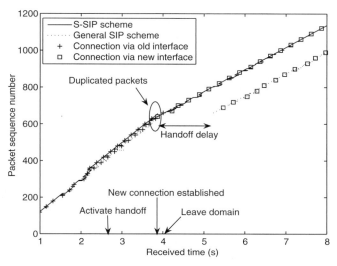

FIGURE 11.4
Handoff events.

interface, the MN terminates the old connection and leaves domain A at time 4 s. Virtually no packets are lost during this handoff procedure.

11.4 Efficient Transport of SIP Traffic over SCTP

SIP is expected to be the dominant signalling protocol for multimedia sessions over both wireless and wired networks in the future. Efficient transport of SIP traffic becomes increasingly important. Currently, SIP traffic is mainly transported over UDP due to latency constraints. For reliability, SIP employs its own retransmission mechanisms. Recent study [30] on transporting SIP traffic over SCTP shows some performance improvements when SIP traffic is heavy. However, transporting SIP over SCTP may not improve the performance when the traffic is light. Previous work suggested that SCTP is not recommended to be used between a SIP user agent (UA) and a SIP server due to the low traffic load, although SCTP gives other benefits, such as fast handoff over heterogeneous mobile networks [21,22]. In this section, we propose a new scheme for transporting SIP over partially-reliable SCTP (PR-SCTP) [31], which chooses the most effective retransmission mechanism between the application and transport layer protocols depending on the traffic condition, thus achieving performance improvements no matter whether the network traffic load is high or low [32]. We compare the performance of SIP over UDP, SIP over SCTP, and SIP over PR-SCTP by simulations.

11.4.1   SIP and SCTP Retransmission Mechanisms

SIP recovers lost messages by retransmitting them in the application layer. SIP messages are classified as request and response messages. A SIP request message is considered by the sender to be lost if the corresponding response message is not received within a specified time period. SIP increases the retransmission timeout period exponentially in successive retransmissions until the expected response for the message has been received or the number of retransmissions has reached the maximum value of six. The initial SIP timeout period T1 is normally set to 500 ms, based on an estimate of the round-trip-time (RTT) of the network. According to the RTT in each specific situation, the value of T1 may be adjusted.

SCTP has similar flow and congestion control mechanisms to those of TCP, although there are some differences between them. SCTP detects packet losses by checking the gaps between blocks in its selective acknowledgment (SACK) chunks. Compared to TCP, SCTP allows a larger number of SACK blocks to be reported in each SACK. SCTP considers a data packet as missing if it is reported missing in four SACK chunks, thus triggering the fast retransmission algorithm. SCTP also retransmits a lost packet when it has not been acknowledged within a retransmission timeout period, and puts the SCTP association into slow start. In order not to generate too much SACK traffic, SCTP may apply the delayed acknowledgment algorithm specified in [33], which generates a SACK after the second unacknowledged data chunk has been received. Generally, a SACK is bundled with data chunks in an SCTP packet returned to the other endpoint. If no SCTP packet is available for the SACK bundling within a certain time limit, a dedicated acknowledgment packet may be sent. The recommended delay limit is 200 ms but should not be more than 500 ms. In addition, the recommended retransmission timeout of SCTP is within the range of 1 second to 60 seconds.

From the above descriptions of the retransmission mechanisms of SIP and SCTP, we can see that SCTP is more effective in loss recovery than SIP when SIP traffic is high, due to fast retransmissions. Thus, SIP over SCTP can yield a good performance. However, when SIP traffic is low, there are less opportunities for SACK bundling and SCTP has to wait between 200 ms and 500 ms to send a dedicated SACK. If the time taken for SCTP error recovery exceeds the SIP retransmission timeout, then redundant SIP retransmissions will be generated. In this case, SIP over UDP, which relies solely on SIP retransmissions to recover errors, can be more effective than SIP over SCTP.

Although SCTP was originally designed to provide transport services for SS7 signalling messages over IP networks, it is useful for the transport of other signalling messages, such as SIP messages. However, the following problems exist in sending SIP messages over SCTP: (1) Transporting obsolete SIP messages may waste network resources. Once a SIP message is included in a SCTP data packet, SCTP will keep resending the packet until it is acknowledged. Since SIP messages are only useful within a period of time, the retransmissions of obsolete SIP messages by SCTP is undesirable for the networks. (2) It gives better network performance than SIP over UDP only when SIP traffic is high.

When SIP traffic is low, SIP over UDP, which recovers errors at the application layer only, can be more effective as discussed above. (3) SCTP transports all SIP messages reliably regardless of needs. SIP provisional response messages, like "180 Ringing" and "100 Trying," are informative and not required to be transported reliably. In SIP over UDP, these messages are not retransmitted by SIP. However, SCTP only provides a reliable service. It treats all data packets in the same way and transports all of them reliably. Our proposed scheme of SIP over PR-SCTP can overcome the above limitations.

11.4.2   Proposed Scheme of SIP over PR-SCTP

By engaging an appropriate PR-SCTP service according to the reliability requirement of a SIP message, SIP messages can be transported over PR-SCTP either with partial reliability or unreliably as appropriate. Moreover, for efficient recovery of lost packets, our proposed scheme chooses the most effective retransmission mechanisms between SIP and PR-SCTP, thus achieving a better performance regardless of whether SIP traffic is high or low.

Our SIP over PR-SCTP scheme adopts the "timed reliability" service of PR-SCTP. Instead of setting the lifetime of a PR-SCTP data packet to a constant value, the lifetime is changed according to the application layer timeout values of the SIP message transported. Besides applying the partial reliable transport service of PR-SCTP within the lifetime of the SIP message, our scheme uses SIP to retransmit the lost message as well. This allows our scheme to benefit from both the flexibility of PR-SCTP and the retransmission mechanism of SIP. When a SIP message is first transmitted, the lifetime of the data packet carrying this SIP message is set to T1, the initial SIP retransmission timeout. PR-SCTP attempts to provide reliable delivery of the data packet within the lifetime using the same error recovery mechanism as in SCTP, and drops the packet if unsuccessful within the lifetime. At this time the SIP timeout period is also reached, and SIP takes over by retransmitting the SIP message dropped by the transport layer. As SIP increases its retransmission timeout to $2 \times T1$, accordingly, the lifetime of the new data packet in PR-SCTP carrying the retransmitted SIP message is also increased to $2 \times T1$. This process continues until the message has been sent successfully or the sixth retransmission of this SIP message has failed. If all the retransmissions have failed, an error message would be reported to the application layer to indicate transmission failure of the SIP message. PR-SCTP sets the lifetime of the data packets containing these messages to a constant when SIP messages that do not need to be transmitted are sent. These messages are then sent once and not retransmitted by PR-SCTP.

In the proposed scheme, the application can require either an ordered or unordered transport service. The unordered service avoids the head-of-line (HOL) queue blocking problem for messages in the same stream. Unordered service is used to send SIP provisional messages, since the loss of these message should not block the delivery of other SIP messages. Other SIP messages can be sent either ordered or unordered. If multiple SIP sessions are mapped into one SCTP stream, unordered service is recommended since it prevents the SIP messages of one session from blocking other SIP messages belonging to different SIP sessions. The schemes for mapping SIP sessions into SCTP streams are discussed in [34]. The lifetime of the PR-SCTP data packets containing the SIP message is set to the SIP retransmission timeout. This will result in the application and transport layers discarding the SIP messages at the same time, thus avoiding the transport of stale messages in the networks. Network resources are therefore used more efficiently.

Comparing SIP to traditional SCTP or SIP over UDP, our proposed SIP over PR-SCTP is more effective in retransmitting lost SIP messages. According to [34], loss recovery is faster using SCTP than SIP when SIP traffic is high. Therefore, when SIP traffic is low, loss recovery at the SIP layer is a better choice. The proposed scheme coordinates the loss recovery at both the RP-SCTP and SIP layers so that the faster PR-SCTP loss recovery takes over if SIP loss recovery fails. This coordination also prevents duplicating retransmissions over both layers. The effect of this coordination is that when SIP traffic is low, loss recoveries are mainly performed by SIP, whereas when SIP traffic is high, loss recoveries are mainly performed by PR-SCTP. Thus, the proposed SIP over PR-SCTP scheme is able to employ the most appropriate loss recovery mechanism in the networks.

11.4.3   Simulation Results and Discussions

Our simulation experiments for SIP signalling are designed to study the performance of the proposed SIP over PR-SCTP scheme. In our simulations, SIP messages with an exponentially distributed inter-arrival time are passed to the transport layer. Figure 11.5 shows a simple network topology that is used to evaluate the performance differences among different schemes. Two SIP agents are connected through two routers. The overall link propagation delay is set to 45 ms, which approximates the propagation delay between two

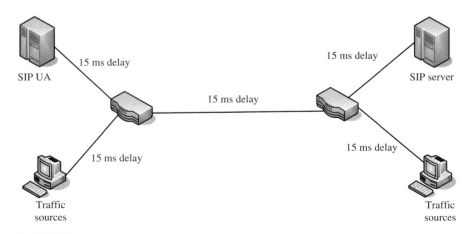

FIGURE 11.5
A simple network topology in simulations.

entities located in North America and Europe. The capacities of the links in the network are chosen so that the only bottleneck in the network is the link between two routers. We simulate scenarios where SIP traffic is low, which is the case between a SIP UA and a SIP server. The two other traffic sources are provided as network background traffic.

The traffic between SIP UA and SIP server has an exponential distribution in its inter-arrival time. Background traffic is generated according to a deterministic rate. In the scenarios of low background traffic, this data rate is set at 32 Kb/s, while in the scenarios of high background traffic, it is increased to 4,000 Kb/s. In the simulations considered in this section, the link capacity between two routers is set to 4.5 Mb/s. we can also see that our proposed dynamic adjustment methods for T1 improve the network performance as the link error rate increases. We dynamically adjust T1 to improve the network performance as the link error rate increases using one of the following methods: method 1 adjusts T1 according to the data packet loss ratio; method 2 adjusts T1 according to the RTT.

Since the transmission delay between two endpoints is a constant (set to 45 ms) that cannot be reduced, we compared the average delays, excluding the 45 ms transmission delay, as shown in Figure 11.6. From this Figure, we can see that, compared to the transport method using SCTP, this average delay is reduced by 57% using PR-SCTP when the link error is 0.5%, and by 83% when the link error is 1%. If the proposed dynamic T1 adjustment methods are applied, the average packet delay is further reduced. Comparing the two dynamic T1 adjustment methods, we can see that the method based on the RTT is more effective than the method based on the packet loss ratio when SIP traffic is low. In the case of low SIP traffic, the real RTT is likely to be less than 500 ms. In the method based on the RTT, T1 can be adjusted to be

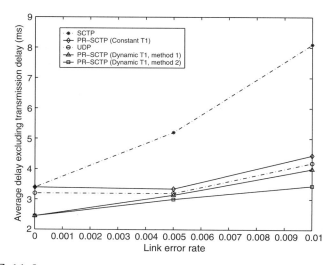

FIGURE 11.6

Comparison of average packet delay in different schemes.

any value between 500 ms and 250 ms, which is compliant with the changes of the real RTT. In the method based on the packet loss ratio, if the packet loss ratio is less than the UPPER LIMIT, which is 10% in our simulation, T1 is set to be 500 ms. When SIP traffic is low, data loss rate is low, thus the T1 is likely to remain at 500 ms. Therefore, in the case of low SIP traffic, the dynamic T1 adjustment method according to the RTT is more effective than the method according to the packet loss ratio.

11.5   Conclusions

One of the main challenges for NG wireless networks to provide "Always Best Connected" services is how to manage vertical handoffs seamlessly. Compared with horizontal handoffs, vertical handoffs require technology-independent mobility management schemes and involve more extensive procedures, thereby leading to longer delays. Vertical handoffs schemes proposed in the literature can be classified according to the layer they operate on. Network layer schemes require additional mobility supporting network entities that may hinder their wide deployment. TCP-based transport layer schemes have the drawback that they require modifications to the TCP operations in all peer nodes. MSCTP proves to be a good candidate for supporting vertical handoffs when SCTP becomes widely adopted in the future. Due to the operation at the highest layer, SIP-based schemes are easy to deploy but also incur long handoff delays. To solve this problem, we have proposed a S-SIP scheme that enables the MN to simultaneously establish two ongoing connections with the CN and realize "seamless-handoff." Moreover, with the help of tracking technologies, the MN can predict potential handoffs and establish the new connection before the current connection losses its signal. Simulation results have been presented to show that the proposed scheme can support seamless inter-domain roaming. In addition, current SIP signalling traffic is transported over UDP with loss recovery performed at the SIP layer using its own retransmission mechanism. This may not work well during vertical handoffs when the wireless links may be subject to a high packet loss rate. Therefore, we have also proposed a new scheme for transporting SIP over PR-SCTP, which chooses the most effective retransmission mechanism between the transport and application layers. Simulation results have been presented to show that the proposed protocol yields excellent performance under various network conditions.

References

[1] Gustafsson, E., and A. Jonsson. 2003. Always best connected. *IEEE Wireless Communication Magazine* 10(1): 49–55.

[2] 3GPP. 2005. 3rd Generation Partnership Project; Technical Specification Group Services and System Aspects; Telecommunication management; Charging management; Charging principles (Release 5), TS 32.200 version 5.8.0, March.

[3] Handley, M., Schulzrinner, H., and Schooler, E. 1999. SIP: Session initiation protocol. *IETF RFC* 2543, March.

[4] 3GPP 2005. 3GPP Systems to Wireless Local Area Network (WLAN) Interworking; System Description (Release 6), 3GPP TS 23.234, March.

[5] Ahmavaara, K., H. Haverinen, and R. Pichna. 2003. Interworking architecture between 3GPP and WLAN systems. *IEEE Communication Magazine* 41(11): 74–81.

[6] Song, J., S. Lee, and D. Cho. 2003. Hybrid coupling scheme for UMTS and wireless LAN interworking. In *Proceedings IEEE Vehicular Technology Conference*, Orlando, FL, 2247–51. October.

[7] Perkins, C.E., 2002. IP mobility support for IPv4. *RFC 3220*. January 41.

[8] Stewart R., and Q. Xie. 2001. Stream control transport protocol. *IETF RFC2960*. October.

[9] Haverinen, H., and J. Malinen. 2001. Mobile IP Regional Paging. Internet draft, IETF, draft-haverinen-mobileip-reg-paging-00.txt. September.

[10] Perkins, C.E., and D.B. Johnson. 2001. Route Optimization in Mobile IP. Internet draft, IETF, draft-ietf-mobileip-optim-11.txt, September.

[11] Hsieh, R., Z.G. Zhou, and A. Seneviratne. 2003. S-MIP: A seamless handoff architecture for mobile IP. In *Proceedings IEEE INFOCOM*. San Francisco, CA, 1774–84, March–April.

[12] Gustaffson, E., A. Jonsson, and C. Perkins. 2001. "Mobile IP Regional Registration", draftietf-mobileip-reg-tunnel-05.txt, September.

[13] Misra, A., S. Das, A. Dutta, A. Mcauley, and S. K. Das. 2002. IDMP-based fast handoffs and paging in IP-based 4G mobile networks. *IEEE Communication Magazine* 40(3): 138–45. March.

[14] Soliman, H., C. Castelluccia, K. El-Malki, and L. Bellier. 2004. Hierarchical Mobile IPv6 Mobility Management (HMIPv6), Internet draft, IETF, draft-ietf-mipshop-hmipv6-02.txt. June.

[15] Perkins, C.E., and K. Wang. 1999. Optimized smooth handoffs in mobile IP. In *Proceedings IEEE Symposium on Computers and Communications*, Red Sea, Egypt, 340–46. July.

[16] Saltzer, J.H., D.P. Reed, and D.D. Clark. 1984. End-to-end arguments in system design. In *ACM Transactions on Computer Sysems.* 2(4): 278–88.

[17] Snoeren, A.C., and H. Balakrishnan. 2000. An end-to-end approach to host mobility. In *Proceedings of International Conference on Mobile Computing and Networking*, Boston, MA. 155–66. September.

[18] Zhang, Q., C. Guo, Z. Guo, and W. Zhu. 2003. Efficient mobility management for vertical handoff between WWAN and WLAN. *IEEE Communication Magazine* 41(11): 102–8.

[19] Guo, C., Z. Guo, Q. Zhang, and W. Zhu. 2004. A seamless and proactive end-to-end mobility solution for roaming across heterogeneous wireless networks. *IEEE JSAC* 22(5): 834–48.

[20] Koh, S.J., M.J. Lee, M. Riegel, L. Ma, and M. Tuexen. 2004. Mobile SCTP for Transport Layer Mobility. draft-sjkoh-sctp-mobility-04.txt, June, work in progress.

[21] Ma, L., F. Yu, V.C.M. Leung, and T. Randhawa. 2004. A new method to support UMTS/WLAN vertical handover using SCTP. *IEEE Wireless Communication Magazine* 11(4): 44–51.

[22] Ma, L., F.R. Yu, and V.C.M. Leung. 2007. Performance improvements of mobile SCTP in integrated heterogeneous wireless networks. To appear In *IEEE Trans. Wireless Comm.* 6(10): 3567–77.

[23] Wu, W., N. Banerjee, K. Basu, and S.K. Das. 2005. SIP-based vertical handoff between WWANs and WLANs. *IEEE Wireless Communications* 12(3): 66–72.

[24] Vali, D., S. Paskalis, A. Kaloxylos, and L. Merakos. 2003. An efficient micro-mobility solution for SIP networks. *Proceedings IEEE Global Communications Conference'03*. San Francisco, CA, 3088–92, December.

[25] Kwon, T.T., M. Gerla, S. Das, and S. Das. 2002. Mobility management for VoIP service: Mobile IP vs. SIP. *IEEE Wireless Communication Magazine* 9(5): 66–75.

[26] Banerjee, N., A. Acharya, and S.K. Das. 2006. Seamless SIP-based mobility for multimedia applications. *IEEE Network Magazine* 20(2): 6–13.

[27] Zhang, J., H.C.B. Chan and V.C.M. Leung. 2007. A SIP-based handoff scheme for heterogeneous mobile networks. In *Proceedings of IEEE Wireless Communication and Networking Conference (WCNC'07)*, Hong Kong, China. 3946–50, March.

[28] Mahy, R., and D. Petrie. 2004. The Session Initiation Protocol (SIP) 'Join' Header. *IETF RFC 3911*. October.

[29] Soh, W., and H.S. Kim. 2001. Dynamic guard bandwidth scheme for wireless broadband networks. In *Proceedings of IEEE Conference on Computer Communications (INFOCOM'01)*, Anchorage, AK. 572–81. April.

[30] Camarillo, G., R. Kantola, and H. Schulzrinne. 2003. Evaluation of transport protocols for the session initiation protocol. *IEEE Network Magazine* 17(5): 40–6. September–October.

[31] Stewart, R., M. Ramalho, Q. Xie, M. Tuexen, and P. Conrad. 2004. SCTP Partial Reliability extension. *IETF RFC 3758*. May.

[32] Wang, X., and V.C.M. Leung. 2005. Applying PR-SCTP to transport SIP traffic. In *Proceedings IEEE Global Communications Conference'05*. St. Louis, MO. November.

[33] Allman, M., V. Paxson, and W. Stevens. 1999. TCP congestion control, *IETF RFC 2581*. April.

[34] Marco, G., D. Vito, and M. Longo. 2003. SCTP as a transport for SIP: a case study. In *Proceedings of World Multiconference on Systemics, Cybernatics and Informatics*, Orlando, FL. 284–89.

[35] Stewart, R., Q. Xie, M. Tuexen, S. Maruyama and M. Kozuka. 2007. "Stream Control Transmission Protocol (SCTP) dynamic address reconfiguration". *IETF RFC 5061*. September.

12

NAT Traversal

Mario Baldi, Fulvio Risso, and Livio Torrero
Technical University of Turin

CONTENTS

12.1 Introduction 277
12.2 NAT Behavior 278
12.3 Problem Statement 282
12.4 Generic NAT Traversal Techniques 284
12.5 SIP-Specific NAT Traversal Techniques 292
12.6 ALEX: Ensuring End-to-End Connectivity for Both SIP and
      Media Flows across NATs 302
     References 305

12.1   Introduction

The wide diffusion of network address translators (NATs) through the Internet is the result of the increasing demand for hosts capable of communicating across the network although they cannot have routable network addresses due to their shortage in IPv4. The increasing number of mobile devices with wi-fi connectivity has made the demand grow further. The idea behind the NAT is that a restricted minority of the hosts belonging to the same network need to effectively communicate with hosts placed in other networks at any point in time: so assigning a public routable network address to each host of the network is a waste. On the contrary, a NAT dynamically "assigns" a public network address (and port) only to hosts (and applications) at the moment in which they are actually communicating with other hosts in the Internet. By doing this, the number of public network addresses actually required for a subnetwork is reduced (as it is much lower than the number of hosts connected to such a subnetwork), thus rationalizing this precious resource.

Furthermore, NATs are considered by network administrators as a (weak) manner to provide network security, hiding the real topology of the network behind a NAT: since public network addresses (and ports) are assigned dynamically, the same host (and application) can be bound to the different network addresses (and ports) in at subsequent times; this makes it difficult for an external attacker to discover that the internal host is the same. Moreover, in the most common NAT operation and deployment modes, a public address is assigned only to hosts initiating a packet exchange; consequently hosts behind a NAT that are not engaged in communications with hosts over the Internet cannot receive packets sent from possible attackers somewhere in the Internet.

However, the presence of NATs between networks is not transparent to applications: NATs have been designed around the client–server approach, considering the server placed in the Internet. By doing this, it is not trivial for a server placed behind a NAT to be reachable from clients over the Internet. This is due to two main reasons: first, the client may not know exactly which public network address (and port) has been assigned to the server (and its service) by the NAT when starting the session; in addition the NAT may be configured to work with an operational mode that blocks sessions started by hosts in other networks. Among applications somehow impaired by NATs, the most important are the peer-to-peer ones: since the Session Initiation Protocol (SIP) defines the user agents (UAs) as entities made up of both a client and a server (each UA can start or receive a media session), SIP UAs fall in this category. The scope of this chapter is to describe in detail how NATs work and to present techniques developed to overcome the connectivity limitations introduced by NATs, known as NAT traversal techniques.

The chapter is structured as follows: the first section introduces definitions concerning NAT behavior, highlighting how a NAT should be configured to make peer-to-peer like applications work. The second section explains how NATs may affect SIP UA's connectivity, thus defining the problem that is the main focus of this chapter. In the third section, the most effective NAT traversal techniques will be introduced, while the fourth section describes how they can be specifically applied to SIP signaling and media sessions establishment. Finally, the last section presents ALEX, an alternative solution that provides dual-stack UA support and NAT traversal, for both signaling and media sessions, through a specific SIP header field.

12.2   NAT Behavior

NATs can be configured by network administrators to operate according to a large variety of possible behaviors, some of which may restrict host connectivity. The BEHAVE IETF working group [1] has defined a set of requirements that have to be satisfied by "application-friendly NATs," i.e., NATs that allow peer-to-peer applications to work properly. Specifically, these requirements refer to hosts communicating using User Datagram Protocol (UDP) as a transport protocol (exactly the way SIP UAs were originally supposed to do). The result of the work by BEHAVE is RFC 4787 [2]. This section, while describing NAT behavior, also analyzes the most relevant requirements defined in the RFC; for detailed information on the requirements not discussed here, please refer to RFC 4787.

12.2.1   Network Address and Port Translation

The network address and port pair uniquely identifying a host and an application running on it, respectively, will be referred in the following as *transport address*. More specifically, the transport address configured on a host behind a NAT will be referred as the *internal transport address*.

A NAT defines a binding between an internal and an external transport address every time a host behind a NAT, .i.e., in the *internal network*, starts sending packets to hosts placed in other networks, called *external networks*. When the packets exit the internal network through the NAT, the source address and port are rewritten, thus replacing the internal transport address with the external one, as exemplified in Figure 12.1.

Host A is placed in an internal network behind NAT A, while host B is placed in an external network, e.g., the open Internet. When an application running on host A starts sending packets to host B, the source address and port address (i.e., the internal transport address) are replaced by an external network address, possibly being assigned to one of the interfaces of the NAT, and an external port, which constitute the external transport address. The NAT stored the binding between the internal transport address and the external one in an internal *binding table*. When host B receives the packets, it believes that the sender has the external network address A_1, which it uses in its replies. The packets sent by host B are received by the NAT, which looks for the target network address and port in its binding table. By doing this, the NAT discovers that this packet is targeted to a specific internal transport address. The NAT rewrites the destination network address and port of the packets and then forwards them in the internal network so that they can reach the proper application running on host A. One widespread application scenario for NAT is when the internal network is deploying private addresses.

NATs can be categorized based on their mapping behavior. The network scenario in Figure 12.2 is deployed in the following to introduce the possible mapping behaviors of a NAT: host A is behind NAT A, which maps the internal transport address (A:a) on the external transport address $(A_1{:}a_1)$ when packets are sent to host B, and to the external transport address $(A_2{:}a_2)$ when packets are sent to host C.

If $(A_1{:}a_1)$ and $(A_2{:}a_2)$ are the same external transport address, the NAT implements an *endpoint independent mapping* behavior. On the contrary if $(A_1{:}a_1)$ differs from $(A_2{:}a_2)$, but $(A_1{:}a_1)$ does not change when sending packets to different ports of host B, an address dependent mapping behavior is adopted; if changing the target port on $B(A_1{:}a_1)$ changes, an address and port dependent mapping behavior is used.

The requirement established in RFC 4787 says that an application-friendly NAT must have an endpoint independent mapping behavior.

12.2.2   Mapping Refresh

Since a NAT temporarily assigns an external transport address to an internal host, the host itself has to refresh this binding by periodically sending

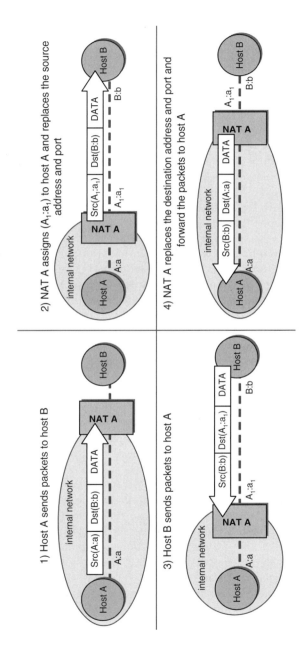

FIGURE 12.1
Generic NAT behavior.

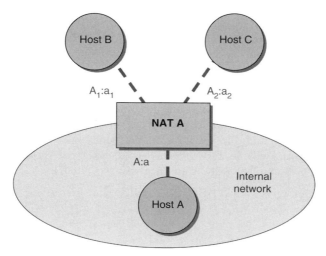

FIGURE 12.2
Host establishing sessions across NAT.

packets to the other host involved in the session, if the hosts are communicating using UDP. This is typically obtained by exchanging keepalive dummy packets between hosts. RFC 4787 states that each binding must stay active without packets traversing the NATs for at least two minutes. However, for specific destination ports in the well-known range, this value may be lower. A suggested default value for the UDP binding timeout is five minutes. Static bindings, to be manually configured by the network administrator, might also be supported.

12.2.3   Filtering Behavior

When an application running on an internal host opens a session through a NAT and is assigned an external transport address, the fact that any other host can send packets to the application by using the external transport address assigned by the NAT depends on the filtering behavior of the NAT. There are three possible filtering behaviors:

- *Endpoint independent filtering behavior*: as soon as the NAT binds an external transport address $(A_1{:}a_1)$ to an internal transport address $(A{:}a)$, the NAT forwards all the packets targeted to $(A{:}a)$, independently from the transport address of the sender. This means that any application running on any external host can communicate with the internal one.

- *Address-dependent filtering behavior*: if an application running on a host behind a NAT using an internal transport address $(A{:}a)$ has been bound to an external transport address $(A_1{:}a_1)$ by the NAT, the NAT will filter out all the packets not targeted to $(A{:}a)$. In addition, packets sent to

(A:a) by an external host will be filtered out if (A:a) has not sent packets to the network address of the sender before (independently of the port where the external sender received the packets).

- *Address- and port-dependent filtering behavior*: if an application running on a host behind a NAT using an internal transport address (A:a) has been bound to an external transport address $(A_1:a_1)$ by the NAT, the NAT will filter all the packets not targeted to (A:a). In addition, an application running on an host outside the NAT using the transport address (X:x) will be allowed to send packets to (A:a) only if packets have been previously sent from (A:a) to (X:x).

The requirement defined by RFC 4787 states that *if the application transparency is most important, it is recommended that NAT have an Endpoint Independent Filtering Behavior. If a more stringent filtering behavior is most important, it is recommended that a NAT have an Address Dependent Filtering Behavior.*

12.2.4 Deterministic Properties

NATs are non-deterministic if they change their filtering or mapping behavior without any configuration change. For example, there are NATs that try to preserve the port when translating the internal transport address to the external one. However, if the NAT finds that the internal port is currently used by another mapping and the NAT cannot use another public network address (because only a limited number of public addresses, possibly one, is available), NAT mapping behavior varies.

RFC 4787 states that a NAT must have a deterministic behavior, i.e., the NAT must not change its behavior at any point in time or under particular conditions.

12.2.5 Hairpinning Behavior

Consider the two communicating hosts A and B in Figure 12.3. Because they are behind the same NAT, they should use their internal transport addresses to communicate. However, since A knows only an external transport address (B_1, b_1) assigned to B by the NAT, it sends packets to B through the NAT using the external endpoint $(B_1, b1)$. This procedure is referred to as hairpinning. RFC 4787 states that a NAT must support hairpinning.

12.3 Problem Statement

SIP is commonly used to establish media sessions among UAs. Each UA wishing to communicate usually performs two basic actions:

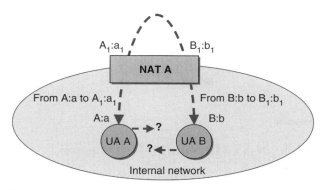

FIGURE 12.3
NAT supporting hairpinning.

- It registers itself in a SIP domain to be reachable by other UAs.

- It then initiates media sessions, possibly creating direct media flows with other participants. In addition, SIP messages are exchanged to manage and eventually terminate the session.

Concerning the registration procedure, the UA sends a REGISTER message that binds the unique address of record (AOR) of the user to a globally routable Universal Resource Identifier (URI), i.e., a URI that uniquely identifies the UA and ensures UA reachability, placed in the Contact header field of the message itself: the URI must ensure reachability because at any time the proxy may have to forward a SIP message to the UA. Typically, this URI includes a transport address rather than a fully qualified name since most UAs are not executed on hosts registered in the Domain Name Service (DNS). When a UA placed behind a NAT registers with an external SIP proxy, the provided URI must contain the external address dynamically assigned by the NAT that shall not change the binding with the corresponding internal address for the whole validity period of the registration. In addition, it must be ensured that incoming messages sent by the proxy are not filtered by the NAT.

When the internal UA attempts to participate in new media session establishment, the problem is further complicated. SIP messages used to establish media sessions must include a globally routable URI that will be used to exchange direct mid-dialog SIP messages and a valid reachable, hence external, transport address for each media flow that will be established: NAT filtering and mapping policies may compromise these media flows. The situation gets even more complex when considering non-deterministic NATs. Currently, SIP connectivity is usually ensured by delivering all mid-dialog SIP messages through intermediate SIP proxies. Each proxy that wishes to stay in the path of the mid-dialog SIP messages simply adds a Record-Route header field containing its URI to the dialog creating messages before forwarding them. The Record-Route header field is defined in RFC 3261 [3], which recommends a moderate usage of this approach to ensure scalability. A stronger effort is required to ensure direct connectivity for media flows; for this reason, NAT traversal for SIP signaling and for media sessions will be discussed separately. However, before doing this, the most-used generic NAT traversal techniques will be introduced.

12.4   Generic NAT Traversal Techniques

NAT traversal techniques are the set of operations required to establish end-to-end connectivity involving hosts behind NATs. More specifically, this section introduces hole punching and relaying methodologies. The first approach is usually preferred when NATs in the packets path satisfy the NAT requirements set by RFC 4787 as discussed previously. The presented techniques are described in the context of applications running over UDP (exactly the way SIP UAs were originally supposed to do). For additional details, please refer to [4].

12.4.1   Hole Punching

Hole punching is a basic technique to ensure end-to-end communication across NATs. If two hosts willing to exchange packets are both behind a NAT, this technique requires the presence of a rendezvous server S that is for both of them in an external network and helps them establish a direct communication channel. Considering a generic scenario where two hosts called A and B wish to communicate directly, first, both hosts register their external and internal transport addresses with the rendezvous server S; the external addresses are devised by S from the received packets or by the hosts using the solutions described later (see Section 12.4.3). Then, if A is the initiator of the direct connection to B, it sends a communication request to S, which in response returns A the internal and the external transport addresses of B and, at the same time, forwards the request to B. Moreover, S sends B the internal and the external transport addresses of A. As soon as both hosts know each other, they *probe* the addresses received from S by attempting to communicate directly using each of them. To understand how the communication is actually established, three main scenarios are considered in the following:

1. Both A and B are placed in the same internal network.

2. A and B are placed in separate internal networks connected by the same external one.

3. A and B are placed behind multiple layers of NATs.

In the first scenario (see Figure 12.4), each host that successfully probes the transport addresses will be successful for the internal addresses as both hosts are in the same internal network, and possibly for the external ones, if the NAT

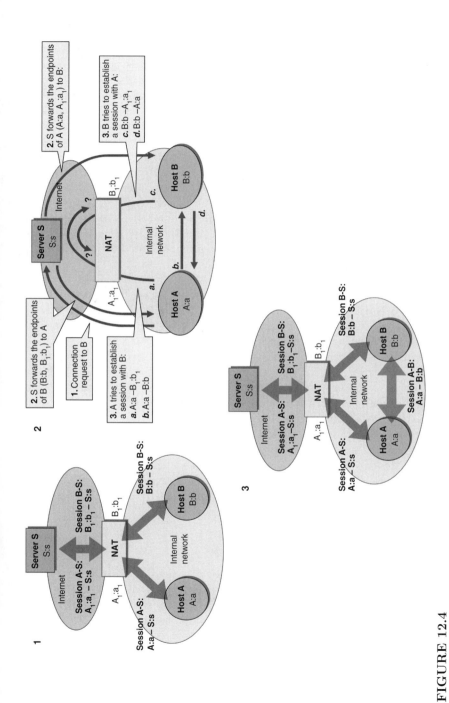

FIGURE 12.4

A NAT traversal scenario showing hosts behind the same NATs.

supports hairpinning. The internal transport addresses will be the preferred choice.

In the second (and most common) scenario (see Figure 12.5), probing of the internal transport addresses fails since the hosts belong to different internal networks. The success of the external transport address probing depends on the timing. If, for example, A starts sending probing packets to B first, when the packets leave the internal network, NAT A will replace the internal transport address of A with an external one. If the NAT implements an endpoint-independent mapping behavior, it will assign A the same external transport address registered with the rendezvous server S. In addition, since the packet is targeted to the external transport address of B, if the NAT implements an address-dependent filtering behavior, future packets coming from host B will not be dropped. If NAT B implements an address-dependent filtering behavior, the probing packets from A will not reach B. However, on the other side, B in turn sends probing packets to A, thus "activating" NAT B to the forwarding of packets going to B from A. Obviously, if both NATs (or at least one of them) implement an endpoint-independent filtering behavior, the communication establishment will be successful as well.

The last scenario involving multiple layers of NAT, as depicted in Figure 12.6, is quite common on the network of many Internet service providers (ISPs) where NAT C is a large ISP NAT, while NAT A and NAT B are small residential NATs. Such a deployment is required because the addressing space of the ISP is private, as well as the addressing space of the customers' networks. It is not unlikely that the customers' address spaces overlap with the addressing space of the ISP and with each other, i.e., external transport addresses assigned by NAT A may collide with the internal ones of NAT B and vice versa. So the most conservative choice would be to have host A and host B communicating using the external transport addresses assigned by NAT C. This requires S to be in the external network of NAT C and NAT C to support hairpinning. Indeed, it is more frequent to have the rendezvous server S placed in the external network of NAT C rather than in the intenal one. If this is the case, host A and host B cannot communicate using the external transport addresses assigned by NAT A and NAT B, since they have no chance to learn these addresses.

12.4.2 Relaying

When discussing the scenarios related to hole punching, we observed that direct connectivity may be impossible if specific types of NAT are on the path. However, if NATs satisfy the requirements defined in RFC 4787, hole punching ensures direct connectivity. Nevertheless, there are still NATs that do not comply with those requirements: for example, NATs with address- and port-dependent mapping behavior or NATs that do not support hairpinning may impair connectivity. Under such conditions, *relays* are needed. Relaying is the only NAT traversal technique that always ensures connectivity, but it comes at the cost of increased latency and reduced scalability.

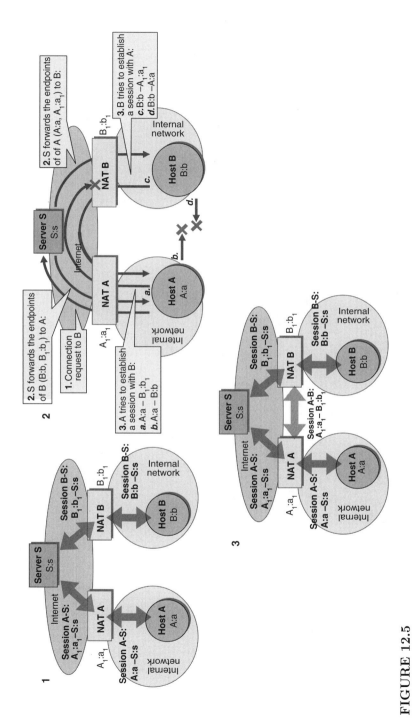

FIGURE 12.5

A NAT traversal scenario including hosts behind different NATs.

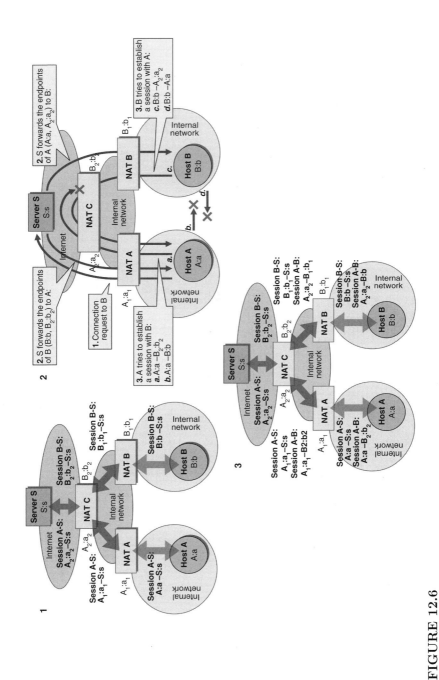

FIGURE 12.6
A NAT traversal scenario including multiple layers of NATs.

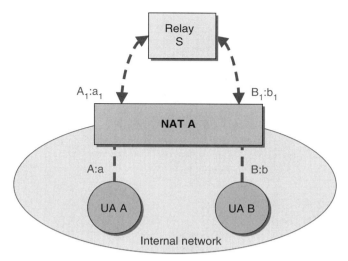

FIGURE 12.7

A scenario including relay node.

The scenario depicted in Figure 12.7 is deployed to explain how relays work. To ensure connectivity, A and B communicate with server S, which will forward packets to the other party. In other words, relaying merges two client server communications into a peer-to-peer one. Relaying is commonly used only as a fallback solution when hole punching fails.

12.4.3 STUN: A Standardized Set of Functionalities to Support NAT Traversal

Session Traversal Utilities for NAT (STUN) is a protocol that provides a set of functionalities used for NAT traversal. STUN is not a complete NAT traversal solution by itself, but must work in conjunction with other protocols. Originally defined in RFC 3489 [5] (now referred as "classic STUN" specification), STUN has been partially redefined by a set of subsequent Internet drafts [6]. Basically, STUN allows a host placed behind NATs to discover its external transport address. In addition, STUN provides a set of extended functionalities, including a keepalive mechanism in order to ensure connectivity for the whole duration of an end-to-end session by preventing NAT bindings from expiring. Several techniques described in the following make use of STUN messages. They are generally referred as "STUN usages."

Figure 12.8 shows a typical STUN usage scenario with the two main STUN entities: the *STUN client* and the *STUN server*. The STUN entities communicate by means of STUN messages consisting of a STUN header and a sequence of zero or more TLV (i.e., type-length-value) encoded attributes. A host behind a NAT willing to discover its external transport address must implement a STUN client and send to the STUN server a STUN binding

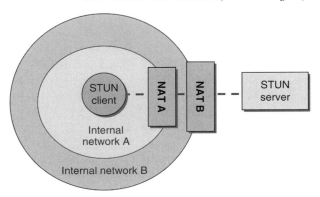

FIGURE 12.8
A typical STUN scenario.

request. The STUN server must be placed in the external network, and possibly on the open Internet in order to be reachable by any host and must be known by any client, e.g., through configuration; the STUN server is typically related to a known DNS entry. The STUN binding request traverses all the NATs up to the STUN server, and because it is generated from the internal network, NATs do not drop it. In the example depicted in Figure 12.8, the transport address is modified twice: a first time by NAT A and then by NAT B, which assigns the message an external transport address valid for the STUN server network. The STUN server reads the source transport address of the packet carrying the STUN message and copies it in the XOR-MAPPED-ADDRESS attribute of a STUN binding response that is sends back to the STUN client. Since the STUN binding response traverses exactly the same NATs of the request, none of them drops it, whatever filtering behavior it is implementing.* Upon receiving the STUN binding response, the STUN client learns its external transport address (i.e., the transport address assigned by the NAT closest to the STUN server) from the XOR-MAPPED-ADDRESS attribute, which has indeed not been modified by the NATs.

"Classic STUN" proposed this approach as a NAT traversal solution, since it provides a manner for the STUN client to discover its external transport address. The issue is that this solution when used alone is not effective with certain types of NATs, such as, for example, the non-deterministic NATs and NATs implementing address- and/or port-dependent mapping. So this STUN usage must be considered as a building block usable in more complete solutions, such as, for example, ICE [7,8], which will be described later on in this chapter.

Besides the STUN binding request/response message pair, STUN defines an indication message that generates no response.

---

*If this were not the case, communication would not be possible between hosts in the network of the STUN client and hosts in the network of the STUN server.

12.4.4   TURN: A STUN Extension to Support Relaying

Traversal Using Relays around NAT (TURN) [9] is an extension to STUN to make STUN servers act as relays for hosts placed behind NATs. TURN introduces a framing method and a few new messages called TURN control messages. The new messages are used to request a transport address and to activate the relaying of data sent to/from external peers. Indeed even when a TURN client obtains a relayed transport address, it cannot be used to send/receive data to/from an external peer until this peer is enabled. Note that the TURN client may use the same relayed transport address to send/receive data from more than one peer. To this purpose, TURN encompasses different channels between the TURN server and the TURN client, each one for data received by the STUN server from a corresponding enabled peer; channel 0 is reserved for TURN control messages. The channels are implemented by adding a so-called *framing header* to data exchanged between a TURN client and a TURN server.

Figure 12.9 shows an example of TURN operation. The client sends an allocate request to the TURN server to obtain a public transport address from the server. The server communicates the assigned address placing it in the RELAY-ADDRESS attribute of the allocate response. Both messages are new STUN messages defined in the context of TURN. The allocate response also includes a XOR-MAPPED-ADDRESS, thus offering the functionality of a STUN binding response. At this point, the TURN client enables the target peer by sending a send indication, another newly defined STUN message containing the channel number chosen by the client (2 in Figure 12.9) together

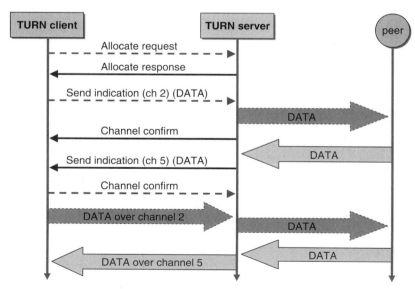

FIGURE 12.9
A TURN operation.

with the transport address of the peer stored in the PEER-ADDRESS-ATTRIBUTE. The send indication can also be used to send encapsulated data to the peer, placing the data to be delivered in a DATA attribute. If UDP is used between STUN client and server, the client continues sending data to the peer encapsulated in send indication messages until it receives a channel confirmation message from the server confirming the channel assignment. After the confirmation, the client can send data to the STUN server simply by adding a framing header in front of the packets. The confirmation, not used if TCP is deployed between STUN server and client, is needed because since UDP messages may be received out of order, the client must make sure that the server knows about a channel before starting to send packets over it. When the enabled peer starts sending data, the server forwards them to the client using data indications until the first one is confirmed by the client (the server reserves channel 5 in Figure 12.9). When the STUN server receives a channel confirmation by the server, it forwards data received by the peer on the assigned channel simply adding a framing header in front of the packets.

12.5   SIP-Specific NAT Traversal Techniques

This section describes how the NAT traversal techniques described above are used to support SIP UAs behind NATs, providing a general overview of the approaches adopted; more details can be found in [10].

12.5.1   Signaling Layer: Ensuring Correct Delivery of a Response to a SIP Request

Every host that forwards a SIP message adds a `Via` header field to the message in order to record the path of SIP entities traversed: the related response will be sent through the same path in order to reach the sender of the SIP request message as each SIP entity will attempt to forward the response to the transport address stored in the topmost `Via` header field (copied from the related request message). Figure 12.10 shows two UAs, called UA A and UA B wishing to set up a media session: no NATs are present. Before forwarding the message, the SIP proxy appends its transport address in a new `Via` header field. When the message reaches UA B, the `Via` headers are copied in the 200 OK response and the response is forwarded to the transport address stored in the topmost `Via` header field: when the 200 OK response reaches the proxy, it removes the topmost `Via` header (storing the transport address of the proxy) and the message is forwarded to UA A.

Now suppose that UA A is placed behind a NAT and that UA A knows only its internal transport address (see Figure 12.11): the transport address placed by UA A in the `Via` header field of the INVITE request will be a

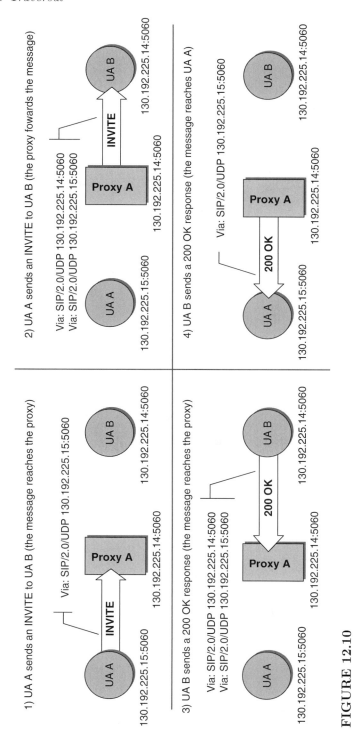

FIGURE 12.10
An INVITE request and 200 OK response routing.

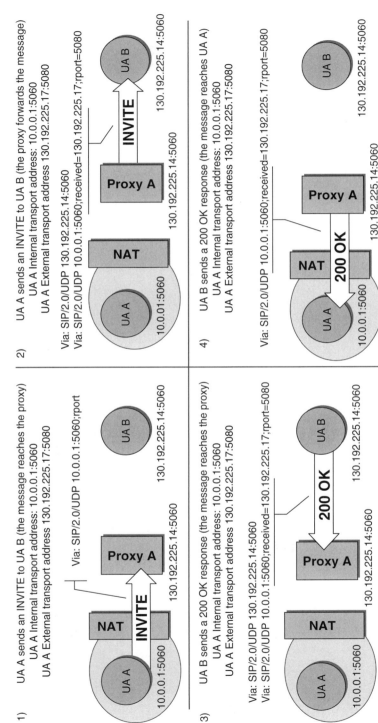

FIGURE 12.11

An INVITE request and 200 OK response routing when NATs are present.

private one. This means that when the 200 OK answer reaches the proxy, the SIP message cannot be forwarded to UA A since the transport address in the `Via` header field is not globally routable. To overcome this limitation, two additional parameters called, respectively, "received" and "rport," have been defined. The first one was defined in RFC 2543, while the second one was defined in RFC 3581 [11]. In Figure 12.11, UA A simply sends the INVITE request, placing an empty "rport" parameter in the `Via` header field. As proxy A receives the INVITE request, the transport address stored in the `Via` header field is compared to the source network address and the port of the packet carrying the request. If they differ, a NAT is on the path and the source network address and port observed by the proxy are the external transport address assigned to UA A by the NAT. Therefore, before adding its own `Via` header field, the proxy amends the one inserted by UA A by adding the "received" parameter, followed by the source network address observed by the proxy. The "rport" parameter is used to store the related external port: without the "rport" parameter, it would have been impossible to record the port assigned by the NAT. When the 200 OK response is sent by UA B, all the `Via` header fields of the request are inserted. When the proxy receives the message, the topmost Via header is removed: the proxy forwards the 200 OK response to the network address stored in the "received" parameter, using the port stored in the "rport" parameter as the target port. Without the "rport" parameter, the response would have been forwarded to the port originally placed by UA A in the `Via` header field. Since the NAT changed the port from 5060 to 5080, it would have been impossible to deliver the request.

12.5.2 Signaling Layer: Ensuring UA Reachability and Supporting Mid-Dialog Requests

As mentioned when discussing NAT traversal-related issues, UAs must be always reachable by their SIP proxy, which is not necessarily the case when the UA is behind a NAT as the NAT binding might expire between the REGISTER message from the client and the first message from the proxy. The IETF Network Working Group has released an Internet draft [12] providing a solution based on a modification of the registration procedure to create a sort of "flow" between the UA and the proxy when the UA sends the REGISTER message. The term flow is used to indicate a bidirectional stream of UDP datagrams or a TCP connection, depending on the transport protocol adopted. The proxy later reuses this flow to send incoming SIP messages to the UA, which is in turn responsible to periodically verify that the flow is active and to re-establish it in case of failure; multiple flows might be established to ensure reachability. The `Contact` field of the REGISTER message is extended with two new subfields: the `instance-id` to store a *Unified Resource Name* (URN) that uniquely identifies the UA and the `reg-id` to distinguish among flows related to a specific UA when it performs multiple registrations. Ideally, a UA registers multiple times, establishing different flows,

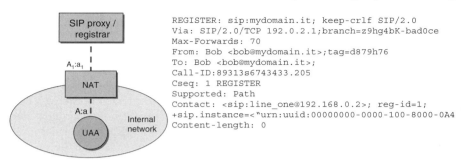

FIGURE 12.12

A scenario with a SIP proxy, including SIP registrar.

and all the registered Contact fields share the same instance-id, while the reg-id vary. UAs have to keep flows alive by periodically sending STUN messages to the proxy if the UDP transport protocol is used. After the registration, UAs send out-of-dialog SIP requests on existing flows; the Contact field in the SIP request should also contain the instance-id used during the registration to uniquely identify the UA.

A few scenarios are discussed in the following to provide a better understanding of the solution.

The first scenario (shown in Figure 12.10), which is the simplest, encompasses a single SIP proxy including a registrar and a UA behind a NAT sending to the proxy/registrar the REGISTER message depicted in Figure 12.12, containing instance-id (introduced by the "+sip.instance" string) and reg-id. The registrar will store both the instance-id and the reg-id values together with the URI placed in the Contact field. When it receives an incoming SIP request for the registered URI, instead of resolving it to devise information to forward the request to the UA, the proxy uses the flow initiated by the UA with the REGISTER message. Note that the network address in the registered URI is an internal one (private in the example in Figure 12.12), hence unusable to deliver messages to the UA.

The next scenario features a single logical proxy/registrar function running on separate hosts, as shown in Figure 12.13; load balancing is ensured by the DNS alternatively resolving the proxy name to SIP proxy/registrar 1 and to SIP proxy/registrar 2. The UA establishes multiple flows with separate SIP proxy/registrars: this helps to ensure connectivity when one of the SIP proxy/registrars fails. For more details, please refer to [12].

The last scenario, depicted in Figure 12.14, encompasses *edge proxies* and a separate registrar. Edge proxies are between a UA and the registrar, while the latter that can access the SIP location service in order to retrieve the URI needed to forward a SIP message to the corresponding UA.

In this scenario, the UA sends the REGISTER request to the registrar through an edge proxy that adds a Path field before forwarding it to the registrar. By doing this, each out-of-dialog SIP request forwarded by the registrar

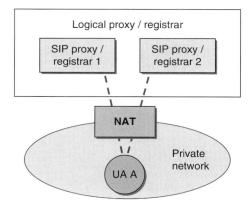

FIGURE 12.13

A logical proxy/registrar function running on separate hosts.

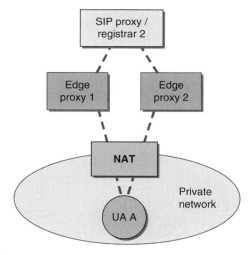

FIGURE 12.14

A scenario including edge proxies.

to the UA will traverse the edge proxy[†]. The flow creation mechanism works between UA and the edge proxy as described before; but the edge proxy places an identifier value unique to the flow in the `Path` field of REGIS-TER request forwarded to the registrar. Such a value will be used by the edge proxy to map future requests to the correct flow. Keepalive messages are sent to the edge proxies. To ensure the correct delivery of mid-dialog SIP requests (e.g., messages like the BYE requests used to terminate an existing session), the edge proxy will have to add a `Record-Route` header field to the dialog-forming request: in other words, the solution proposed consists

---

[†]For further details on the path header filed, please refer to RFC 3327 [13].

in having the edge proxies relaying all the SIP messages. When the edge proxies are not present, the SIP proxy/registrar itself must add a `record route` header field to the dialog forming request, since this specification needs to have mid-dialog requests routed over the same flow established through the registration.

12.5.3   Traversal for Media Flows: ICE

Interactive Connectivity Establishment (ICE) [7] (developed by the Multipart Multimedia Session Control working group [8]) is used for NAT traversal of UDP-based multimedia sessions using an offer/answer model. ICE makes use of STUN and TURN and basically extends the Session Description Protocol (SDP) with the insertion of multiple network addresses and ports into SDP offers and answers. Consequently, ICE can be applied to all protocols making use of the SDP offer/answer mechanism, among which SIP is one of the most important. However, ICE is not intended for NAT traversal in SIP. On the contrary, it assumes that SIP messages can be exchanged by UAs. Hence, in case NATs are present, UAs have to ensure SIP connectivity as discussed above.

Figure 12.15 depicts a typical scenario for ICE-capable UAs. Before trying to establish a media session, each UA gathers all its relevant transport addresses in order to insert the MIN the SDP payloads as candidates. The first available candidate is obviously the one related to a physical or logical interface of the UA, which is called a *host candidate*. Multi-homed hosts will have multiple host candidates. Typically, each UA will request at least a public transport address to a TURN server, which is called a *relayed candidate*. A UA also tries to discover its external transport address using a STUN server. The resulting address is called a *server reflexive candidate* as it is a transport

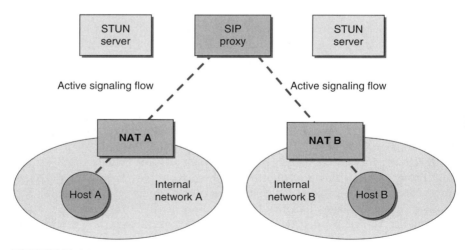

FIGURE 12.15
An ICE deployment scenario.

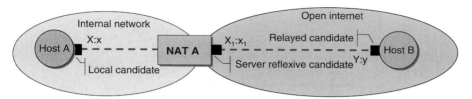

FIGURE 12.16

Transport address relationships.

address as seen by a STUN server. Note that since TURN servers include a STUN server, a UA can gather at the same time from the same server both a relayed candidate and a server reflexive one. Since STUN requests are sent from a local interface, each server reflexive candidate is related to the internal transport address used as the source for the STUN binding request; the corresponding host candidate will be referred to as the *base* for the server reflexive one.

Figure 12.16 shows a sample scenario that describes the relationships among the above-mentioned addresses. X:x is the base candidate, while $X_1:x_1$ and Y:y are the server reflexive candidate and the relayed candidate, respectively; obtained by sending a STUN binding request and a TURN allocate request, respectively, from the base candidate.

After collecting its candidates, each UA orders them using a numeric priority field that will be sent along with each candidate: the entire candidate list will be included in the SDP payload of the dialog-forming message using multiple ad hoc SDP attribute entries: for such purpose the "a = candidate" lines have been added to the SDP definition. In addition, default candidates, i.e., transport addresses that can be used by UAs that are not ICE-capable, are placed in the m and c lines of the SDP payload. As an example, given a media session including a RTP flow and a RTCP flow, the default candidate for RTP consists in the network address be placed in the c line and in the port stored in the m line; the RTCP default candidate address is placed in the rtcp attribute of SDP if present, otherwise the default candidate consists of the same network address used for RTP and the port in the m line increased by one. Candidates of the same type (e.g., server-reflexive candidates rather than relayed ones) get the same priority values; candidates corresponding to more direct routes (i.e., with a lower number of relays in the path) will receive higher priority values.

Each candidate entry will also include its own *foundation*: two candidates have the same foundation if they are of the same type (e.g., both server-reflexive), and have been obtained from the same base network address; if the candidates are server-reflexive or relayed ones, they have the same foundation if they have been obtained from the same STUN or TURN server. The need for the foundation stems from observing that a media session is usually made up of different components (where a component is intended to be a piece of media session that requires a single transport address). For example, an audio

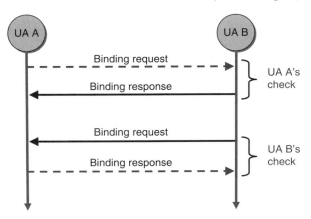

FIGURE 12.17
Connectivity checks.

session typically has two components: an RTP flow and an RTCP flow. If the UA obtains two server reflexive candidates, one for the RTP flow and one from the RTCP flow sending two STUN binding requests from the same base, they will get the same foundation. Since the components of a media session are strongly correlated (for example, it is possible to leverage information about a component if it is known that the other component works), the foundation has been introduced to express such relation also among the corresponding transport addresses.

When two UAs have successfully exchanged their candidate lists, they organize them into a list of candidate pairs and start connectivity checks to verify whether candidate pairs provide end-to-end connectivity. Specifically, each UA sends a STUN binding request message and waits for a related STUN binding response, performing the four-way handshake depicted in Figure 12.17.

The candidate pairs are sorted according to a priority value obtained by combining the priority values of the candidates; higher priority candidate pairs are checked first. Furthermore, each candidate pair has a foundation obtained by combining the foundations of its candidates. Initially, only one candidate pair per foundation is tested, while the others are placed in a frozen state. Considering an audio session as an example, only the RTP flows are checked first, while the RTCP ones are in a frozen state. As soon as the connectivity check for a candidate pair is successful, the other candidate pairs with the same foundation are unfrozen. The checks are performed exchanging STUN messages directly between the UAs. In order to do this, both UAs must integrate a STUN client and a STUN server listening on the ports that will be used by the flows as the STUN messages used for the checks must be exchanged using the same transport addresses that the media flows are going to use. Hence, it is necessary to distinguish STUN messages from normal packets in order to pass the STUN messages to the appropriate STUN client/server. This is obtained by setting the first two bits of each STUN message to zero and by placing a magic cookie on top of the header of the STUN messages.

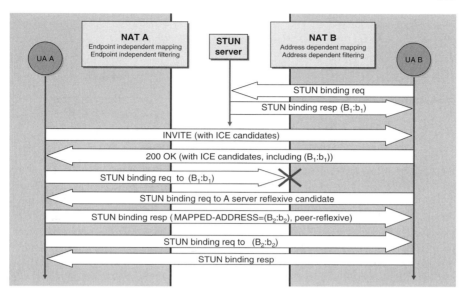

FIGURE 12.18

A NAT traversal using ICE.

STUN messages are usually exchanged also after the checks to ensure that the bindings on the NATs stay active even when no data packets are exchanged for a while, which is referred as the keepalive usage.

Figure 12.18 shows an interesting scenario from the point of view of the connectivity checks. NAT A implements an endpoint-independent mapping and filtering behavior, while NAT B implements an address-dependent mapping and filtering behavior. When UA B receives an INVITE message from UA A, it inserts in its 200 OK answer a candidate list that includes the server reflexive candidate from the public STUN server. When UA A starts connectivity checks, the STUN binding request sent to the UA B server reflexive candidate is dropped by NAT B because of its address-dependent filtering policy. Instead, a STUN binding request from UA B is successfully delivered to UA A as arriving from the transport address $(B_2:b_2)$ assigned by NAT B (since the NAT changes mapping depending on the target). UA A replies with a STUN binding response, while at the same time discovering this new external address for UA B, which is referred to as *peer-reflexive candidate* since it is discovered directly from the peer. Later, UA A checks the peer reflexive candidate of B by sending a STUN binding request to the transport address $(B_2:b_2)$.

A basic approach for choosing a winning candidate pair is to declare a winner as soon as all the checks for all the components of the session are successful. However, ICE defines more sophisticated techniques that continue the execution of connectivity checks even in case of success. For further detail, please refer to the ICE Internet draft [7,8]. To conclude the procedure, once the winning candidate pairs are identified, if they differ from the default candidates, i.e., the entries stored in the m and c lines of the original messages used to exchange the lists, they are exchanged again by the UAs.

12.6 ALEX: Ensuring End-to-End Connectivity for Both SIP and Media Flows across NATs

The address list extension (ALEX) [14] is a NAT traversal solution for both SIP and media flows developed by the NetGroup at Politecnico di Torino. The solutions presented so far address NAT traversal for media flows aiming at establishing direct connectivity (e.g., using ICE), while leaving to SIP proxies the forwarding of all SIP messages exchanged by the UAs. However, in most cases, there it is not mandatory to have intermediate nodes on the path of SIP messages, since NAT traversal techniques used for media flows can be easily extended to SIP signalling as well. If the SIP session does not rely on intermediate nodes to deliver both SIP messages as happens for media flows, the overall network robustness and scalability are increased. Otherwise, fault tolerance and scalability have to be addressed through the deployment of redundant proxies (see, for example, the scenario depicted in Figure 12.14) together with mechanisms to redirect SIP sessions in case of failure of a proxy, which adds complexity to the network infrastructure. If end-to-end connectivity is available for SIP signalling, it is up to the UAs to detect communication failures and establish a valid communication channel again as soon as possible. Moreover, a direct channel established across NATs to deliver SIP messages can be used also to provide hints to choose optimal transport addresses for media flows, thus reducing the overall number of messages exchanged during connectivity checks.

The core of ALEX is the `ALEX-item`, a new header field used to announce transport addresses, one in each field, together with a priority value. Hence, while ICE announces its candidates in specific attributes of the SDP payload (i.e., outside SIP), ALEX provides this information directly in the SIP header. As a consequence, ALEX also announces transport addresses to be used for SIP, not only for media flows.

The ALEX operation consists in four phases, executed by each UA during session establishment, as exemplified in Figure 12.19.

1. *Gathering transport addresses*: both UAs collect all the transport addresses they can use to send and receive IP packets, including relayed transport addresses gathered from TURN servers and server reflexive transport addresses gathered from STUN servers.

2. *Exchanging transport addresses* by placing them in the ALEX items inside SIP messages. By doing this, the UA presents itself as a sort of multihomed virtual UA that will decide at runtime the best transport

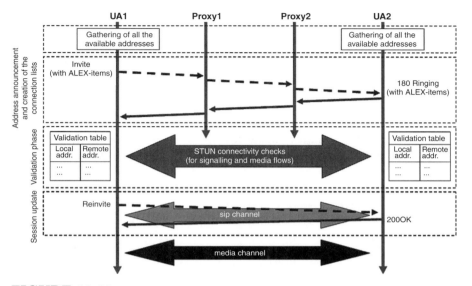

FIGURE 12.19

SIP and media session establishment with ALEX.

addresses to use for each flow during the session. In fact, ALEX was originally developed as a methodology to allow multi-homed dual-stack UAs
to choose at runtime the best (IPv6 or IPv4) network addresses to use
for each flow, including SIP. In Figure 12.19, UA B sends ALEX-item
fields related to the SIP channel directly in the 180 RINGING provisional response; UA B may choose to send ALEX-item fields related
to media flows in the 180 RINGING response as well, or to postpone
their transmission in the 200 OK answer. However, all the ALEX-item
fields must also be present in the 200 OK message because they must
be exchanged in a reliable way.

3. *Creation of validation tables*: each UA pairs each one of its transport
   addresses with the ones in the ALEX-item fields received by the other
   UA. Pairs are referred to as *candidate channels* and are the entries of the
   validation tables. Validation tables on both UAs contain the same data.

4. *Channel validation*: candidate channels are checked using STUN messages to verify connectivity, beginning from the candidate channels for
   the SIP session. The successful channels with the highest priority,
   obtained by combining the priority of the component transport addresses,
   are referred to as *optimal channels*. The time required by the checks,
   especially when multiple retransmissions of STUN messages are needed
   to confirm failed checks (i.e., to make sure that the transport address
   cannot be reached rather than packets being dropped due to congestion),
   might not be negligible. In order to reduce the delay in beginning SIP

and media sessions, the validation phase has been split into two sub-phases:

a. The *fore validation sub-phase* consists of a non-thorough procedure aiming at discovering in a short time, if possible, one optimal channel during which STUN messages are retransmitted quickly—up to three times in case of missing answers. The fore validation sub-phase begins the probing procedure from the default channels (i.e., channels made up of default transport addresses) and stops as soon as an optimal channel is discovered or when a timer fires.

b. The *background validation sub-phase* follows, while concurrent SIP and media sessions start using the best channels successfully validated in the previous sub-phase, to probe again high-priority channels that may have not been successfully checked in the fore validation sub-phase and to probe the remaining channels. The validation procedure during the background validation sub-phase is less aggressive in order to limit network overhead due to connectivity checks, e.g., retransmission intervals of STUN messages are longer than in the fore validation sub-phase.

In the most likely case, the fore validation sub-phase concludes with the discovery of an optimal SIP channel before the 200 OK SIP message is sent. The information related to the selected channel can be used to reduce the number of STUN messages needed to detect optimal channels for the media flows by properly setting the priority sub-field of the `ALEX-item` fields related to media flows placed in the 200 OK answer. For example, if the optimal SIP channel is based on a relayed address, the the `ALEX-item` containing a relayed address is assigned highest priority for each media flow.

ALEX stores the optimal channels discovered in the *ALEX cache*. The transport addresses stored in the cache will be checked first during subsequent session establishments since they probably have a high probability of success, i.e., of being reachable. Each entry of the ALEX cache contains an additional tag to keep its current state. Initially, the state of an optimal channel is set to `live` and the UA periodically sends a keepalive message on the channel to ensure that it is still working. If no keepalive messages are sent for a given period, the state of the channel is set to `probed`: the channel is no longer active, but it is kept in the cache as a hint to speed up the establishment of subsequent sessions.

ALEX has been implemented in a modified version of the OpenWengo NG UA [15] and has been tested in some significant scenarios including different NAT behaviors [2]. The tests prove ALEX to be effective in providing end-to-end connectivity for both SIP and media flows at a very low cost in terms of the increase of the SIP message size and in terms of the number of STUN messages exchanged during the validation step.

References

[1] Behavior Engineering for Hindrance Avoidance (behave), IETF Working Group. http://www.ietf.org/html.charters/behave-charter.html. Accessed November 15, 2007.

[2] Audet, F., and C. Jennings. 2007. Network Address Translation Behavioral Requirements for UDP. *RFC 4787*, January, http://www.ietf.org/rfc/rfc4787.txt. Accessed November 15, 2007.

[3] Rosemberg, J., et al., 2002. SIP: Session Initiation Protocol. IETF Network Working Group, June, http://www.ietf.org/rfc/rfc3261.txt, *RFC 3261*. Accessed November 15, 2007.

[4] Ford, B., P. Srisuresh, and D. Kegel. Peer-to-peer Comminucations Across Network Address Translators (NATs). http://www.bford.info/pub/net/p2pnat/. Accessed November 15, 2007.

[5] Rosenberg, J., J. Weinberger, C. Huitema, and R. Mahy. 2003. STUN—Simple Traversal of User Datagram Protocol (UDP) Through Network Address Translators (NATs). *RFC 3489*, March, http://www.ietf.org/rfc/rfc3489.txt. Accessed November 15, 2007.

[6] Rosenberg, J., R. Mahy, P. Matthews, and D. Wing. 2007. Session Traversl Utilities for (NAT) (STUN). Internet Draft, November 17, http://tools.ietf.org/html/draft-ietf-behave-rfc3489bis-13.

[7] Rosenberg, J., 2007. Interactive Connectivity Establishment (ICE): A Protocol for Network Address Translator (NAT) Traversal fro Offer/Answer Protocols. Internet Draft, October 29, http://tools.ietf.org/html/draft-ietf-mmusic-ice-19.

[8] Multiparty Multimedia Session Control (mmusic), IETF Working Group, http://www.ietf.org/html.charters/mmusic-charter.html. Accessed November 15, 2007.

[9] Rosenberg, J., R. Mahy, P. Matthews, and D. Wing. 2007. Traversal Using Relays around NAT (TURN): Relay Extensions to Session Traversal Utilities for NAT (TURN). Internet Draft, November 15, http://tools.ietf.org/html/draft-ietf-behave-turn-05.

[10] Boulton, C., J. Rosenberg, and G. Camarillo. 2007. Best Current Practices for NAT traversal for SIP. Internet Draft, July 9, http://tools.ietf.org/ html/draft-ietf-sipping-nat-scenarios-07.

[11] Rosenberg, J., and H. Schulzrinne. 2003. An Extension to the Session Initiation Protocol (SIP) for Symmetric Response Routing. *RFC 3581*. August, http://www.ietf.org/rfc/rfc3581.txt.

[12] Jennings, C. and R. Mahy. 2007. Managing Client Initiated Connections in the Session Initiation Protocol (SIP). Internet Draft, November 18, http://tools.ietf.org/html/draft-ietf-sip-outbound-11.

[13] Willis, D. and B. Hoeneisen. 2002. Session Initiation Protocol (SIP) Extension Header Field for Registering Non-Adjacent Contacts. *RFC 3327*, December, http://www.ietf.org/rfc/rfc3327.txt.

[14] Baldi, M., L. De Marco, F. Risso, and L. Torrero. 2008. Providing End-to-End Connectivity to SIP User Agents Behind NATs. *IEEE International Conference on Communications (ICC 2008) Advances in Networkds & Internet Symposium*, Beijing (China), May.

[15] The OpenWengo project. Available at http://www.openwengo.org. Accessed November 15, 2007.

13

Multipoint Extensions for SIP

Sureswaran Ramadass and Omar Abouabdalla
University of Science, Malaysia

CONTENTS

13.1 Introduction 307
13.2 Multipoint Session Initiation Protocol (MSIP) Entities 308
13.3 Control Messages 310
13.4 Messages Flow 311

13.1 Introduction

In the distributed network entities environment, a message-passing distributed algorithm is needed. Session Initiation Protocol (SIP) is point-to-point protocol. In this chapter, we are going to describe a Multipoint Session Initiation Protocol (MSIP). This protocol is an application-layer control (signaling) protocol for creating, modifying, and terminating sessions with one or more participants. The protocol is a signaling protocol based on distributed network entities architecture, and the use of the server is mandatory.

The server here is the main entity, and it communicates with the other network entities involved in the system (mainly the client entities) using the standard Internet communication protocol (TCP/IP). There are some terminologies that are going to be used in this chapter. These terminologies are:

- MSIP server: is a software that runs on a central machine and its main functions are to control and manage the session.

- MSIP client: is a software that runs on end users' devices and is used by users to start a new session, to join a session, or to participate in a session.

- User: is a person who is registered to a MSIP server. He may or may not be taking part in a session run by the MSIP server.

- Chairman: is a user who starts a new session and invites other users to join the session.

- Participant: is a user in a session who can actively participate or contribute to the session.

- Observer: is a user in a session who is allowed to only view and listen to the session but is not allowed to take active part in it.

Based on the above, the chairman, participants, and observers are all users who are participating in a session via MSIP client entities with the session being managed by the MSIP server entity.

13.2   Multipoint Session Initiation Protocol (MSIP) Entities

The general network architecture contains MSIP server, MSIP client, and MSIP reflector. The architecture SHOULD contain at least one MSIP server and at least two MSIP clients. The MSIP reflector is optional. Figure 13.1 below shows the general architecture of MSIP entities.

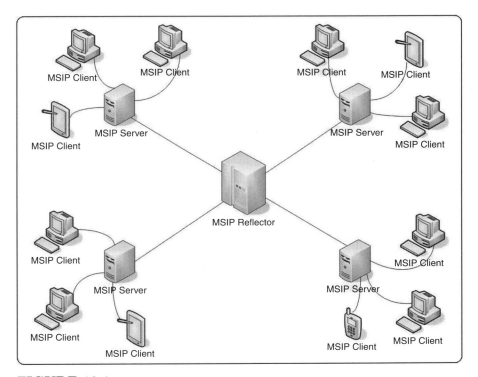

FIGURE 13.1
The general architecture of MSIP.

13.2.1   MSIP Reflector Behavior

The MSIP reflector is in charge of reflecting all session control messages to all MSIP servers involved in the session. Once it is started, it waits for connections from MSIP servers. The MSIP reflector must only accept connections from MSIP servers. The MSIP reflector holds information on MISP servers as well as information on sessions handled by MSIP servers. The information is stored in two different tables. The tables are:

- Session data list
- Remote MSIP servers list

13.2.2   MSIP Server Behavior

The MSIP server is the heart of the session. It controls all the entities involved in the session. The MSIP server holds information on the users and sessions handled by the MSIP server. It also holds information on the MSIP reflector that can be used. The information is stored in three different tables. The tables are:

- Session data list
- MSIP reflectors list
- Users info list

The MSIP server takes full control of all entities involved. The input and output transmissions from the MSIP server consist of control communications (TCP/IP). Once started, the MSIP server searches for the MSIP reflector (if available) and opens a connection to it. The main MSIP Server is the MSIP server that the chairman (the user who starts the session) is logged into. This is the MSIP server that will manage the entire session.

13.2.3   MSIP Client Behavior

The main functions of the MSIP client entity are as the control and session monitor, which does all the coordination and synchronization for the MSIP client entity. It provides controls for the following:

- Login/logout
- Create session and join session
- Switch status (available, busy, or away)

The MSIP client provides a communication interface with the MSIP server. It also provides a users list and servers list. Finally, it provides feedback of MSIP client and MSIP server messages to users. The local MSIP clients are the client entities connected to the local MSIP server's local-area network (LAN), while the client entities from different LANs are known as remote MSIP clients.

13.3   Control Messages

Since MSIP is an application layer signaling protocol, it uses the data field in TCP/IP to send its own messages. The first field of the message is the MSIP header (version). A header is used to avoid possible conflict with other applications using the same communication port. The second message field contains the message type. Figure 13.2 shows the general format of MSIP messages.

There are essentially six phases in a session, each of which has varying message formats or types. The six phases are registration, session initiation, session establishment, joining a session, controlling the session and session updates, and finally, terminating a session.

13.3.1   Registration

During registration, the user notifies the local MSIP server at which terminal he is located. Users can also change their passwords or change their status after registration.

13.3.2   Session Initiation

In this phase, the user (chairman) requests to create a new session.

13.3.3   Session Establishment

In this phase, the MSIP server sends an invitation message to all users chosen by the chairman to be involved in the session.

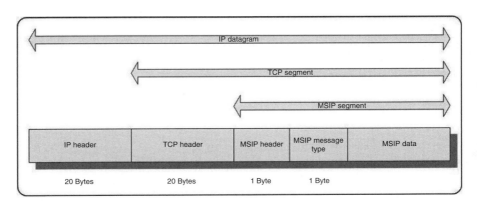

FIGURE 13.2
MSIP general message format.

13.3.4　Joining a Session

A user can join a session when he receives (through a MSIP client) an invitation message by responding to it. He can also join a session by sending a message to the MSIP server requesting the list of sessions he has been invited to.

13.3.5　Controlling the Session and Session Updates

This phase continually occurs while a session is running. A control criterion is used to control the flow of the session. In this phase, a participant can request to be an active site (the MSIP server will then put him in the first empty cell in the session queue), release the active site status, withdraw from the session queue, or logout from the session. In this phase, the chairman can invite new users, remove users from the active site, change user status (from participant to observer or vice versa) or drop a user completely from the session.

13.3.6　Terminating a Session

In this phase, the chairman client terminates the session by sending a terminate message to the MSIP server. The MSIP server informs all MSIP clients about the termination of the session by sending an update message to all MSIP clients. After that, the MSIP server frees the session resources within the MSIP server.

Within the session, there are three levels of participation: chairman, participant, and observer. The chairman has the authority to cut short a lengthy active site, that is, to "kill" a currently active site or to drop him completely from the session.

The chairman is also the only one who has the authority to completely terminate the session. Individual participants may leave or rejoin the session as and when they wish, as long as the chairman still keeps the session going. Participants can be active or passive during the session, but observers are limited to passive mode only. An observer or participant can be upgraded or downgraded by the chairman at any time during the session.

13.4　Messages Flow

The control messages exchanged between all entities mostly flow between MSIP client and MSIP server (MSIP client → MSIP server and MSIP server → MSIP client). If more than one MSIP server is involved in the session, the basic structure of the messages flow will be as in Figure 13.3.

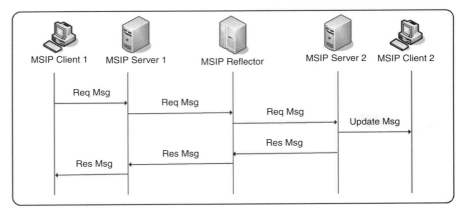

FIGURE 13.3
Basic structure of message flow.

13.4.1   Session Messages Flow

The message flow is between the MSIP server and MSIP client and vice versa.
When the MSIP server starts up, it performs the following tasks:

- Initialize session data list

- Initialize users info list

- Wait for any incoming message

Once a message is received, the MSIP server checks the message header. If
the message header belongs to MSIP, it checks the message and acts accordingly. As mentioned before, the first field of the message is the MSIP header
(version). A header is used to avoid possible conflict with other applications
using the same communication port. The second message field contains the
message type. There are essentially six phases in the session, each of which
have varying message formats or types.

13.4.1.1   Registration Phase

The first message that should be received from the MSIP client is C_USER_
LOGIN. This message contains the user information. The MSIP server will
check the user authentication and update user table by adding a new record
with the user information (if successful). The MSIP server then sends back
an S_USER_LOGIN message to the MSIP client with login successful or not.
It also sends S_USER_UPDATE to other clients (if they log into the MSIP
server).

After a user logs into the MSIP server, the user can change his password by
sending a C_PASS_CHG message to the MSIP server. The message format is
as shown in Figure 13.4. The VER field is for MSIP version. The type field is

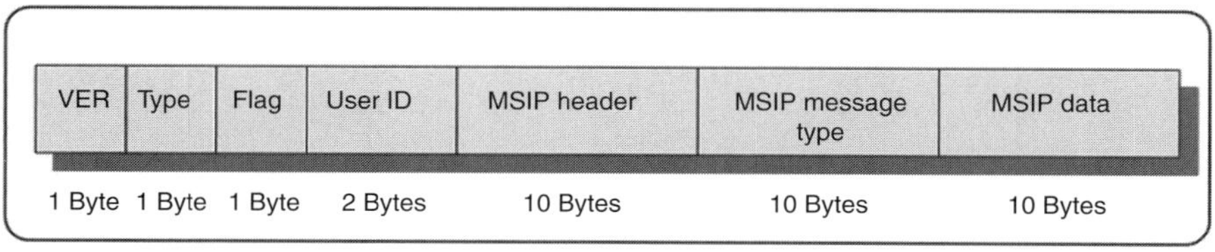

VER	Type	Flag	User ID	MSIP header	MSIP message type	MSIP data
1 Byte	1 Byte	1 Byte	2 Bytes	10 Bytes	10 Bytes	10 Bytes

FIGURE 13.4
Change password message format.

for message type; in this case it is either C_PASS_CHG if it is a request sent by the MSIP client, or S_PASS_CHG if it is a response from the MSIP server. The flag field is used to return the operation's result to the client. The user ID, user name, user old password, and user new password constitute the rest of the fields.

When the MSIP server receives C_PASS_CHG from a client, it compares the old password with the one in the user table stored in the server. After that, the MSIP server will change the password in the table with the new password (if correct), then send back an S_PASS_CHG message to the client with a flag. The flag will explain the operation. The flag can be one of the following:

- PASS_WRG $\rightarrow$ 0 × 11 if a wrong password is received from the client

- PASS_OK $\rightarrow$ 0×10 if a correct password is received and change password is successful

- FILE_ERROR $\rightarrow$ 0 × 12 if writing the new password in the user's file could not be done for any reason

Figure 13.5 shows an unsuccessful attempt to change the password followed by a successful one. The Figure shows the communications between the MSIP server and the MSIP client.

The MSIP client can also send a C_STS_CHG message to the MSIP server if one of his flags has changed (i.e., video flag, audio flag). The MSIP client can also use a C_STS_CHG message to announce the change of his status (in a session, available, busy, etc.)

The message format is as shown in Figure 13.6. The VER field is for the MSIP version. The type field is for message type, in this case it is C_STS_CHG. The flag field is used to determine which device is changed (video, audio, client status, etc.) Other fields are the user ID and user status.

When the MSIP server receives C_STS_CHG from a client, it updates the user table based on the information received, and then it sends an S_USER_UPDATE message to the current user and all other involved users. Involved users are users communicating with the user who changes his status. Figure 13.7 shows the message flows between the MSIP server and three MSIP clients involved in the session. The messages showed in Figure 13.7 are from the MSIP server start-up and before starting a session.

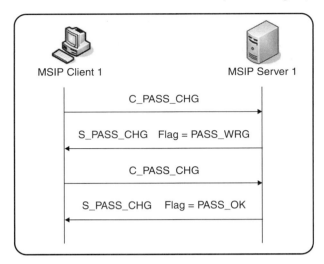

FIGURE 13.5
Change password messages flow.

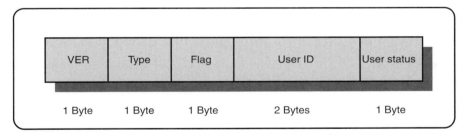

FIGURE 13.6
Change status message format.

13.4.1.2   Session Initiation Phase

In this phase, the user (chairman) requests to initiate a new session. Before the MSIP client starts a session, it sends a C_GET_USERLIST to the MSIP server, the MSIP server replies with S_START_USERLIST, followed by S_CONT_USERLIST if required. The S_CONT_USERLIST is required if the number of users is very big and their information cannot fit in one message (message buffer size = 202 byte).

The MSIP server will also send a S_USER_UPDATE message to the MSIP client (one for each user in the user list). The S_USER_UPDATE message contains the latest updates for the users. The MSIP client will then send a C_GET_GROUPRLIST message to the MSIP server, asking for the list of groups. The MSIP server replies with S_START_GROUPLIST, followed by S_CONT_GROUPLIST, if required. The S_CONT_GROUPLIST is required

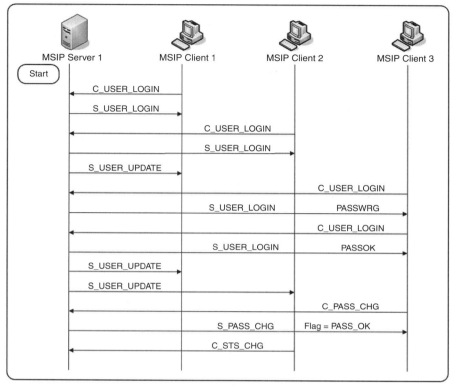

FIGURE 13.7

Messages flow from MSIP server start-up and before starting a session.

if the number of groups is very big and their information cannot fit in one message. After the MSIP client receives all the information, it builds the users list, which contains all the information about all users having accounts in the MSIP server. User names are in the form of name@server.

The MSIP client can also send C_GET_CONFLIST to the MSIP server, to ask for a list of the sessions it is invited to. The MSIP server replies with S_START_CONFLIST followed by S_CONT_CONFLIST, if required. The S_CONT_CONFLIST is required if the number of sessions is very big and their information cannot fit in one message. Figure 13.8 shows the messages flow for pre-create while Figure 13.9 shows the messages flow for a pre-join session.

After the MSIP client builds user information, it can start a session by sending a C_CREATE_CONF message to the MSIP server. The C_CREATE_CONF message contains the session name, session priority, session control criteria, and chairman info (i.e., chairman name, chairman id, chairman group id, and chairman flags). When the MSIP server receives this message, it checks for an available session slot. If no slot is available, it sends back an

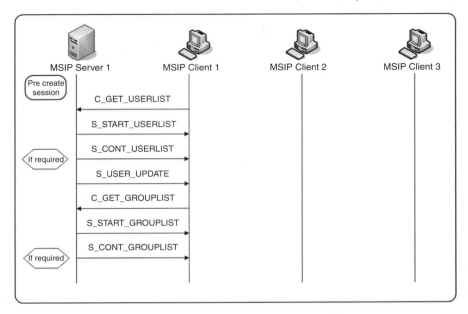

FIGURE 13.8
Messages flow for the pre-create session stage.

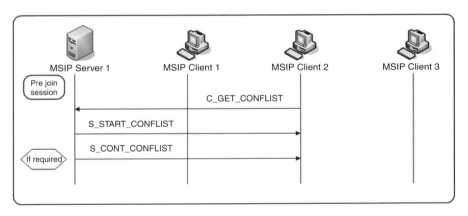

FIGURE 13.9
Messages flow for pre-join session stages.

S_CREATE_CONF message with session priority field = CREATE_CONF_ MAX (0x21).

If the maximum number of sessions is not reached, the MSIP server will initiate all session parameters based on the information received in a C_CREATE_ CONF message, and then add a new session record. After that, the MSIP server sends back an S_CREATE_CONF message with session priority field = CREATE_CONF_OK (0x20).

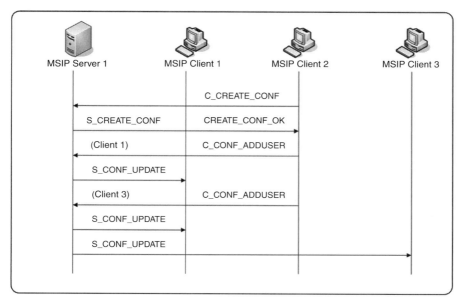

FIGURE 13.10
Messages flow for establishing a session.

13.4.1.3   Session Establishment Phase

After the MSIP client receives CREATE_CONF_OK, it will send multiple C_CONF_ADDUSER messages (one for each user invited to the session). This message contains the session ID, session operation, user ID, user group ID, user MSIP server ID, operation type, and user name.

Once the MSIP server receives the message, it adds the user to the session invited list, then sends an S_CONF_UPDATE message to all users involved in the session and to the user invited with Update_Operation field = ADDPAR. Figure 13.10 shows the messages flow for session establishment between three clients. Client 2 is considered the session chairman.

13.4.1.4   Joining a Session Phase

A user can join a session when he receives an invitation message by responding to it. A user should know that he is invited to a session when he receives an S_CONF_UPDATE message that contains session operation field = add participant (ADDPAR), and the participant ID is equal to his ID.

A user can also join a session by sending a C_GET_CONFLIST message to the MSIP server, requesting for the list of sessions he has been invited to. The MSIP sever will send an S_START_CONFLIST message back to the client, followed by S_CONT_CONFLIST (if required). After the MSIP client receives the full list, he chooses the session he wishes to join and sends a C_REQJOIN_CONF message to the MSIP server. The MSIP server will check session availability and reply with an S_REQJOIN_CONF message. The reply message includes a flag that determines whether the client can join the session or not. The flag value can be:

- NO_INVITE—if the client is not invited to the session

- CONF_ENDED—if session was terminated before the client requested to join

- JOIN_PAR—if the client is invited as a participant

- JOIN_OBS—if the client is invited as an observer

The difference between participant and observer is that observer is only allowed to listen to the conversation and not allowed to take part in it. Technically, the observer is only allowed to receive audio, video, and data streams but not allowed to send audio, video, and data streams.

When the flag equals JOIN_PAR or JOIN_OBS, the S_REQJOIN_CONF message also includes information about the session such as chairman name, chairman group, and control criteria. Here the client can join a session by sending a C_JOIN_CONF message. If an error occurs, the MSIP server will reply to the client with an S_ERRORB4JOIN message. Otherwise, it will send back a series of messages as follows:

The first message will be S_START_PARLIST, followed by S_CONT_PARLIST (if required). These two messages contain all users invited to the session and their status (participant or observer). After that, the MSIP server will send an S_ACTIVE_LIST message to the client that contains the current active users. Lastly, the MSIP server sends an S_START_QLIST message followed by an S_CONT_QLIST message (if required). The last two messages contain information about the users in the session queue (if any) who are waiting for their turn to become active. The joining client uses all the information to build up its session table.

The MSIP server sends an S_CONF_UPDATE message to the other clients involved in the session to update them on newly joined clients. Figure 13.11 shows the messages flow for joining a session between three clients. Client 2 is considered the session chairman, and he invited client 1 and client 3 to a session. Client 1 joined immediately, while client 3 joined later, and found out that the session has already ended.

13.4.1.5 Controlling the Session and Session Updates Phase

This phase continually occurs while a session is running. A control criterion is used to control the flow of the session. In this phase, a participant can request to be an active site by sending C_CONF_UPDATE message to the MSIP server with Update_Operation field = REQ.

If one of the active site cells is empty, the participant will be active, if not, the MSIP server will put the participant ID in the first empty cell in the session queue then send an S_CONF_UPDATE message with updated information to

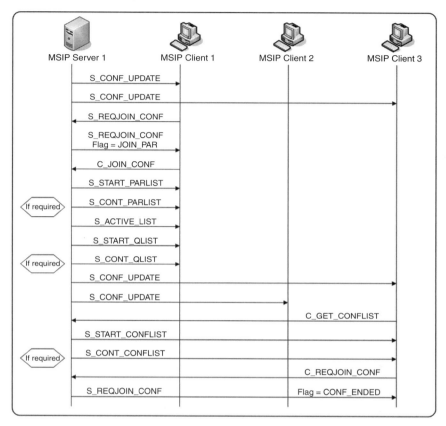

FIGURE 13.11

Messages flow for joining a session.

all clients involved so that the clients can also update their session table. The number of active cells can be one, two, three, or more (based on the control criteria used for session control).

Participants can also release the active site status or withdraw from the session queue by sending C_CONF_UPDATE message to the MSIP server with the Update_Operation field = REL or UNQ. The MSIP server will update the session queue and the session table accordingly, and then send an S_CONF_UPDATE message with updated information to all clients involved. By doing this, all involved clients are up to date. Figure 13.12 shows message flows for requesting and releasing the active site from MSIP client 2.

The S_CONF_UPDATE message contains the update type (Update_Operation Field), user ID, and session ID, so when the current active participant releases, the participant who is next at the top of the session queue will know that he is now an active site. Participants are able to recognize themselves since each user has a unique ID in the MSIP server. The S_CONF_UPDATE message is also sent to all clients when a new user joins a session.

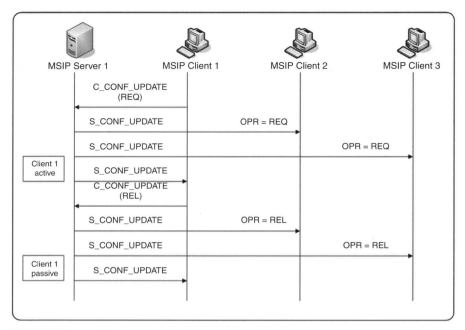

FIGURE 13.12
Messages flow for request and release active.

If a participant wants to logout from a session (hang up), his client will send a C_LEAVE_CONF message to the MSIP server. Once the MSIP server receives this message, it will update the session table by changing the user's Join_Status field in the session table to LEFT_CONF. Then it sends an S_CONF_UPDATE message to all MSIP clients involved in the session with the Update_Operation field = LEAVE.

During this phase, the chairman can upgrade observers to participants or conversely, using the C_CONF_UPDATE message. The chairman will send the message to the MSIP server with the Update_Operation field = CHGSTS. When the MSIP server receives the message, it will check and confirm if the message is received from the session chairman or not. If so, the MSIP server will update the user information in the session table and send an S_CONF_UPDATE message to all MSIP clients involved in the session with The Update_Operation field = CHGSTS.

The session chairman can also cut short a lengthy active site, that is to "kill" a currently active site by sending a C_CONF_UPDATE message to the MSIP server with the Update_Operation field = KILL, with which the MSIP server will then follow the procedure for releasing the active site.

During the session, the chairman can invite new participants or drop an existing participant. To invite a new participant, the chairman sends aC_CONF_ADDUSER message to the MSIP server. Please refer to Section 13.4.1.3 (session establishment phase) for message contents and MSIP server behavior

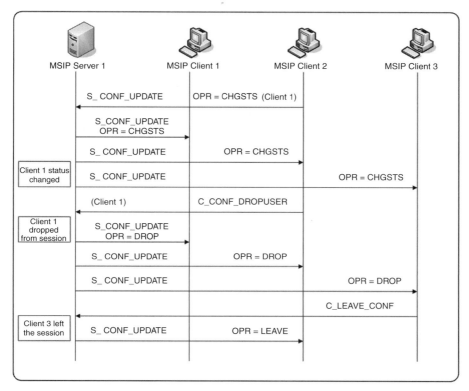

FIGURE 13.13
Messages flow for change status, drop client, and leave session.

upon receiving it. The session chairman could also drop any participant at any time during the session.

To drop a participant, the chairman sends a C_CONF_DROPUSER message to the MSIP server. The server will remove the dropped participant from the session invited users, update the session table, and then send an S_CONF_UPDATE message to all involved clients with Update_Operation field = DROP. Once a user is removed from the invited participant list, he would not be able to join the session again even if the session is still running. Figure 13.13 shows the message flow for some of the session update messages between three clients. Client 2 is considered the session chairman. The chairman changes client 1's status first, then drops it completely from the session. After that, Client 3 leaves the session.

13.4.1.6 Terminating a Session Phase

In this phase, the chairman's MSIP client terminates the session by sending a C_END_CONF message to the MSIP server. This message contains only two fields; VER, and type, which in this case, is C_END_CONF. The MSIP server

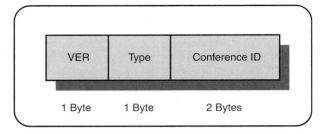

FIGURE 13.14

The S_END_CONF message format.

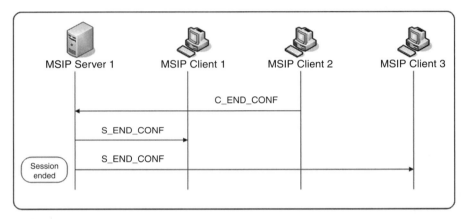

FIGURE 13.15

Message flow for session termination.

first confirms that the message is received from the session chairman by checking the sender ID and comparing it with the chairman ID in the session table.
If so, it will inform all the MSIP clients that are involved in the session about
the session termination by sending an S_END_CONF message to all of them.
The S_END_CONF contains three fields; VER, type, and conference ID. The
conference ID is used by the MSIP server to identify which session is terminated. The S_END_CONF message format is as shown in Figure 13.14.

After that, the server disconnects all clients from the session and then frees
the session resources within the server. Figure 13.15 shows the messages flow
for terminating a session. The session chairman is the only one allowed to
terminate the session. If the MSIP server detects a network disconnection
with the chairman, it will terminate the session automatically.

14

Narrowcasting in SIP: Articulated Privacy Control

Sabbir Alam, Michael Cohen, and Julián Villegas
University of Aizu

Ashir Ahmed
Kyushu University

CONTENTS

14.1 Introduction 323
14.2 Conferencing 325
14.3 Media Privacy: Narrowcasting Concept 328
14.4 System Design and Implementation 332
14.5 Conclusion and Future Research 339
    Acknowledgements 343
    References 343

14.1 Introduction

In multimedia conferencing, media streams are exchanged between participants upon session establishment by setting up communication channels within a group. By default, each participant receives a combined stream obtained by mixing other participants' media. Situations arise when one wants to select a subset of the conference participants to whom one's media is sent or from whom streams are received. Media filters are necessary to configure privacy of the participants in the conference. In analogy to broad-, multi-, any-, and swarm-casting, narrowcasting is a technique for limiting and focusing information streams. Narrowcasting systems extend broad- and multicasting systems by allowing media streams to be filtered—for relevancy control, privacy, and user interface optimization. We describe four narrowcasting commands: `mute`, `deafen`, `select`, and `attend`—to provide distributed privacy in SIP-based conferencing.

Extensive development has been carried out in the area of conference and floor control [1,2]. Conventional features regarding media privacy in conferences are typically limited to scheduling and selecting the speaker. Advanced conferencing features such as adding/deleting participants, changing user agents (UAs) or modes (like switching from a desktop to a mobile phone),

TABLE 14.1

Three Different Mute Operations

	Self-mute	Pbx mute	Narrowcasting mute
Media vector	(diagram of P_1, P_2, P_3, P_4 with media connections)		(diagram of P_1, P_2, P_3, P_4 with media connections)
Media distribution	$P_1 \leftarrow (P_2 + P_3 + P_4)$ $P_2 \leftarrow (P_3 + P_4)$ $P_3 \leftarrow (P_2 + P_4)$ $P_4 \leftarrow (P_2 + P_3)$		$P_1 \leftarrow (P_3 + P_4)$ $P_2 \leftarrow (P_1 + P_3 + P_4)$ $P_3 \leftarrow (P_1 + P_2 + P_4)$ $P_4 \leftarrow (P_1 + P_2 + P_3)$
Semantics	Self control. P_1 mutes himself by turning off his mic so that no media goes to the media server, or P_1 can send a "self-mute" signal to the application server so that the media server simulates self-censorship.	Control by admin. P_1 is muted by moderator. P_1's media is not mixed in the media server. P_1 is in a listen-only, or "lurker" (stealth) mode.	P2P Control. P_1 mutes P_2. P_2 may speak to everyone, but P_1 won't hear his voice.

changing media, authenticating or authorizing participants, granting privileges, controlling presentation of media, sidebars, passive participants, whisper/private messages, audio-only, and lecture mode are described in RFC 4597 [3]. Media privacy features allow participants to control their own information and to distribute their attention, based on secrecy, anonymity, and solitude [4].

Mute is a popular feature for media privacy. It has three different varieties, juxtaposed in Table 14.1: self-mute, PBX-mute, and narrowcasting mute. Self-mute allows a user to withhold his media streams from other participants. In PBX-mute, a moderator disables a participant's outgoing media to other participants. Narrowcasting mute refers to P2P control with which a participant (controller) can select another participant (controllee) to disallow the controllee's media towards the controller.

A Call Whisper [5] feature allows a participant to talk privately to one or more participants in a group. This walkie-talkie-like feature creates a one-way voice or video communication channel. The session terminates when the controller releases the push-to-talk (PTT) button, so such a system is not practical when two-way communication is necessary. Voice Chat [6] allows participants to create one or more private audio conferences: although the communication channel in the private voice chat group provides duplex communication, participants can hear the main conference at low volume. Private Conversation [7] offers a private video, voice, and text conversation session inside a main conference. It is similar to a Call Whisper feature, but adds duplex

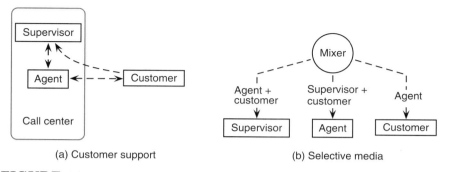

FIGURE 14.1

Media privacy: A call center application scenario.

communication capability and text messaging. In a WebEx audio conference, a conference chairperson can selectively disable microphones to allow only certain attendees to speak. An "audio-only" option allows a moderator to revoke and restore speaking privileges to attendees, so that muted attendees can only lurk, listen but not speak. WebEx participants can have a private chat with someone during a meeting. Whisper Coaching (www.audiocodes.com) allows a supervisor to listen to a main conference conversation while talking to a selected set of participants. The privacy control allowed by these applications is rather blunt. In order to better control media privacy, we are exploring the concept and practical applications of narrowcasting [8–10].

A call center scenario provides an example of media privacy: in instances when a first-tier agent cannot answer a customer's questions, the agent might have a private side-channel communication with a supervisor as back-up for realtime customer support, as shown in Figure14.1a. Even though the supervisor can overhear the customer, privacy control is invoked so that the supervisor's media goes only to the agent, not to the customer, as shown in Figure 14.1b. Traditional conferencing systems do not generally provide such features. In this chapter, we describe a mechanism and instance of "Media Server Component Model" architecture for policy-based media mixing with a centralized media mixer, using the standard SIP [11] framework for multimedia conferencing systems. We have defined privacy commands and developed a policy evaluation algorithm: media mixing and delivery factoring policy configured by conference participants.

14.2 Conferencing

A conference server and the participants are two major components of a centralized conference system, as shown in Figure 14.2. A SIP conference server comprises a focus, policy server, and media mixer. The focus handles the

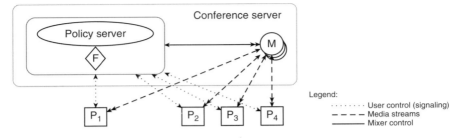

FIGURE 14.2

Typical conference architecture: F is the focus, M are media mixers, and P_i are the participants.

conference control—creating, modifying, and terminating conferences. Conference connectivity is managed by the policy server, which can configures the media server. Mixing and distribution of media streams are the main functions of a media mixer, such as a "voice switch" for audio conferencing, which transmits some composite signals to the respective terminals, as suggested by the multiple arrowheads on the (dashed) return vectors. Value-added services—such as monitoring conference status, participant status, and billing—can be implemented inside or outside of this framework.

14.2.1 SIP Conference Model

There are two generic conference models: loosely and tightly coupled. In a loosely coupled model, there is neither a central point of control nor a conference server, whereas in a tightly coupled model, a centralized conference control server manages the conferences. A tightly coupled conferencing model can be further classified into six different types depending on the location of the focus and the mixer, as illustrated in Table 14.2, including the Media Server Component Model used for our proof-of-concept. These models are detailed by J. Rosenberg [12] and Y. Cho et al. [13].

14.2.2 SIP Conference Control

Conference control refers to the ability to manipulate the state of a session. A conference is represented by a unique Uniform Resource Identifer (URI), usually a SIP URI, that identifies the focus of a conference. A conference URI can be e-mailed, sent in an instant message, linked on a web page, or obtained from some other mechanism. Conference control includes three primary functions:

- Creation: A participant joins a conference by sending an INVITE request to its focus ("dial-in") or by the focus sending an INVITE request to the participant ("dial-out"), citing the conference URI.

TABLE 14.2

Conferencing Models: "P" indicates participant, "F" indicates focus, "M" indicates media mixer, and (in the last model) 'PF' indicates primary focus. Dotted lines indicate signaling, dashed lines indicate media transmission, and solid lines indicate mixer control.

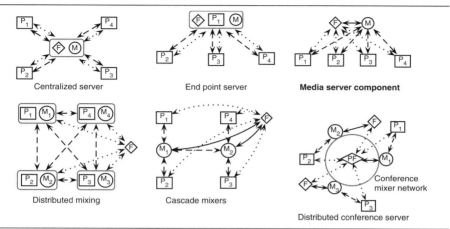

| Centralized server | End point server | Media server component |
| Distributed mixing | Cascade mixers | Distributed conference server |

Note: P = participant; F = focus; M = media mixer. In the last model, PF = primary focus. Dotted lines = signaling; dashed lines = media transmission; solid lines = mixer control.

- Modification: A participant or focus can modify a session in a conference using a re-INVITE. For instance, when an audio conference extends to video, the focus re-INVITEs each participant adding a video media stream. A participant or focus may also put media streams on hold or take them off hold. Narrowcasting commands are applied to a session by selectively enabling the media streams.

- Termination: A privileged participant (typically a moderator or conference creator) closes a session by sending a BYE request to the focus. The focus then distributes a BYE request to all other participants in the conference, terminating the session.

14.2.3 Conference Privacy

Privacy has two interpretations. The first association, with sources, is that of avoiding "leaks," protecting secrets. But a second interpretation, with sinks, means freedom from disturbance; in the sense of solitude, not being bothered by irrelevance or interruption, as suggested by Figure 14.3. Our distributed interface features narrowcasting operations that manage privacy in both senses, by filtering duplex media flow through an articulated conferencing model that limits and focuses information streams.

FIGURE 14.3
Privacy: Freedom from disturbance. (© The New Yorker Collection 1996 Sam Gross from `cartoonbank.com`. All rights reserved.)

14.3 Media Privacy: Narrowcasting Concept

In traditional conferencing systems, participants have little or no privacy, as their voices are by default shared with all others in a session. Such systems cannot offer participants options for muting and deafening other members. The concept of narrowcasting can be applied to make these kinds of filters available in multimedia conferencing systems. A symmetric model treats media sinks (such as, listeners) as full citizens, peers of the media sources (conversants' voices), and we defined therefore duals of mute & select: deafen & attend respectively block a sink or focus on it to the exclusion of others. Figure 14.4 shows a famous carving which informally illustrates multimodal narrowcasting. Three monkeys—Kikazaru (blocked ears), Iwazaru (covered mouth), and Mizaru (with covered eyes)—manifest the notion of

FIGURE 14.4
Media privacy (narrowcasting features).

limiting media vectors. Kikazaru can not hear but can speak and see; Iwazaru can not speak but can see and hear; Mizaru can not see but can hear and speak.

For modern groupware situations like teleconferences, in which everyone can have a presence across the global network, users want to shift and distribute attention (apathy) and accessibility/availability/exposure (privacy), and narrowcasting provides a formalization of such filters. The narrowcasting predicate calculus [14], shown in Figure 14.5, is an appropriate basis for such a permission scheme.

Users are represented by objects in an interface which have attributes corresponding to narrowcasting state. To distinguish between operations reflexive

---

The general expression of inclusion is:

`active (object x)` = ¬`exclude (x)`∧
(∃ *y* (`include` (*y*)∧ (`self` (*y*) ⇔ `self` (*x*))) ⇒ `include` (*x*)).   (1)

So, for `mute` and `select` (`solo`), the relation is:

`active (source x)` = ¬`mute` (*x*) ∧
(∃ *y* (`select` (*y*) ∧ (`self` (*y*) ⇔ `self` (*x*))) ⇒ `select` (*x*)),   (1a)

`mute` explicitly turning off a source, and `select` disabling the complement of the selection (in the spirit of "anything not mandatory is forbidden"). For `deafen` and `attend`, the relation is:

`active (sink x)`= ¬`deafen` (*x*) ∧
(∃ *y* (`attend` (*y*) ∧ (`self` (*y*) ⇔ `self` (*x*))) ⇒ `attend` (*x*)).   (1b)

FIGURE 14.5
Formalization of narrowcasting and selection functions in predicate calculus notation, where '¬' means "not," '∧' means conjunction (logical "and"), '∃' means "there exists," '=>' means "implies," and '⇔' means mutual implication (equivalence).

(invoked on oneself) and transitive (on others), the attributes include a `self` flag, applied to both sources and sinks. The duality between source and sink operations is tight, and the semantics are analogous: an object is inclusively enabled by default unless it is explicitly excluded (with $\overbrace{\texttt{mute}}^{\text{source}}$ or $\overbrace{\texttt{deafen}}^{\text{sink}}$), or, peers of the same `self`/`non-self` class are explicitly included (with $\overbrace{\texttt{select [solo]}}^{\text{sources}}$ or $\overbrace{\texttt{attend}}^{\text{sinks}}$) when the respective object is not. Narrowcasting attributes are not mutually exclusive, and the dimensions are orthogonal. Because a source or sink is active by default, invoking `exclude` and `include` operations simultaneously on an object results in its being disabled. For instance, a sink might be first `attended`, perhaps as a member of some non-singleton subset of a chatspace's sinks, then later `deafened`, so that both attributes are simultaneously applied. (As audibility is assumed to be a revocable privilege, such a seemingly conflicted attribute state disables the sink, whose receptivity would be restored upon resetting its `deafen` flag.) Symmetrically, a source might be `selected` and then `muted`, akin to inclusion on a "short list" but relegated to back-up.

Our system allows each user to send or receive data streams to/from a specific recipients in a session. For easier understanding, we consider only audio streams in this chapter, but the design applies to other media types. Narrowcasting audio commands are listed and their characteristics arrayed in Table 14.3.

In this section, we formally define four narrowcasting commands. In the following expressions, P_a denotes the actor (controller), P_o the object (controllee), P_i a sender of the media (source), P_j a receiver of the media (sink), for $a, i, j, o \in \{1, \ldots, n\}$, where n is the total number of participants.

14.3.1 Mute

The narrowcasting command `mute` blocks media coming from a source. The mute in traditional systems is a self-mute function which allows a user to withhold his/her media from other participants, but the modern `mute` is a control function that can select another participant (or a group of participants) to disallow media towards the controller, still allowing other participants to hear the controllee. The Σ operator composites media from the respective participants.

$$
P_j \leftarrow \begin{cases} \sum_{i=1}^{n} P_i - P_j - P_o & \text{when } \overbrace{P_j = P_a}^{\text{transitive}} \text{ or } \overbrace{P_a = P_o}^{\text{reflexive}}, \\ \sum_{i=1}^{n} P_i - P_j & \text{otherwise.} \end{cases} \tag{14.1}
$$

The example modeled by the matrix in the first column of Table 14.3 illustrates when P_1 `mutes` another participant P_2. In this example, $n = 4$, $P_a = P_1$

TABLE 14.3

Narrowcasting Commands

	P_1 mutes P_2	P_1 deafens P_2	P_1 selects P_2	P_1 attends P_2
Media vectors				
Semantics	Block the media stream coming from a source.	Block media streams going to a sink.	Limit projected sound to particular sources.	Limit received sound to particular sinks.
Situation	Participant wants to block media from specific participants.	Participant wants to block media to specific participants.	Participant wants to receive media only from particular participants.	Participant wants to send media to specific participants.
Figurative avatars				
Mobile icons	$\bar{\triangle}$	$-\triangle-$	$\overset{+}{\triangle}$	$+\triangle+$
Media distribution	$\begin{bmatrix} \times & 1 & 1 & 1 \\ 0 & \times & 1 & 1 \\ 1 & 1 & \times & 1 \\ 1 & 1 & 1 & \times \end{bmatrix}$	$\begin{bmatrix} \times & 0 & 1 & 1 \\ 1 & \times & 1 & 1 \\ 1 & 1 & \times & 1 \\ 1 & 1 & 1 & \times \end{bmatrix}$	$\begin{bmatrix} \times & 1 & 1 & 1 \\ 1 & \times & 1 & 1 \\ 0 & 1 & \times & 1 \\ 0 & 1 & 1 & \times \end{bmatrix}$	$\begin{bmatrix} \times & 1 & 0 & 0 \\ 1 & \times & 1 & 1 \\ 1 & 1 & \times & 1 \\ 1 & 1 & 1 & \times \end{bmatrix}$

(the controller), and $P_o = P_2$ (the controllee). Due to this operation, P_1 will not receive any media from P_2. (This is actually a simplification of the evaluation performed by our system, since our model supports multipresence, the designation by a single human user of possibly multiple iconic representatives in an interface. Such complicated subtleties are beyond the scope of this chapter.)

14.3.2 Deafen

Deafen is a sink-related media privacy command that blocks media streams to a selected participant. For example, if Bob (P_1) wants to share his media with everyone in a conference except Alice (P_2), then Alice will not receive any streams from Bob if Bob deafens Alice. (Transposing the participants suggests an equivalent operation via reciprocity of sources and sinks, P_2 mutes P_1.) The second column of Table 14.3 shows the media relationship among four participants when `deafen` is invoked.

$$P_j \leftarrow \begin{cases} \sum_{i=1}^{n} P_i - P_j - P_a & \text{when } \overbrace{P_j = P_o}^{\text{transitive}}, \\ \phi & \text{when } \overbrace{P_a = P_o}^{\text{reflexive}}, \\ \sum_{i=1}^{n} P_i - P_j & \text{otherwise.} \end{cases} \quad (14.2)$$

Again in this example, $n = 4$, $a = 1$, and $o = 2$.

14.3.3 Select (Solo)

The privacy command `select` limits received media to particular sources. For instance, students might `select` a teacher to avoid distractions. P_1 will receive media only from P_2 if P_1 selects P_2, implicitly muting the complement of the selection. The third column of Table 14.3 shows the media relationships among four participants when `select` is invoked; two vectors are disabled in this case.

$$P_j \leftarrow \begin{cases} P_o & \text{when } P_j = P_a \text{ and } \overbrace{P_a \neq P_o}^{\text{not reflexive}}, \\ \sum_{i=1}^{n} P_i - P_j & \text{otherwise.} \end{cases} \quad (14.3)$$

14.3.4 Attend

`Attend` is the other including command for narrowcasting, limiting received media to a particular recipient. If Alice `attends` Bob, only Bob will hear Alice, since other participants are implicitly `deafened`. The rightmost column of Table 14.3 shows the media relationship among four participants when `attend` is invoked; again two media vectors are suppressed.

$$P_j \leftarrow \begin{cases} \sum_{i=1}^{n} P_i - P_j & \text{when } P_j = P_o \text{ and } \overbrace{P_a \neq P_o}^{\text{not reflexive}}, \\ \sum_{i=1}^{n} P_i - P_j - P_a & \text{otherwise.} \end{cases} \quad (14.4)$$

14.4 System Design and Implementation

The main required functions for media control are policy configuration, policy evaluation, media mixing, and media distribution. The Media Server Component Model (top right of Table 14.2) selected for our implementation comprises a centralized focus (collocated with the policy server), a centralized mixer, and

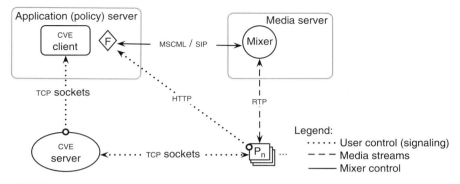

FIGURE 14.6
Media Server Component Model with collaborative virtual environment integration.

distributed participants. The architecture, elaborated in Figure 14.6, embeds policy configuration, media mixing, and a collaborative virtual environment (CVE) interface within a SIP framework. All the components in this architecture are standard SIP UAs extended with additional user interfaces needed for media policy configuration and control. Communication protocols XCAP (Extensible Markup Language Configuration Access Protocol) and MSCML (Media Server Control Markup Language) [16] are IETF standards, considered in the following descriptions.

14.4.1 Policy Configuration

In an extended SIP framework, conference participants could configure privacy by sending requests to the policy server using XCAP [15], a standardized way to use HTTP to store, retrieve, and manipulate configuration and application data in XML format. In our proof-of-concept, participants set policies using GUIs to invoke narrowcasting commands on specified controllees, and control is distributed via TCP sockets or HTTP directly (without XCAP).

14.4.2 Policy Evaluation

An application server performs three major functions to evaluate policy:

Evaluating policies configured by each participant: The policy from each participant can be logically compiled into a matrix, as shown in Table 14.4, where entry c_{ij} of the matrix represents connectivity of source i to sink j, and the main diagonal is populated by "don't care"s. Each participant $(P_1, P_2, \ldots, P_n)$, where n is the total number of participants, logically sets permissions in authorized cells. Since a media relationship ultimately factors at least two participants, a source and a sink, each cell contains policies from both. For example, $P_1 \rightarrow P_2$, i.e., media sourced at P_1 and sunk at P_2, has

TABLE 14.4
Policy Matrix $P = [\mathrm{p}_{ij}]$

	P_1	P_2	$\ldots$	P_n
P_1		$\mathrm{P}_1(\mathrm{P}_1 \to \mathrm{P}_2)$ $\mathrm{P}_2(\mathrm{P}_1 \to \mathrm{P}_2)$	$\ldots$	$\mathrm{P}_1(\mathrm{P}_1 \to \mathrm{P}_n)$ $\mathrm{P}_n(\mathrm{P}_1 \to \mathrm{P}_n)$
P_2	$\mathrm{P}_2(\mathrm{P}_2 \to \mathrm{P}_1)$ $\mathrm{P}_1(\mathrm{P}_2 \to \mathrm{P}_1)$		$\ldots$	$\mathrm{P}_2(\mathrm{P}_2 \to \mathrm{P}_n)$ $\mathrm{P}_n(\mathrm{P}_2 \to \mathrm{P}_n)$
$\vdots$	$\vdots$	$\vdots$	$\ddots$	$\vdots$
P_n	$\mathrm{P}_n(\mathrm{P}_n \to \mathrm{P}_1)$ $\mathrm{P}_1(\mathrm{P}_n \to \mathrm{P}_1)$	$\mathrm{P}_n(\mathrm{P}_n \to \mathrm{P}_2)$ $\mathrm{P}_2(\mathrm{P}_n \to \mathrm{P}_2)$	$\ldots$	

policy involvement of both P_1 and P_2: P_1 sets permissions about whether or not to send media to P_2, and at the same time, P_2 sets permissions about whether or not to receive such media. The policy then is evaluated depending on the combined relationship between P_1 and P_2.

Responding to participants regarding changes made in the policy: A policy evaluation report (confirming success or alerting failure of a configuration request) might be sent to participants via standard XCAP response codes.

Sending requests to a media mixer for necessary media mixing: After compiling the media policies, the system determines which media streams need to be mixed and delivered to whom. Using MSCML or some equivalent, the policy server instructs the media mixer to perform the necessary mixing.

14.4.3   Media Mixing and Distribution

The media server receives MSCML requests from a policy configuration server. According to the accumulated state, it performs the necessary mixing and delivers these streams to subscribed participants. The maximum number of mixes, the power set of the participants excluding the empty and universal sets, is

$$\sum_{i=1}^{n-1} nC_i = 2^n - 2. \tag{14.5}$$

Therefore, for $n = 3, 4, 5$, the maximum number of mixes would be 6, 14, and 30, respectively. However, depending on participants' media privacy requests, the actual number of mixes might be less.

Figure 14.7 illustrates narrowcasting media distribution among four participants when P_1 mutes P_2 and deafens P_4. All participants send their media

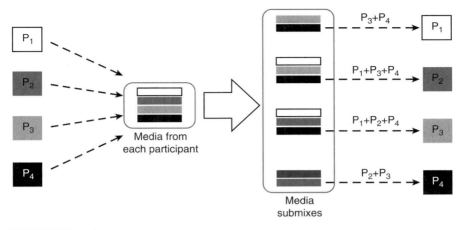

FIGURE 14.7
Media mixing and delivery (P_1 mutes P_2 and deafens P_4).

to the media server. The media server mixes only the necessary streams and delivers them back to the appropriate recipients.

14.4.4 Sample Mixing Configuration

Our prototype environment comprises a SIP server (BEA WebLogic SIP Server), an application server (BEA WebLogic Workshop), a media server (Dialogic/ Cantata Snowshore IP Media Server), and four SIP clients (X-lite). We implemented narrowcasting commands mute, deafen, attend, and (partially) select, integrating these filter functions into the application server. Figure 14.8 shows the control and media streams among a participant, application server, and media mixer when applying a narrowcasting command.

The MSCML configuration and audibility are shown in Table 14.5 when P_1 deafens P_2. P_1 makes a private group with P_3 and P_4, so P_1, P_3, and P_4 can hear each other, but P_2 cannot hear P_1. The application server evaluates the policy and configures the media server.

14.4.5 Narrowcasting Interfaces

A bridge between CVE clients and SIP-based narrowcasting allows distributed multimodal interfaces [19]. The results of narrowcasting operations are expressed aurally by the SIP-based mixer and visually by graphical interfaces. In contrast to general multimedia systems, virtual environments are characterized by the explicit notion of the position (location and orientation) of the perspective presented to respective users. Often such vantage points are modeled by the standpoints and directions of representative objects in a virtual space. These representatives might be more or less symbolic (abstract icons) or figurative (avatars), but act as delegates of human users. Icons and avatars can

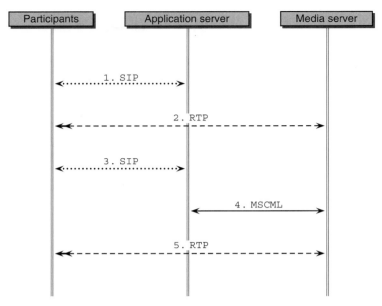

FIGURE 14.8
Communication flow between SIP entities: A default configuration (1) establishes a normal session (2), but it can be adjusted (3) to reconfigure (4) the mixes returned to the participants (5).

TABLE 14.5
MSCML configuration: P_1 deafens P_2

Participant	ID	Team Members	Mixmode	Hears
P_1	P_1	P_3,P_4	Private	$P_2 + P_3 + P_4$
P_2	P_2	None	Full	$P_3 + P_4$
P_3	P_3	P_1,P_4	Full	$P_1 + P_2 + P_4$
P_4	P_4	P_1,P_3	Full	$P_1 + P_2 + P_3$

reify embodied virtuality, treating abstract presence as user interface objects, symbols and manifestations of sources and sinks. We have developed workstation and mobile interfaces [16] to manipulate narrowcasting attributes in virtual spaces via a Java3D [23] interface. This "virtual reality"-style interface features perspective displays of virtual rooms and spaces with figurative avatars, each of which can be associated with an audio source, corresponding to the voice of a user. A participant can rearrange the locations of avatars in virtual spaces and designate sinks, through whose ears the resulting spatialized soundscape is heard. Also, each participant can apply narrowcasting attributes to the avatars, altering the sound mix. Recalling the monkeys in Figure 14.4, Figure 14.9 illustrates the visual cues used for narrowcasting, including a hand covering the mouth of a muted avatar and hands

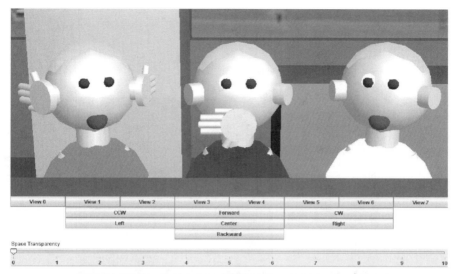

FIGURE 14.9

Narrowcasting control in virtual environment: P_1 (avatar 0, right) mutes P_2 (avatar 1, middle) and deafens P_3 (avatar 2, left).

clapped over the ears of a deafened avatar. Manipulations by participants are communicated using a CVE client/server architecture, which framework allows multimodal clients to exchange status data through the network. Clients currently include sound spatializers, telepresence applications, panoramic and turnoramic browsers [10], music visualizers, motion platforms, and mobile interfaces.

Besides the previously-described web- and workstation-based interfaces, we have also built a mobile narrowcasting display and control, shown in Figure 14.10, although we have not yet integrated the mobile audio stream, so it is currently more useful as a collocated "remote control" than as a truly mobile application. Symbolic representations of narrowcasting operations were developed for mobile interfaces by flattening figurative 3D avatars to 2.5D icons, as seen in the second-last row of Table 14.3. In the mobile application, narrowcasting attributes graphical displays are triply encoded—by position (before the "mouth" for mute and select, straddling the "ears" for deafen and attend), symbol ('+' for include & '−' for exclude), and color (green for

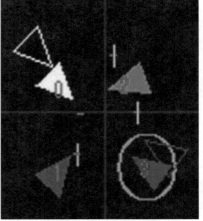

(a) DoCoMo platform   (b) Sample screen

FIGURE 14.10

Mobile narrowcasting interface. Isosceles triangles iconfiy users. #0 has surged forward and #3 has swayed to the left, but the "ghost icons" show that their local state has not yet been flushed out to the network. #1 has been muted and selected. #2 has been deafened and attended. #3 is in the (currently singleton) selection set, toggled for yawing (spinning).

assert & red, yellow, and orange for inhibit—by self, other, and implicitly, respectively).

The bridge between the interfaces and the SIP-based backend is a 'read-only' CVE client embedded in the SIP application server. When the policy server is launched, the embedded client connects to a CVE session server and opens a duplex channel for each member in the conference. Every time a user enables or disables one of the narrowcasting attributes, the action is relayed to the embedded CVE client. As each message is received, the client invokes the necessary methods to reflect the changed status in the SIP conference.

14.4.6 System Performance

The narrowcasting control is basically lightweight: the commands are typically infrequent, and each of them is easily processed by an application server. For excluding narrowcasting commands (deafen and mute), the time complexity is constant ($O(1)$), independent of the number of participants in a session.

For including narrowcasting commands (`select` and `attend`), in which the connectivity state of the complement of the selection needs to be adjusted, the time complexity is O(n), linear in the number of participants. The configuration for the IP media server used in our experiments supports up to a hundred clients. Even though our laboratory testbed uses a much smaller user pool, typically about four, there is no reason not to assume that the signaling protocol can keep up with practical realtime demands and support the same number of session participants.

14.5 Conclusion and Future Research

The Media Server Component Model architecture can be deployed for policy-based media-mixing and narrowcasting within the standard SIP framework for multimedia conferencing systems. Narrowcasting privacy command interpretation includes a policy evaluation algorithm, a media mixing and delivery mechanism that considers fine-grained policy configured by participants. The policy can be displayed and controlled via interfaces in which hands and other attributes (megaphones and ears trumpets) clapped over figurative avatars' mouths and ears represent audio stream filters. Applications like "Second Life" and virtual environments represent a fertile platform for conferences, distance learning, meetings, and recreational activity like chatspaces.

14.5.1 Practical Conferencing

In ordinary conversation, participants generally observe turn-taking, as in a CSMA/CD (carrier sense multiple access/collision detection) protocol with discretionary backup. That is, an utterance that collides with another will cause one or both of the simultaneous speakers to stop and wait until a break before repeating.

One might wonder what happens to such conversational turn-taking in the presence of asymmetric media filters and the absence of a moderator. Narrowcasting features—like blocklists, side channels, and call-within-a-call—complicate teleconferences, since a deafened conversant might not be aware that another is talking and multiple sources might speak at once. If some participants in a conference are muted or deafened to some other participants, without formal floor control there is a likelihood of some "talking on top of" others. In the absence of common floor control, won't private chats and decentralized control lead to cacophonous anarchy? Without "traffic signals," how can collisions be avoided?

In fact, such parallel conversation streams are not a problem. For example, if two participants set up a private side-conference using narrowcasting commands, even though their utterances might collide with others', they wouldn't expect or want others to stop conversing. Rather they "listen with

FIGURE 14.11
Theme-based discussion in articulated chatspace. (© The New Yorker Collection 1961 Robert J. Day from cartoonbank.com. All rights reserved.)

one ear" to ongoing conversations while enjoying their own caucus. Listeners can still untangle conversational threads, by context, voice quality, etc., as suggested by Figure 14.11. Just as in real social situations, including informal gatherings like parties, multiple simultaneous speakers are analyzable. Even "linear" conversations like formal meetings might have some subsets of conversants whispering among themselves while a main speaker is talking. Narrowcasting audio interfaces are even more useful when extended by spatial sound and attenuation based on mutual virtual position (source projection, sink bearing, and distance) [20], distributing the respective voices across a soundscape.

14.5.2 Event Notification Framework for Exchanging Narrowcasting Control Status Information

A narrowcasting conference allows one to influence the media streams of other participants, as well as those of oneself. As a result, each session member can send or receive media streams to and from specific groups in a conference.

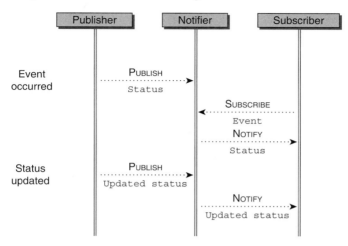

FIGURE 14.12
Event notification mechanism (RFC 3265).

It is potentially useful for participants to be informed of who is control-ling media using a narrowcasting command (mute, deafen, select, or attend). For example, teachers may want to be notified whether any stu-dents have muted them during a lecture. Depending on the conditions, such role-based transparency could be appropriate or not. A parent might insist upon the ability to override a teenager's 'ignore' filter: "How dare you mute me?" In the current narrowcasting implementation, a notification mechanism has not been incorporated. This section proposes such an event notification mechanism.

The SIP event notification framework (RFC 3265) [22] defines general mech-anisms for subscribing to, and receiving notifications of, events within SIP networks. The framework uses standard SIP methods PUBLISH, SUBSCRIBE, and NOTIFY to deliver event-related information. Figure 14.12 shows an event information exchange flow.

In order to allow conference participants to be notified of narrowcasting control status, an event template package can be introduced. A specialized template event package, which might be called "NARROWCAST" or "NC," needs to be standardized within the IETF community. Such a template event package could be used in combination with the "CONFERENCE" event package (defined in RFC 4575 [23]) to inherit the state of a conference. As shown in Figure 14.13, the publisher (a narrowcasting conference participant), the notifier (presence server), and the subscriber (another narrowcasting conference participant) must support "CONFERENCE.NC," a combination of the CONFERENCE event and NARROWCAST template event packages. In this example, P_2 subscribes to P_1 and mentions CONFERENCE.NC in the "Event" header in the subscription body. As a result, P_2 will be notified of the narrowcasting control informa-tion of P_1.

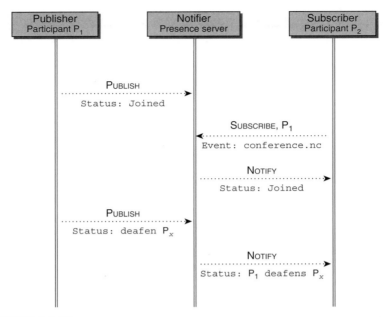

FIGURE 14.13
Narrowcasting control status notification.

14.5.3   Architectural and Interface Refinement

Future research includes allowing selection of multiple sources and sinks for narrowcasting commands to support arbitrary multiuser narrowcasting configurations. A basic challenge is that the SIP side-conference model is not rich enough to support arbitrary multiparty narrowcasting with orthogonal attributes. A related engineering issue is that MSCML is not expressive enough to convey such detailed, articulated channel crossbar control. We are also considering other conference models with multiple policy servers or media mixers. For instance, as multimedia processing becomes less of a specialized service and more of a commodity, a grid computing paradigm could be used instead of a centralized server architecture to mix and deliver media streams to distributed narrowcasting-enabled terminals. `Muffle` (partial `deafen`) and `muzzle` (partial `mute`) will enrich the narrowcasting state space [24,25]. We will also generalize policy determination in metasessions with multiple simultaneous chatspaces, in which one has presence across multiple virtual spaces, each with multiple conversants, including "multipresence," allowing designation of multiple instances of "`self`" [26].

14.5.4   Convergence

Besides wireline-connected workstation-based interfaces, narrowcasting might find an even more fertile platform in mobile devices. The "four-play" convergence of telephony, television/video, internet, and wireless is driving a proliferation of new devices and services. Mobile terminals, almost as intimate as clothing, are a kind of wearable computer, and a diversity of next-generation functionalities and form factors for smartphones is emerging, including mobile stereotelephony, inspired by cyberspatial audio [27] and augmented audio models. Meanwhile, location-based services—along with seamless handoff, FMC (fixed-mobile convergence), and heterogeneous roaming via MIMO (multiple input/multiple output) smart antennas leading to software-defined radio (SDR) and cognitive radio—leverage geolocation and portable GPS/GIS. Such advanced sensing enables ubicomp and ambient intelligence, including an awareness of user status and availability, and articulated models of privacy, like narrowcasting, which allow users to distribute their attention, availability, and virtual presence. Multipresence and persistent channels, encouraged by ABC (always best connected) networks, will extend the way people communicate.

Acknowledgements

This research has been sponsored in part by grants from Kyushu University Research Superstar Program and the Japan Science and Technology Foundation. We also thank Dialogic/Cantata and BEA for providing media and application servers for our experiments.

References

[1] Koskelainen, P., H. Schulzrinne, and X. Wu. 2002. A SIP-based conference control framework. In *Proc. NOSSDAV: 12th Int. Wkshp. on Network and Operating Systems Support for Digital Audio and Video* 53–61. New York, NY: ACM Press.

[2] Dommel, H.-P., and J. Garcia-Luna-Aceves. 1995. Floor control for activity coordination in networked multimedia applications. In *Proc. APCC: 2nd Asian-Pacific Conference on Communications.* June, Osaka, Japan.

[3] Even, R., and N. Ismail. 2006. RFC 4825—Conferencing Scenarios. July.

[4] Spinello, R. A. 2004. *Cyberethics Morality and Law in Cyberspace.* Jones and Bartlett Publishers, Sudbury.

[5] Berc, L., H. Gajewska, and M. Manasse. 1995. Pssst: Side conversations in the argo telecollaboration system. In *Proc. UIST: 8th Annual ACM*

Symposium on User Interface and Software Technology 155–56. November, New York, NY: ACM Press.

[6] Yankelovich, N., J. McGinn, M. Wessler, J. Kaplan, J. Provino, and H. Fox. 2005. Private communications in public meetings. In *Proc. CHI: Human Factors in Computing Systems,* 1873–76. April, Portland, OR: ACM Press.

[7] Wang, C. C., J. Cahnbley, and J. W. Richardson. 2003. US Patent—Method and system for providing a private conversation channel in a videoconference system. June. Publication No. WO 2003/053034 A1.

[8] Fernando, O. N. N., K. Adachi, U. Duminduwardena, M. Kawaguchi, and M. Cohen. 2006. Audio narrowcasting and privacy for multipresent avatars on workstations and mobile phones. *IEICE Trans. on Information and Systems* E89-D(1): 73–87.

[9] Alam, M. S., M. Cohen, and A. Ahmed. 2007. Narrowcasting—Controlling media privacy in SIP multimedia conferencing. In *Proc. IEEE CCNC: 4th Consumer Communications and Networking Conf.* January, Las Vegas.

[10] Alam, M. S., M. Cohen, and A. Ahmed. 2008. Narrowcasting: Implementation of privacy control in SIP conferencing. *JVRB: Journal of Virtual Reality and Broadcasting* 4(9).

[11] Rosenberg, J., H. Schulzrinne, G. Camarillo, A. Johnston, J. Peterson, R. Sparks, M. Handley, and E. Schooler. 2002. RFC 3261—SIP: Session Initiation Protocol. June.

[12] Rosenberg, J. 2006. RFC 4353—A framework for conferencing with the session initiation protocol. February.

[13] Cho, Y.-H., M.-S. Jeong, J.-T. Park, and W.-H. Lee. 2005. Distributed management architecture for multimedia conferencing using SIP. In *Proc. DFMA: 1st Int. Conf. on Distributed Frameworks for Multimedia Applications.* February, Besancon, France.

[14] Cohen, M. 2000. Exclude and include for audio sources and sinks: Analogs of mute & solo are deafen & attend. Presence: Teleoperators and Virtual Environments 9(1): 84–96. February. www.u-aizu.ac.jp/~mcohen/welcome/publications/ie1.pdf.

[15] Rosenberg, J. 2007. RFC 4897—The Extensible Markup Language (XML) Configuration Access Protocol (XCAP). May.

[16] Dyke, J. V., E. Burger, and A. Spitzer. 2006. RFC 4722—Media Server Control Markup Language (MSCML) and Protocol. November.

[17] Walsh, A. E., and D. Gehringer. 2002. *Java 3D API Jump-Start.* Prentice-Hall. Upper Saddle River, NJ, USA.

[18] Sowizral, H., K. Rushforth, and M. Deering. 2000. *The Java 3D API Specification.* 2nd ed. Addison-Wesley. Upper Saddle River, NJ, USA.

[19] Alam, M. S., M. Cohen, and A. Ahmed. 2007. Articulated narrowcasting for privacy and awareness in multimedia conferencing systems and design for implementation within a SIP framework. In *Proc. ICME: Int. Conf. on Multimedia & Expo.* July, Beijing.

[20] Cohen, M., and N. Koizumi. 1998. Virtual gain for audio windows. *Presence: Teleoperators and Virtual Environments* 7(1): 53–66. February. issn 1054-7460.

[21] Cohen, M., and E. M. Wenzel. 1995. The design of multidimensional sound interfaces. In *Virtual Environments and Advanced Interface Design,* ed. W. Barfield and T. A. Furness III, Chapter 8, 291–346. Oxford University Press. New York, New York, USA.

[22] Roach, A. B. 2002. RFC 3265—Session Initiation Protocol (SIP)—specific event notification. June.

[23] Rosenberg, J., H. Schulzrinne, and O. Levin. 2006. RFC 4575—A Session Initiation Protocol (SIP) event package for conference state. August.

[24] Cohen, M. 2003. Emerging exotic auditory interfaces. In *114th Convention of the AES*, Amsterdam, March. Preprint # 5819.

[25] Cohen. M. 1993. Throwing, pitching, and catching sound: Audio windowing models and modes. *IJMMS: The Journal of Person-Computer Interaction* 39(2): 269–304. August. www. u-aizu.ac.jp/~mcohen/welcome/publications/tpc.ps.

[26] Cohen, M. 1998. Quantity of presence: Beyond person, number, and pronouns. In *Cyberworlds*, eds. T. L. Kunii and A. Luciani, Chapter 19, 289–308. Tokyo: Springer-Verlag. www.u-aizu.ac.jp/~mcohen/welcome/publications/bi1.pdf.

[27] Cohen, M., J. Herder, and W. L. Martens. 1999. Cyberspatial audio technology. *Journal of the Acoustical Society of Japan* 20(6). November.

15

Q-SIP/SDP for QoS-Guaranteed End-to-End Real-Time Multimedia Service Provisioning on Converged Heterogeneous Wired and Wireless Networks

Young-Tak Kim and Young-Chul Jung
Yeungnam University

CONTENTS

15.1 Introduction 347
15.2 Background and Related Work 349
15.3 Session Establishment for QoS-Guaranteed Real-Time Multimedia
     Service Provisioning on Broadband Convergence Network (BcN) 352
15.4 Experimental Implementations and Performance Evaluation of QoS
     Provisioning in Heterogeneous Wired and Wireless Convergence
     Networks 363
15.5 Conclusions 370
     References 371

15.1   Introduction

In next generation Internet, converged heterogeneous broadband wired and wireless subnetworks and various terminal equipments will be interconnected [1]. As an example, the backbone transit network may be implemented by IP/T-MPLS (Transport Multiprotocol Transport Switching) [2], IP/Carrier Ethernet [3], or IP/WDM (wavelength-divisioning multiplexing) technology with xDSL or PON (passive optical network) access loop, while the home/office intranet may be implemented based on IEEE 802.3 Fast/Gigabit Ethernet or IEEE 802.11 wireless local area network (LAN) (i.e., WiFi) technology [4]. Each subnetwork may have a different control plane function, such as signaling protocol for connection establishment or resource reservation with connection admission control (CAC). Since most home/office intranets are currently implemented with connectionless and contention-based Ethernet or wireless LAN, the quality of service (QoS)-guaranteed service provisioning is strictly limited without well-designed CAC.

Also, the end terminal may have different capabilities in multimedia data processing, display size and resolution. For example, the sender may be a desktop PC-based multimedia device with high capability in video data encoding/decoding and large display size, while the receiver may be a PDA-based multimedia device with limited capability and small display size. In this case, an end-to-end real-time multimedia session should be established that can be supported by both the end terminal nodes. As a result, we must resolve two major problems in order to provide QoS-guaranteed end-to-end real-time multimedia service on such heterogeneous networking environments: (1) multimedia session establishment and negotiation that adjusts the difference in the capability of multimedia data processing at the end terminal nodes, (2) QoS-guaranteed connection establishment or resource reservation with connection admission control (CAC) in each heterogeneous subnetworks along the path.

For the multimedia session establishment and negotiation among heterogeneous multimedia terminals with different capability in audio/visual data processing, QoS-extended SIP (Q-SIP)/Session Description Protocol (SDP) should be used. Session Initiation Protocol (SIP) is a signaling protocol on the application-layer for managing sessions with one or more participants [5]. The session may be VoIP (Voice-over-Internet Protocol) telephone calls, real-time video conferences, or a real-time multimedia group game. SDP is used for describing multimedia sessions for the purpose of session announcement, session invitation, and other forms of multimedia session initiation [6]. In order to establish a multimedia communication session, SIP/SDP messages must be exchanged among participants to check the availability of participants, capability of terminals, determination of media type, transport protocol, format of media, related network and port addresses [7]. After determination of parameters for a multimedia session, QoS-guaranteed connection for the session must be established between the participant's terminals.

When the QoS parameters and traffic parameters for the multimedia session are determined via the exchanges of SIP/SDP messages between the end terminals, a network layer connection establishment in the connection-oriented subnetwork and a resource reservation with CAC should be provided. For the IP/MPLS transit network, the resource ReSerVation Protocol with Traffic Engineering extension (RSVP-TE) [8] can be used to establish an MPLS LSP for the multimedia session. In the non-connection-oriented subnetworks, such as IEEE 802.3 Fast/Gigabit Ethernet or IEEE 802.11 wireless LAN, however, there is no explicit connection establishment. Instead, we must provide the bandwidth allocation by configuration of Fast/Gigabit Ethernet Switch ports with proper CAC function for the guaranteed QoS provisioning. The end-to-end connection establishment or resource reservation in network on such heterogeneous networking environment, should be tightly cooperated with the session layer signaling (e.g., SIP/SDP).

In this chapter, we propose a session and connection management architecture for the QoS-guaranteed real-time multimedia service provisioning on converged heterogeneous wired and wireless network, with Q-SIP/SDP, RSVP-TE and CAC functions. The detailed interaction scenario and related algorithms for QoS-guaranteed real-time multimedia session, resource reservation, and connection establishment are proposed. From the experimental implementation of the proposed scheme on a small-scale converged hetero-geneous wired and wireless network testbed, we verified that the proposed architecture is feasible for the QoS-guaranteed real-time multimedia service provisioning. We analyze the network performance and end-to-end QoS performance parameters.

The rest of this chapter is organized as follows. Section 15.2 briefly explains the related work, including heterogeneous networking technologies in broadband convergence network (BcN), SIP/SDP, and RSVP-TE. Section 15.3 explains the functional model of the proposed session- and network-layer signaling with related resource reservation and CAC. Section 15.4 explains the performance analysis of QoS provisioning with the proposed scheme based on experiments on a testbed. Finally, Section 15.5 concludes this chapter.

15.2 Background and Related Work

15.2.1 Heterogeneous Subnetwork Technologies in Broadband Convergence Network (BcN)

In the BcN, various broadband networking technologies, such as IP/MPLS or IP/WDM transit network, vDSL, or PON access loop, and IEEE 802.3 Giga-bit/Fast Ethernet or IEEE 802.11 wireless LAN (WiFi) home/office intranet, will be integrated [1]. Each subnetwork may have a different signaling mechanism for connection establishment or resource reservation with connection admission control (CAC). Figure 15.1 shows the end-to-end session layer signaling and per-domain subnetwork management for QoS-guaranteed multimedia service provisioning.

The IP/MPLS or IP/WDM backbone transit network will be equipped with (G)-MPLS signaling, such as RSVP-TE, that provides constraint-based LSP (label switched path) establishment in the MPLS network domain. The transit network may be implemented by DiffServ-over-MPLS virtual overlay networks, in order to efficiently provide QoS-guaranteed differentiated multimedia services.

The access network interconnects commercial backbone transit network and home/office intranet. Currently, most access loops are implemented by vDSL or PON technologies that provide a fixed bandwidth channel. Since most access loops do not include packet switching functions, bandwidth-guaranteed channel allocation with simple CAC is a practical scheme for the QoS-guaranteed multimedia service provisioning in the access loop.

The IEEE 802.3 Fast/Gigabit Ethernet and IEEE 802.11 wireless LAN (WiFi) are the most popular home/office intranet technologies because of the simplicity and low installation cost. The CSMA/CD-based Ethernet and CSMA/CA-based WiFi, however, are contention-oriented networking technology where bandwidth and QoS cannot be guaranteed by itself. Well-designed

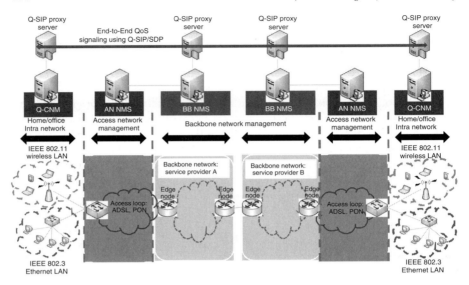

FIGURE 15.1

End-to-end session layer signaling and per-domain subnetwork management for QoS-guaranteed multimedia service provisioning.

CAC should be provided for the Ethernet switch or WiFi access point node that may be included in the home/office gateway.

15.2.2   SIP/SDP

SIP is a session-layer signaling protocol that is used for establishment, modification, and release of sessions among participants [5]. The session may be VoIP, video conference, multimedia distribution, and IMS (IP Multimedia Subsystem) applications. SIP supports five facets of establishing and terminating multimedia communications: user location, user availability, user capabilities, session setup, and session management. SIP may run on top of several different transport protocols, such as UDP and TCP.

SIP is composed of three major functional elements:

   i) *SIP user agents (UAs)* originate SIP requests to establish media sessions and to exchange media. A user agent can be a SIP client software running on a desktop PC, laptop, PDA, or other available devices.

  ii) *SIP servers* are intermediary devices that are positioned within the network and assist user agents in session establishment and other functions. There are three types of SIP servers defined in [5]: proxy, redirect, and registrar servers.

    • *A SIP proxy server* is used to help the routing request to the user's current location, to authenticate and authorize users for services, to implement provider's call-routing policies, and to provide features to users.

- *A redirect server* receives a request from a user agent or proxy, and returns a redirection response, indicating where the request should be redirected.

- *A registrar server* provides a registration function that receives the SIP registration request and updates the user agent's information into a location server or other database.

iii) *SIP location databases* support the location information of users, such as URLs or IP addresses, scripts, features and other preferences.

SIP invitations are used to create sessions and carry session descriptions that allow participants to agree on a set of compatible media types. SDP [6] has been designed to convey session description information, such as session name and purpose, the type of media (e.g., video, audio, etc.), the transport protocol (e.g., RTP/UDP/IP or H.320), and the media encoding scheme (e.g., H.261 video, MPEG video, etc.). SDP also includes information associated with QoS (e.g., bandwidth, delay and jitter) and security (e.g., encryption key). For a multicast session, multicast IP address and transport layer port for media are delivered; for a unicast session, the remote contact addresses (e.g., IP address and port number) are delivered.

When a multimedia communication session is accepted by the SIP/SDP signaling, the related traffic and QoS parameters should be determined. The traffic parameters for real-time multimedia services are committed information rate (CIR), committed burst size (CBS), excess burst size (EBS), and peak information rate (PIR). The QoS parameters include end-to-end transfer delay, delay variation (jitter) tolerance, packet error rate (PER), and packet loss rate (PLR). Based on the determined QoS and connection parameters, a network-layer connection establishment is requested using User–Network Interface (UNI) and Network–Node Interface (NNI) signaling [10].

When the subnetwork access mechanism is based on connectionless or contention mode, however, the network layer should allocate the required resources with the appropriate CAC mechanism. Some examples of contention-based subnetworks are IEEE 802.3 Fast/Gigabit Ethernet and IEEE 802.11 wireless LAN. In following section, we propose a detailed procedure of the network resource allocation scheme for QoS-guaranteed service provisioning.

15.2.3 RSVP-TE for IP/(T-)MPLS Subnetwork

After determination of the QoS and traffic parameters for a multimedia session using the SIP/SDP session layer signaling, QoS-guaranteed per-class-type connection must be established among the participant end terminal nodes [10]. In the network layer, connection establishment is accomplished by UNI and NNI signaling. For UNI signaling between the user terminal and the ingress edge router, RSVP-TE can be used to carry the connection request [8]. MPLS [9] NNI signaling standard for inter-AS (autonomous system) domain network, however, is not yet well standardized to establish per-class-type QoS-guaranteed connections across multiple AS domain networks of different network operators. As a consequence, an efficient alternative transit networking scheme must be provided so as to configure a per-class-type QoS-guaranteed Differentiated Service (DiffServ) provisioning and to support the scalable CAC [10]. In order to support the per-class-type DiffServ provisioning, RSVP-TE must provide traffic engineering extensions so as to deliver the traffic and QoS parameters.

Recently, ITU-T and IETF are jointly developing transport MPLS (T-MPLS), which uses a subset of the existing MPLS standards and is designed specifically for application in transport networks [2]. T-MPLS is designed as a connection-oriented packet switched application. It offers a simpler implementation by removing features that are not relevant to connection-oriented packet switching, and adding mechanisms that provide support of critical transport functionality. T-MPLS uses the same architectural principles of layered networking that are used in synchronous digital hierarchy (SDH) and optical transport network (OTN), and provides a reliable packet-based technology on the circuit-based reliable transport networking.

The user agent in the multimedia terminal must provide the RSVP-TE client function, while the ingress router must support the RSVP-TE server function. Since RSVP-TE establishes only unidirectional connection, two PATH-RESV message exchanges should be implemented to establish a bidirectional path between the user terminal and the ingress router [10]. The RSVP-TE UNI signaling message carries the QoS and connection parameters for the specified application data flows. RSVP-TE NNI signaling message is used by routers to deliver QoS requests to all nodes along the path in the IP/MPLS subnetwork, in order to establish and maintain the state to provide the requested QoS-guaranteed service. The routers should be able to reserve network resources according to the requested QoS and traffic parameters [11].

15.3   Session Establishment for QoS-Guaranteed Real-Time Multimedia Service Provisioning on Broadband Convergence Network (BcN)

15.3.1   Q-SIP/SDP Interaction for End-to-End QoS Session Negotiation

Figure 15.2 depicts the overall functional block diagram and interaction model of session and connection management in QoS-guaranteed multimedia service provisioning [10,12]. The user terminal will first discover available service from the service directory of the network operator that provides service profiles. The service subscription will define the service level agreement (SLA) of the

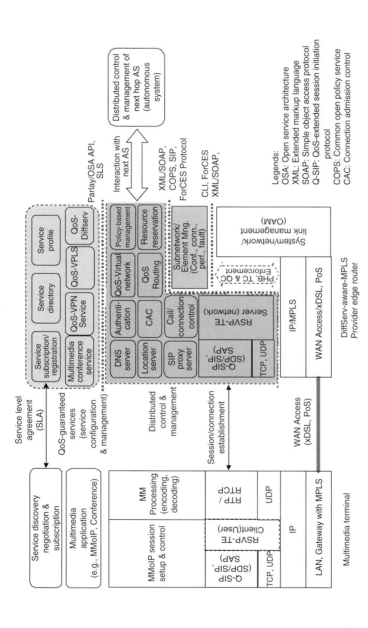

FIGURE 15.2

Functional block diagram and interaction model of session and connection management in QoS-guaranteed multimedia service provisioning.

service, and service level specification (SLS)/traffic level specification (TLS) will further specify the detailed traffic parameters and QoS parameters. For QoS-guaranteed service provisioning, the network operator must configure DiffServ-over-MPLS virtual overlay networks for differentiated services. The scalability can be achieved by using a virtual overlay network to configure the predefined paths for different traffic classes.

If a specific application service is agreed via SLA, a session setup and a connection establishment for the media application should be initiated using QoS-extended SIP (Q-SIP)/SDP and RSVP-TE signaling. Q-SIP/SDP will be used to find the current location of the destination, and to determine the availability and capability of the RoS-aware station (RSTA) terminal for the requested multimedia application. The Q-SIP proxy server may utilize a location server to provide some value-added service based on presence and availability management functions. The sending Q-SIP/SDP user agent will offer the list of possible media types and encoding standards, and the receiving Q-SIP/SDP user agent, considering its processing capacity, will select some acceptable media type and its encoding format, and reply with the acceptable list to the sender. Finally, the sending Q-SIP/SDP user agent will determine the selected media type and the encoding format to be used in the multimedia session.

Once a multimedia session description is agreed upon by the participants, a QoS-guaranteed per-class-type end-to-end connection establishment will be requested through UNI signaling. If the multimedia communication service is using a separated connection for each media type, multiple connections should be established for each media-type packet flow. As an end-to-end connection establishment signaling, RSVP-TE can be used in the backbone transit IP/MPLS network. The edge node (router) of the ingress provider network should take the role of connection control and management functions for on-demand connection establishments [10].

When the user's terminal is attached to a contention-based intra network, such as IEEE 802.3 Fast/Gigabit Ethernet or IEEE 802.11 wireless LAN, or the participant's terminal does not support RSVP-TE signaling, the end-to-end connection cannot be established; instead, per-class-type packet flow should be registered and controlled by the connection management function of the ingress provider edge (PE) node with CAC [4]. The customer network management (CNM) system may support the procedure of per-class-type packet flow registration [10]. In the following Section 15.3.3 and Section 15.3.4, we explain the packet flow management in contention-based intranet for QoS-guaranteed multimedia service provisioning.

Figure 15.3 depicts the Q-SIP/SDP message exchanges for QoS-guaranteed session and connection establishment in real-time multimedia service provisioning [7,12]. User agent A (the caller) invites user agent B (the callee) through an *INVITE* request message that is relayed by the SIP proxy server to the destination (callee). When the INIVITE message is created by the caller, the message body includes the information of services that the caller can provide. This information is contained in SDP, which describes the details of media, QoS, bandwidth, and security related attributes. When the callee

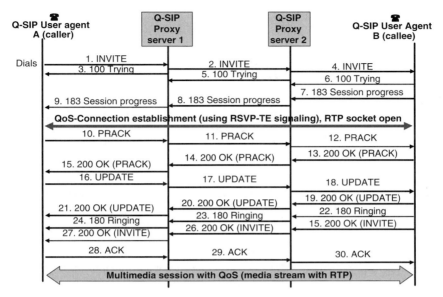

FIGURE 15.3
Q-SIP/SDP messages for QoS-guaranteed multimedia session establishment.

receives the INVITE request from the caller, the callee checks the requested QoS parameters in the SDP message, checks the availability of resource and capacity, and sends a response message (183 Session Progress Response) to the caller. In this response message, the callee specifies the acceptable media types, QoS parameters, and bandwidth and traffic parameters. When the caller receives the 183 message, it checks the parameters that have been selected by the callee, and sends a Provisional Response ACKnowledge (PRACK) message to the callee as the answer to the 183 message. When the callee receives the PRACK message, it sends a 200 OK message as the acknowledgement to the the the PRACK. Table 15.1 presents the Q-SIP messages used in multimedia session establishment.

After receiving the 200 OK message, the caller initiates the connection establishment procedure using network layer signaling, such as RSVP-TE UNI signaling in IP/MPLS subnetwork. Details of the connection establishment procedure in the IP/MPLS subnetwork using RSVP-TE UNI/NNI signaling are explained in the next section.

When the two connections are successfully established for each direction, the caller then sends an UPDATE message to inform that the multimedia stream may be turned on to deliver the multimedia audio/video data. By receiving the UPDATE message, the called party finally accepts the multimedia call/session by replying a 200 OK (INVITE) message to the caller. The caller sends an ACK message that informs the successfully established call/session. Bidirectional QoS-guaranteed multimedia communication can be provided at this moment. When either the caller or the called party sends

TABLE 15.1
Q-SIP Messages Using Multimedia Session Establishment

SIP Message	Direction	Major Contents
INVITE	A (caller) → B (callee)	Media information of the caller, QoS or security information
100 Trying	A (caller) ← B (callee)	Typically does not contain a *To* tag
183 Session Progress	A (caller) ← B (callee)	Information about the progress of the session (call state)
PRACK	A (caller) → B (callee)	Acknowledges receipt of reliably transported provisional response (1xx)
UPDATE	A (caller) → B (callee)	Used to change media session parameters in early dialogs (before the final response to the initial *INVITE*)
180 Ringing	A (caller) ← B (callee)	QoS or security information or ring tone or animations from user agent server (UAS) to the user agent client (UAC)
200 OK	A (caller) ← B (callee)	Request has completed successfully
ACK	A (caller) → B (callee)	Acknowledgment of final response to *INVITE*

a BYE message, then the bidirectional connection will be released, and the call/session will be closed. The VOVIDA project provides a good example of Q-SIP/SDP implementation, and VoIP applications [13].

15.3.2 Connection Establishment in the IP/MPLS Transit Network

RSVP-TE is required as the UNI signaling to request QoS-guaranteed connection establishment [11]. In our implementation of RSVP-TE, we expanded the objects for PATH message and RESV message to contain the traffic parameters (such as peak data rate, peak burst size, committed data rate, committed burst size, and excess burst size) and service class-type that implicitly specifies the QoS parameters (such as end-to-end delay, jitter boundary, packet error ratio, packet loss ratio, service availability, protection mode, and service reliability). Table 15.2 describes the RSVP-TE messages with their major contents.

To establish a bidirectional QoS-guaranteed connection, a coordinator module of the caller provides the related traffic/QoS parameters to the RSVP client. These parameters are based on session parameters that have been negotiated during the previous Q-SIP/SDP INVITE procedure between the caller and the callee via SIP proxy servers. In the current implementation of the multimedia terminal, as shown in Figure 15.4, the MM (multimedia) service coordinator is managing the overall session status and connection establishments.

TABLE 15.2

Functionality of RSVP-TE Messages

RSVP-TE Message	Description	Major Contents
PATH	Used to set up and maintain reservations	SENDER_TEMPLATE, SENDER_TSPEC, ADSPEC
RESV	Sent in response to *PATH* messages to set up and maintain reservations	FLOWSPEC, FILTER_SPEC
RESV-Conf	Optionally sent back to the sender of a *RESV* message to confirm that a given reservation actually got installed	FLOWSPEC, FILTER_SPEC, ERROR_SPEC
PATH-Error	Sent by a recipient of a *PATH* message who detects an error in that message	ERROR_SPEC, POLICY_DATA
RESV-Error	Sent by a recipient of a *RESV* message who detects an error in that message	ERROR_SPEC, POLICY_DATA
PATH-Teardown	Analogous to *PATH* messages, but used to remove reservations from the network	SENDER_TSPEC, ADSPEC
RESV-Teardown	Analogous to *RESV* messages, but used to remove reservations from the network	FILTER_SPEC

A PATH message is delivered to the destination, and the destination terminal sends a RESV message to request the resource reservation and connection establishment. Since Q-SIP/SDP has been used to configure the bidirectional session, the PATH/RESV message exchange is executed for each direction. We assume that the QoS-parameters are defined in the service level agreement (SLA)/service level specification (SLS), and traffic engineered label switched paths (TE-LSPs) are established among Provider Edge (PE) pairs for simplified traffic grooming. A multimedia application may require the multiple QoS-guaranteed connections for multiple media streams.

Since the RSVP-TE is using "soft state" connection management [11], the end user's terminal should periodically exchange the PATH and RESV message to keep the connection in normal operational mode. When the required network resource is changed for the multimedia session, the traffic parameters of the connection must also be updated by the PATH/RESV message exchange.

Another consideration in the UNI and NNI signaling is the scalability of connection status management. Since each QoS-guaranteed connection must

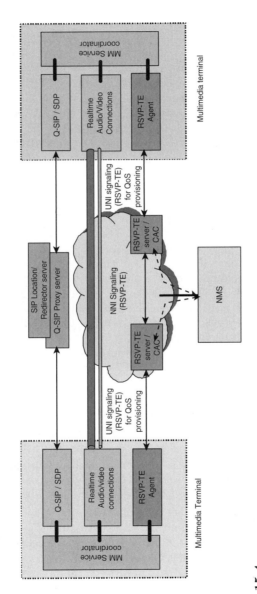

FIGURE 15.4
Proposed QoS-guaranteed multimedia service system.

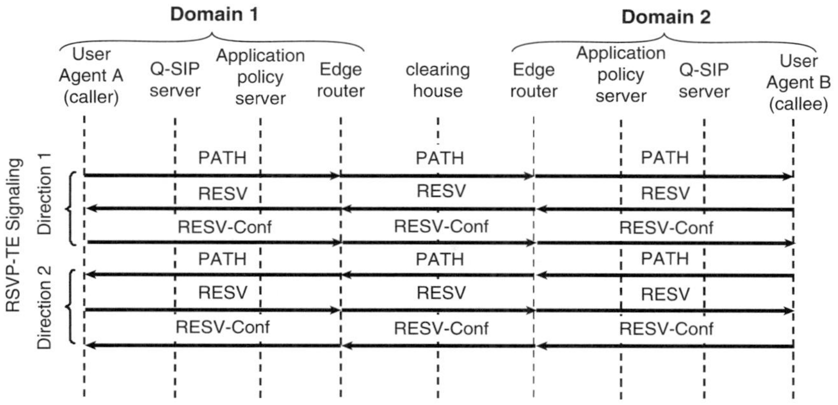

FIGURE 15.5
An RSVP-TE signaling procedure to establish TE-LSP in an IP/MPLS subnetwork.

be maintained by the provider network, an efficient traffic grooming scheme must be used. In the current implementation, we are using edge-to-edge TE-LSP and node-to-node traffic engineering (TE)-link for each QoS class-type based on the virtual overlay networks [12]. In this hierarchical traffic grooming, the ingress edge router is managing the detailed connection status of each QoS-guaranteed connection; the transit routers are not keeping detailed information for each connection, but maintain the aggregated traffic parameters of each QoS class-type. Figure 15.5 shows the procedure of MPLS TE-LSP establishment using RSVP-TE signaling.

15.3.3 Resource Reservation and CAC on IEEE 802.3 Fast/Gigabit Ethernet Switch

Currently, most broadband end user terminals are attached to the home network or office network via IEEE 802.3 Fast/Gigabit Ethernet or IEEE 802.11 WLAN. In order to provide an end-to-end QoS-guaranteed multimedia service across a heterogeneous network environment, as shown in Figure 15.1, each subnetwork should be able to provide QoS-guaranteed packet delivery with appropriate resource allocation and CAC [4].

In IEEE 802.3 Fast/Gigabit Ethernet, the legacy dummy hubs cannot provide guaranteed QoS; while some recent Fast/Gigabit Ethernet switches, such as Cisco Catalyst 3550 and 2950, can provide QoS scheduling and queuing capability with strict priority queue and weighted random early detection (WRED) functions [14,15]. In Cisco Catalyst 3550, each port has four different output queues, and one of the queues can be configured as a strict priority queue, while the remaining queues are configured as non-strict priority queues and are serviced with the use of weighted round-robin (WRR). The strict priority queue is specially designed for delay/jitter-sensitive traffic, such as voice

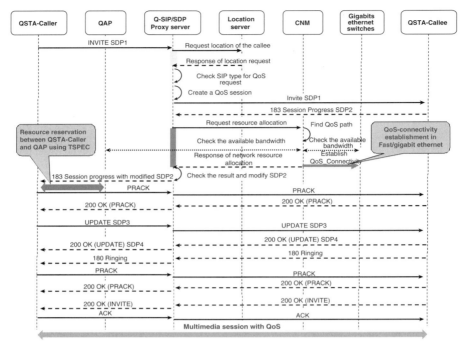

FIGURE 15.6

Interactions between a Q-SIP/SDP proxy server and Q-CNM for QoS-connectivity establishment.

and real-time video. Since the strict priority queue is always emptied first regardless of the status of other queues, it may eventually cause starvation of the other WRR queues.

Since the currently available Gigabit Ethernet switch supports neither Q-SIP/SDP session layer signaling or RSVP-TE network layer signaling, the QoS-aware customer network management (Q-CNM) for home/office intra network must provide the QoS-scheduling and the queue configuration functions with CAC. Figure 15.6 depicts the procedure of QoS-connectivity establishment in an intranet with IEEE 802.11e Wireless LAN and Fast/Gigabit Ethernet switch.

When a caller QoS-aware Station (QSTA) initiates a session establishment with an INVITE message, the Q-SIP/SDP proxy server firstly checks the location of the called party (callee), determines the next hop, and relays the Q-SIP/SDP message. When the callee accepts the request of session establishment by replying with a *183 Session Progress* message, the Q-SIP/SDP proxy server receives it, and sends a resource allocation request message to the Q-CNM that manages the intranet resources. Based on the CAC function, the Q-CNM first checks the availability of the requested bandwidth in the Fast/Gigabit Ethernet switch, and confirms that the addition of the new QoS-connectivity does not deteriorate the QoS provisioning of the existing flows. When the new QoS flow is admitted, the Q-CNM configures the QoS-scheduler and the strict priority queue.

15.3.4 Resource Reservation and CAC in IEEE 802.11e WLAN

Guaranteed QoS provisioning on currently available IEEE 802.11 wireless LAN (WLAN) products is very limited, since the data delivery on most WLANs is based on contention-based medium access control, i.e., carrier sensing multiple access, collision avoidance (CSMA/CA). IEEE 802.11e [16] has been standardized to support QoS provisioning with an enhanced channel access mechanism: Enhance DCF Channel Access (EDCA) and HCF Controlled Channel Access (HCCA). The QoS provisioning on IEEE 802.11e is based on the enhanced distributed channel access by EDCA and the centralized channel access by HCCA. These two channel access functions are managed by a centralized controller called Hybrid Coordinator (HC), which is a module in the QoS Access Point (QAP). The IEEE 802.11e EDCA is a contention-based medium access method, and is realized with the introduction of traffic categories (TCs). The EDCA provides differentiated distributed accesses to the wireless medium with eight access priorities from stations. The EDCA defines the access category (AC) mechanism that provides support for the priorities at the stations. Each station may have up to four ACs (AC_VO, AC_VI, AC_BE, and AC_BK) to support eight user priorities. One or more user priorities are assigned to one AC.

Even though EDCA provides differentiated access categories, it does not guarantee the QoS parameters of hard real-time applications, i.e., jitter and delay [17]. In order to provide real-time services with guaranteed QoS-parameters, HCCA that has been designed for parameterized QoS support with a contention-free polling-based channel access mechanism must be used. The QAP scheduler computes the duration of polled-transmission opportunity (TXOP) for each QSTA based upon the TSPEC parameters of an application flow. The scheduler in each QSTA then allocates the TXOP for different TS queues according to the priority order. In IEEE 802.11e, traffic specification (TSPEC) is used to describe the traffic characteristics and the QoS requirements of a traffic stream (TS) to and from a station.

In practical QoS-guaranteed multimedia service provisioning on IEEE 802.11e WLAN, there are two problems yet to be solved: (1) IEEE 802.11e WLAN access point (AP) node and network interface card that supports both EDCA and HCCA are not commercially available, (2) the IEEE 802.11e WLAN access point (AP) is limited and does not provide the Q-SIP/SDP session layer signaling and RSVP-TE network layer signaling. So, we have to use the Q-CNM that can manage the resource utilization in WLAN and interacts with SIP proxy server. Similar to the networking scenario in the QoS provisioning in Fast/Gigabit Ethernet, the Q-CNM nodes in the source and destination intranets receive the *QoS-connectivity* requests from the Q-SIP proxy server, and check the available network resources as CAC operations. If the new QoS flow is accepted, the Q-CNM controls the configuration of the QoS scheduling on appropriate AC in the IEEE 802.11e WLAN AP. The actual QoS-connectivity establishment is requested by the QSTA via IEEE 802.11e MAC protocol with TSPEC parameters. The Q-CNM can manage the overall aggregated throughput of each AC in the intranet by adjusting the beacon interval, and the proportion of contention-free periods (CFPs) and contention periods (CPs).

15.3.5 QoS-Aware CNM (Q-CNM)

In order to support QoS-aware service provisioning in a Fast/Gigabit Ethernet and IEEE 802.11 WLAN environment, the QoS-aware customer network management (Q-CNM) system should provide QoS-aware resource allocation with CAC function. In this study, we designed and implemented Web-based Enterprise Management (WBEM)-based Q-CNM [4]. The Q-CNM provides QoS-aware resource allocation and CAC in the intranet, configures the details of the per-port queue management and scheduling of the Fast/Gigabit Ethernet switch. It also configures the policy-based operational parameters in IEEE 802.11e WLAN. The Q-CNM maintains the detailed operation status of each QoS-aware Basic Service Set (QBSS) in the IEEE 802.11e WLAN and Fast/Gigabit Ethernet subnetwork.

For the QoS-guaranteed multimedia service provisioning on IEEE 802.11e-based intranet, we developed a Q-CNM as shown in Figure 15.7 [4]. The Q-CNM is based on WBEM architecture for extensibility and interoperability

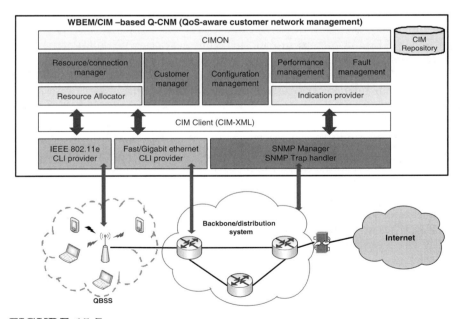

FIGURE 15.7
QoS-aware customer network management (Q-CNM).

with Network Management Systems (NMSs) in other domains. Q-CNM configures the intranet with Fast/Gigabit Ethernet and IEEE 802.11e WLAN. Since Fast/Gigabit Ethernet switch and IEEE 802.11e WLAN AP do not provide any network layer signaling, the Q-CNM handles the resource management and resource allocation. Through the Fast/Gigabit Ethernet command line interface (CLI) provider and IEEE 802.11e provider, the Q-CNM configures the details of QoS provisioning in Fast/Gigabit Ethernet and IEEE 802.11e WLAN. The Q-CNM has the detailed information of each QSTA in the intranet, including the location and current attached QBSS, the availability of IEEE 802.11e functions. Q-CNM also maintains the aggregated allocated bandwidth at each QBSS and at each port of the Fast/Gigabit Ethernet switch node.

15.4 Experimental Implementations and Performance Evaluation of QoS Provisioning in Heterogeneous Wired and Wireless Convergence Networks

15.4.1 Testbed Network of Converged Heterogeneous Wired and Wireless Networks

Figure 15.8 depicts the testbed network of converged heterogeneous wired and wireless network for the performance evaluation of the overall QoS provisioning of multimedia real-time services. In order to analyze the performance of QoS provisioning in a practical intranet environment, we configured two kinds

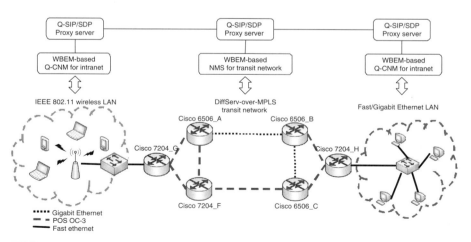

FIGURE 15.8

Testbed network of converged heterogeneous wired and wireless network.

of intranets with commercially available products: (1) *wired intranet* with a
Cisco 2950 IEEE 802.3 Fast/Gigabit Ethernet switch, (2) *wireless intranet*
with a Cisco Aironet 1242 AG IEEE 802.11e Wireless LAN access point and
a Cisco Aironet 802.11a/b/g PI21AG network interface card (NIC). The backbone core network is configured by IP/MPLS network with Cisco 7204 and
6505 IP/MPLS routers.

According to the request of a TE-LSP establishment from the Q-SIP/SDP
proxy server and the WBEM-based NMS for the transit network, a bidirectional TE-LSP between the boundary edge Label Edge Routers (LERs) in the
IP/MPLS backbone network is established using the RSVP-TE signaling of
Cisco IP/MPLS routers. The scheduling of TE-LSP (e.g., bandwidth, class-
type) is controlled by the NMS with a traffic engineering function. As we
explained in Section 15.3, the QoS-connectivity establishments in intranets
are controlled and managed by the Q-CNM of each intranet.

For performance analysis of the end-to-end QoS provisioning for real-time
multimedia services, two kinds of end-to-end video telephony service sessions
are configured: (1) a QoS-guaranteed video telephony session, and (2) a best-
effort video telephony session. For the QoS-guaranteed video telephony session, an EDCA AC_VI is used in the IEEE 802.11e WLAN, a strict-priority
queue is used in the Fast/Gigabit Ethernet switch port, and a TE-LSP is configured in the IP/MPLS transit network. For the best-effort video telephony
session, an EDCA_BE is used in IEEE 802.11e WLAN, a WRED queue is used
in the Fast/Gigabit Ethernet switch port, and a best effort LSP is configured
in the IP/MPLS transit network.

15.4.2   Operations of Q-SIP/SDP and Q-CNM for End-to-End QoS-Guaranteed Service Provisioning

15.4.2.1   Connectivity Establishment for QoS Provisioning at Fast/Gigabit Ethernet

In order to provide QoS-guaranteed multimedia service in the Fast/Gigabit
Ethernet subnetwork, the Q-CNM configures two kinds of service categories
(i.e., best effort and video) in the access network, according to the traffic
parameters that has been informed via Q-SIP/SDP proxy server. Figure 15.9
depicts the system configuration functional architecture for QoS-guaranteed
service provisioning on Fast/Gigabit Ethernet. For the experiments of QoS
provisioning, eight end-user stations are configured to produce video traffic
of two service categories; each service category uses four stations to generate individual traffic flows. To guarantee QoS-guaranteed services for AC_VI
service, the Q-CNM configures classification, metering, marking, WRED, priority queuing, and priority scheduling at the Fast/Gigabit Ethernet switch
and the edge IP/MPLS router. In the experiment, an aggregated bandwidth
of 60 Mbps at IP-layer is allocated for the AC_VI service in the access network
switch. Consequently, AC_VI service can receive the QoS-guaranteed service
unless its data rate does not exceed 60 Mbps.

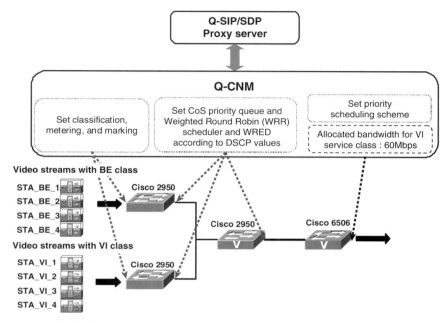

FIGURE 15.9

Configurations for QoS-guaranteed service provisioning on Fast/Gigabit Ethernet.

15.4.2.2 Connectivity Establishment for QoS Provisioning at IEEE 802.11e Wireless LAN

The QoS-provisioning in IEEE 802.11e WLAN requires configuration of QoS-aware Access Point (QAP) to handle MAC frames according to the access category (i.e., AC_VO, AC_VI, AC_BK, and AC_BE) in EDCA. The RTS/CTS and fragmentation mechanisms are not used in most practical WLAN environments with default settings. In the case of IEEE 802.11e, we differentiated and classified the QoS-guaranteed video service, and map the QoS required multimedia services into the AC_VI in IEEE 802.11e EDCA mechanism by using Type of Service (ToS) field in the IP header. The other best effort traffic is processed as AC_BE in IEEE 802.11e EDCA. The QoS-aware traffic and background traffic are generated in the same traffic pattern. The packet size is 1,400 bytes, and the packet inter-arrival time is adjusted according to the target throughput. The data rate shown in this paper is measured at IP-layer, not at physical layer.

As shown in Figure 15.10, the Q-CNM configures the QoS-provisioning on IEEE 802.11e WLAN according to the request from Q-SIP/SDP proxy server [4]. Q-CNM includes the resource management of the WLAN with CAC for QoS-guaranteed services, traffic classification based on service level agreement (SLA), and configures the QAP to provide mapping between classified traffic and IEEE 802.11e EDCA access categories. Q-CNM also configures

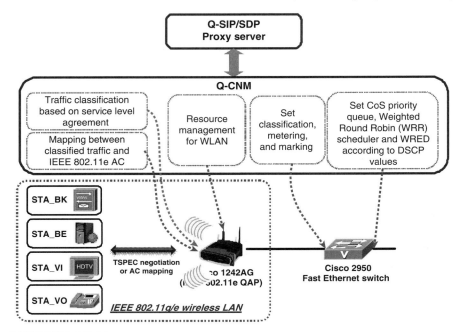

FIGURE 15.10
Configurations for QoS-guaranteed service provisioning on IEEE 802.11e Wireless LAN.

the packet classification, metering, and marking at the access switch in the intranet, and sets the CoS priority queue, WRR scheduler, and WRED queue for each class type.

15.4.2.3   Connection Establishment for QoS Provisioning at IP/MPLS Backbone Network

The real-time multimedia service will require guaranteed QoS provisioning and traffic parameters according to its applications [18]. The highly interactive real-time transaction service, such as user-to-user signaling and highly interactive transaction service, will require less than 50 ms of end-to-end delay. Highly interactive real-time constant bit rate (CBR) conversation service, such as VoIP, will require end-to-end delay of 100 ms and jitter of 50 ms [18], and can be supported by expedited forwarding (EF) class-type in DiffServ standard. Highly interactive real-time variable bit rate (VBR) conversation service, such as multimedia phone, will require the same QoS parameters, but may be provided by assured forwarding (AF) a class-type that efficiently handles the VBR characteristics.

Multimedia conference will require extended end-to-end delay of 400 ms, and jitter of 50 ms [18]. Interactive transaction data, such as a telnet session, will require end-to-end delay of 400 ms without jitter constraints, while Web search or bulk data transfer will require much looser time constraints, such as more than one sec of end-to-end delay. Finally, the best effort service is a class type for legacy Internet service that does not guarantee time constraints or traffic parameters.

As explained above, each class-type requires different QoS requirements; some service classes requires tight end-to-end delay of 100 or 400 ms and limited jitter, while the other non-real-time service classes do not require tight end-to-end delay. The different service class-types will require different fault restoration capabilities; highly interactive real-time transaction or conversational services will require fast restoration with $1+1$, 1:1 or M:N backup path, while non-real-time services will require less strict fault restoration with 1:N or sometimes may allow no restoration.

In order to simplify the required functions of connection establishment for QoS-guaranteed DiffServ provisioning, traffic grooming in the transit networks should be supported. Configuration of scalable per-class-type virtual networks in an intra-AS domain network is one of the key traffic grooming functions in QoS-guaranteed DiffServ provisioning.

The MPLS TE-LSP establishment in an AS domain network may be easily provided by the proprietary MPLS signaling that is supported by the IP/MPLS routers. The NNI signaling for inter-AS TE-LSP establishment is not well standardized yet, and transport MPLS (T-MPLS) is under development as a joint working effort between ITU-T and IETF [2]. As a intermediate solution for the inter-AS TE-LSP establishment, as shown in Figure 15.11, the network management system (NMS) in each AS domain network may configure multiple virtual overlay networks for each DiffServ class-type considering the QoS parameters [10].

In a per-class-type virtual overlay network, multiple QoS-guaranteed TE-LSPs are established among provider edge (PE) routers to configure connectivity of full mesh topology. The DiffServ-over-MPLS AS domain network provides the network-view information of IP/MPLS routers, data links among routers, and constraint-based shortest path first (CSPF) routing function. For open service provisioning, the QoS-guaranteed per-class-type virtual overlay networking function must be provided as the connectivity management application programming interface (API) of Parlay/Open Service Architecture (OSA) standard.

In order to provide QoS-guaranteed service while maintaining the network in optimal resource utilization level, we need to configure the IP/MPLS transit network to provide multiple per-class-type virtual overlay networks with different operation modes. In order to provide QoS-guaranteed differentiated services, we use the concept of multiple virtual overlay network on an MPLS network, where virtual networks for network control traffic (NCT), expedited forwarding (EF), and assured forwarding (AF) class-types are configured and managed separately [10].

In the performance analysis of end-to-end QoS-provisioning across the IP/MPLS transit network, we used only a single AS domain network with six IP/MPLS routers, as shown in Figure 15.8. In total eight end terminal nodes

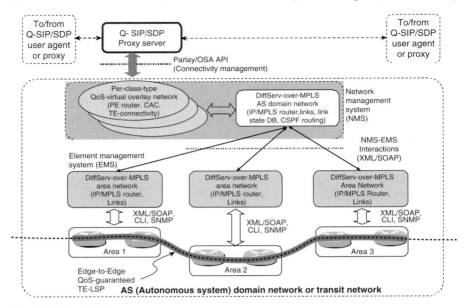

FIGURE 15.11
Configurations for QoS-guaranteed service provisioning on IP/MPLS transit network.

are generating real-time multimedia traffic (as QoS-required traffic) and best effort traffic. The aggregate traffic load increases continuously and is evenly distributed among eight stations. In the same way as in previous scenarios, four stations are assigned for each category. We configured two MPLS TE-LSPs for AC_VI and AC_BE with allocated bandwidth limits of 60 Mbps each. The TE-LSP for AC_VI is configured with higher priority than AC_BE.

15.4.3　Performance Analysis of End-to-End QoS Provisioning

Figure 15.12 depicts the end-to-end performance analysis of QoS-provisioning on the testbed network with IEEE 802.11e WLAN, IP/MPLS transit network, and Fast/Gigabit Ethernet. The end-to-end real-time application is a video conference with H.323 video coding at 10 frames per second, 320 × 240 pixels display size, and the target average data rate of 384 Kbps at IP-layer. In order to compare the end-to-end quality of video conference according to the service provisioning scenario, we measured the end-to-end QoS performances of two networking scenarios: QoS-aware transport networking, and best-effort transport networking. In the QoS-aware transport networking, the IEEE 802.11e EDCA AC_VI is used in the WLAN subnetwork, and the strict-priority queue is used in the Fast/Gigabit Ethernet switch. In the best-effort transport networking, the IEEE 802.11e EDCA AC_BE is used in the WLAN subnetwork,

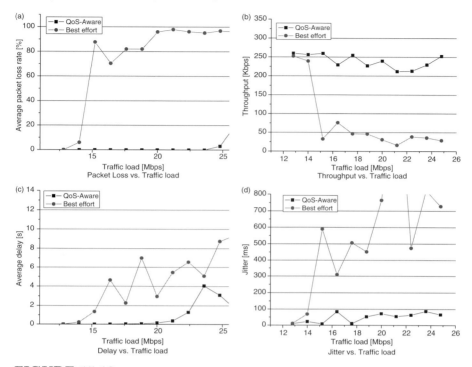

FIGURE 15.12

Traffic load vs. end-to-end throughput, packet loss, delay, and jitter.

while the non-strict priority WRED queue is used in the Fast/Gigabit Ethernet switch. The backbone transit network is configured to provide TE-LSPs for both cases.

From Figure 15.12, we can see that the end-to-end packet delivery of QoS-aware video traffic is guaranteed until 22 Mbps, while the best-effort video traffic is suffering from massive packet loss. The average delay of QoS-aware video traffic is guaranteed to be less than 100 ms until the aggregated traffic at IP-layer is less than 20 Mbps, while the end-to-end delay of best-effort video traffic is suffering a longer delay, which is more than 2 s. Since the currently available IEEE 802.11e WLAN AP and NIC do not support the HCCA mechanism, QoS-guaranteed real-time multimedia service provisioning is not possible if the aggregated traffic load is beyond 22 Mbps, which is around 60% of the practically available throughput of IEEE 802.11 g WLAN. From these experiments, we found that the QoS-aware connectivity provisioning in WLAN is the most critical factor in the end-to-end QoS-guaranteed real-time service provisioning in broadband convergence network with heterogeneous networking technologies. Figure 15.13 shows two examples of captured images in the experiment: QoS-guaranteed and best-effort video conferences.

(a) QoS-guaranteed   (b) Best effort

FIGURE 15.13
Capture images of a video conference.

15.5   Conclusions

In this chapter, we proposed a Q-SIP/SDP for establishment of a QoS-guaranteed end-to-end session, and Q-CNM for resource allocation with CAC in Fast/Gigabit Ethernet and IEEE 802.11e WLAN subnetworks. Based on the proposed session and connection management architecture, the QoS-guaranteed real-time multimedia service provisioning is possible in a converged heterogeneous network environment where various contention-based intranets with IEEE 802.11e WLAN and Fast/Gigabit Ethernet are attached to the IP/(T-)MPLS based connection-oriented transit networks. Detailed interaction scenarios among Q-SIP/SDP, intranet Q-CNM, IEEE 802.11e WLAN AP, and Fast/Gigabit Ethernet switch have been proposed. Based on a small-scale BcN testbed network, the performances of the QoS provisioning for real-time video conference have been analyzed at the IEEE 802.11e-based WLAN intranet, Fast/Gigabit Ethernet-based intranet, and IP/MPLS transit network. Also, the end-to-end QoS-performance has been analyzed.

We performed a series of experiments to analyze the performance of QoS provisioning in wired and wireless intranet subnetwork, access loop, and IP/MPLS-based transit network. The QoS-guaranteed service provisioning in IP/MPLS transit network is relatively easy to configure, since the transit networks are mostly well-equipped with network-layer signaling, such as MPLS with RSVP-TE, and traffic engineering with TE-LSPs. The access loop, such as vDSL, is also not a critical bottleneck, since most wired access loop technologies are expected to provide more than 25 Mbps without a big problem. The QoS-aware real-time service provisioning in Fast/Gigabit Ethernet can be easily configured by priority-based scheduling and differentiated queuing at each port.

From these experiments, we found that the QoS-aware connectivity provisioning in WLAN is the most critical factor in the near future end-to-end QoS-guaranteed real-time service provisioning in broadband convergence network with heterogeneous networking technologies. Even in the IEEE 802.11 g

WLAN where the maximum physical layer transmission rate is 54 Mbps, the practical limitation of QoS-aware service provisioning with IEEE 802.11e EDCA mechanism is around 22 Mbps at IP packet layer. In order to provide QoS-guaranteed multimedia service provisioning in IEEE 802.11e WLAN, the HCCA/DLS scheme must be implemented.

Another key factor in QoS-guaranteed real-time multimedia service provisioning is the end-to-end session layer Q-SIP/SDP signaling with tight cooperation with the network-layer signaling and the resource management function with CAC, in each subnetwork. In this paper, we proposed the functional architecture of this tight collaboration between the session-layer and the network-layer. As future work, we are developing the IEEE 802.11e distributed CAC scheme and channel allocation algorithm to optimize the WLAN and Fast/Gigabit Ethernet resource utilization, while providing the end-to-end QoS provisioning according to the user's request.

References

[1] Kim, Y.-T., Y.-C. Jung, and S.-W. Kim. 2007. QoS-guaranteed realtime multimedia service provisioning on broadband convergence network (BcN) with IEEE 802.11e Wireless LAN and Fast/Gigabit Ethernet, *Journal of Communications and Networks (JCN)* 9 (4): 511–23.

[2] ITU-T Rec. G.8110.1, Architecture of Transport MPLS (T-MPLS) layer network, 2007.

[3] Metro Ethernet Forum, http://metroethernetforum.org/.

[4] Jung, Y.-C., B.-K. Kim, and Y.-T. Kim. 2008. Home/office intranet resource management for QoS-guaranteed realtime stream service provisioning on IEEE 802.11e WLAN, in *Proceedings of NOMS 2008*, pp. 959–62.

[5] Rosenberg J., H. Schulzrinne, G. Camarillo, A. Johnston, J. Peterson, R. Sparks, M. Handley, and E. Schooler. 2002. SIP: Session Initiation Protocol, *IETF RFC 3261*, June.

[6] Handley, M., and V. Jacobson. 1998. SDP: Session Description Protocol, *IETF RFC 2327*, April.

[7] Camarillo, G., W. Marshall, and J. Rosenberg. 2002. Integration of Resource Management and Session Initiation Protocol (SIP), *IETF RFC 3312*, October.

[8] Awduche, D., L. Berger, D. Gen, T. Li, V. Srinivasan, and G. Swallow. 2001. RSVP-TE: Extensions to RSVP for LSP Tunnels, *IETF RFC 3209*, December.

[9] Le Faucheur, F., L. Wu, B. Davie, S. Davari, P. Vaananen, R. Krishan, P. Cheval, and J. Heinanen. 2002. Multiprotocol Label Switching (MPLS) support or differentiated services, *IETF RFC 3270*, May.

[10] Kim, Y.-T., H.-S. Kim, and H.-H. Shin, 2005. Session and connection management for QoS-guaranteed multimedia service provisioning on IP/MPLS networks. In *Proceedings of ICCSA2005 (LNCS 3481)*, May.

[11] Braden, R., L. Zhang, S. Berson, S. Herzog, and S. Jamin. 1997. Resource ReSerVation Protocol (RSVP) Version 1 functional specification, *IEEE RFC 2205*, September.

[12] Kim, Y.-T. 2005. Inter-AS Session and Connection Management for QoS-guaranteed DiffServ Provisioning. In *Proceedings of SERA2005*, August, 325–30.

[13] VOCAL—Vovida Open Communication Application Library, Software version 1.4.0, http://www.vovida.org.

[14] Cisco—QoS Scheduling and Queuing on Catalyst 3550 Switches, Cisco, July 2007.

[15] IEEE Std 802.11e, Part 11: Wireless LAN Medium Access Control (MAC) and Physical layer (PHY) specifications: Amendment 8: Medium Access Control (MAC) Quality of Service Enhancements, 2005.

[16] Mangold, S., S. Choi, G. Hiertz, O. Klein, and B. Walke. 2003. Analysis of IEEE 802.11e for QoS support in Wireless LANs, *IEEE Wireless Communications*, 19 December, 40–50.

[17] Choi, S., J. del Prado, S. Shankar, and S. Mangold. 2003. IEEE 802.11e Contention-based Channel Access (EDCF) Performance Evaluation. In *Proceedings of ICC 2003*, 1151–56.

[18] ITU-T Recommendation Y.1541, Network Performance Objectives for IP-based Service, May 2002.

16

SIP Modeling and Simulation

Yanlan Ding, GuiPing Su, and Huaxu Wan
Graduate University of Chinese Academy of Sciences

CONTENTS

16.1 Introduction 373
16.2 General Formal Modeling Methodologies and Tools for SIP 375
16.3 Modeling and Simulation of SIP through Timed Hierarchical
       CPN and CPN Tools 382
16.4 General Network Simulators for SIP 390
16.5 Summary 393
     References 393

16.1   Introduction

16.1.1   Background

Although Session Initiation Protocol (SIP) is becoming widely used, it is still a developing protocol with a lot of extensions proposed day-by-day to solve existing problems and to provide new services. From the academic research of SIP ideas to SIP-based software development, a complex network environment, cost inefficiency, and impossibility or impracticality of direct implementation are reasons for the utilization of modeling and simulation methods and tools to simulate new network architectures, to solve security problems, and to improve performance of the SIP scenario, etc. Modeling and simulation ideas are necessary, and already many methods are proposed to deal with the realistic problems of SIP. The following list contains some popular and hot SIP research topics and related modeling and simulation methods.

- SIP in wireless network—NS-2 [1]

- Interactions between SIP and H.323, PSTN, etc.—FSM [2,3]

- Performance evaluation of SIP such as SIP traffic, call delay, mobility, etc.—FSM [4], NS-2 [5], HSPA [6]

- Optimizing extensions for next generation services with SIP—FSM [7], OPNET [8], NS-2 [9]

- SIP problems such as SIP firewall traversal, security problems, etc.—CPN [10], EFSM [11]

- SIP services or application development, e.g., VoIP, Instant Messaging and Presence, IP Multimedia Subsystem (IMS), multiparty conferencing—CPN [12], UML [13]

- P2PSIP—P2PSim [14]

16.1.2 Modeling and Simulation

Modeling and simulation are always work done prior to system implementation. Modeling provides architects and others with the ability to visualize entire systems, assess different options, and communicate designs more clearly before taking on the risks—technical, financial, or otherwise—of actual construction [1]. SIP is a protocol and SIP modeling can be classified as two types: one is the modeling of the protocol itself, which focuses on call flowing, states while running, timer mechanisms, and some SIP extensions. Some formal modeling methodologies are suitable for this modeling, such as finite state machine (FSM) used in request for comments Standard IETF RFC Documents (RFCs) containing language-specific modeling [15] and graphical-specific modeling. The other puts SIP into a network environment, tests the whole network (such as Voice-over-Internet Protocol (VoIP) network, wireless mobile network, etc.), tests the interactions between different protocols, and evaluates performances. Specific network simulators are used for this complex modeling, e.g., NS-2 [16], OPNET [17], and QualNet [18]. Not only do they provide standard SIP modules but also performance evaluations, statistic collection, and analysis functions. The key to modeling is the ability to closely match the generated model map to the real system. And for SIP, standard RFCs are the benchmark of the purpose.

Simulation has been defined in various ways. In this chapter, simulation is considered to be the running of a model by use of simulation toolkits. Simulation is the basis for making decisions. Decisions are formulated based on the information resulting from the simulation [19], which based on the level of understanding of the problem, the correctness of the model and the interpretation of running results. Simulation and modeling can be developed by the same simulator which provides such functions. After modeling and setting the characteristics and properties of the system, simulation is simplified as running tracing, data collection, and result analysis. However, complex network architectures and topologies cause the burden of "trial and error."

16.1.3 Our Work

Based on our previous modeling and simulation trial for SIP problems, we focused on general formal methodologies and general network simulators for modeling and simulation of SIP, which are separately discussed in Section 16.2 and Section 16.4. As one of our research projects is about the security problems of SIP, some of the examples are illustrated by such problems. The main method for the modeling and simulation work of SIP is timed Hierarchial Coloured Petri Nets (HCPN), which is proposed in Section 16.3.

16.2 General Formal Modeling Methodologies and Tools for SIP

There are a lot of general formal methodologies and tools for SIP modeling and simulation. Different tools are suitable for specific research purposes of SIP. Although there are programming languages like C, JAVA is a programming language to define and develop SIP modules and implementation languages like XML to test compatibility and protocol stacks of SIP [20], it is always cost-inefficient to use these methods. In this section, we focus on some simple but useful graphical modeling methods and tools. As SIP is a control-plane protocol, SIP modeling is discussed from finite state machine (FSM), which is used in standard RFCs followed by unified modeling language (UML) for SIP application development to the Petri Net cluster for complex modeling and further simulation.

16.2.1 Finite State Machine (FSM)

Seen from the standard RFCs of SIP, most models are represented by FSM. Even an FSM framework is established for SIP extensions [21]. FSM is a simple graphical tool that focuses on states and events and is useful to describe the control flow of a system. Protocol processes can often be modeled as a collection of communicating finite state machines [11]. For SIP modeling, it consists of four elements [21].

S: States that define behavior and can produce actions

T: State transitions that are changing from one state to another

R: Rules or conditions that are to be met to allow a state transition

E: Events that are either externally or internally generated, which can possibly trigger rules and lead to state transitions

For SIP modeling, transaction procedures can be represented as one or more state machines. The formal model of a communicating finite state machine plays an important role in the formal validation of protocol, protocol synthesis, and its conformance testing [11]. Figure 16.1 is an FSM model with the formal definition of the INVITE client transaction procedure defined in RFC 3261.

As we know, SIP should be used in conjunction with other protocols in order to provide complete services to the users. It is easy to use FSM to model channels and interaction between every two protocols, which helps to analyze problems such as security problems in the interactions boundary. Each protocol establishes an FSM with synchronous transitions for interaction. See

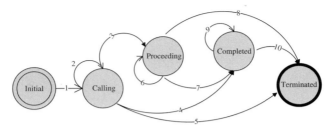

S	s_0 = Initial, s_1 = Calling, s_2 = Proceeding, s_3 = Completed, s_4 = Terminated
R	r_1 = +INVITE, r_2 = (+INVITE) && (t_{wait}>=2*T_1)) r_3 = (+2xx-2xx) ‖ (t_{wait}>=64*T_1) ‖ (+Error)), r_4 = +1xx-1xx r_5 = +(300-699)-ACK, r_6= r_4, r_7 = r_5 r_8 = +2xx-2xx, r_9 = r_5 && t_{wait} <=32s, r_{10} = -Error
E	e_1 = TU-INVITE, e_2 = Timer A (T_1) e_3 = (TU+2xx) ‖ (Timer B(64*T_1)) ‖ (Error), e_4 = TU+1xx e_5 = TU+(300-699)+ACK, e_6 = e_4, e_7 = e_5 e_8 = TU+2xx, e_9 = e_5 e_{10} = (TU+Error) ‖ (Timer D(64*T_1))
T	t_1 = s_0-> s_1, t_2 = s_1-> s_1, t_3 = s_1-> s_4, t_4 = s_1-> s_2, t_5 = s_1-> s_3 t_6 = s_2-> s_2, t_7 = s_2-> s_3, t_8 = s_2-> s_4, t_9 = s_3-> s_3, t_{10} = s_3-> s_4
	Notes: "+" means receive message, "-" means send message

FIGURE 16.1

FSM model for INVITE client transaction procedure.

Figure 16.2 for a simple model of interactions between SIP and the public switched telephone network (PSTN), which is introduced in [22].

In [21], an FSM framework is defined for the SIP extensions (for current as well as future SIP extensions).Various FSMs within a modelled system can interact with each other to achieve the basic functionality.

Further, from [23] we learn about an extended finite state machine (EFSM), which extends FSM with input and output parameters, context variables, operations, and predicates. These extensions are used to describe the data flow and the context of the communication. Figure 16.3 is a simple EFSM model for a CANCEL attack prevention method for SIP, which is proposed in [10]. The intruder sends spurious a CANCEL message to the callee to end the calling from the caller. And then the caller would continue waiting for a 200OK message, which causes a (Denial of Service) attack. In order to prevent

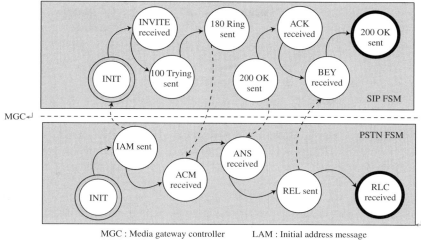

MGC : Media gateway controller   LAM : Initial address message
ACM : Address complete message   ANM : Answer
REL : RELease   RLC : Release complete

FIGURE 16.2
FSM model for interaction between SIP and PSTN.

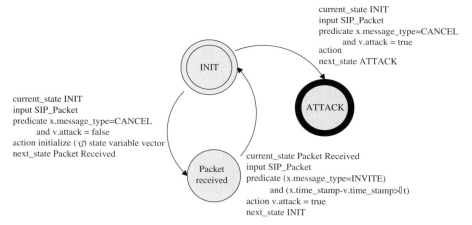

FIGURE 16.3
EFSM model for CANCEL attack prevention method.

such an attack, the caller fires a timer for retransmitting INVITE. If the time between a CANCEL and an INVITE message from the same caller is a short period, the callee would ignore the next CANCEL.

Another modeling tool called CRESS diagram is similar to FSM and it also deals with states of a system. CRESS is a notation and a set of tools for graphical description and analysis of services [24]. In [24], it introduces CRESS and tools for specification and analysis of SIP services.

FSM, EFSM, and CRESS are state-based modeling tools. From examples above, these graphical tools are useful to help model SIP processing and interactions. However, they are limited for network analysis and simulation.

16.2.2   Unified Modeling Language

UML, which has proven useful to improve the quality and productivity of software design and development, would be suitable for the SIP-based application. Consider the importance of SIP in the development of a next generation telecommunications network, it's necessary to use a model-driven method through UML to save time, reduce cost, and improve efficiency for SIP-based application design and development . Besides that, SIP itself is an excellent candidate for UML modeling.

UML supports model-driven development, which helps narrow the gap between a model and the corresponding system it represents to make it accurate. The diagrams of recent UML edition are of two categories [25]: structure diagrams (e.g., class, component diagrams) to show the static structure of the system being modeled and behavior diagrams (e.g., activity, use case, sequence diagrams), which show the dynamic behavior between the objects in the system.

We model those UML SIP models through UML software design tools such as Rational Rose Platform, Microsoft Visio and the SIP Modeling Toolkit. Specifically, the SIP Modeling Tookit is a plugin for IBM Rational Software Architect that adds SIP-specific extensions to the UML modeling and development platform and supports the analysis and design of applications using the Session and Initiation Protocol [25]. The toolkit provides call flow modeling, servlet design, and transformation before application development. By using the model templates and a train of modeling elements of the toolkit, it is easy to create a SIP design and a SIP servlet model. The powerful function of the toolkit is transformation, which translates servlet models and sequence diagrams into standard JAVA code that would be deployed in applications.

While defining the activities and collaboration of a SIP application, the sequence diagram [25], which describes the object that has completed a certain course and the information of time sequence between objects, is fit for call flow modeling. SIP designs are often expressed as sequence diagrams. In fact, SIP in RFCs is always described as a simple form of sequence diagram. Based on the SIP modeling toolkit and the platform, Figure 16.4 is the sequence diagram model of the general SIP dialogue in RFC 3261 and Figure 16.5 is the message content of the INVITE message in Figure 16.4.

While designing and before coding, the class diagram helps to design the class structure of SIP application. Important classes like SIP type, call flow messages, etc., are bases of SIP programming. Sees Figure 16.6 for some defined SIP classes in IBM Rational Software Architect.

Also, the state chart diagram in UML can describe those states of a SIP system and those events that lead to the transformation of states, which helps

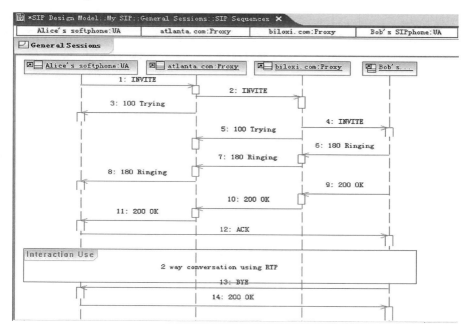

FIGURE 16.4
Sequence diagram of general SIP dialogue.

Request URI

sip:bob@biloxi.com

Headers

Header	Content
Via	SIP/2.0/UDP pc33.atlanta.com;branch=z9hG4bKnashds8
Max-Forwards	70
To	Bob <sip:bob@biloxi.com>
From	Alice <sip:alice@atlanta.com>;tag=1928301774
Call-ID	a84b4c76e66710
CSeq	314159 INVITE
Contact	<sip:alice@pc33.atlanta.com>
Content-Type	application/pkcs7-mime; smime-type=enveloped-data;name=smime.p7m
Content-Disp...	attachment; filename=smime.p7mhandling=required
Content-Type	application/sdp

[Add...] [Copy...] [Edit...] [Remove] [Update Content Length]

Message Body

```
v=0
o=alice 53655765 2353687637 IN IP4 pc33.atlanta.com
s=-
t=0 0
c=IN IP4 pc33.atlanta.com
m=audio 3456 RTP/AVP 0 1 3 99
a=rtpmap:0 PCMU/8000
```

FIGURE 16.5
Message content of INVITE in Figure 16.4.

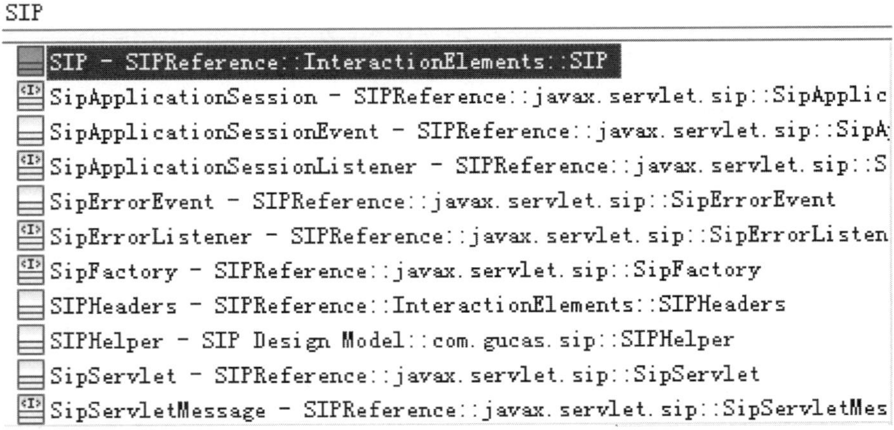

SIP

```
SIP - SIPReference::InteractionElements::SIP
SipApplicationSession - SIPReference::javax.servlet.sip::SipApplic
SipApplicationSessionEvent - SIPReference::javax.servlet.sip::SipA
SipApplicationSessionListener - SIPReference::javax.servlet.sip::S
SipErrorEvent - SIPReference::javax.servlet.sip::SipErrorEvent
SipErrorListener - SIPReference::javax.servlet.sip::SipErrorListen
SipFactory - SIPReference::javax.servlet.sip::SipFactory
SIPHeaders - SIPReference::InteractionElements::SIPHeaders
SIPHelper - SIP Design Model::com.gucas.sip::SIPHelper
SipServlet - SIPReference::javax.servlet.sip::SipServlet
SipServletMessage - SIPReference::javax.servlet.sip::SipServletMes
```

FIGURE 16.6
SIP classes defined in IBM rational software architect.

to keep track of the whole process and confirm the quality of the design. Such a diagram is similar to FSM.

16.2.3 Petri Net Cluster

Here, we have to emphasize another important formal graphical tool for modeling and simulation, Petri Net. As there are various kinds of Petri Nets with different features, we call these Petri Nets as Petri Net Cluster. Petri Net Cluster contains Hierarchical Petri Net, Colored Petri Net, Temporal Petri Net, and so on, each of which can be utilized for specific SIP modeling and simulation.

Hierarchical Petri Net (HPN) introduces a hierarchical modeling method, which gives a high abstraction ability to synthesize and simplify relevant objects to simulate a realistic system. The top-down and bottom-up methods are often used for hierarchical modeling with an idea of substitution transition. The idea of *substitution transition* is to allow the user to relate a transition and its surrounding arcs to a more complex Petri Net, called subnet, which could give a more precise and detailed description of the activity represented by the substitution transition [26].

While modeling a large and complex system such as a SIP system, when designing a low hierarchy model, you don't need to consider complex realizations just like a black box. What's more, the overall system can be reduced to the generation of smaller models that integrate related activities [10]. Figure 16.7 shows the first hierarchy of a simple SIP model and a subnet of UA1.

(A) The first hierarchy of SIP

(B) The subnet model of UA1

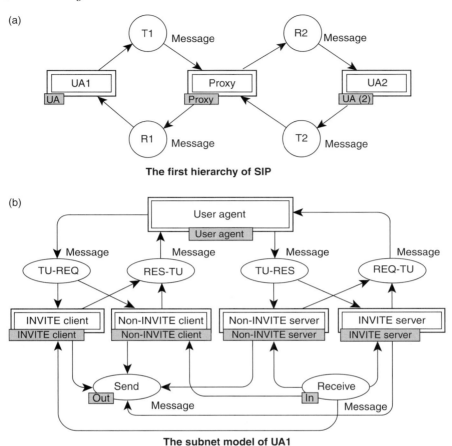

FIGURE 16.7
Simple SIP model though HPN.

From the model, there are some places in the subnet UA1 called *port places* containing *input, output* and *I/O* socket places as the interface with which the subnet could communicate with its surroundings. Figure 16.7b is the subnet of the substitution transition UA1 in Figure 16.7a and the place *Send* which refers to place T1 is the output socket labeled *Out* and the place *Receive,* which refers to place R1 is input socket labeled *In.* As SIP works with SIP agents, e.g., SIP UA, SIP proxy server and SIP interacts with different protocols, it's necessary to use HPN to model the interaction in high level and the functions of SIP agents in subnets.

Colored Petri Net (CPN) [27], introduced by Kurt Jensen, is well-known for its capability in modeling discrete event systems [26]. The concept of color set, trigger functions, and other properties are suitable for modeling protocol and folding a large model. Color sets are tokens carrying complex information

TABLE 16.1
Formal Definition of SIP Messages through CPN

---

colset Message = with INVITE | ACK | BYE | CANCEL | OK200 | Temp1xx |
    Error | Trying100 | NotFound404 | NULL timed;
colset STATE = with STATE_IDLE | STATE_TALKING | STATE_
    WAITRING | STATE_WAIT200 | STATE_WAITBYEOK | STATE_
    WAITACK | STATE_CALLCOMING | CALLING | Terminated |
    Completed | Proceeding | Confirmed;

---

and data. The passing scenario of color sets describes the control flow of a system. As SIP is a text-based protocol and SIP messages are of different kinds between server and client, color set is suitable to model a SIP message. The Table 16.1 is a simple color set example of SIP messages and system states through CPN.

CPN is capable of describing the dynamic behavior of process systems and handling the hierarchy [26]. Combined with HPN and CPN into Hierarchical Colored Petri Net (HCPN), it allows the modeler to construct a SIP model by using a number of small CPNs, which are related to each other in a well-defined way.

Temporal Petri Net [28], just as its name implies, timed factors are added to Petri Net to consider time in modeling and simulation. Learned from RFC 3261, we know that timer is an important factor for SIP service, which deals with overtime and retransmission. So, timed modeling of SIP is necessary and important.

In the chapter, we added a time factor to HCPN into timed HCPN and utilize a tool called CPN tools [29] for practical modeling and simulation. In next section, we focus on the formal definition of the timed HTPN by use of which to model and simulate SIP in order to solve concrete problems.

16.3 Modeling and Simulation of SIP through Timed Hierarchical CPN and CPN Tools

16.3.1 Formal Definition of Timed HCPN and CPN Tools

In order to further understand timed HCPN and the modeling method of SIP, We give the formal definition of timed HCPN. As the base of timed HTCPN is CPN, the following is the formal definition of CPN from [27].

Definition 16.1 *A Colored Petri Net (CPN) is a nine-tuple* $\text{CPN} = (\sum, \text{P}, \text{T}, \text{A}, \text{N}, \text{C}, \text{G}, \text{E}, \text{IN})$*, satisfying:*

$\sum$ *is a finite set of non-empty types, called colored sets.*
P *is a finite set of places.*

T is a finite set of transitions.

A is a finite set of arcs that $P \cap T = P \cap A = T \cap A = \emptyset$

N: $A \rightarrow P \times T \cup T \times P$ *is a node function.*

C: $P \rightarrow \sum$ *is a color function denotes tokens in a place.*

G is a guard function, which is defined from T into expressions such that:
   $\forall t \in T$: $[\text{Type}(G(t)) = \text{Bool} \wedge \text{Type}(\text{Var}(G(t))) \subseteq \sum]$.

E is an arc expression function, which is defined from A into expressions such that: $\forall a \in A$: $[\text{Type}(E(a)) = C(p(s))_{\text{MS}} \wedge \text{Type}(\text{Var}(E(a))) \subseteq \sum]$.

IN is an initialization function. It is defined from P into closed expressions such that: $\forall p \in P$: $[\text{Type}(\text{IN}(p)) = C(p(s))_{\text{MS}} \wedge \text{Var}(\text{IN}(p)) = \emptyset]$

where Type(expr) denotes the type of an expression and Var(expr) denotes the set of variables in an expression.

An HCPN consists of a set of subnets, each of which is a non-hierarchical CPN. We don't care about the formal definition of a whole HCPN, as hierarchical method here only contributes to the modeling process.

For SIP modeling, as we mention above, $\sum$ denote SIP messages and system states which are stored in P mapping through C. IN defines the initialized state of SIP such as UA owns a SIP message token INVITE. A, G and E master the whole control flows of SIP. G is similar to R of FSM. By formally defining CPN, we can apply the properties and related analysis tools of CPN for analysis of the SIP model, which will be presented in 16.3.4.

The timed factor in the HCPN model can be represented to two ways. The first is to add time to $\sum$ that each token could take time which indicates the time it runs. The other is to add time to T that each transition consumes time to deal with the input tokens and to finish defined actions. And then, the whole system is labeled with time.

Here, we introduce the tool we use to model and simulate SIP through timed HCPN, which gives powerful functions for our purpose—CPN tools [29]. CPN tools support hierarchical modeling of SIP with a set of tools called hierarchical tools, such as the "Move to Subpage" tool to relate a transition with a subpage and "Set Subpage" tool to assign subpages to the substitution transitions on a superpage. CPN tools also support timed modeling, which provides temporal inscription. It introduces a new color set called timed color set, which appends the keyword "timed" to the color set declaration, then tokens representing the color set would take time and change the value as model runs. Also, each transition owns a function to add time by attaching an INT time to keyword "@+." With the definition of timed color set, there are also time simulator functions for query and analysis [29].

One of the main advantages of CPN tools is the powerful simulation, performance, and state space analysis tools. Simulation tools can run a simulation of an existing model, manually choose bindings, change marking during simulation, send feedback, and give reports. Performance tools collect, calculate statistics, and output simulation data into log files. The state space analysis tools automatically calculate state space and SCC graph of a model, report state space results and make state space query through CPN ML functions for analysis. The CPN ML language is used in CPN tools for declarations and net inscriptions, which are important for our analysis.

16.3.2 Modeling Methods for SIP

From the related work on modeling for SIP through CPN, there are always two ways for SIP modeling, the whole SIP modeling for all-sided test and the simplified SIP modeling for specific interest to the study.

16.3.2.1 Whole SIP Modeling

In order to test SIP functions and the extensions of SIP, sometimes it has to model the whole complex SIP based on RFCs. As SIP is a transactional protocol, models shown in Figure 16.7 are a top SIP and a top UA based on an INVITE transaction described in RFC 3261. The expanding subnets INVITE-Client and INVITE server of the UA model in Figure 16.7 are shown in Figure 16.8 and Figure 16.9.

Modeling of the whole SIP always depends on structures in RFC 3261, such as SIP structure, SIP calling sequence, security mechanisms etc. The model in the paper is based on the layered SIP structure and the INVITE transaction

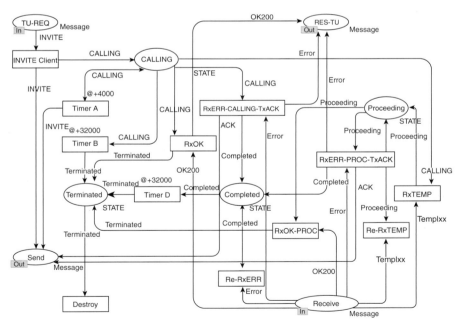

FIGURE 16.8
INVITE client model.

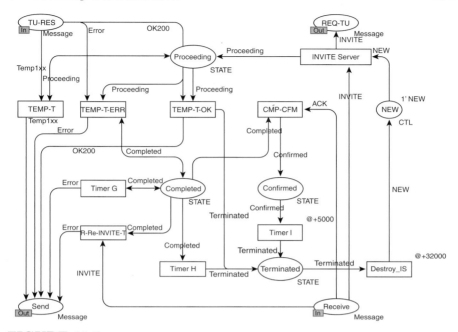

FIGURE 16.9

INVITE server model.

structure. In [12], a SIP-based dynamic discovery protocol, modeling also relies on the SIP transaction structure of client and server.

No matter the model in the section, or the large models shown in other papers, such as [10,12], the whole SIP modeling is complex, which would even become the resistance of analysis. For example, if we would like to model the interaction between SIP and H.323, a modeling interaction based on a whole SIP model would become onerous and the state space would be too large for analysis. So, for specific analysis purposes, a simplified modeling method would be more suitable.

16.3.2.2 Simplified SIP Modeling

In order to simplify the modeling procedure and analysis of the model, it is always useful to pay particular attention to those that are of interest to the study [15] and select the necessary hierarchies and functions for modeling. That is modeling on-the-fly. Although SIP is a simple protocol, transaction of layers, interaction of different protocols, and security problems are complex to model integrally. Even if the whole system is modeled, it is too large for further analysis.

Simplified SIP modeling helps to solve the problem. As an example of the method, Figure 16.10 shows a simple BYE attack with interactions between SIP and RTP.

(A) Top level model

(B) Intruder model

(C) UA1 model

(D) UA2 model

In the illustration, the SIP model is simplified from models in Figure 16.8 and Figure 16.9. RTP here is modeled by two places called RTP. Then interaction between SIP and RTP is simply modeled by the passing scenario of an RTP color set with two values of RTPCONN and RTPNONE indicating if receiving or sending RTP packages into the Internet or not.

Although the model is simple, it describes the security problem of the interaction boundary between SIP and RTP, specifies the BYE attack of SIP, and presents an intrusion detection mechanism of the attack. Here, we only focus on the interactions between SIP and RTP to find out the detection method of the BYE attack and ignore the concrete content of SIP and RTP. Modeling and simulation of SIP are for a specific purpose: the on-the-fly modeling helps simplify modeling if it doesn't require the whole modeling.

16.3.3   Timed SIP Model

By use of timed color set and useful time attributes in tokens defined in CPN tools, timer and time-related problems of SIP can be easily modeled and analyzed. Timer is an important part for SIP, especially the INVITE transaction. There are some ways in CPN tools to model timer, overtime, and retransmission of SIP. The following example is a small model for Timer A and Timer B of the INVITE client transaction in RFC 3261 (Figure 16.10).

```
colset Message=with INVITE|ACK|BYE|CANCEL|OK200|Ring180Busy486|
       Trying100| NotFound404 |Timeout408|NULL timed;
colset time_cons=INT timed; var time_pass, time_cal: time_cons;
val T1=500; fun OT(t:time_cons):bool=if t>64*T1 then
          true else false;
```

As seen in Figure 16.11, retransmission is controlled by transition "Timer A and B" with T1, an interval 2*T1 and Place A. Timeout is controlled by declared overtime function OT of 64*T1 for Timer B with transition "Timer A and B" and Place B. Time transition inscription of "Timer A and B" indicates time consuming by attaching time to "@+." Another way to realize time modeling is to calculate consuming time of a token by code segment of transition inscription (see Figure 16.12).

The ModelTime and time () are defined timed functions for timed net modeling. By calculating the *starcal* function, *time_cal* carries the time consuming by "Timer A and B" and also indicates the time attribute of each token, which is an important value for further calculation and analysis.

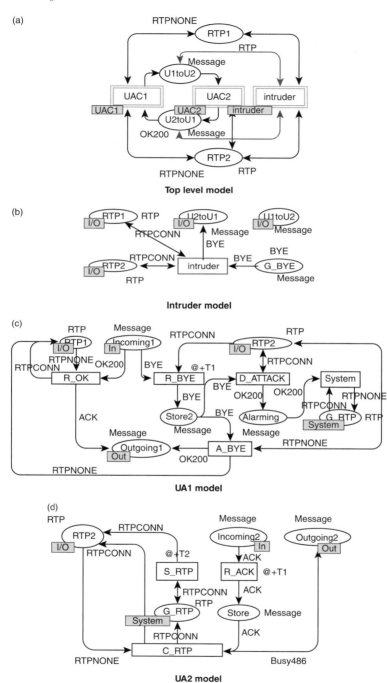

FIGURE 16.10
BYE attack model through simplified modeling method.

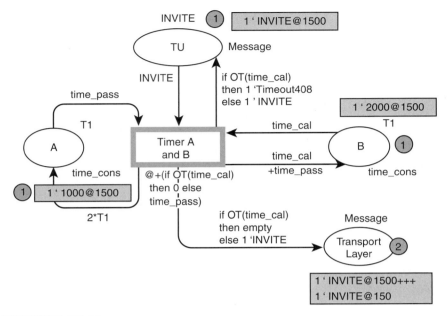

FIGURE 16.11
Timer modeling for INVITE client transaction.

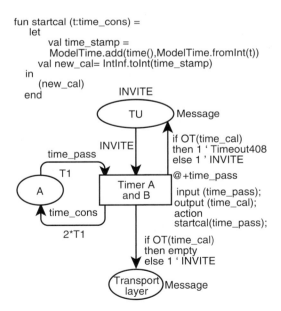

FIGURE 16.12
Timer modeling with code segment of time transition.

16.3.4   Simulation and Analysis

As CPN tools is based on CPN, it offers specific tools to analyze properties of modeled net, such as boundness and liveness properties shown in the simulation report, automated state space calculation, supported query functions of CPN ML, simulation, and performance tools, etc.

Using the BYE attack detection model shown in Section 16.3.2 as an example, Figure 16.13 is the corresponding report and result of state space simulation.

As state space is calculated, CPN ML query functions can be utilized for further analysis. CPN ML language is used for declarations and net inscriptions. The declarations of SIP message shown in Figure 16.14 are of standard CPN ML, which also provides many functions for state space, monitoring, and performance analysis. The following function is a query of a BYE attack

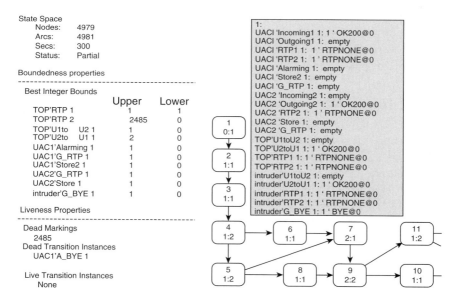

FIGURE 16.13

State space simulation report and result for model in Figure 16.11.

```
fun AuthViolation():Node list
= PredAllNodes(fn n=>
cf(OK200, Mark.UAC1'Alarming 1 n)>0
andalso
cf(RTPCONN, Mark.UAC2'RTP2 1 n)>0
);

AuthViolation();
```

FIGURE 16.14

CPN ML query functions for BYE attack.

model, which means that if alarming occurs when there is still an RTP package from UAC2, the BYE attack exists.

The monitoring and performance tools are useful for simulation of models, which would store simulation data for further analysis. They are not mentioned here. See details in [29].

16.4   General Network Simulators for SIP

Although general formal modeling and simulation tools are useful in analysis of SIP, they lack the simulation of a network environment. Now a lot of network simulators supporting SIP can help to solve the problem. Although SIP is an application-layer protocol, it always interacts with other lower-layer protocols, such as TCP, UDP, etc. What's more, systems such as VoIP based on SIP do not only consider SIP itself, the network environment, traffic and QoS problems are also important for a whole system, which should be tested and simulated through simulations.

16.4.1   NS-2 and SIP Modeling

Network Simulation version 2 (NS-2) is a discrete event simulator targeted at networking research from the Virtual Internetwork Testbed project VINT. NS-2 is free and provides substantial support for simulation of TCP, routing, and multicast protocols over wired and wireless (local and satellite) networks [16]. In order to simulate a SIP-based system, we use NS-2 with the addition of the SIP module, SIP Patch.

We tested some versions of SIP Patch for the NS-2 network simulator, containing the simple 2.1b9a SIP Patch from [16] and the recent 2.26 SIP Patch from [30]. SIP Patch gives the original support for SIP-based simulation and most SIP modules in it can simulate bidirectional dialogue between two SIP agents that send SIP requests, establish topology, and trace SIP events during the simulation through NAM-The Network Animator; see Figure 16.15 as a media test example in SIP Patch 2.26.

SIP modules can run over UDP (unreliable), TCP or SCTP (reliable). Figure 16.15 is an example of SIP over UDP. In the patch, *sip_udp.cc* and *sip_udp.h* define the standard SIP over UDP and the file *trace.cc* has been modified to allow the tracing of SIP events during the simulation.

The nodes in the topology graph are first defined as *ns nodes* and then attached to significant objects. For example, the node 4 and 5 in Figure 16.15 are SIP proxy servers, parameters of which should be set, especially time parameters. There is also an initial setting for the whole simulation environment, such as the *simtime,* which controls the time limitation for simulation. All these definitions are written in *tcl* files and the trace results are written

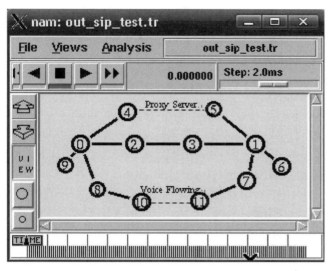

FIGURE 16.15
Topology of SIP over UDP media test in NAM.

r 1.010004	4	0	udp	170	-------	0	4.0	5.0	2	9	RINGING	0
+1.020827	7	11	exp	160	-------	2	10.0	11.0	3	7	INVITE	0
- 1.020827	7	11	exp	160	-------	2	10.0	11.0	3	7	INVITE	0
r 1.020836	7	11	exp	160	-------	2	10.0	11.0	3	7	INVITE	0
+1.021657	4	0	udp	577	-------	0	4.0	5.0	3	11	INVITE	1
- 1.021657	4	0	udp	577	-------	0	4.0	5.0	3	11	INVITE	1
+1.025031	1	5	udp	170	-------	0	4.0	5.0	2	9	RINGING	0
- 1.025031	1	5	udp	170	-------	0	4.0	5.0	2	9	RINGING	0
r 1.026687	4	0	udp	577	-------	0	4.0	5-0	3	11	INVITE	1
r 1.030039	1	5	udp	170	-------	0	4.0	5.0	2	9	RINGING	0
+1.03926	4	0	udp	170	-------	0	4-0	5.0	4	13	2000K	0

FIGURE 16.16
Trace report during SIP file compiler.

in *tr* files based on which the NAM draws topology and performs simulation. See trace example in Figure 16.16.

As NS-2 is an open source simulator, we can develop a new *tcl* program or even modify and recompile existing files for specific simulation goals. Based on the 2.1b9a SIP Patch, we simulate a simple INVITE flooding attack. As we know, when SIP runs on UDP, the client retransmits the INVITE until the "100 trying" response is received and the server retransmits the "100 trying" each time it receives a retransmission of the INVITE. Timer A and Timer B work as shown before. The INVITE flooding attack means the intruder lunches INVITE flooding to overwhelm a single proxy server within a short period of time and causes DoS. We edit only a SIP agent and a SIP proxy server for our simulation. In order to simplify the procedure, the time for retransmission of INVITE is shortened by changing the value of the parameter for *resched()* function with an *Invitetimer* object *inv_tmr*, which is initialized as 0.5 *s*. The intruder ignores the "100 Trying" response to simulate INVITE flooding by deleting the code to close the timer for retransmission when *sh−>type* is of *SIP_RSP_TRYING* type.

We can even develop a mechanism for detection of this attack. A normal SIP Agent A sends INVITE to SIP proxy server, if the SIP proxy server suffers INVITE flooding, the queue to deal with INVITE is too long to deal with the request of A. So, A is always in timeout status. What's more, by testing the number of INVITE messages the server received in a short period in trace file, we may know about the time the attack starts.

16.4.2   Other Simulators Supporting SIP

16.4.2.1   OPNET and QualNet

Besides NS-2, there are also some powerful network simulators supporting the SIP module, especially OPNET and QualNet, which are popular within academia, commercial, and industrial communities.

OPNET, the primary product of MIL3 Inc. provides a comprehensive development environment for the specification, simulation, and performance analysis of communication networks, computer systems and applications, and distributed systems [17]. OPNET provides a SIP model for VoIP, which contains SIP, RTP and H.323 to allow VoIP simulations based on the standard model library of OPNET Modeler. There are also SIP agent modules such as proxy server and UAC modules that contribute the topology of SIP simulation scenarios. Before simulation, the attributes of them should be configured, such as the Maximum Simultaneous Calls, Proxy Type, Area Name of Proxy Server and the Proxy Server Connect Timeout, Domain Name, and Current Domain parameters of UAC. After configuration of agent parameters and establishment of topology, the SIP scenario is run for statistics collection, performance evaluation, and result analysis.

QualNet network simulation software has been developed by Scalable Network Technologies [31], which provides a comprehensive set of tools with many components for custom network modeling and simulation. QualNet has a range of wired and wiresless models. However, the main strength of QualNet is in the wireless area. SIP Module is defined in the Multimedia & Enterprise Models section in Standard QualNet Model Libraries which consists of H323, SIP, VoIP Traffic Generator, and VoIP Jitter Buffer. SIP is enabled as the signaling protocol after setting SIP configurations, which recommend all nodes of a simulated network are SIP intelligible [32]. Although SIP is independent of the underlying transport layer protocol and can run over TCP, UDP, and SCTP, only TCP is used in current implementation with two types of call model-Direct and Proxy-routed. During simulation, a set of statistics about UAC, UAS, proxy and overall data can be collected and printed for further analysis, such as total number of INVITE, ACK, and BYE requests sent in UAC and Trying, Ringing, and OK responses received by a SIP node.

16.4.2.2   P2P Simulators for P2P-SIP

Currently, an increasing number of people are beginning to pay attention to P2P-SIP (peer-to-peer SIP). Therefore, the P2P simulators are also considered in this section. The concept behind P2P-SIP is to leverage the distributed nature of P2P to allow for distributed resource discovery in a SIP network, eliminating (or at least reducing) the need for centralized servers [33]. P2P can reach huge dimensions like millions of nodes joining and leaving continuously, which is very difficult to simulate. So P2P simulators are developed with special features to solve these problems.

Peersim, or Peer-to-Peer Simulator, written in Java Language, is composed of many simple extendable and pluggable components, with a flexible config-uration mechanism and a scalable simulation testbed [34]. P2PSim as part of IRIS project is a free, multi-threaded, discrete event simulator to evaluate, investigate, and explore P2P protocols [35]. They both support various P2P protocols like Chord, Accordion, Koorde, Kelips and so on. Besides such P2P simulators, there are specific simulators, e.g., the P2P File-sharing simulator, which is designed as an implementation of "Incentive Resource Distribution in P2P Networks" scheme and the Progressive Filling (PF) algorithm [36]. Now there are extensions to SIP regarding utilization of SIP for different applica-tions, such as SIP based file-sharing service, which could be modeled by use of these simulators.

16.5   Summary

In the paper, only some general formal modeling methodologies and specific simulators are enumerated and discussed through related examples. After all, no matter what kind of methodologies we used for SIP, the key point is the purpose of modeling and simulation which is the motivation, the standard to select methods, and to evaluate the results of our work.

References

[1] Rajaram, V.S. 2006. Session initiation protocol for wireless channels. Master's thesis, Office of Graduate Studies of Texas A&M University.

[2] Wang, L., A. Agarwal, and J.W. Atwood. 2004. Modeling and verifica-tion of interworking between SIP and H.323, *Computer Networks: The International Journal of Computer and Telecommunications Network-ing*, Elsevier North-Holland, Inc.

[3] Wang, L., J.W. Atwood, and A. Agarwal. 2003. Validation of SIP/H.323 Interworking Using SDL/MSC. SDL Forum, 335–51.

[4] De Marco, G, G. Iacovoni, and L. Barolli. 2005. A technique to analyse session initiation protocol traffic, *Proceedings of the 11th International Conference on Parallel and Distributed Systems-Workshops (ICPADS'05)*, IEEE Computer Society.

[5] Ohta, M. 2006. Overload protection in a SIP signaling network, *International Conference on Internet Surveillance and Protection (ICISP'06)*, IEEE Computer Society.

[6] Fan, R., M. Wang, M. Ericson, and S. Wanstedt. 2007. Evaluation and analysis of SIP and VOIP performance with presence traffic over HSPA, Personal, *IEEE 18th International Symposium on Indoor and Mobile Radio Communications* (PIMRC 2007).

[7] Chan, K.Y., and G.V. Bochmann 2003. Methods for designing SIP services in SDL with fewer feature interactions, *Feature Interactions in Telecommunications and Software Systems VII*, 59–76. Amsterdam, IOS Press.

[8] Bo Rong, J. Lebeau, M. Bennani, M. Kadoch, and A.K. Elhakeem. 2005. Modeling and simulation of traffic aggregation based SIP over MPLS network architecture, proceedings. *38th Annual Simulation Symposium*, IEEE Computer Society 305–11.

[9] Cerqueira E., A. Neto, E. Monteiro, L. Veloso, M. Curado, and P. Mendes. 2006. Multi-user session control in the next generation wireless system, *MOBIWAC Proceedings of the 4th ACM International Workshop on Mobility Management and Wireless Access*, ACM.

[10] Ding, Y. and G. Su. 2007. Intrusion detection system for signal based SIP attacks through timed HCPN, *Second International Conference on Availability, Reliability and Security, 2007. ARES 2007*, IEEE Computer Society, 190–7.

[11] Sengar, H., D. Wijesekera, H. Wang, S. Jajodia. 2006. VoIP intrusion detection through interacting protocol state machines, *2006 International Conference on Dependable Systems and Networks (DSN 2006)*, IEEE Computer Society, 393–402.

[12] Gehlot, V. and A. Hayrapetyan. 2006. A CPN model of a SIP-based dynamic discovery protocol for webservices in a mobile environment, *Proceedings 7th Workshop and Tutorial on Practical Use of Coloured Petri Nets and the CPN Tools*. Denmark: Aarhus.

[13] Cernosek, G. and E. Naiburg. 2004. The Value of Modeling, IBM Software Group.

[14] Harjula, E. 2007. *Peer-to-Peer SIP in Mobile Middleware Intercommunication*. University of Oulu, Department of Electrical and Information Engineering, Oulu, Finland, Master's thesis.

[15] Garzia, R.F. 1990. *Network Modeling, Simulation, and Analysis*. Marcel Dekker, Inc. New York, NY, USA.

[16] Network simulator v.2, available at www.isi.edu/nsnam.

[17] OPNET Technologies Ltd. WWW-page, http://www.opnet.com.

[18] QualNet Technologies Ltd. WWW-page, http://www.qualnet.com.

[19] Daly, F. 1999. The Simulation. PhD thesis. Ireland: University of Limerick.

[20] Ranganathan, M., O. Deruelle, and D. Montgomery. 2003. Testing SIP call flows using XML protocol templates, *IFIP International Conference on Testing of Communicating Systems*. Berlin: Springer-Verlag.

[21] Kaul, A. 2003. Finite State Machine (FSM) framework for SIP extensions, Internet-draft, draft-kaul-sip-fsm-framework-00.txt, Network Working Group.

[22] Zhang, Y. 2002. SIP-based VoIP network and its interworking with the PSTN, *Electronics & Communication Engineering Journal*, 273–82, Institution of Electrical Engineers, London, ROYAUME-UNI (1989–2002) (Revue).

[23] Lee D., and M.Yannakakis. 1996. Principles and methods of testing finite state machines—a survey, *Proceedings IEEE*, 84, 1090–1123.

[24] Turner, K.J. 2002. Modelling SIP services using Cress, In *Proceedings Formal Techniques for Networked and Distributed Systems (FORTE XV)*, eds. Moshe Vardi and Doron Peled, 162–77. *LNCS 2529*. Berlin: Springer-Verlag, November.

[25] Conallen, J. 2007. Introduction to the SIP Modeling Toolkit for IBM Rational Software Architect, IBM Product Report.

[26] Németh E., R. Lakner, K. M. Hangos, and I. T. Cameron. 2003. Hierarchical CPN model-based diagnosis using HAZOP knowledge, Research Report SCL-009/2003, Systems and Control Laboratory. Hungary: Computer and Automation Research Institute, Budapest.

[27] Jensen, K. 1995. Coloured Petri Nets. *Basic Concepts, Analysis Methods and Practical Use*, 2. Berlin: Springer-Verlag.

[28] Racloz, P. 1994. Properties of petri nets modellings: The temporal way, *7th International Conference on Formal Description Techniques for Distributed Systems Communications Protocols*.

[29] CPN tools. The CPN Group, Department of Computer, Science, University of Aarhus, Denmark, http://www.daimi.au.dk/designCPN/libs/commscpn/.

[30] Network simulator v.2 SIP Module, patch-ns2_2.26-SIP-SCTP-UDP, http://www.tti.unipa.it/~fasciana/materiale.htm.

[31] Begg, L., W. Liu, K. Pawlikowski, S. Perera, and H. Sirisena. 2006. Next generation networks for studying service availability and resilience, Department of Computer Science & Software Engineering University of Canterbury Christchurch, New Zealand, Technical Report.

[32] QualNet 3.9.5 Model Library: VoIP, Scalable Network Technologies, Inc., 2006.

[33] P2P-SIP websites, http://www.p2psip.org/.

[34] PeerSim available at http://peersim.sourceforge.net/.

[35] P2PSim available at http://www.pdos.lcs.mit.edu/p2psim/.

[36] The P2P File-sharing simulator available at http://www.cs.uiowa.edu/~rbriggs/gt/.

17

SIP-Based Mobility Management and Its Performance Evaluation

Nilanjan Banerjee
IBM India Research Lab

Kalyan Basu
Treveni Systems

CONTENTS

17.1 Summary 397
17.2 Introduction 397
17.3 Mobility Management 399
17.4 Protocols 403
17.5 SIP-Based Mobility Management: A Quantitative Analysis 408
17.6 Handoff Delay Mitigations 424
17.7 Summary 428
     References 429

17.1   Summary

Session Initiation Protocol (SIP) was originally designed as an Internet Protocol for establishing and maintaining multimedia sessions. It, however, gained prominence recently as a Voice-over-Internet Protocol (VoIP) signaling protocol. In addition to its signaling capabilities, SIP can also act as an application layer mobility protocol supporting various flavors of mobility. In this chapter, we first argue in favor of the virtues of SIP in handling mobility in heterogeneous networks and then present a comprehensive performance study of a SIP-based vertical handoff scenario in third generation (3G) networks, where we show that performance-wise SIP may not meet the expectations of the 3rd Generation Partnership Project (3GPP)-specified standards. We conclude the chapter with covering some of the auxiliary schemes that help SIP meet the standard specified requirements.

17.2   Introduction

SIP [1] is gaining acceptance as an application-layer signaling protocol for Internet multimedia and telephony services, as well as for wireless Internet applications. SIP is free of many of the drawbacks of network or transport layer mobility, such as suboptimal routing and protocol stack modification. The fact that SIP has been accepted as the signaling standard by 3GPP and is also capable of providing mobility support made us choose SIP as the representative application-layer mobility solution. Another advantage is that the application layer, architecturally the highest layer of operation, can function across networks, a capability that will be tremendously useful in next-generation heterogeneous wireless networks.

SIP is a control protocol that allows creation, modification, and termination of sessions with one or more participants. It is used for both voice and video calls, either for point-to-point or multiparty sessions. It is independent of the media transport which for example, typically uses Real-time Transport Protocol (RTP)/Real-time Transport Control Protocol (RTCP) over User Datagram Protocol (UDP) [2]. It allows multiple endpoints to establish media sessions with each other. This includes terminating the session, locating the endpoints, establishing the session, and modifying the media session after the session establishment has been completed. Recently, SIP has gained widespread acceptance and deployment among wireline service providers for introducing new services such as VoIP, telepresence, video conferencing; within the enterprises for instant messaging and collaboration; and among mobile carriers for push-to-talk services. Industry acceptance of SIP as the protocol of choice for converged communications over IP networks is thus highly likely. As shown in Figure 17.1, a SIP infrastructure consists of user agents, registration servers, location servers, and SIP proxies deployed across a network.

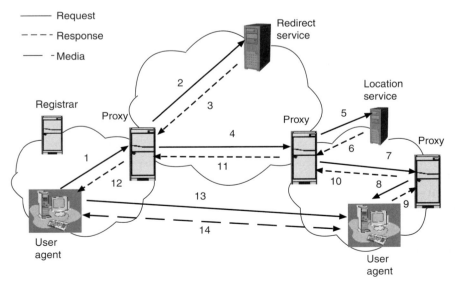

FIGURE 17.1
SIP architecture.

A user agent is a SIP endpoint that identifies services such as controlling session setup and media transfer. User agents are identified by SIP Uniform Resource Identifiers (URIs), which are unique HTTP-like URIs of the form sip:user@domain. All user agents register their IP addresses with a SIP registrar server (which can be colocated with a SIP proxy). Details of SIP can be found in [1]. SIP defines a set of messages, such as INVITE, REGISTER, REFER, etc., to setup sessions between the user agents. These messages are routed through SIP proxies that are deployed in the network. The Domain Name Server (DNS) service records help in finding SIP proxies responsible for the destination domain.

A session or dialog is setup between two user agents following a client–server interaction model, where the requesting user agent client (UAC) interacts with the target user agent server (UAS). A logical entity formed by concatenating a UAC and a UAS that keeps all the dialog information and intercepts all the messages within a dialog to participate in the same, is known as a back-to-back user agent (B2BUA). All requests from an originating UAC, such as an INVITE are routed by the proxy to an appropriate target UAS, based on the target SIP URI included in the Request-URI field of the INVITE message. Proxies may query location and redirect servers for SIP service discovery or to determine the current bindings of the SIP URI. Signaling messages are exchanged between user agents, proxies, and redirect/location servers to locate the appropriate services or endpoints for media exchange. For reasons of scalability, multiple proxies are used to distribute the signaling load [3]. A session is setup between two user agents through SIP signaling messages comprised of an INVITE (messages 1, 2, 4, 7, and 8 in Figure 17.1), an OK response (messages 9–12 in Figure 17.1) and an ACK (message 13 in Figure 17.1) to the response [1]. The call setup is followed by media exchange using RTP. The session is torn down through an exchange of BYE and OK messages.

The performance of SIP comprises the performance of call set-up and tear-down and performance of SIP mobility. Of these performance challenges of SIP, the mobility performance challenge is most critical. In this chapter, we will focus only on the challenge of SIP mobility and its performance. If the reader understands the underlying messages of SIP mobility and its performance modeling, the performance modeling of call set-up and tear-down will be pure applications of these models. The other aspects of SIP performance like reliability and robustness are not addressed in this chapter.

17.3 Mobility Management

Mobility is the most important feature of wireless networks that makes continuous service delivery possible in ubiquitous environments. Seamless service is usually achieved by supporting *handoff* or *handover* mechanisms. Handoff is the process of changing parameters (e.g., channel) associated with the current connection. It is often initiated by the movement of a mobile device or mobile host (MH), either within a network or across different networks. The former scenario initiates *horizontal handoff*, whereas the latter initiates *vertical handoff*. Handoff is again divided into two broad categories: *hard* and *soft* handoffs. They are characterized by "break-before-make" and "make-before-break" connections. In hard handoffs, current resources are released before new resources are used while in soft handoffs, both existing and new resources are used during the handoff process. For a soft handoff, the MH should be capable of communicating to multiple antennas through multiple interfaces.

Existing mobility protocols have mainly been designed for network, transport, and application layers, and the majority of existing literature refers to the inherent mobility support provided as the subnetwork layer mobility. We will evaluate candidate protocols in each layer for their merits and demerits with respect to such important design issues as the scope of mobility, the set of applications supported, and the objectives of mobility protocols.

A layer-by-layer classification of existing mobility protocols is shown in Figure 17.2, which also shows the scope or range of operation of these protocols. Only the transport and application layer protocols maintain the end-to-end semantics of a connection between communicating hosts. Most of the mobility protocols are limited to operate on a single layer only, so they are transparent to the other layers. In the following, the protocol design issues in each layer are discussed in more detail.

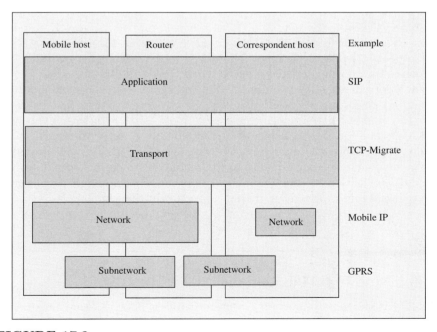

FIGURE 17.2
A layer-by-layer classification of the mobility protocols.

17.3.1   Subnetwork Layer

Subnetwork layer mobility is transparent to network and upper layers because an MH changes its point of attachment using solely layer two (link layer) mechanisms. This makes the deployment of these mechanisms easy from the perspectives of the network and end-to-end hosts, since all affected components are confined to the mobile host and its immediately adjacent subnetwork, with no changes to routing or correspondent hosts. But the obvious disadvantage of this solution is that the scope is a single subnetwork. It is the responsibility of the operating system to provide transparent layer two connections to higher layers of the protocol. Subnetwork mobility protocols provide mobility "invisibly." Unmodified network layer, transport layer, and application layer protocol entities continue to function without knowledge of the underlying link layer mechanisms.

Subnetwork layer mobility is inherently supported by all wireless cellular systems, such as intra-cell mobility in a circuit-switched cellular network. Inter-cellular mobility in cellular networks is achieved with the help of sophisticated switching mechanisms in the mobile switching centers (MSCs) that control the base stations. The main objective of this type of mobility protocol is to provide limited mobility to subnetworks only, which can work in conjunction with higher-layer protocols to provide mobility outside the scope of a subnetwork.

17.3.2   Network Layer

Network layer mobility finds its utility exactly where the subnetwork layer mobility falls short with respect to the scope of operation. The principal goal has been to provide network-level transparency. Specifically, in the Internet, hosts are identified solely with their network-level point of attachment (e.g., IP address). Thus, when a network layer mobility protocol is operating in a Transmission Control Protocol (TCP)/IP-based network, the upper layers do not have to bother about IP address change due to host mobility. Although the notion of network layer mobility started with the Mobile IP proposal [4], a host of different network layer mobility protocols have been proposed, each attempting to improve performance with respect to certain parameters. These protocols can be broadly classified into two classes—micro-mobility and macro-mobility—based on their scope of operation with respect to the administrative domains in the Internet. Micro-mobility protocols work within an administrative domain, while macro-mobility protocols operate across domains.

The legacy applications can still be used unchanged with a mobility protocol in this layer. But this protocol requires operating system level modification in the MH operating system to implement the functions specific to network-layer mobility (e.g., the registration messages in Mobile IP).

Architecturally, the network layer is the "right layer" to provide mobility for all transports and all applications. For a variety of reasons, such as ingress filtering or non-deployment of mobility agents, network-layer mechanisms have not been widely deployed to-date, although many of these issues have been addressed in the Mobile IPv6 proposal [5]. The availability of workable micro-mobility mechanisms may, however, increase the attractiveness of Mobile IP in the future.

17.3.3 Transport Layer

The transport layer maintains the true end-to-end connection, whereas the lower layer is completely ignorant of this end-to-end semantic. As mobility protocol implementation moves upward in the protocol stack, its binding with applications grows stronger. The interface between the application and transport layers gets changed with the introduction of a transport-layer mobility protocol. Of course, there are proposals like TCP Migrate [6] to keep the interface unchanged, but they involve substantial change in the network, such as the introduction of a proxy server and changes in the MH kernel. Thus, they are neither scalable nor deployable on a global basis. However, there are two ways to get around the interface change problem. One way is to recompile all the applications with the modified interface to the transport layer; the other is to put a wrapper around the transport layer libraries that the applications use. Both solutions are fairly costly and hence not widely accepted or deployed.

Transport-layer mobility solutions suffer from an additional disadvantage: they require modified transport layer implementations on correspondent hosts. This means users who are not mobile must install transport implementations to communicate with mobile users. Transport-layer mobility solutions are also transport-specific. Thus, TCP-oriented solutions are appropriate for TCP-based applications, but provide no mobility solution for UDP-based and RTP-based applications.

17.3.4 Application Layer

Most of the transport- and network-layer mobility solutions entail considerable changes in the MHs operating system. This is one of the primary reasons behind their low acceptance and deployment. Application-layer mobility management schemes overcome such drawbacks. As the mobility protocol is implemented in the application layer, the scope of operation is not limited to only network boundaries; this solution, like transport-layer mobility, also achieves the true end-to-end semantics. The overhead associated with these solutions is the delay involved with the application layer processing of longer protocol messages. Moreover, lower-layer support in detecting network change is also required.

Thus, the main motivation behind an application-layer mobility protocol has been to alleviate the major drawbacks of lower-layer solutions, such as suboptimal routing [7], network architecture change, and host operating system change, while providing a way to make an MH always reachable for packet forwarding. However, such a protocol incurs considerable (and unacceptable) delay for real-time services due to application-level processing.

Because these mobility protocols operate at the application layer, they cannot maintain a transport layer connection. Hence, they are somewhat decoupled from the actual applications, which work on the basis of a transport level connection (e.g., a TCP/IP application communicates over a TCP connection).

We believe that application-specific mobility mechanisms may be sufficient for the wireless Internet of the future. To gauge this, we must find out how many applications require mobility. "Portability" often suffices, whether or not mobility is available. Today's Internet application suite is heavily weighted toward client–server applications (e.g., the Web, e-mail, and server-based instant messaging). Client–server applications do not require the same level of mobility support as peer-to-peer applications. In fact, the majority of the applications are client–server-based and can be supported using application-layer mobility solutions with the exception of applications like sustained file downloads during movement. Today's servers do not usually move, nor do they initiate communication with a client that has not already located the server. If inherently peer-to-peer applications are widely used, mobility mechanisms operating at a lower layer of the protocol stack will be more attractive.

17.4   Protocols

In this section, we select a representative protocol in each layer and evaluate its merits and demerits with respect to the above discussions. The selected protocols either have been standardized for handling the mobility in the corresponding layer, or require minimal changes in the network stack and legacy infrastructure. Understanding of the working of the protocols is necessary to develop the performance models. Also, as the mobility in a real network will involve multiple mobility protocols, we selected the representative protocols to cover the scope of our models.

17.4.1   Link-Layer (Subnetwork-Layer) Mobility

One example of subnetwork-layer mobility, often termed link-layer mobility, is the access mobility provided in General Packet Radio Service (GPRS) [8] or Universal Mobile Telecommunication System (UMTS) [9] networks. Packets are forwarded from the core network to the MH or user equipment (UE) in the radio access networks by setting up GPRS Tunneling Protocol (GTP) in two segments from the gateway GPRS support node (GGSN) to the MH/UE. The serving GPRS support node (SGSN) handles inter-radio network controller (RNC, functioning as the base station controllers of second generation (2G) or 2.5G cellular networks) mobility, and the GGSN manages inter-SGSN mobility. When the serving RNC (i.e., the one to which the MH/UE is attached) changes and it is within the scope of the same SGSN, a redirection of the GTP tunnel takes place between the SGSN and the RNC. This changeover takes place in the subnetwork layer, and the solution works fine until the MH/UE leaves the GPRS or UMTS network. This type of mobility management mechanism has its own location update, registration, and paging scheme for maintaining connection to the MH/UE. However, seamless mobility between the GPRS/UMTS network and other data networks still remains elusive.

17.4.2   Network-Layer Mobility

The value of network-layer mobility is open to question with the wide deployment and popularity of Dynamic Host Configuration Protocol (DHCP) [10], Post Office Protocol (POP3) [11], roaming user profile support for Web browsers through Lightweight Directory Access Protocol (LDAP) [12], and the general move to client–server applications where the MH initiates communication. However, to-date the network layer remains the most dominant of all the layers in the protocol stack for mobility protocols. As mentioned earlier, network-layer mobility protocols can be broadly classified into macro- and micro-mobility protocols depending on the scope of their operation.

1) Macro-mobility Protocols: Macro-mobility refers to user mobility that is infrequent and also spans considerable space, often between several administrative domains. We have chosen Mobile IP (in short, MIP) as the representative network layer (or macro) mobility solution since it has been standardized by the Internet Engineering Task Force (IETF).

MIP was proposed with the objective of supporting mobile users with application transparency and the possibility of seamless roaming. This was done by adding IP tunneling to a unified IP routing infrastructure. Under MIP, it is assumed that the MH has a home network from which it obtains its statically allocated IP address and that the MH roams from the home network among various foreign networks. Typically, an MH is a host or router that changes its point of attachment from one network to another while keeping its IP address constant. An entity, called the home agent (HA), maintains reachability information for the MH and forwards the datagram sent to the MH through an IP tunnel to any other visited foreign network. The foreign agent (FA) acts as a router at the visited network, and cooperates with the HA to forward datagrams to any visiting MHs. MIP consists of three important functions:

(1) Mobile agent discovery

(2) Registration with HA

(3) Delivery of datagrams using tunneling

The mobility agents make themselves known by sending advertisement messages. On receiving agent advertisements, the MHs get to know whether they are in the home network or in a foreign network. When the MH moves to a foreign network, it obtains a care-of address (CoA) on the foreign network by either listening to the advertisement from the FA or contacting a local DHCP server or Point-to-Point Protocol (PPP) access server. The MH then registers its new CoA with the HA. The datagrams sent by the correspondent host (CH), destined for the MH, are intercepted by the HA and forwarded to the MH after encapsulation using a tunnel, with the endpoint at either the FA or the MH itself. The encapsulated datagrams are decapsulated and delivered to the MH. Mobile IP works fairly well, except when the distance between the visited foreign network and the home network of the MH is sufficiently large, thereby inducing large signaling delay for these registrations. After an MH roams into a new network, no CH traffic will reach the MH until the MH updates its HA. This outage may be noticeable to applications and hence to end users. To reduce the effect of this outage time, MIP was enhanced to incorporate route optimization and smooth handoff [7] by enabling the MH to directly bind to the CH and keep it updated on its current location. The regional registration scheme [13] has been proposed to reduce the number of signaling messages to the home network, and hence the signaling delay, when an MH moves from one FA to another. This proposal employs a hierarchy of FAs to handle MIP registration updates within a visited network.

2) Micro-mobility Protocols: The limitations of MIP, discussed above, cause severe quality of service (QoS) degradation for the user. Therefore, some kind of localized micro-mobility protocols are needed. These protocols operate in a restricted administrative domain and provide the MHs within that domain with connections to the core network, while keeping signaling cost, packet loss, and handover latency as low as possible. Several micromobility protocols have been proposed, including HAWAII [14], Cellular IP (CIP) [15], Terminal Independent Mobile IP (TIMIP) [16], and Intra Domain Mobility Management Protocol (IDMP) [17].

17.4.3 Transport-Layer Mobility

A TCP connection is uniquely identified by a 4-tuple: <source IP address, source port, destination IP address, destination port>. Once an MH changes its point of attachment and acquires a new IP address, any existing TCP connection will be interrupted. Transport-layer mobility protocols maintain TCP connectivity, which means a TCP session broken due to the mobility of an MH can be resumed after the MH acquires a new IP address. One solution to this problem is TCP-Migrate [6], which we have chosen as the representative transport layer mobility protocol. TCP-Migrate requires minimal change in the network infrastructure, whereas other solutions [14] require the introduction of an additional network entity such as a proxy, to split up the TCP connection. Introduction of such additional entities is not only costly but also fairly impractical. However, TCP-Migrate involves the following modifications to TCP:

- A token contained in the Migrate option is negotiated between both ends during the initial connection establishment. Thus, a TCP connection can be identified by a triple: <source IP address, destination IP address, token>. A new `Migrate TCP` option included in SYN flags identifies a SYN packet as part of a previously established connection, rather than a request for a new connection. The previously broken TCP connection can be resumed after the token information exchange between the client and the server.

- To track an MH, the hostname to address (A-record) mapping in the DNS must be updated dynamically once the MH changes the point of attachment. This scheme relies on some secure DNS update protocols [18]. Also, some DHCP servers can issue a DNS update at the client boot time when assigning a new IP address to a known MH, according to the MAC-to-DNS table.

- It is required to modify the transport layer implementation for all the hosts, regardless of whether they are mobile or static. This is difficult considering the number of hosts already in operation today. In addition to this major problem, there exist several other problems with this approach. For example, during the DNS update period, the packets destined to the MHs may be lost, especially for the hosts that move frequently. A caching time to live (TTL) is associated with each DNS resource record, forcing a choice between long TTLs and long periods of inaccessibility, or short TTLs and increased load on the DNS servers due to ineffective resolver caching.

17.4.4   Application-Layer Mobility

SIP is capable of handling terminal, session, personal, and service mobility [19]. Since we are discussing terminal mobility support that allows a device to move between subnets while being reachable to other hosts and maintaining any ongoing session, we focus on the terminal mobility feature of SIP only. Further details on terminal mobility can be found in Section 17.5.1.

SIP has one obvious disadvantage: it cannot support seamless mobility for any other applications, even other TCP-based applications. A proposed architecture [19] supports the complete range of applications by using SIP for real-time communication and MIP for TCP connections. However, recently there have been attempts to support TCP connection with SIP [19] by adding a middleware for spoofing constant endpoints for mobile TCP connections. But most often, the TCP connection duration (e.g., POP or IMAP mail retrieval, HTTP transactions) is short enough to make the cost of reconnection relatively small on average. Another drawback of SIP is the large handoff delay incurred in application-layer messaging during mid-call mobility management.

A synopsis of the strong and weak points of the mobility solutions, based on the above discussions, is presented in Table 17.1.

Thus, our qualitative studies [20–22] have shown that SIP is an eligible candidate as a mobility management protocol for next generation heterogeneous

TABLE 17.1
Comparison of Mobility Management Schemes [20–22]

Layer	Protocols	Advantages	Disadvantages
Subnetwork	GPRS or UMTS access mobility	Mobility within a subnetwork provider's service area requires no changes to the network, transport, or application layers	Mobility is restricted to the subnetwork domain — the user cannot move to another subnetwork without using additional mobility solutions
Network	Mobile IP	Mobility transparent to upper layers	Requires deployment of foreign agents in visited networks; Requires deployment of home agent in home network; Requires changes to mobile host network-layer implementation; Triangular routing can add considerable latency, based on the distance between the mobile host and home agent; Signaling overhead for location update
	IDMP	Provides faster handoff for intra-domain mobility, reduces the signaling load	Specific architectural requirements; Characteristic routing protocols; Changes in MH
Transport	TCP Migrate	Supports session continuity for any TCP connection	TCP layer needs modification; Requires dynamic DNS to track MH; Supports mobility for one host at a time; Session cannot be recovered if disconnection time is too long; Does not provide mobility for applications built on other transports
Application	SIP	No change in TCP and IP layers for both mobile host and correspondent host; Supports personal, terminal, service and session mobility for a specific application	Requires deployment of application-specific proxy servers; Does not support session continuity for any other application; Large hand-off delay

Source: Banerjee, N. et al., *Computer Communications* 27, 697–707, 2004; Banerjee, N., et al., *IEEE Wireless Communications Magazine* 10, 54–61, 2003; Das, S.K. et al., *IP Mobility Protocols for Wireless Internet*, Kluwer, 2003.

wireless networks. Apart from the advantage of being an application-layer protocol, it has been shown in [20,23] that it can meet the stringent delay bounds of real-time multimedia traffic belonging to the conversational class. However, SIP uses TCP or UDP to carry its signaling messages and hence, is limited by the performance of TCP or UDP over wireless links. In addition, SIP entails application-layer processing of the messages, which may introduce considerable delay. These are the prime factors behind the handoff delay while using SIP as the mobility management protocol.

17.5   SIP-Based Mobility Management: A Quantitative Analysis

The European Telecommunications Standards Institute (ETSI) has quantitatively defined four different classes of performance — *best, high, medium, and best effort* — for voice traffic (e.g., VoIP) and streaming media over IP networks (e.g., streaming video for applications such as videoconferencing) [24]. The first two performance classes specify the type of IP telephony services that have the potential to provide a user experience better than the Public Switched Telephone Network (PSTN). The medium class has the potential to provide a user experience similar to common wireless mobile telephony services. The best effort class includes the type of services that will provide a usable communications service but may not provide performance guarantees. The specification for the end-to-end media packet delay for the best and high classes of services is less than 100 ms, while for the medium and best effort classes, the delay is less than 150 ms and 400 ms, respectively. In fact, a handoff delay of more than 200–250 ms makes voice conversations annoying. Clearly, the handoff delay, being a component of the total end-to-end delay, should also abide by these delay limits. Thus, it is evident that for QoS-sensitive streaming multimedia traffic belonging to either the *best* or *medium* class, the handoff delay should be preferably less than 100 ms.

In the rest of this chapter, we investigate the performance of SIP as a mobility management protocol from the above perspective in a heterogeneous access networking environment akin to fourth generation (4G) wireless networks. In particular, we perform a case study of SIP-based handoff delay analysis using SIP to handle terminal mobility in an IP-based network. Two different types of access technologies, viz. UMTS and IEEE 802.11b-based WLAN, have been considered for the IP-based network.

17.5.1   SIP Mobility Support

Apart from the session setup function, SIP inherently supports personal mobility and can be extended to support service and terminal mobility [19,25]. Personal mobility enables a user to be found independent of the location and network device. Terminal mobility, on the other hand, enables a user to change location or IP address during the traffic flow of an ongoing session. It can be explained with an example of an ongoing session between a mobile host (MH) and a correspondent host (CH) as follows. Each MH belongs to a home network with a SIP server providing a registrar service. Each time the MH changes location, it registers with the home network's registrar service. This is in principle similar to Mobile IP home registration. For ongoing sessions, the MH sends a re-INVITE message to the corresponding CH using the same call identifier as in the original setup. The former procedure takes care of *pre-call mobility*, while the latter enables *mid-call mobility*. High-level messaging of SIP-based mid-call mobility management is depicted in Figure 17.3. The new contact information (e.g., URI for future contact) is put in the `Contact` field of the SIP message to redirect the subsequent SIP messages to the current location. The data traffic flow is redirected by updating the transport address field in the Session Description Protocol (SDP) [26] part of the re-INVITE message. For mid-call mobility, the CH starts sending data to the new location as soon as it gets the re-INVITE message. Hence, the handoff

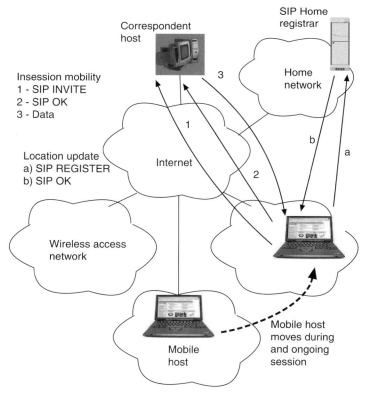

FIGURE 17.3
SIP-based mid-call terminal mobility management.

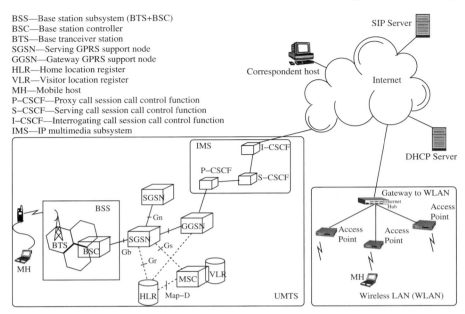

BSS—Base station subsystem (BTS+BSC)
BSC—Base station controller
BTS—Base tranceiver station
SGSN—Serving GPRS support node
GGSN—Gateway GPRS support node
HLR—Home location register
VLR—Visitor location register
MH—Mobile host
P–CSCF—Proxy call session call control function
S–CSCF—Serving call session call control function
I–CSCF—Interrogating call session call control function
IMS—IP multimedia subsystem

FIGURE 17.4
4G Architecture considered for case study.

delay is essentially the one-way delay for sending an INVITE message from the MH to the CH.

17.5.2   System Architecture for Performance Evaluation

For our case study, we have considered an architecture conceptually similar to IP-based 4G networks in terms of heterogeneity in access network technologies. A logical view of the architecture considered is shown in Figure 17.4. The architecture is primarily focused on wireless mobile multimedia networking and is constructed around an IP core network (the Internet) with two different types of the access networks, viz. UMTS and WLAN. The UMTS Release 5 multimedia architecture [9] has been proposed by 3GPP to provide multimedia-based services in an all-IP environment.

UMTS Release 5 defines GPRS/EDGE* radio access network (GERAN) as its access technology. We have assumed only GPRS access network due to its wide acceptance. GPRS networks are built on existing Global System for Mobile communications (GSM) networks by adding a new class of network nodes called the GPRS support nodes (GSN). A *serving GPRS support node* (SGSN) is responsible for mobility and link management, and delivering packets to the MH under its service area. A *gateway GPRS support node*

---

*Enhanced data rates for GSM evolution

(GGSN) acts as an interface between the GPRS network and the external packet data networks (the Internet, in this case). Home Location Register (HLR) and Visited Location Register (VLR) are two databases to keep user location information for mobility management. These databases are derived from legacy GSM architecture. A location register in the SGSN keeps track of the current VLR for a user.

A salient feature of UMTS Release 5 standardization is the new subsystem, known as the IP Multimedia Subsystem (IMS) that works in conjunction with the Packet Switched Core Network (PS-CN) for supporting legacy telephony service as well as new multimedia services. The IMS enables an IP-based network to support both IP telephony services as well as the multimedia services. SIP is the signaling protocol used between the MH or User Equipments (UE) and the IMS as well as with its internal components. As far as the SIP signaling is concerned, the main component of the IMS involved is the Call Session Control Function (CSCF), which is basically a SIP server. The CSCF performs a number of functions such as multimedia session control and address translation (i.e., evolution of digit translation function). In addition, the CSCF must perform switching function for services, voice coder negotiation for audio communication, and handling the subscriber profile (analogous to the VLR). The CSCF play three roles, viz. the Proxy CSCF (P-CSCF) role, the Interrogating CSCF (I-CSCF) role and the Serving CSCF (S-CSCF) role. P-CSCF is the mobile's first point of contact with the IMS network; I-CSCF is responsible for selecting the appropriate S-CSCF based on load or capability; and S-CSCF is responsible for mobile session management.

The other access network technology considered is IEEE 802.11-based WLAN that consists of several access points (AP) providing the radio access to the MH. The APs are connected to the backbone IP network with an ethernet switch. A Dynamic Host Configuration Protocol (DHCP) [10] server is used to assign an IP address to a visiting MH.

We assume that an MH moving between a UMTS network and WLAN has separate network interfaces to connect to these networks. The MH, after moving to a UMTS network or a WLAN, switches to the respective interface in order to attach to the corresponding access network infrastructures. The switching-over operation is initiated by the reception of the GPRS pilot signal in a UMTS network and the characteristics beacon in a WLAN.

1) In-session or Mid-call handoff with SIP: For pre-call mobility, as described before, the MH re-registers its new IP address with its "home" by sending a REGISTER message, while for mid-call mobility, the terminal needs to intimate the correspondent host (CH) or the host communicating with the MH, by sending an INVITE message about the terminal's new IP address and updated session description. In principle, this is similar to the Mobile IP route optimization [7] strategy. The CH starts sending data to the new location as soon as it gets the re-INVITE message. Hence, the handoff delay is essentially the one-way delay for sending an INVITE message from the MH to the CH. Here the "home" refers to the redirect or SIP server in the home network of the MH. The MH needs to register with the redirect server in the home network for future calls. However, in mid-call mobility management, before sending the SIP re-INVITE message, there are some procedures that need to be completed to get the MH attached to the wireless access network infrastructure. For example, an MH attaches to the GPRS radio access of a UMTS network using the GPRS attach and Packet Data Protocol (PDP) context activation procedure, while for the wireless LAN it uses DHCP to attach to the WLAN.

Now, mobility in such a heterogeneous networking environment can give rise to the following four cases: (1) MH moves from a UMTS network to another UMTS network, (2) MH moves from a UMTS network to a WLAN network, (3) MH moves from a WLAN network to a UMTS network, and (4) MH moves from a WLAN network to another WLAN network. Since we are concerned with the analysis of the delay incurred in the *vertical* handoff procedure, the above four cases can be mapped to only two cases of interest:

- MH moving to a UMTS network

- MH moving to a WLAN

This is because the handoff delay is caused mainly by the message exchange that occurs while an MH attaches to a new access network (either UMTS or WLAN in our case) followed by the location update. These two cases are discussed in more details as follows.

2) Mobile host moving to a UMTS network: When an MH moves to a UMTS network from another UMTS network or a wireless LAN, it performs two key functions to initiate a handoff:

- Data connection setup that involves the execution of two procedures known as GPRS attach and the PDP context activation. This establishes the data path required to carry the SIP-related messages to the proxy-CSCF through the GGSN, which acts as the gateway for the proxy-CSCF.

  The messages involved in the GPRS attach and the PDP context activation procedures are shown in Figure 17.5 and Figure 17.6. The steps are described as follows: as a part of the GPRS attach procedure, the MH sends an attach message (1) to the SGSN (responsible for mobility management, logical link management, and authentication and charging functions in a UMTS network) with the MH's International Mobile Subscriber Identity (IMSI). The SGSN uses the IMSI to authenticate (messages 2, 3, 4 and 5) the MH with its HLR. Successful authentication is followed by the SGSN sending a location update to the HLR (messages 6 and 7). The SGSN finally completes the attach procedure by sending an attach complete message (8) to the MH. Thus, a logical association is established between the MH and the SGSN. Once an MH is attached to an SGSN, it must activate a PDP address (or IP address) to begin packet data communication. Activation of the PDP address creates an association between the MH's current SGSN and the GGSN (acting as the interface between the GPRS/UMTS backbone network

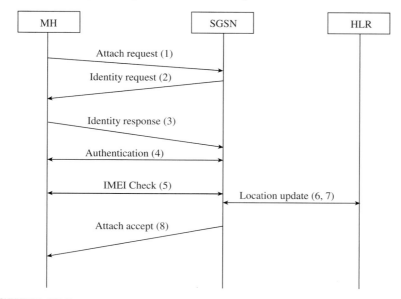

FIGURE 17.5
GPRS attach procedure.

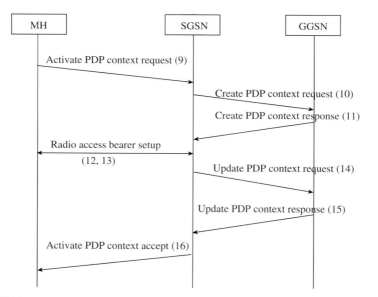

FIGURE 17.6
PDP aontext activation procedure.

and the external packet data networks) that anchors the PDP address. A record of such an association is known as the PDP context. The PDP context transfer is initiated by the MH by sending a PDP context activation message (9) to the SGSN. The SGSN, after receiving this activation message, discovers the appropriate GGSN (messages 10 and 11). It selects a GGSN capable of performing functions required for the SIP-related activities. The SGSN and the GGSN create special paths for the transfer of SIP messages to the P-CSCF, which is identified by the GGSN. The corresponding IP address of the P-CSCF is sent along with the activation accept message (messages 12–16).

- The SIP message exchange for re-establishing the connection is shown in Figure 17.7. The MH re-invites the CH to its new temporary address by sending a SIP INVITE message (1) through the P-CSCF, S-CSCF and the I-CSCF servers. The INVITE message uses the same call identifier as in the original call setup and contains the new IP address at the new location. Once the CH gets the updated information about the MH, it sends an acknowledge message (2) while starting to send data.

3) Mobile host moving to a wireless LAN: When an MH moves to a WLAN from another wireless LAN or a UMTS network, it goes through the following major steps to update its location with the CH:

- The MH goes through a DHCP registration procedure to secure a new IP address for its new location. The message exchanged in the registration procedure is shown in Figure 17.8. When the MH identifies the presence of a WLAN after receiving the characteristics beacons, it broadcasts a DHCP DISCOVER message (1) to discover the DHCP server willing to lend it with registration service. The appropriate DHCP server sends out a DHCP OFFER message (2) to offer service to the requesting MH.

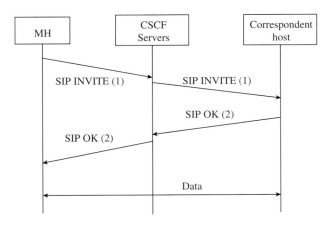

FIGURE 17.7
Messages involved in SIP-based mid-call terminal mobility management.

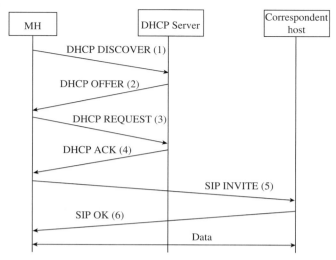

FIGURE 17.8
DHCP registration procedure.

The MH, on receiving this OFFER message, sends a DHCP REQUEST message (3) to the DHCP server to confirm the offer made. The DHCP server then sends the MH DHCP ACK message (4) with the new IP address to be assigned to the MH.

- The SIP message exchange to re-establish the connection is similar to that for UMTS networks, where the MH, after acquiring the new IP address, re-invites the CH to its new address by sending a SIP INVITE message (messages 5, 6, and 7).

17.5.3 Handoff Delay Analysis

In this section, we derive the handoff delay introduced due to the wireless link in the UMTS and WLAN access networks and the queueing delay in the different servers in the signaling path. We show that the handoff delay due to the wireless access is beyond the maximum tolerable limit for streaming media in low-bandwidth UMTS networks and not in WLANs. Let $D_{Handoff}$ denote the handoff delay during the mid-call terminal mobility, which can be divided into two parts: (1) the delay occurred during the attachment procedure and (2) the delay due to location update using the SIP INVITE message. During each of these procedures, messages are transported over the wireless access link, which introduces major delays in comparison with the queuing and transmission delay introduced by the high-speed backbone networks.

To compute the delay for transmitting messages over the wireless links in the access networks, we have used the delay models for frame and packet transmission over a wireless link under various link error conditions, as proposed in [27]. The outdoor operation of GPRS radio access networks makes it more vulnerable to noise, thus increasing the bit error rate (BER) for the wireless channel. To improve the BER performance for transmission of control packets over wireless links, a semi-reliable link-layer retransmission mechanism like the Radio Link Protocol (RLP) is used on top of the physical layer. RLP encapsulates the MAC layer frames into RLP frames by adding control information that determines the retransmission mechanism. When RLP at the receiving end finds a frame in error or a frame is missing, it sends back a NAK (negative acknowledgement) requesting for retransmission of this frame and starts a retransmission timer. When the timer expires for the first attempt, RLP resets the timer and sends back NAK again. This NAK triggers a retransmission of the requested frame from the sender. In this way, the number of attempts per retransmission increases with every retransmission trial. However, due to much higher bandwidth and the indoor operation of WLAN, no such retransmission scheme is used. So we need to consider two types of wireless delay models for our case study.

Transmission delay with RLP (for GPRS radio access): The analysis considers the following parameters:

- p = Probability of an RLP frame being in error in the air link

- k = Number of frames in a packet transmitted over the air

- D = End-to-end frame propagation delay over the air link (typical values of the order of 100 ms).

- τ = Interframe time of RLP (typical values of the order of 20 ms for GPRS).

The effective packet loss P_f seen at the transport layer, with RLP operating underneath, is given as:

$$P_f = 1 - p + \sum_{j=1}^{n}\sum_{i=1}^{j} P(C_{ij})$$

$$= 1 - p(p(2-p))^{\frac{n(n+1)}{2}}$$

(17.1)

where n is the maximum number of RLP retransmission trials and C_{ij} (representing the first frame received correctly at the destination) is the ith retransmission frame at the jth retransmission trial. The number of trials (n) is usually less than four. We assume three trials in our analysis. For $n=3$ (typical value), the packet loss rate with RLP for packets with k frames is given as:

$$q = 1 - (1 - p(p(2-p))^6)^k$$

Considering the RLP retransmissions, the transport delay in transmitting a packet over the RLP is given by:

$$D' = D + (k-1)\tau + \frac{k(P_f - (1-p))}{P_f^2}$$

$$\times \left(\sum_{j=1}^{n} \sum_{i=1}^{j} P(C_{ij}) \left(2jD + \left(\frac{j(j+1)}{2} + i \right) \tau \right) \right) \qquad (17.2)$$

Interested readers can find further details on the derivations of these expressions in [27].

Next, we determine the value of k corresponding to different types of messages involved in the GPRS attach and PDP context activation procedures. The maximum size of the messages exchanged in the GPRS Attach procedure is 43 bytes [28].

Now, for a 9.6 Kbps channel, there are $9.6 \times 10^3 \times 20 \times 10^{-3} \times \frac{1}{8} = 24$ bytes in each frame. Therefore, the number of frames (k) to be transferred for a single message in the attach procedure is $43/24 \approx 2$. Similarly, for a 19.2 Kbps or higher bandwidth (e.g., 128 Kbps) channel, the number of frames per message is $k = 1$. On the other hand, the maximum size for the PDP context activation messages is 537 bytes [28]. So the number of frames for PDP messages are $k = 537/24 \approx 23$, 12 and 2 for 9.6 Kbps, 19.2 Kbps and 128 Kbps channels, respectively.

Using the expression for delay model in Equation 17.2, the delays corresponding to the GPRS attach and the PDP context activation procedures can be determined as: $D_{Attach} = 8D'$ and $D_{PDP} = 4D'$. This is because, as shown in Figure 17.5 and Figure 17.6, GPRS attach and PDP context activation procedures requires eight and four message exchanges, respectively.

Transport delay without RLP (for WLAN): In this case, there is no RLP retransmission. Instead, due to packet loss, retransmission may be done by upper-layer protocols like TCP or DHCP until there is a successful transmission. Let N_m be the number of such retransmissions. The DHCP packet loss rate in this case is $q = 1 - (1-p)^k$. The average delay for successfully transmitting a TCP or DHCP packet with no more than N_m retransmissions is given as,

$$D'' = (k-1)\tau + \frac{D}{(1-q^{N_m})(1-2q)} + \frac{1-q}{1-q^{N_m}} D \left[\frac{q^{N_m}}{1-q} - \frac{2^{N_m+1} q^{N_m}}{1-2q} \right] \qquad (17.3)$$

Note that the DHCP messages have a maximum length of 548 bytes. Also the IEEE 802.11 standard specifies that the WLAN frame duration is 3.5 ms. So, using similar calculations as for the case with RLP, we get $k = 1$ for both 2 Mbps and 11 Mbps WLAN. Also the end-to-end transmission delay, D, of the wireless channel $= 0.27$ ms and 0.049 ms for 2 Mbps and 11 Mbps channel, respectively. The interframe time, $\tau = 0.001$ secs, is independent of the bit rate. Now, the delay due to DHCP registration is given by $D_{DHCP} = 4D''$, since it is shown in Figure 17.8 that DHCP registration requires four message exchanges.

1) Handoff delay to UMTS network from another UMTS or WLAN network: Since SIP is an application-layer protocol, the processing of SIP messages in the intermediate and destination servers may take considerable time due to the queuing of messages that need to be accounted for. Rough estimates of the queuing delays can be obtained using the classical queueing theory-based waiting time formulas.

The major delays occur in the MH, the P-CSCF, I-CSCF, S-CSCF and the destination server due to the queuing of the SIP messages. This is shown in Figure 17.9.

To compute the queueing delay we have assumed an M/M/1 queueing model for MH as well as CSCF servers, and a priority based M/G/1 model for the destination server. The rationale behind these assumptions is that while the MH and the CSCF servers perform dedicated jobs, the destination server may be busy with a variety of jobs other than serving the SIP messages and thus, may have arbitrary service time distribution. Table 17.2 lists the parameters used in the analysis and their meanings. Although, it has been shown that the Internet delay varies between 100 ms to 1 s [29], emerging high-speed technologies like Generalized Multiprotocol Label Switching (GMPLS) [30] provide efficient traffic engineering to reduce this Internet delay to a nominal fraction of the minimum allowed end-to-end delay (on the order of few ms). Hence, the major concern is with the delay introduced by the wireless links in the access networks.

Let us now determine the queueing delay of a SIP message at the MH, the intermediate CSCF servers, the destination and the transport delay over the wireless access. We assume that multiple MHs are served by the CSCF servers that also support load balancing functions. Hence $\lambda M < \lambda$ or λM is a fraction of λ. The SIP message transmission delay, $D_{SIP\text{-}UMTS}$, for GPRS radio access of a UMTS network can be computed as:

$$D_{SIP\text{-}UMTS} = D_{MH} + D_{RLP} + D_{P\text{-}CSCF} + D_{I\text{-}CSCF} + D_{S\text{-}CSCF} + \Delta_I + D_{Dest}$$
$$(17.4)$$

Using the results from queuing theory [31] and the parameters presented in Table 17.2, the delay components are estimated as follows.

$$D_{MH} = \frac{1}{\mu - \lambda_M} \qquad (17.5)$$

$$D_{P\text{-}CSCF} = D_{I\text{-}CSCF} = D_{S\text{-}CSCF} = \frac{\rho_s}{\lambda(1 - \rho_s)} \qquad (17.6)$$

$$D_{Dest} = \frac{\frac{1}{\mu_s}(1 - \rho_0 - \rho_s) + R}{(1 - \rho_0)(1 - \rho_0 - \rho_s)} \qquad (17.7)$$

where $R = \frac{\lambda_0 \bar{X}_1^2 + \lambda_s \bar{X}_s^2}{2}$ while $\bar{X}_1^2$ and $\bar{X}_s^2$ are the second moments of μ_0 and μ_s, respectively. The expression for D_{Dest} is obtained by using the result of

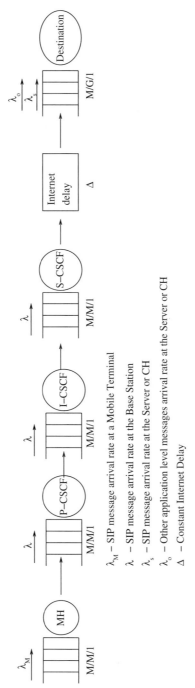

FIGURE 17.9

Queuing model for analyzing delay in SIP-based handoff to UMTS network.

TABLE 17.2

System Parameters

Parameters	Symbols
ΛM	SIP message arrival rate at the UE/MH
Λ	SIP message arrival rate at the CSCF servers
μ	Processing delay for each SIP message in the UE/MH
ρ_s	Destination and the CSCF server load
λ_s	SIP message arrival rate at the destination
μ_s	Processing delay for each SIP message at the destination
ρ_o	Load at the destination for messages other than SIP
λ_0	Arrival rate at the destination for messages other than SIP
μ_0	Processing delay at the destination for messages other than SIP
Δ_I	Internet delay in transmitting of SIP messages
D_{MH}	Queueing delay at the MH
D_{RLP}	Delay in transmitting a packet over an RLP link in UMTS network
$D_{P\text{-}CSCF}$	Queueing delay at the P-CSCF server
$D_{I\text{-}CSCF}$	Queueing delay at the I-CSCF server
$D_{S\text{-}CSCF}$	Queueing delay at the S-CSCF server
D_{Dest}	Queueing delay at the destination (CH)
D_{SIP}	Queueing delay at the P-CSCF server
D_{GW}	Queueing delay at the gateway to the WLAN
D'	Transport delay with RLP
D''	Transport delay without RLP

a non-preemptive priority-based M/G/1 queue [31]. Since our objective is to estimate the SIP message processing delay, we have considered only those messages having higher priority than SIP messages and ignored other lower priority messages. The derivation of the delay, D_{RLP}, requires us to adopt a transport-layer-based delay model over wireless links. Now, SIP messages work with both TCP and UDP. Since we are dealing with wireless links and TCP is a reliable protocol, the SIP messages are assumed to be sent over TCP. So a delay model for TCP transmission over wireless links is required.

According to the model used and the results reported in [27], the delay to transmit a TCP segment consisting of k frames over a radio link with RLP operating over it, is given by

$$D_{RLP} = D + (k-1)\tau + \frac{k(P_f - (1-p))}{P_f^2}$$
$$\times \left(\sum_{j=1}^{n} \sum_{i=1}^{j} P(C_{ij}) \left(2jD + \left(\frac{j(j+1)}{2} + i \right)\tau \right) \right)$$
$$+ \frac{2Dq(1-q)}{1-q^{N_m}} \left[1 + \frac{4q(1-(2q)N_m - 2)}{1-2q} - \frac{q(1-qN_m - 2)}{1-q} \right] \quad (17.8)$$

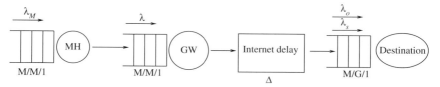

λ_M — SIP message arrival rate at a Mobile Terminal

λ — SIP message arrival rate at the Base Station

λ_s — SIP message arrival rate at the Server or CH

λ_o — Other application level messages arrival rate at the Server or CH

Δ — Constant Internet Delay

FIGURE 17.10

Queuing model for analyzing delay in SIP-based handoff to WLAN.

where $n = 3$ is the maximum number of RLP retransmission trials, and N_m is the number of TCP retransmissions. The parameters τ, p, P_f, and C_{ij} are the same as defined earlier, and $q = 1 - (1 - p)^k$ is the packet loss rate.

To derive the value of k, we have assumed that a TCP segment is carried in one packet. We assume that the air link frame duration is 20 ms. As derived earlier, a 9.6 Kbps radio channel can afford 24 bytes in each frame. Also, we assume that the size of one SIP message is 500 bytes. Therefore, the number of air link frames in a SIP message is $k = \frac{500}{24} \approx 21$. For 19.2 Kbps and 128 Kbps, channels the number of frames are $k = 11$ and $k = 2$, respectively.

2) Handoff delay to a WLAN network from another WLAN or UMTS network: For the WLAN network, the queueing delays are shown in Figure 17.10. Different parameters used are also listed in Table 17.1. The corresponding transmission delay for a SIP message, $D_{SIP\text{-}WLAN}$, can be calculated in the same manner as $D_{sip\text{-}umts}$ and is given as follows:

$$D_{SIP\text{-}WLAN} = D_{MH} + D'' + D_{GW} + \Delta_I + D_{Dest}$$

Here D_{GW} is essentially the same as the queueing delay at any of the CSCF servers and is given as follows.

$$D_{GW} = \frac{\rho_s}{\lambda(1 - \rho_s)}$$

For both 2 Mbps and 11 Mbps WLAN networks, the number of frames corresponding to a SIP message (500 bytes) is $k = 1$. Although, typically in a WLAN, the SIP-based control messages and the data use the same channel, we have assumed that the control messages would have higher preemptive priority than the data frames and would not wait for pending data frame transmission.

Once we have the estimates for all the components, we can determine the total handoff delay for UMTS networks as

$$D_{Handoff} = D_{Attach} + D_{PDP} + D_{SIP\text{-}UMTS} \qquad (17.9)$$

TABLE 17.3

System Parameter Values

Parameters	Values
μ	4×10^{-4} sec
ρ_s	$\dfrac{\lambda}{\mu}(\lambda < \mu)$
ρ_o	0.7
N_m	10

and for WLAN as

$$D_{Handoff} = D_{DHCP} + D_{SIP\text{-}WLAN} \qquad (17.10)$$

3) Numerical results: In this section, we present the results for the handoff delay computation in a SIP-based multimedia session, using the delay models described in the previous section. The values used for different system parameters are given in Table 17.3.

We have assumed the SIP message arrival rate (λ) and the processing rate at the CSCF servers and the destination server, are the same (i.e., $\mu_s = \mu$). Also, we have assumed $\lambda_M = 0.1\lambda$. The derivation of D_{Dest} involves the second moment of the processing delay at the destination, which can be derived once the mean and variance are given. For our analysis, we have assumed the standard deviation of the processing delays at the destination is 5% of the mean. Now $\bar{X}_1^2 = E[X_1^2]$ and $\bar{X}_s^2 = E[X_s^2]$. Also, $E[X_1^2] = \sigma_1^2 + (E[X_1])^2$ and $E[X_s^2] = \sigma_s^2 + (E[X_s])^2$, where σ_1^2 and σ_s^2 are the respective variances. Substituting μ_0 and μ_s for $E[X_1]$ and $E[X_s]$ and the values for the variances, we get $R = 0.501[\rho_o^2 + \rho_s^2]$.

As mentioned before, due to the varying nature of the Internet delay and the computing power of the intermediate servers, it is difficult to characterize the end-to-end handoff delay. With proper traffic engineering (e.g., GMPLS), the Internet delay can be made to suit the application requirements. Hence, we focus on the component of the handoff delay introduced due to the wireless access networks to estimate the minimum handoff delay. Subsequently, we have also estimated the end-to-end handoff delay assuming a constant value for the Internet delay and some representative values for the computing capabilities of the servers as shown in Table 17.3. Figure 17.11 shows the increase of the handoff delay component due to the wireless access only, with the increase of channel Frame Error Rate (FER) for channel bandwidth of 9.6 Kbps, 19.2 Kbps, and 128 Kbps, when the MH moves to a UMTS network. Table 17.4 shows the corresponding end-to-end handoff delay including the queuing delay at different servers and the transmission delay over the Internet. Figure 17.12 shows the handoff delay component due to wireless access with the increase of SIP-based session request rate. The request rate of the SIP-based session in an MH is assumed to be $\lambda_M = 50$ requests/s when the channel FER is varied. On the other hand, the channel FER is kept constant at 0.05 when the arrival rate (λ_M) for a SIP-based session is varied.

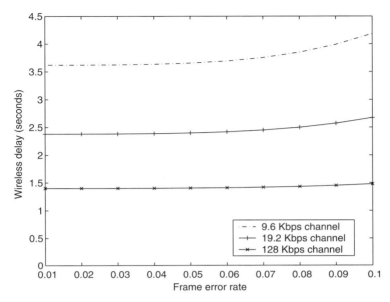

FIGURE 17.11
Wireless delay vs. FER for MH moving to UMTS network from another UMTS network or a WLAN.

The corresponding variation of handoff delay with the channel FER and SIP session request rate, for the case when the MH moves to WLAN, are given in Figure 17.13 and Figure 17.14. Table 17.5 shows the end-to-end handoff delay for an MH moving to a WLAN access network.

Observe that the component of handoff delay due to wireless access for a 128 Kbps GPRS radio access of a UMTS network is 1.404818 s, where the channel FER is 0.05 and the SIP-based multimedia session arrival rate is 50/s. Whereas for the 11 Mbps WLAN, the handoff delay is only 0.267 ms. As mentioned earlier, to ensure QoS for streaming multimedia, the maximum handoff delay should be ideally less than 100 ms and not more than 200 ms. Moreover, such abrupt variation in end-to-end delay due to handoff essentially means unacceptable delay jitter for multimedia streaming applications. Clearly, this requirement cannot be satisfied for a UMTS network even with a channel data rate of 128 Kbps. Figure 17.15 shows the exponential decrement of the total handoff delay with the increase of access bandwidth available for handoff. With higher speed data access (on the order of Mbps) as promised in the emerging 4G networks [32], the handoff delay will be within the stipulated requirement for streaming media and SIP would be able to meet the performance requirement. For example, with WLAN access networks, this is not as much of a problem because the wireless component of the delay is only around 0.2 ms.

As shown in Table 17.5, excluding the Internet delay, the end-to-end handoff delay is only around 1.9 ms. This leaves a leverage of about 98 ms for the Internet delay. As mentioned before, with appropriate traffic engineering deployed

TABLE 17.4
Handoff Delay Components (MH Moving to a UMTS Network)

Channel bandwidth	Channel FER	Processing delay	Wireless delay
9.6 Kbps	0.010000	0.214408	3.620081
	0.020000	0.214408	3.621036
	0.030000	0.214408	3.624850
	0.040000	0.214408	3.634785
	0.050000	0.214408	3.655453
	0.060000	0.214408	3.692894
	0.070000	0.214408	3.754647
	0.080000	0.214408	3.849831
	0.090000	0.214408	3.989220
	0.100000	0.214408	4.185324
19.2 Kbps	0.010000	0.214408	2.380042
	0.020000	0.214408	2.380538
	0.030000	0.214408	2.382520
	0.040000	0.214408	2.387681
	0.050000	0.214408	2.398418
	0.060000	0.214408	2.417867
	0.070000	0.214408	2.449945
	0.080000	0.214408	2.499389
	0.090000	0.214408	2.571796
	0.100000	0.214408	2.673663
128 Kbps	0.010000	0.214408	1.400010
	0.020000	0.214408	1.400137
	0.030000	0.214408	1.400650
	0.040000	0.214408	1.401998
	0.050000	0.214408	1.404818
	0.060000	0.214408	1.409945
	0.070000	0.214408	1.418423
	0.080000	0.214408	1.431517
	0.090000	0.214408	1.450721
	0.100000	0.214408	1.477774

in the Internet the end-to-end handoff delay can be restricted well within the stipulated maximum limit of 100 ms. Table 17.4 and Table 17.5 show the components of the end-to-end delay for two handoff cases considered earlier.

17.6 Handoff Delay Mitigations

As discussed above, the handoff delay has been a major deterrent for the deployment of a SIP-based mobility solution, particularly in low bandwidth

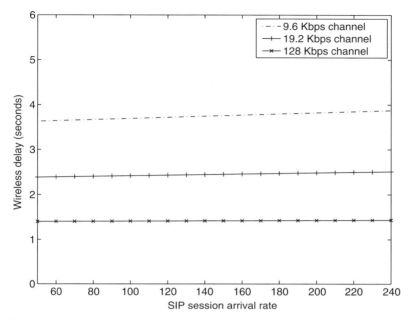

FIGURE 17.12
Wireless delay vs. session arrival rate for MH moving to UMTS network from another UMTS network or a WLAN.

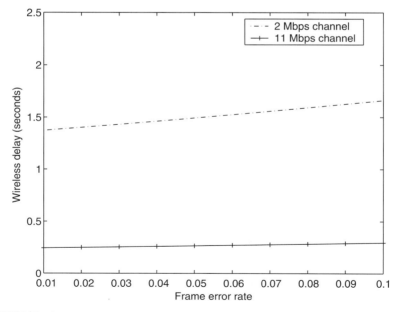

FIGURE 17.13
Wireless delay vs. FER for MH moving to WLAN from another WLAN or a UMTS network.

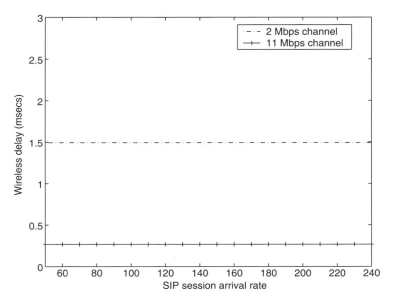

FIGURE 17.14

Wireless delay vs. session arrival rate for MH moving to WLAN from another WLAN or a UMTS network.

TABLE 17.5

Handoff Delay Components (MH Moving to WLAN)

Channel bandwidth	Channel FER	Processing delay	Wireless delay
2 Mbps	0.010000	0.201908	0.001373
	0.020000	0.201908	0.001402
	0.030000	0.201908	0.001431
	0.040000	0.201908	0.001462
	0.050000	0.201908	0.001493
	0.060000	0.201908	0.001525
	0.070000	0.201908	0.001558
	0.080000	0.201908	0.001592
	0.090000	0.201908	0.001626
	0.100000	0.201908	0.001661
11 Mbps	0.010000	0.201908	0.000246
	0.020000	0.201908	0.000251
	0.030000	0.201908	0.000256
	0.040000	0.201908	0.000262
	0.050000	0.201908	0.000267
	0.060000	0.201908	0.000273
	0.070000	0.201908	0.000279
	0.080000	0.201908	0.000285
	0.090000	0.201908	0.000291
	0.100000	0.201908	0.000297

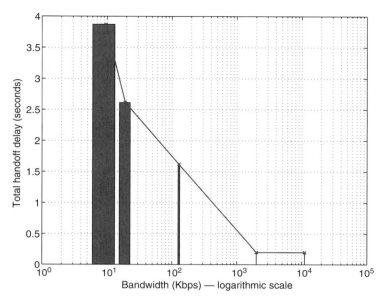

FIGURE 17.15
Variation of handoff delay with the increase of wireless access bandwidth.

access networks and hence, there have been several proposals to minimize the handoff delay and thereby minimizing packet loss during the transient handoff period.

In [33], three different types of delay components in a SIP-based handoff have been identified. They are: link layer delay, network layer delay and the SIP layer delay. The proposal is to detect the movement before the link layer handoff starts and start the network layer and SIP layer handoff procedures to minimize the effect of the corresponding delay components.

SIP mobility in a IPv6 network has been investigated in [34], where kernel modifications based on the design practices of MIPL IPv6 [35], such as Aggresive Router Selection, have been done to reduce some key delay components involved in the handoff. The modifications have been shown to pay off for mid-call mobility support for real-time communication. However, it is still not as fast as the network-layer mobility implemented by MIPL IPv6. The extra delay in the case of SIP mobility has been suspected due to the application-layer processing of SIP based handoff messages in contrast to the operating system level message processing in case of the network layer mobility implemented with MIPL IPv6.

Several other attempts in the higher layers have been made to minimize packet loss during the handoff period. RTP translators have been used in conjunction with SIP servers to forward packets during the transient handoff period [36]. The SIP servers however, need to take up additional responsibility to configure the RTP tranlators, in this case. Alternatively, a B2BUA has been used to establish a transient supplementary connection during the handoff period to minimize packet loss. Similarly, a transparent B2BUA functionality compounded with the packet filters and replicators has been proposed [40] in the base stations of cellular networks, again to minimize the packet loss during SIP-based handoff in cellular networks.

As mentioned before, SIP was originally designed for UDP-based multimedia-based sessions. Hence, SIP-based mobility provides mobility support to UDP-based applications only. There are, however, numerous already existing applications that are based on TCP, and SIP alone could not provide mobility support to them. There have been a number of proposals to solve this problem. For example, SIP has been complemented with Mobile IP [37] to handle mobility for both UDP- and TCP-based applications. The combination of SIP and Mobile IP has been further leveraged to reduce packet loss and handoff latency in [38], where the two separate registration databases maintained by SIP and Mobile IP have been integrated to minimize the location update delay and hence packet loss.

Alternatively, SIP has been combined with HIP [39] for a hybrid mechanism called the SHIP to solve the same problem. It has been reported [39] that the SHIP performs better in terms of handoff delay when compared to the SIP and Mobile IP scheme. In the latter scheme, handoffs need to be processed in both Mobile IP and SIP, and home agents redirect the packets until SIP re-INVITE process is completed. SHIP avoids the re-INVITE message in SIP and therefore, its signaling overhead is smaller. In addition, SHIP provides multi-homing support, which does not exist in hybrid Mobile IP and SIP.

17.7 Summary

The selection of an appropriate protocol for the mobility management framework warrants careful consideration with respect to several factors. In this chapter, a framework of qualitative comparative study of the mobility protocols from the network layer perspective has been performed with the conclusion that SIP is the most appropriate mobility protocol for next generation heterogeneous wireless networks. However, SIP suffers from some handoff performance issues that have detrimental effects to multimedia applications. We have investigated the handoff performance of SIP in a heterogeneous access networking environment and validated this claim. In particular, we have performed a case study of SIP-based handoff delay analysis using SIP to handle terminal mobility in a IP-based network. Two different types of access technologies, viz. UMTS and IEEE 802.11b-based WLAN, have been considered for the IP-based network. Analytical results show that for WLAN networks, the handoff delay is suitable for streaming media but for low bandwidth UMTS networks, the minimum handoff delay does not meet the specification of 100 ms. More precisely, handoff to a UMTS network from either another UMTS network or a WLAN, introduces a minimum delay of 1.4048 s for

128 Kbps channel, while a handoff to a WLAN access network from another WLAN or a UMTS network, the minimum delay is 0.2 ms. Abrupt and large end-to-end delay causes unacceptable packet loss for multimedia applications. Clearly, in the former case, the minimum delay is unacceptable for streaming multimedia traffic and requires the deployment of auxiliary techniques to reduce the packet loss to a minimum. Considering SIP mobility, a number of proposals now are in place to reduce the SIP hand-off delay. This framework and the underlying modeling techniques and assumptions can be easily extended to these SIP upgrade scenarios to evaluate their performance.

References

[1] Rosenberg, J., H. Schulzrinne, G. Camarillo, A. Johnston, J. Peterson, R. Sparks, M. Handley, and E. Schooler. 2002. SIP: Session Initiation Protocol, *draft-ietf-mobileip-optim-11.txt. IETF RFC 3261*, June.

[2] Schulzrinne, H., S. Casner, R. Frederick, and V. Jacobson. 2003. RTP: A transport protocol for real-time applications, *IETF RFC 3550*, July.

[3] Jiang, W., J. Lennox, H. Schulzrinne, and K. Singh. 2001. Towards junking the PBX: Deploying IP telephony. *NOSSDAV* 177–85.

[4] Perkins, C. 2002. IP Mobility support for IPv4, *IETF RFC 3344*, August.

[5] Johnson, D., C. Perkins, and J. Arkko, 2004. Mobility support in IPv6, *RFC 3775 Internet Engineering Task Force*, June.

[6] Snoeren, A.C., and H. Balakrishnan. 2000. An end-to-end approach to host mobility. *MobiCom*, 155–66.

[7] Perkins, C., and D. Johnson. 2001. "Route optimization in mobile IP," *draft-ietf-mobileip-optim-11.txt*, Expired March 2002.

[8] 3GPP TS 23.060.2001. General Packet Radio Service (GPRS), Service Description, Stage 2, December.

[9] The Third Generation Partnership Project (3GPP), http://www.3gpp.org

[10] Droms, R. 1997. Dynamic host configuration protocol, *RFC 2131*, March.

[11] Myers, J., and M. Rose. 1996. Post office protocol—version 3, *RFC 1939*, May.

[12] Hodges, J., and R. Morgan 2002. Lightweight directory access protocol (v3): Technical Specification, *RFC 3377*, September.

[13] Fogelstroem, E., A. Jonsson, and C. Perkins. 2007. Mobile IPv4 Regional Registration, *RFC 4857*, June.

[14] Ramjee, R., T. La. Porta, L. Salgarelli, S. Thuel, K. Varadhan, and L. Li. 2000. IP-based access network infrastructure for next-generation wireless data networks. *IEEE Personal Communications* 7 (4): 34–41.

[15] Wan, C.Y., A.T. Campbell, and A.G. Valko. 2000. Design, implementation and evaluation of cellular IP. *IEEE Personal Communications* 7(4): 42–49.

[16] Grilo, A., P. Estrela, and M. Nunes. 2001. Terminal independent mobility for IP (TIMIP). *IEEE Communication Magazine* 39(12): 34–41.

[17] Das, S., A. Mcauley, A. Dutta, A. Misra, K. Chakraborty, and S. K. Das. 2002. IDMP: An intra-domain mobility management protocol for next-generation wireless networks. *IEEE Wireless Communications* 9(3): 38–45.

[18] Eastlake, D.E. 1997. Secure domain name system dynamic update, *IETF RFC 2137*, April.

[19] Wedlund, E., and H. Schulzrinne. 1999. Mobility support using SIP. *ACM/IEEE International Workshopp for Wireless and Mobile Multimedia*, 76–82.

[20] Banerjee, N., W. Wu, K. Basu, and S.K. Das. 2004. Analysis of SIP-based mobility management in 4G wireless networks. *Computer Communications* 27(8): 697–707.

[21] Banerjee, N., W. Wu, S.K. Das, S. Dawkins, and J. Pathak. 2003. Mobility support in wireless internet. *IEEE Wireless Communications Magazine* 10(5): 54–61.

[22] Das, S.K., N. Banerjee, W. Wu, S. Ganeshan, and J. Pathak. 2003. *IP Mobility Protocols for Wireless Internet*, Kluwer Academic Publisher, USA, June.

[23] Wu, W., N. Banerjee, K. Basu, and S.K. Das. 2005. SIP-based vertical handoff between WWAN and WLAN. *IEEE Wireless Comminucations Magazine, Special Issue: Toward Seamless Internetworking of Wireless LAN and Cellular Networks* 12 (3): 66–72, June.

[24] Telecommunications and Internet Protocol Harmonization over Networks, *QoS Class Specification*, TS 101329-2, http://www.3gpp.org.

[25] Schulzrinne, H., and E. Wedlund, Application-layer mobility using SIP. *Mobile Computing and Communication Review* 1 (2): 47–57.

[26] Handley, M., and V. Jacobson. 1998. SDP: Session description protocol. *IETF RFC 2327*, April.

[27] Das, S.K., E. Lee, K. Basu, and S.K. Sen. 2003. Performance optimization of VoIP calls over wireless links using H.323 protocol. *IEEE Transactions on Computers* 52(6): 742–52.

[28] ETSI, GSM 03.60 Digital cellular telecommunications system (Phase 2+): General Packet Radio Service, Service Description Stage.

[29] Paxson, V. 1997. End to end Internet packet dynamics. *SIGCOMM* 139–52.

[30] Berger, L. 2003. Generalized multi-protocol label switching (GMPLS) signaling resource ReserVation Protocol-Traffic Engineering (RSVP-TE) Extensions. *RFC 3473 Internet Engineering Task Force*, January.

[31] Kleinrock, L. 1975. *QUEUING SYSTEMS Volume I: Theory*, John Wiley & Sons, USA.

[32] Varshney, U., and R. Jain. 2001. Issues in emerging 4G wireless networks. *IEEE Computer Magazine* 34(6): 94–96, June.

[33] Kim, W., M. Kim, K. Lee, C. Yu, and B. Lee. 2004. Link layer assisted mobility support using SIP for real-time multimedia communications. *MobiWac* 127–29.

[34] Nakajima, N., A. Dutta, S. Das, and H. Schulzrinne. 2003. Handoff delay analysis and measurement for SIP based mobility in IPv6. *IEEE International Conference on Communications* 2: 1085–89.

[35] Tuominen, A.J., and H. Petander. 2001. MIPL Mobile IPv6 for Linux in HUT campus network MediaPoli, *Proceedings of Ottawa Linux Symposium*, June.

[36] Dutta, A., S. Madhani, W. Chen, O. Altintas, and H. Schulzrinne. 2004. Fast-handoff schemes for application layer mobility management. *PIMRC* 3: 1527–32.

[37] Lee, H., J. Song, S. Lee, S. Lee, and D. Cho. 2005. Integrated mobility management ethods for mobile IP and SIP in IP based wireless data networks. *Journal of Wireless Personal Communications* 35(3): 269–87.

[38] Jung, J.W., R. Mudumbai, D. Montgomery, and K. Hyun-Kook. 2003. Performance evaluation of two layered mobility management using mobile IP and session initiation protocol. *GLOBECOM* 3: 1190–94.

[39] So, J.Y.H., W. Jidong, and D. Jones. 2005. SHIP Mobility Management hybrid SIP-HIP Scheme. *Sixth International Conference on Software Engineering, Artificial Intelligence, Networking and Parallel/Distributed Computing, 2005 and First ACIS International Workshop on Self-Assembling Wireless Networks. SNPD/SAWN 2005*, 226–30.

[40] Banerjee, N., A. Acharya, and S.K. Das 2006. Seamless SIP-based mobility for multimedia applications. *IEEE Network* 20(2): 6–13.

Part III

SECURITY

18

SIP Security: Threats, Vulnerabilities and Countermeasures

Dimitris Geneiatakis, Georgios Kambourakis and Costas Lambrinoudakis
University of the Aegean

CONTENTS

18.1 Introduction 435
18.2 SIP Threats and Vulnerabilities 436
18.3 Attacks against the SIP 437
18.4 SIP Security Requirements 447
18.5 SIP Security Mechanisms and Services 448
18.6 Conclusions 452
     References 453

18.1   Introduction

Until the early 1990s telecommunication providers were employing dedicated closed networks, like the Public Switch Telephone Network (PSTN), for transmitting voice. By this way the establishment of communication and voice transmission in PSTN was exploiting only 10–25% of the total available resources [1]. Even though significant effort has been devoted to improve resource utilization and to minimize the cost of the provided services, very little has been achieved to date. The Internet has supported the revolution in telecommunications and specifically in the provision of voice services, voice is transmitted over the Internet Protocol (VoIP) utilizing the Internet as a backbone network. Despite the fact that a plethora of novel, sophisticated, low-cost services can be offered in this way (e.g., conference rooms, click to dial etc), telecommunication providers are now facing security problems that have not been previously encountered.

Contrariwise to PSTN, which is based on a closed network architecture and thus it neither experiences a big variety of attacks nor a high attack frequency [2], VoIP utilizes an open network architecture and thus VoIP providers and end users should deal not only with the Internet's native vulnerabilities and attacks but also with VoIP oriented ones. Several researchers [3–6] have identified various security flaws in VoIP signaling protocols, like Session

Initiation Protocol (SIP) [7], H.323 [8] and IAX [9]. These security flaws can be exploited by attackers in order to achieve either unauthorized access or to cause Denial of Service (DoS). It is stressed that the majority of the published research work focuses on SIP, because SIP has been adopted by various standardization organizations as the predominant protocol for both wireline and wireless in the Next Generation Networks (NGN) era. The rather straightforward way to access the communication channel combined with SIP's text-based message structure, makes SIP an attractive target for attackers. Without doubt, in order to gain global acceptance, it is necessary for VoIP to offer at least similar levels of security, availability and reliability to those of PSTN.

This chapter focuses on SIP security issues. Particularly, it presents and thoroughly analyzes existing vulnerabilities, threats and attacks against SIP. Moreover it identifies specific security requirements that the SIP-dedicated security mechanisms and the corresponding Intrusion Detection Systems (IDS) must fulfill in order to eliminate the respective security flaws.

18.2   SIP Threats and Vulnerabilities

Prior to the design and development of any new service, critical or not, the identification of potential risks and vulnerabilities should pave the way towards the adoption of the appropriate security countermeasures; in accordance to the well-accredited statement *"It is very important to know what you are trying to protect against"* [10]. For example, in a confined network you can generally trust a participant to share your data with, while in an open environment this can be only done through the employment of security mechanisms that establish an acceptable level of trustworthiness with other network entities. This is exactly the case for SIP-based VoIP services. PSTN, VoIP's predecessor, relies on a closed network architecture where risks are limited and well controlled [1]. On the other hand, SIP-based VoIP services are offered through an open network architecture, like the Internet, facing a hostile environment and thus having to cope with a variety of existing and new threats. The attackers, in order to accomplish their goals, try to either exploit a well-known Internet inherited vulnerability or some other vulnerability that stems directly from SIP's architecture.

Among the most common threats that communication systems have to deal with is that of unauthorized access. In the PSTN, unauthorized access concerns only voice data and not signaling. Despite PSTN's simplistic nature, it is very difficult to achieve unauthorized access (to voice data) because it requires physical access to the medium. On the downside, in a SIP-based VoIP service unauthorized access, to both voice data and signaling, is considered relatively simple because: (a) aggressors can easily gain access to the communication channel, (b) SIP uses a text-based message structure and thus everything can be effortlessly tracked, and (c) there are restrictions regarding the employment of confidentiality protection mechanisms in SIP that mainly arise from the fact that intermediate nodes must have access to specific SIP message headers in order to correctly process and route messages to their final destination. Clearly this makes eavesdropping attacks an easy task for malicious users. Furthermore, the lack of integrity mechanisms allows malicious users to manipulate SIP signaling data (e.g., man-in-the-middle, MITM, attack) and tamper them (e.g., with SQL code).

On top of the privacy and integrity issues VoIP providers should also take into account the client or/and service masquerading threat which was not applicable to PSTN. In the first case, the malicious user tries to act on behalf of a legitimate user (victim) by sending SIP requests that identify the victim's ID instead of the malicious user's one. For example, the malicious user may replace his ID with the victim's ID in the SIP REGISTER message that is used during the registration procedure of every SIP VoIP based service. During a service masquerading incident, a malicious user responds on behalf of the service, taking advantage of the fact that the SIP server is not properly authenticated by the other side and thus the attacker gains access to call information without being identified.

Another important threat for all telephony services is toll fraud. Without doubt, all users expect accurate billing procedures. In the PSTN a malicious user could make free calls, without affecting a legitimate user's billing account. This was achieved by notifying, through specific tones, the call center that the call had been finished while the conversation was actually still in progress [11]. Similar incidents can occur in SIP based VoIP services, affecting, in financial terms, not only the provider but also the legitimate users [12].

With the continuously increasing demand for VoIP services, service corruption is considered by most VoIP service providers as the most severe threat they have to cope with. Similarly to other Internet applications and services, in a SIP-based VoIP service a malicious user may exploit implementation errors or generate a large number of SIP requests (e.g., SIP INVITE, SIP REGISTER) in order to cause DoS and/or gain access in the VoIP provider's core network [3,6,13]. Furthermore, a malicious user can easily modify the signaling data in order to illegally terminate either an established session or a session in progress [3]. Last but not least, critical services like VoIP suffer from deception incidents exploiting social engineering methods [14]. In this case the malicious user reveals sensitive private information by deceiving innocent users who are aware of the common security practices. Table 18.1 summarizes the potential threats, the vulnerabilities exploited by each threat and the most common attacks in a SIP-based VoIP environment.

18.3   Attacks against the SIP

According to Oxford Reference an attack is defined as *"An attempt to overcome the security provisions of a computer network"* [15]. More specifically, an attack epitomizes a threat by exploiting one or more vulnerabilities in order

TABLE 18.1

Potential Threats, Vulnerabilities, Attacks and Their Impact in an SIP-Based VoIP Environment

Threat	Vulnerability	SIP attack	Affects
Unauthorized access to communication data	Lack of confidentiality mechanisms	Eavesdropping SIP traffic	Confidentiality
	Easy access to the medium		
Unauthorized data modification	Lack of integrity mechanisms	SIP message tampering	Integrity, availability
	Easy access to the medium	SIP malformed messages	
Client impersonation	Inadequate authentication mechanisms	SIP signaling attack	Confidentiality, authenticity
		SIP Man-in-the-Middle	
Server impersonation	Lack of server authentication	SIP Man-in-the-Middle	Authenticity, integrity
Toll fraud	Inadequate authentication mechanisms	SIP signaling attacks	Authenticity, integrity
	Lack of integrity mechanisms	SIP message tampering	
Service disruption	Parsers' implementation Eerrors	SIP malformed messages SIP flooding	Availability
	Inadequate authentication and integrity mechanisms	SIP signaling attacks	
Social engineering	Lack of education	Phising	Confidentiality, authenticity

for the attacker to achieve his goals. The effectiveness of the attack, or the rate of success, depends on the combination of three factors: (a) the security countermeasures and repelling mechanisms that have been employed, (b) the existence of specific vulnerabilities in the underlying infrastructure, protocols etc, and (c) the attacker's particular skills and the tools used to realize the attack.

Although real attack incidents in SIP environments have not yet been published, research work [3–6,12,13] has highlighted SIP's security flaws in test environments. A description of the possible attack categories against a SIP infrastructure is provided in the following sections. It is stressed that attacks stemming directly from the Internet, which may (also) affect SIP, are well documented in the literature and are thus outside the scope of this chapter.

```
REGISTER sip:dgentele.com SIP/2.0
Via: SIP/2.0/UDP 81.0.7.124:5070
From: <sip:3400001586@dgentele.com;user=phone>;tag=3199572059
To: <sip:3400001586@dgenele.com;user=phone>
Call-ID: 3021094946@81.0.7.124
CSeq: 2 REGISTER
Contact: sip:3400001586@81.0.7.124:5070;user=phone;transport=udp>;
expires=300
User-agent: Cisco ATA 186 v3.1.0 atasip (040211A)
Authorization: Digest username="3400001586",realm="dgentele.com",
        nonce="426302039afdf717c6687e28f6c7d39c4fdb9f08",
        uri="sip:dgentele.com",response="af0d725596c8f06f370f8c80ade67b05"
Content-Length: 0
```

FIGURE 18.1
A captured SIP REGISTER message (utilizing Wireshark).

18.3.1   Eavesdropping

One of the most prevalent attacks against a communication system is eavesdropping. Conversely to PSTN where only voice data are subject to this type of attack, in SIP-based VoIP services signaling is a common target too. SIP messages, which in fact are signaling data, contain information about the user's identity, contact addresses, security keys and other useful data required for establishing a session among two or more SIP participants. Normally this information should be kept private. However, the rather straightforward way to access the medium (whether wired or wireless), the existence of several sophisticated eavesdropping tools, like Wireshark (http://www.wireshark.org), that are widely available on the Internet and the SIP's text-based message structure makes such an attack a very easy task to accomplish. For instance, during an eavesdropping attack the attacker may capture a SIP REGISTER message (see Figure 18.1) and thus become fully aware of the user's private information. On top of that, a group of SIP REGISTER captured messages could be exploited at a later time by the attacker for deferred traffic analysis [16], in an attempt to guess the password of the user; this can be realized by utilizing the data contained in the authorization header of each SIP REGISTER message. Even if the traffic is protected by a confidentiality protection mechanism, the attacker may try to discover the corresponding cryptographic keys and gain access to the signaling data. Summarizing, eavesdropping is an attack mainly violating privacy and confidentiality.

18.3.2   Parsing and Message Injection Attacks

Every time that a SIP entity receives a SIP message it must parse it in order to convert it to the appropriate internal structure that will be used for further processing by the corresponding SIP proxy or the end device. Figure 18.2 illustrates the typical steps taken by a SIP proxy in order to process an incoming message; depending on the SIP proxy vendor this procedure may slightly vary.

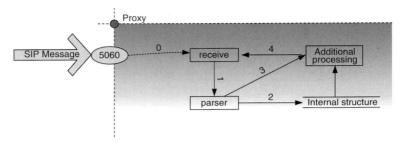

FIGURE 18.2
SIP proxy general architecture.

In most cases parsers are designed having in mind only the processing of normal well-formed messages (see Figure 18.1). As a result, any alteration of the aforementioned normal procedure may severely affect the system stability and availability. This fact could be exploited by an adversary who may generate a large amount of malformed messages trying to cause: (a) DoS, (b) unstable operation, e.g., paralyze the proxy or the end device, (c) gain unauthorized access. A SIP message is considered as malformed if it does not fully conform to the SIP grammar defined in RFC 3261 [7]. For example, an aggressor may generate the malformed message depicted in Figure 18.3 and repeatedly send it to a SIP entity in order to crash it. The parser whenever receives such a message will try to parse the values of *From* and *To* headers respectively, which are intentionally both set to the NULL value. Any attempt of the parser to represent this NULL value to the corresponding internal structure, may cause instability or in the worst case a DoS.

As mentioned elsewhere an attacker may craft a perfectly valid message in an attempt to hamper proper parsing [6]. For instance he may try to cause instability by including unnecessary headers and having a long message body. Alternatively the attacker may embed malicious SQL code [17] in the *Authorization* header of the SIP message in order to illegally modify legitimate users' accounting or registration data. An example of such a message is depicted in Figure 18.4. Further details for these types of attacks can be found elsewhere [17,18]. The reader may also refer to Wieser, Laakso and Schulzrinne for the demonstration of a testing framework utilizing SIP malformed messages against well-known SIP servers [13].

```
REGISTER I am trying to crash a sip proxy SIP/2.0
Via: SIP/2.0/UDP 81.0.7.124:5070
From: NULL
To: NULL
Call-ID: 3021094946@81.0.7.124
CSeq: 2 REGISTER
```

FIGURE 18.3
An example of SIP REGISTER malformed message.

```
REGISTER sip:dgentele.com SIP/2.0
Via: SIP/2.0/UDP 81.0.7.124:5070
From: <sip:3400001586@dgentele.com;user=phone>;tag=3199572059
To: <sip:3400001586@dgentele.com;user=phone>
Call-ID: 3021094946@81.0.7.124
CSeq: 2 REGISTER
Contact: <sip:3400001586@81.0.7.124:5070;user=phone;transport=udp>;expires=300
User-Agent: Cisco ATA 186 v3.1.0 atasip (040211A)
Authorization: Digest username="3400001586'; UPDATE accounting set duration='200'
                where username =340001586",
            realm="dgentele.com",
            nonce="426302039afdf717c6687e28f6c7d39c4fdb9f08",
            uri="sip:voztele.com",response="af0d725596c8f06f370f8c80ade67b05"
Content-Length: 0
```

FIGURE 18.4

An example of a SIP REGISTER message injected with SQL code.

18.3.3 Flooding Attacks

The ultimate goal of a flooding attack is to exhaust victim's resources and thus make a service unavailable to its legitimate users. This type of attacks form a severe threat to Internet services since it is really difficult to distinguish the attack traffic from the normal one. In general, a malicious user will try to overwhelm system's resources by generating a vast number of useless but well-formed requests. This can be realized by employing either single or multiple source message generators; in a way similar to attack architectures employed against web servers (Figure 18.5 and Figure 18.6). Note that in SIP architectures both servers and end-user terminals are susceptible to any kind of flooding attack.

End user devices are more susceptible to single source flooding attacks than SIP servers. This is because their resources and processing capabilities are limited; for instance, SIP phones are only able to process a small number of sessions simultaneously. The adversary may generate hundreds of SIP invitation requests, over a specific time interval, and send them to the end-terminal (either directly to the terminal's IP address or indirectly via the corresponding proxy). This will not only cause the terminal to ring continuously but it will also disrupt its normal operation by eventually crashing it. A single source

FIGURE 18.5

Single source attack architecture.

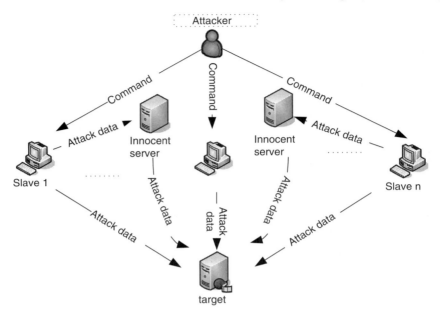

FIGURE 18.6
Multiple source attack architecture.

attack can be also launched against a SIP server even though it is far more difficult to consume its resources by utilizing only a single source message generator.

Alternatively, the aggressor may generate spoofed SIP invitation messages, containing a new URI (which corresponds to a new session), in an attempt to gradually consume the recourses of a stateful SIP proxy. This kind of attack exploits the stateful operational mode of the proxy, or in other words its need to store each transaction's context consuming three Kbytes per request [6], i.e., for every new session, and keep it alive until the transaction ends. Thus, depending on the implementation and configuration, a stateful proxy should be able to maintain a transaction active for a duration of a few seconds up to some minutes. Consequently, the concurrent existence of a large number of uncompleted transactions could eventually render the provided service unavailable. In addition, as highlighted by Sisalem, Kuthan, and Ehlert, a rather simple way to disrupt the operation of a SIP server is to employ a single source attack architectures and include irresolvable URIs in the SIP invitation messages [6]. Every time a SIP server attempts to parse such a message it has to wait for the reply of the DNS resolving the URI before completing the operation. As a result, the combination of single source invitation attacks with unresolvable addresses could cause rapid resource consumption, disrupting the availability of the offered services. All the aforementioned scenarios could lead to more severe attacks by utilizing multiple source flooding architectures, similar to those illustrated in Figure 18.6.

It should be noticed that the aforementioned flooding attacks are similar to the well known TCP/SYN attack, mainly because of the similarities among the SIP and TCP handshake phases. In the TCP/SYN attack the attacker includes a spoofed destination address in a SYN message, mandating the server to reserve resources until the appropriate TCP timer expires [19]. A variation of the TCP/SYN attack appeared in January 2002. According to this attack, innocent TCP servers were exploited as amplifiers (also known as reflectors) to paralyze a specific server or even an entire network. In order to do that the attacker generates a large number of TCP/SYN requests including spoofed (with the victim's IP) IP addresses. Upon that, the innocent servers respond by generating the respective SYN/ACK messages that are forwarded to the victim. This type of attack is known as Reflection Distributed DoS (RDDoS) [20]. Normally, such an attack consumes all the available bandwidth or exhausts the target server's resources.

Similar attacks may be launched against a SIP server or network. Specifically, the attacker crafts spoofed SIP INVITE messages, placing the address of the victim in the requested URI, and sends them to the innocent SIP servers which will eventually forward them to the victim (see Figure 18.7). By maintaining a continuously growing number of incomplete INVITE transactions, the server will very soon become unavailable. Alternatively, the SIP servers could be replaced by innocent User Agents that respond to the spoofed INVITE requests, forwarding the responses to the actual victim, in a way similar to the TCP/SYN reflection attack. This attack can take various forms in SIP, depending on the type of the spoofed SIP message crafted by the

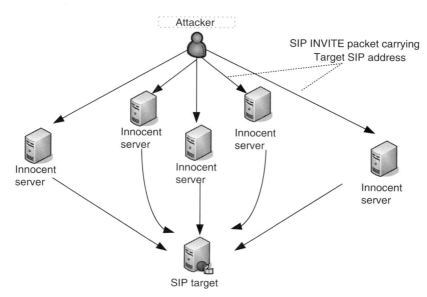

FIGURE 18.7
The INVITE reflection syndrome.

attacker. For instance, an alternative to the "SIP INVITE forwarding" situation, described above, is to add the victim's address in the "via" header, thus having the reflectors to send their responses to the final destination via the victim. Finally, another way to pass the attack traffic through a specific SIP server is to embed the Record-Route header in the SIP spoofed message.

Up to now the focus was on flooding attacks against proxies and end-user terminals since they are considered as the most important SIP networks elements. Nevertheless, similar attack scenarios could be employed against SIP redirect and registrar servers. Especially the registrar server should be considered as a very probable target since it executes expensive cryptographic operations for authentication purposes. Therefore, a malicious user may send against any registrar a huge number of SIP REGISTER requests in order to incapacitate the registration service.

18.3.4 Signaling Attacks

As signaling attack is considered any attempt to illegally modify and manipulate the signaling messages in order to: (a) gain unauthorized access to the provided services and/or (b) modify the state of an established session. SIP's signaling attacks are linked to threats like Toll Fraud, Client and Server Impersonation and Service Disruption (see Table 18.1), while they take advantage of the fact that many SIP architectures feature inappropriate authentication and integrity mechanisms [3]. In order to launch such an attack a malevolent user should eavesdrop the session parameters that are required for matching the SIP messages with the corresponding session.

The first point of attention for a signaling attack is the registration service. The malicious user will act as a MITM trying to impersonate the legitimate user. Under normal circumstances a legitimate user generates a SIP REGISTER message and sends it to the appropriate SIP Registrar. Depending on the Registrar's configuration, i.e., it may require authentication or not, the negotiation procedure unfolds as depicted in Figure 18.8.

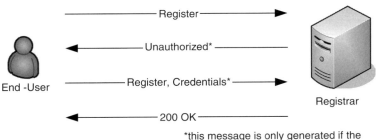

End -User

Register →

← Unauthorized*

Register, Credentials* →

← 200 OK

Registrar

*this message is only generated if the Registrar requires authentication

FIGURE 18.8
Registration procedure in SIP.

Let us now assume that a malicious user can intercept and modify the messages of the standard registration procedure by performing a MITM attack (this situation is illustrated in Figure 18.9). Specifically, the malicious user will either alter the contact address of the SIP REGISTER message, in order to be able to receive and put calls on behalf of the legitimate user, or he will set the value of the expiration header to "0", in order to violently disconnect the legitimate user from the service. Note, that the attacker may even force the legitimate user to create the appropriate SIP REGISTER message, including his credentials, by forwarding to him the unauthorized message generated by the Registrar. This is the case depicted in Figure 18.9, except that the first SIP REGISTER message, generated by the legitimate user, is missing.

As illustrated in Figure 18.10, an attacker could follow a similar approach in order to impersonate an entity when a SIP session is in progress. According to this scenario the malicious user modifies the SIP INVITE message, i.e., its

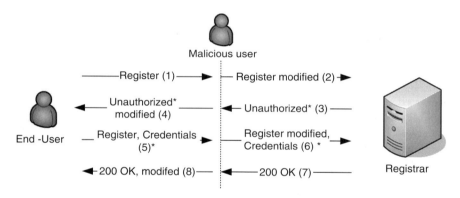

FIGURE 18.9
Man-in-the-Middle attack during registration phase.

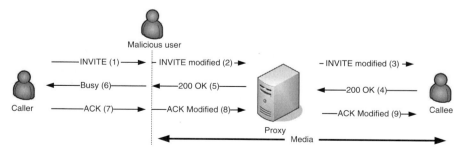

FIGURE 18.10
Man-in-the-Middle attack during call establishment.

contact header, and forwards it to the SIP proxy. The proxy sends the request to the callee who, assuming that he accepts the call, generates a "200 OK" response message which is finally passed to the malicious user. Upon receiving it, he replaces it with a "busy" message and forwards it to the legitimate user who acknowledges the spoofed busy response and terminates the session. However, the conversation between the malicious user and the callee has been successfully established.

Similar attack techniques that are known as Fake Busy, Bye Delay and ByeDrop, are discussed in detail elsewhere [12]. The difference between the aforementioned scenario and the Fake Busy one is the existence of another malicious user acting on behalf of the callee. This aggressor intercepts the SIP messages and generates a "200 OK" response message in order to make his IP address available and thus establish a media session with the malicious user placed on the side of the caller. In this case the (legitimate) callee is unaware of the incoming invitation.

For the rest of the attack scenarios, i.e., Bye Delay and ByeDrop, the MITM attacker drops the SIP BYE message sent by the legitimate user (caller or callee) to the proxy and sends back to the (legitimate) user a spoofed 200 OK response message. This will give the impression to the service provider that the session is still active, whereas the legitimate user thinks that the session has been successfully terminated.

In the same way a malicious user may impersonate the redirect server, forcing the legitimate one to pass its traffic via unauthorized proxies, in order to gain access to the signaling data.

Moreover, an attacker may deliberately inject signaling messages for terminating or modifying [3] the state of an established session or, even, affect the accounting service. Figure 18.11 illustrates an example of an illegally terminated session. The attacker eavesdrops on session parameters and after the session has been successfully established he forwards to the proxy a session termination message (SIP BYE) on behalf of the callee. The proxy forwards the

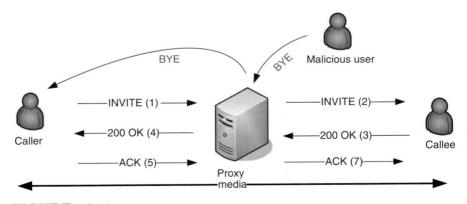

FIGURE 18.11
The SIP BYE signaling attack.

SIP BYE to the caller who, eventually, terminates the media session without realizing that the callee has not generated the SIP BYE message.

The same procedure could be followed for: (a) illegally canceling an SIP session in progress by injecting in the network a SIP BYE or CANCEL message before the caller sends its final ACK (message 5 in Figure 18.11) and (b) modifying SIP session parameters by generating a SIP re-INVITE or UPDATE message—for instance if the contact address is modified the aggressor may forward the session to a malicious user.

18.3.5   Social Attacks

Social attacks, also known as social engineering, are methods exploited by attackers in order to manipulate naive users' trust or goodwill for collecting confidential information. For example, an attacker may take control of a service's e-mail account and send, on behalf of this service, an e-mail asking the user to change his password through a designated link. This link may prompt to a malicious web site that captures the passwords.

This specific scenario may also occur in a SIP environment where the legitimate user is prompted to submit his credentials to a fake web site that impersonates the official web site of the service provider. This is an issue of major importance since nowadays many SIP providers offer accounting and billing services through web interfaces.

18.4   SIP Security Requirements

The identification of the security requirements for each of the offered services should be one of the main concerns of the service providers. During the design and implementation phases it is necessary to employ security measures that satisfy the requirements set and thus ending up with secure and robust systems. In general, security requirements can be classified into three categories: (a) confidentiality, (b) integrity, and (c) availability [21]. However, in the case of SIP-based VoIP services, this categorization must be extended to also include authenticity (this is clearly demonstrated in Table 18.1 by looking at the "security properties" violated in case of a successful attack). In order to avoid misinterpretation and/or confusion, it is necessary to explicitly define the aforementioned security requirements in the framework of a SIP domain.

18.4.1   SIP Security Requirement 1: Confidentiality

Confidentiality can be defined as a data hiding mechanism which ensures that only authorized entities [21] can access (read) the data. In this context, confidentiality services in SIP should guarantee that only authorized entities (proxies, registrar, redirect, end-user) have access to signaling data.

18.4.2   SIP Security Requirement 2: Integrity

Integrity services provide assurance that only authorized entities can modify the data in transit [21]. In the case of SIP, this definition must be further elaborated due to the fact that various SIP entities may require access to the signaling data in order to process and deliver them to its final destination. Consequently, an integrity service in SIP should guarantee that SIP signaling data can be manipulated only by SIP-authorized entities across the entire signaling path. Note that a signaling path may include several intermediate entities (servers) and end-users. Therefore, integrity services should be provided in both an end-to-end and a hop-by-hop fashion.

18.4.3   SIP Security Requirement 3: Authenticity

The requirement of authenticity can guarantee data originality [21]. For the SIP case, data originality means that the incoming data are indeed the data sent by some other entity. In other words, authenticity supports the genuineness of the SIP entity that generates the corresponding signaling data.

18.4.4   SIP Security Requirement 4: Availability

Last but not least, availability provides assurance that whenever a resource is required for the completion of a task the system can instantly provide it [21]. Specifically for SIP, availability could be interpreted as the user's ability to invoke SIP-based VoIP services at any given moment.

18.4.5   SIP-specific Security Restrictions

Apart from the security requirements listed above, during the design phase of a service it is also important to take into account the following SIP-specific security restrictions:

1. Intermediates nodes must (at least) have access to specific SIP message headers, in order to process and route the messages to the proper destination. Therefore, it is not generally acceptable to encrypt the entire SIP message.

2. SIP requests like CANCEL and ACK and naturally SIP responses must not be challenged for authentication since they cannot be resubmitted.

18.5   SIP Security Mechanisms and Services

According to RFC 3261 [7], SIP security basically relies on well-known security mechanisms that have been successfully applied in existing, widely used

FIGURE 18.12
Security services modes in SIP.

Internet applications and services. Furthermore, SIP security services should be provided both in a hop-by-hop (intermediate servers) and an end-to-end (end users) fashion (see Figure 18.12).

In the above context, SIP designers recommend at least the employment of the following security mechanisms:

HTTP digest is a challenge–response authentication protocol [22] employed for message authentication and identity validity in SIP. It is based on a pre-arranged trusted environment for password distribution, so it can be applied only under a specific domain. An example of a successful HTTP digest authentication transaction is depicted in Figure 18.8. Specifically, whenever a user sends a SIP request to the SIP server (e.g., a SIP REGISTER message), the server responds back either with an *Unauthorized (401)* or *Proxy Authentication required (407)* message, depending on the type of the SIP server. Upon the reception of the response, the user computes the appropriate credentials, based on the challenge (a random number) included in the response, and sends a new SIP message containing a *WWW-Authenticate header* with the credentials to the server. The server checks the validity of the message by first (re)computing and then comparing the credentials with those included in the WWW-Authenticate header. If these two values match, it replies with a "200 OK" response message, otherwise it responds back with an authentication error. Note that the HTTP digest does neither provide message integrity, nor complete authentication (mutual and end-to-end). Moreover, considering that both plaintext (challenge) and ciphertext (computed credentials) can be easily captured by an aggressor, the entire procedure is considered vulnerable to "chosen plaintext" attacks [10]. Another problem is that HTTP digest does not take into account specific SIP restrictions, i.e., some requests (e.g., ACKs) and responses must not be challenged. It is therefore clear that certain HTTP digest limitations can be exploited for launching MITM and impersonations attacks.

IP Security (IPSec) [23] is an independent and transparent network level protocol in the Internet architecture, providing services for the protection of IP packets against any possible attack like IP spoofing, session hijacking, traffic analysis and many more [24]. IPSec operates either in transport or in tunnel mode, offering integrity, data origin authentication and confidentiality services by utilizing Encapsulating Security Payload (ESP) and Authentication Header protocols. Consequently, IPSec could be utilized as an alternative safeguarding mechanism for SIP, provided that some sort of trust (e.g., pre-shared keys) has been established between the communicating parties beforehand in a hop-by-hop fashion. However, as IPSec is implemented at the operating system level, most of the existing SIP clients do not support it. Thus, IPSec can be utilized to shield signaling traffic between SIP servers but not between servers and user agents.

Transport Layer Security (TLS) [25] provides security services at the transport layer of the Internet architecture. Similarly to IPSec, TLS offers authentication, integrity, and confidentiality services among SIP entities, without mandating a pre-established trust relationship (as IPSec does) between the communicating parties. In addition, TLS is employed for providing end-to-end security through the support of the SIP Secure (SIPS) protocol. If implemented properly, the latter mechanism guarantees that the requested resource is reached securely. This is achieved by constructing TLS secure tunnels among each hop in the path until the final recipient. However, until today no universal mechanism exists which can guarantee that TLS is applied by all the intermediate nodes along the whole path to the final destination. Obviously, such a scenario will mandate the existence of a central PKI authority for SIP services as well as the deployment of cross-certification procedures between VoIP providers at a large scale. Moreover, the utilization of TLS protocol by SIP servers comes at a considerable cost as the maintenance of many open TLS connections simultaneously may significantly reduce SIP servers' performance. Moreover, TLS does not support connectionless protocols, like UDP, that are actually implemented by most SIP clients today. In fact only a few clients support TLS today; among them are KPhone (http://www.kphone.org), Minisip (http://www.minsip.org) and hardware phones from Snom (http://www.snom.com).

Secure Multipurpose Internet Mail Extensions (S/MIME) [26] is considered one of the most complete security suites for the application layer of the Internet architecture. It supports (a) authentication, (b) message integrity, (c) non repudiation of origin, and (d) confidentiality. According to RFC 3261 [7], SIP messages could also include MIME protected parts. This means that the part of the SIP message that must be protected against any potential manipulation, is replicated in the MIME part (which is the Message Body) of the SIP message, in which S/MIME is applied to offer integrity, authentication or encryption services. However, as a result of specific SIP restrictions [7] the S/MIME can be only applied to the parts of the SIP message that are not processed/modified by intermediate nodes (see also SIP-specific Security Restrictions). Furthermore, it is essential to stress that S/MIME introduces considerable overheads due to the cryptographic functions that it mandates. Currently, we are not aware of any commercial SIP clients supporting S/MIME.

Summarizing, HTTP digest can only be utilized for client authentication without providing confidentiality or integrity. On the other hand, IPSec and TLS do offer confidentiality, integrity and authenticity services. However, they can be only employed for communication between SIP servers without providing any guarantee for true end-to-end security. The S/MIME is the only

TABLE 18.2

Common Security Mechanisms Supported in SIP

	HTTP digest	IPSec	TLS	S/MIME
Confidentiality	No	Hop-by-hop	Hop-by-hop	End-to-end (partial)
Integrity	No	Hop-by-hop	Hop-by-hop	End-to-end (partial)
Authenticity	One-way	Hop-by-hop	Hop-by-hop	End-to-end
Availability	No	No	No	No

mechanism that could be employed for end-to-end security. Nevertheless until today it is not implemented by commercial SIP clients. Moreover, the need of intermediate nodes to modify some parts of the SIP message restricts S/MIME from providing only "partial" security services. This means that S/MIME can protect only certain parts of the SIP message; only these that remain unmodified along the whole path from the caller to the callee and vice versa. A final observation is that SIP specifications do not suggest any mechanism for minimizing the threat of service unavailability. Table 18.2 summarizes the common security mechanisms and protocols currently supported by SIP.

The security mechanisms presented in this section can substantially restrict, or at best prevent, only some of the potential attacks and threats emerging in SIP architectures. In order to develop a truly resilient SIP architecture, additional security mechanisms, acting complementary to the existing ones, are necessary. During the last years this necessity has been identified and manifested by various researchers [26–37], who either propose the design of new security services or the enhancement of existing ones, in all cases aiming to satisfy the identified SIP security requirements.

Ono and Tachimoto define a set of security requirements for providing both integrity and confidentiality for SIP signaling along the entire message path, ensuring that only trusted intermediate nodes can access the data [26]. This is known as end-to-middle security, since not all the intermediate nodes are trusted for processing all sorts of signaling information.

Other authors propose various schemes for the improvement of SIP authentication services, in an attempt to eliminate the impersonation threat and consequently confine signaling attacks [27–29]. For example, Cao and Jennings focus on ways to guarantee the authenticity and integrity of SIP response messages [27]. The proposed solution is particularly effective in cases of signaling attacks where a rogue SIP proxy (acting as MITM) generates a "spoofed" response to redirect the session to an unauthorized user. A similar approach is presented by Geneiatakis and Lambrinoudakis where a lightweight scheme ensures authenticity and integrity across the entire signaling path, for both requests and responses [28]. Yang et al. suggest the replacement of the HTTP digest by a scheme based on Diffie-Hellman key agreement and pre-shared secrets aiming to eliminate the HTTP digest-oriented security flaws [29]. Elsewhere monitoring systems are introduced that are able to detect signaling attacks and thus prevent modifications to the HTTP Digest mechanism that is currently supported by most SIP clients [30,31].

TABLE 18.3
Security Requirements Fulfilled by the Security Mechanisms
Proposed by the Research Community

Reference	[26]	[27,28]	[29]	[30,31]	[18,32]	[33–37]
Confidentiality	Yes	No	No	No	No	No
Integrity	Yes	Yes	No	No	No	No
Authenticity	No	Yes	Yes	No	No	No
Availability	No	No	No	No	Yes	Yes

Lately, significant effort is being spent towards the development of dedicated intrusion detection and prevention systems that can maintain a high availability for the SIP-based services [18–35]. Geneiatakis et al. present a framework that can protect SIP parsers against malformed messages. Also, Zhang et al. propose a solution for protecting SIP proxies against flooding attacks that are attempted through falsified DNS messages containing irresolvable addresses [32]. It is important to stress that any unavailability caused to these auxiliary systems, e.g., parser, DNS lookup service etc., affects all underlying SIP services, turning them unavailable as well.

Focus on the identification of flooding attacks that exploit INVITE messages against SIP proxies is given by the work presented elsewhere [33–37]. Reynolds and Ghosal [33] introduce an application layer attack sensor, based on the cumulative sum method [38] and the correlation between SIP INVITE and "200 OK" response messages, in order to detect a flooding attack. Another interesting approach for detecting flooding attacks is described by Sengar et al. [34]. This time the detection method is based on the correlation of SIP INVITE, "200 OK" response and SIP BYE (or SIP CANCEL) messages, and the Hellinger distance among them. Two slightly different approaches, based on the original finite-state machine for SIP transactions, are presented [35]. Finally, Fiedler et al. [37] present a scalable SIP-based security architecture, known as VoIP defender, capable of processing more than 50.000 SIP requests per second is introduced. This is the only architecture that takes into account performance requirements as well.

Table 18.3 highlights the security requirements fulfilled by each of the security mechanisms described above. It is stressed that none of these mechanisms can assure a complete end-to-end security and thus they must be employed in conjunction with existing SIP security mechanisms.

18.6 Conclusions

This chapter has presented and analyzed security issues of the SIP. It has been demonstrated that it is crucial for SIP service providers to take into serious consideration the emerging and highly sophisticated SIP-oriented threats and vulnerabilities that could be exploited by malicious users to launch an attack. In order to eliminate such threats it is very important to identify the actual security requirements and fulfill them through the appropriate security measures. However, it has been demonstrated that none of the existing SIP security mechanisms can satisfactorily deal with all security requirements. Consequently, SIP providers should also monitor the new developed SIP security mechanisms, in an attempt to explore ways for combining them with existing security measures.

References

[1] Bates, R. J. 2002. *Broadband Telecommunications Handbook.* 2nd ed. New York, USA: McGraw-Hill Professional.

[2] Sicker, D. C., and T. Lookabaugh. 2004. *VoIP Security: Not an Afterthought.* QUEUE ACM, September.

[3] Geneiatakis, D., T. Dagiuklas, G. Kambourakis, C. Lambrinoudakis, S. Gritzalis, K. S. Ehlert, and D. Sisalem. 2006. Survey of security vulnerabilities in session initiation protocol. *Communications Surveys and Tutorials IEEE* 8(3): 3rd. Qtr. 68–81.

[4] Butcher, D., X. Li, and J. Guo, 2007. Security challenge and defence in VoIP infrastructures. *IEEE Transactions on Systems, Man and Cybernetics, part C: Applications and Reviews* 37 (6): 1152–62. November.

[5] VOIPSA. 2005. *VoIP Security and Privacy Threat Taxonomy,* Public Release 1.0. Available online: http://www.voipsa.org/Activities/taxonomy.php, October.

[6] Sisalem, D., J. Kuthan, and S. Ehlert. 2006. Denial of service attacks targeting a SIP VoIP infrastructure: Attack scenarios and prevention mechanisms. *Network, IEEE* 20 (5): 26–31. September–October.

[7] Rosenberg, J., H. Schulzrinne, G. Camarillo, A. Johnston, J. Peterson, R. Spark, M. Handley, and E. Schooler. 2002. *Session Initiation Protocol.* RFC 3261, June.

[8] Hersent, O., J. P. Petit, and D. Gurle. 2005. *IP Telephony Deploying Voice over IP Protocols.* New York, USA: John Wiley & Sons.

[9] Spencer, M., B. Capouch, and E. Guy. 2007. F. Miller, K. Shumard ed. *IAX2: Inter-Asterisk eXchange Version 2.* Internet Draft, April.

[10] Ferguson, N., and B. Schneier. 2003. *Practical Cryptography.* New York, USA: John Wiley & Sons.

[11] Shannons, M. L. 1998. *Phone Book: The Latest High-Tech Techniques and Equipment for Preventing Electronic Eavesdropping, Recording Phone Calls, Ending Harassing Calls, and Stopping Toll Fraud.* Colorado, USA: Paladin Press.

[12] Ruishan, Z., X. Wang, X. Yang, and X. Jiang. 2007. Billing attacks on SIP-based VoIP systems. In *Proceedings of First USENIX Workshop on Offensive Technologies.*

[13] Wieser, C., M. Laakso, and H. Schulzrinne. 2003. Security testing of SIP implementations. Available online: http://compose.labri.fr/documenta tion/sip/Documentation/Papers/Security/Papers/462.pdf.

[14] Mitnick, D. K., W. L. Simon, and S. Wozniak. 2003. *The Art of Deception: Controlling the Human Element of Security.* New York, USA: John Wiley & Sons.

[15] Oxford Reference Online Premium, http://www.oxfordreference.com/.

[16] Fu, X., B. Graham, R. Bettati, and W. Zhao. 2003. Active traffic analysis attacks and countermeasures. Available on line: http://students.cs. tamu.edu/xinwenfu/paper/ICCNMC03_Fu.pdf.

[17] Geneiatakis, D., G. Kambourakis, C. Lambrinoudakis, T. Dagiuklas, and S. Gritzalis. 2005. SIP message tampering: The SQL code INJECTION attack. 13th IEEE *International Conference on Software, Telecommunications and Computer Networks (SoftCOM '05).* N. Rozic et al. eds. 176–81. Split, Croatia.

[18] Geneiatakis, D., G. Kambourakis, C. Lambrinoudakis, T. Dagiuklas, and S. Gritzalis. 2007. A framework for protecting SIP-based infrastructure against Malformed Message Attacks. *Computer Networks* 51 (10): 2580–93.

[19] Eddy, W. 2007. TCP SYN flooding attacks and common mitigations, *Internet Engineering Task Force, RFC 4987,* August.

[20] Gibson, S. 2002. DRDoS distributed reflection denial of service. Available on line: http://grc.com/dos/drdos.htm.

[21] Stamp, M. 2005. *Information Security: Principles and Practice.* John Wiley & Sons, New York, USA: Wiley-Interscience.

[22] Franks, J., P. Hallam-Baker, J. Hostetler, S. Lawrence, P. Leach, A. Luotonen, and L. Stewart. 1999. HTTP Authentication: Basic and digest access authentication. *Internet Engineering Task Force, RFC 2617,* June.

[23] Thayer, R., N. Doraswamy, and R. Glenn. 1998. IP Security Document Roadmap. *Internet Engineering Task Force, RFC 2411,* November.

[24] Bellovin, S. M. 1996. Problem Areas for the IP Security Protocols. In *Proceedings of the Sixth USENIX UNIX Security Symposium,* San Jose. Available on line: ftp://ftp.research.att.com/dist/smb/badesp.ps., July.

[25] Dierks T., and E. Rescoral. 2006. *The Transport Layer Security (TLS) Protocol Version 1.1. RFC 4346*, April.

[26] Ono, K., and S. Tachimoto. 2005. *Requirements for End-To-Middle Security for the Session Initiation Protocol (SIP). RFC 4189*. October.

[27] Cao, F., and C. Jennings. 2006. Providing response identity and authentication in IP telephony. In *Proceedings of the First International Conference on Availability, Reliability and Security* 20–22 April, 8.

[28] Geneiatakis, D., and C. Lambrinoudakis. 2007. A lightweight protection mechanism against signaling attacks in a SIP-based VoIP environment. *Telecommunication Systems*, 36(4): 153–59.

[29] Yang, C., R. Wang, and W. Liu. 2005. Secure authentication scheme for session initiation Protocol. *Computers & Security* 24 (5): 381–86.

[30] Wu, Y., S. Bagchi, S. Garg, and N. Singh. 2004. SCIDIVE: A stateful and cross protocol intrusion detection architecture for voice-over-IP environments. In *Proceedings of International Conference on Dependable Systems and Networks*, 433–42.

[31] Ding, Y., and G. Su. 2007. Intrusion detection system for signal based SIP attacks through timed HCPN. *The Second International Conference on Availability, Reliability and Security (ARES)*. 190–97, 10–13.

[32] Zhang, G., S. Ehlert, T. Magedanz, and D. Sisalem. 2007. Denial of service attack and prevention on SIP VoIP infrastructures using DNS flooding. *Principles, Systems and Applications of IP Telecommunications (IPTComm2007)*. New York, USA.

[33] Reynolds, B., and D. Ghosal. 2003. Secure IP telephony using multi-layered protection. In *Proceedings of the Network and Distributed System Security Symposium (NDSS)*.

[34] Sengar, H., H. Wang, D. Wijesekera, and S. Jajodia. 2006. Fast detection of denial-of-service attacks on IP telephony. In *Proceedings of 14th IEEE International Workshop on Quality of Service*, 199–208.

[35] Chen, E. Y. 2006. Detecting DoS attacks on SIP systems. In *Proceedings of 1st IEEE workshop on VoIP Management and Security*. 3: 53–58.

[36] Sengar, H., D. Wijesekera, H. Wang, and S. Jajodia. 2006. VoIP intrusion detection through interacting protocol state machines. In *Proceedings of International Conference on Dependable Systems and Networks*.

[37] Fiedler, J., T. Kupka, S. Ehlert, T. Magedanz, and D. Sisalem. 2007. VoIP defender: Highly scalable SIP-based security architecture. *Principles, Systems and Applications of IP Telecommunications (IPTComm2007)*. New York, USA.

[38] Brodsky, B. E., and B. S. Darkhovsky. 1993. *Nonparametric Methods in Changepoint Problems*. Kluwer Academic Publisher.

19

SIP Vulnerabilities for SPIT, SPIT
Identification Criteria, Anti-SPIT
Mechanisms Evaluation Framework and
Legal Issues

G. F. Marias
Athens University of Economics and Business

L. Mitrou
University of the Aegean

M. Theoharidou and J. Soupionis
Athens University of Economics and Business

S. Ehlert
Fraunhofer Institute for Open Communication Systems

D. Gritzalis
Athens University of Economics and Business

CONTENTS

19.1 Introduction 457
19.2 Background 459
19.3 SPIT Vulnerability Analysis 461
19.4 SPIT Identification Criteria 465
19.5 Anti-SPIT Mechanisms 468
19.6 Anti-SPIT Mechanisms Evaluations 473
19.7 Anti-SPIT Mechanisms and Legal Issues 476
19.8 Conclusions 477
      References 478

19.1 Introduction

Since 2004, spam over Internet Telephony (SPIT) attack has been officially reported. The first incident was recorded in Japan at a major VoIP provider called SoftbankBB, with Mio-subscribed customers*. Actually, spam history started 30 years ago, when the first e-mail spam was sent to 600 addresses. In the late 1970s and early 1980s there was not any real public Internet, and thus the growth of e-mail spam was not as exponential as it is today. Today, there is widespread use of the Internet and, since 2004, there has been significant growth in mass-market VoIP services over broadband Internet access services, in which subscribers make and receive calls as they would over the Plain Old Telephone Service (POTS) or Public Switched Telephone Network (PSTN). Thus, due to the current status of the Internet and the penetration of VoIP services, including the foreseeable IP Multimedia Subsystem (IMS) paradigm shift, it is likely we will see SPIT spread rapidly around the world. In fact, it is accepted to happen more sharply in the VoIP world than the e-mail world, since "vishing" attacks, i.e., acquiring sensitive information, such as usernames, passwords and credit card details by masquerading as a trustworthy entity using VoIP services, are expected to increase by 50% during 2008, according to industry reports [1].

The explosion in the adoption and usage of VoIP together with the knowledge gained so far by spammers in the e-mail spam domain, and of course their potential profits of spam business, makes it more challenging to design and deploy an anti-SPIT framework. Additionally, VoIP services, and especially Session Initiation Protocol (SIP) when used as the signaling bearer for VoIP session management, significantly differ from classic SMTP e-mail services. Therefore, some of the existing and reliable mechanisms for prevention, detection and management of spam might not be applicable. Finally, even if we assume that spam and SPIT threats are equivalent, the SIP protocol itself might illustrate some new, probably more fatal, vulnerabilities and weak points. In this case, SPITters might find themselves luckier than their predecessor spammers.

If SPIT prevalence becomes proportional to spam, then the acceptance of VoIP will be encumbered. This might be a problem from the service providers' point of view but the real problem will appear to the end-user, who will suffer simultaneously from spamming and spitting.

Even though SPIT is not yet a dominant Internet threat, several mechanisms and frameworks have been introduced to detect and counter the threat. Most of them combine or adapt existing and proved ideas from the anti-spam domain, while some introduce innovative new concepts. Until now as no practical amount of SPIT is present, their efficacy and efficiency could only be estimated in laboratory conditions, although a concrete evaluation framework is still missing.

Unsolicited communications infringe privacy, primarily in its narrow and visible sense, i.e., through the illegal intrusion into the private realms of the person. SPIT as a specific form of spam is considered to be an invasion of privacy [2]. VoIP spam (in the form of "call spam", IM spam, Presence spam),

---

*http://www.voipsa.org/pipermail/voipsec_voipsa.org/2006-March/001326.html. According to Columbia University officials (http://www.voipuser.org/forum_topic_10383.html) a recent SPIT accident was recorded also in the University during a VoIP pilot rollout.

referred to as SPIT, differs from other "traditional" forms of electronic communications (e.g., e-mail) in that is significantly more obtrusive and intrusive, as a phone will actually ring with every SPIT message, possibly even after midnight [3].

The fundamental right of privacy, anchored in various national constitutional texts as well as in the European Convention of Human Rights (Art. 8), encompasses informational privacy, relational privacy and freedom of communication in the meaning of privacy/secrecy of communications. Informational privacy (or right to informational self-determination) relates to the individual's right to decide autonomously, whether and which personal information they can communicate to others and/or processed by them. Affected is also the so-called relational aspect of privacy, i.e., the right to determine, which communications one wishes to receive or not [4].

Communication partners may reasonably expect that their communications and related data will not be used in an unlawful way by third parties. However, not only does SPIT constitute a threat to privacy, the protection against unsolicited communications through the use of various prevention and detection mechanisms raises a lot of questions concerning their compatibility with fundamental rights. Informational and communicational privacy and confidentiality are relevant with regard to preventing SPIT. Concerns are expressed in particular for existing practices to inspect communications in order to prevent and eliminate spam. Use of detection mechanisms and blocking of incoming calls/ messages can restrict peoples' ability to communicate and therefore be an impairment of freedom of speech [4–6] and the right to receive and impart information, which is also recognized as an integral part of the freedom of information.

In this chapter we are discussing SPIT. We provide first a framework to defile and classify SPIT. Then, in Section 19.3, we identify some of the SIP vulnerabilities that might be exploited by potential SPITters. We then stress the criteria that enable any anti-SPIT framework to identify which session and call-establishment attempts should be classified as SPIT. In Section 19.4 we present the results of our survey about the anti-SPIT frameworks already presented in the literature, as well as their classification as prevention, identification or handling approaches. Next we illustrate our proposed quantitative and qualitative criteria for assessment and evaluation of the proposed anti-SPIT frameworks, together with a compliance study. In Section 19.7 we discuss the legal issues pertaining to SPIT detection mechanisms focusing on protection of communications secrecy and privacy. Finally this chapter summarizes its findings and proposals in the corresponding conclusion section.

19.2 Background

19.2.1 SPIT Definitions

Spam over Internet Telephony is defined as a set of bulk, unsolicited calls or instant messages. A SPITter uses the existing IP infrastructure to target users, in order to initiate SIP calls or send SIP instant messages and generate spam. The phenomenon is already known from both the e-mail and the traditional telephony context.

SPIT can be identified in three different forms:

a) *Call SPIT*, which is defined as bulk, unsolicited session initiation attempts in order to establish a multimedia session.

b) *Instant Message SPIT*, which is defined as bulk, unsolicited instant messages and it is known as SPIM (spam over Instant Messages).

c) *Presence SPIT*, which is defined as bulk, unsolicited presence requests so that the malicious user becomes a member of the address book of another user or potentially of multiples users.

SPIT is a relative new identified threat; some first SPIT messages have been reported but the phenomenon has not yet reached the massive volume of the e-mail spam. The two types of spam (e-mail spam and SPIT) illustrate several common characteristics, like the use of the IP protocol to send e-mails and initiate calls or IMs, respectively. Attacks that are common in the e-mail context are applicable and expected to be seen in the SIP context too. Examples include harvesting addresses, dictionary attacks, zombies or bots.

19.2.2   Motivation

Despite the similarities that one can observe on the two spam types, there are also elements that differentiate the threats and vulnerabilities and justify the need for discovering new countermeasures to mitigate SPIT or adopt existing techniques cautiously. Firstly, SIP communication is not asynchronous as the e-mail one. An SIP call is held in real time and is sensitive to delays. Only at the call establishment phase some delay may be accepted and this renders the application of existing content analysis techniques to identify SPIT invalid. E-mail spam is mainly textual and may be combined by images, video or sound. SPIT takes the form of audio, video and maybe some instant text messages. The cost of sending a SPIT call is higher that sending an e-mail, in terms of resources and it can cause overload on a network. The annoyance of the user is considered higher when he receives a call as opposed to receiving an e-mail. Thus, one has to carefully examine the applicability of existing countermeasures to the SIP environment.

In order to design an anti-SPIT framework one needs to also identify the points where anti-SPIT countermeasures can be placed. These can firstly be applied on the domain of the SPITter (i.e., outgoing proxy server) and act proactively in the sense that the SPIT message or call does not leave the boundaries of the caller's domain. In addition, mechanisms can be deployed in the domain of the user that is the target of SPIT (i.e., incoming proxy server), which act reactively and aim at identifying an incoming SPIT call or message. When SPIT is detected, handling mechanisms can be applied, which may vary from flagging the call or message as potential SPIT to blocking it, depending on the domain's policies or the target's preferences.

To identify the overall amount of SPIT risk when using SIP for voice services, a vulnerability analysis is crucial. This study will report which of the SIP protocol specifications are vulnerable, whenever these specifications are mandatory or optional, if the interoperability with other protocols produces flaws, and, finally, if generic security weaknesses can be used for SPIT attacks. This is considered essential since potential anti-SPIT mechanism should be aware of these vulnerabilities in order to mitigate their impact.

Additionally, as several anti-SPIT mechanisms have recently been proposed in the literature, it is useful to survey their main functionality to identify metrics and criteria for their evaluation, in respect to SPIT. Actually assessment and evaluation of the existing anti-SPIT mechanisms is spread in many domains, such as efficiency (i.e., whenever it avoids SPIT false alarms or identifies the actual SPIT successfully), performance (i.e., overhead in the network, number of calls examined per minute, etc.), and finally, qualitative (i.e., scalability issues, adoption, availability, etc.). Another important dimension when evaluating existing or even forthcoming anti-SPIT mechanism is related to the legal framework and any anti-SPIT mechanism should be compliant to EC or national legislation.

19.3 SPIT Vulnerability Analysis

In the VoIP context, a SPITter might perform various SPIT attacks by first building a set of active SIP URIs which contain callees' SIP address. Moreover, any call (i.e., INVITE) automation method can be useful for SPITters, as well. As a conclusion, the identified SIP threats are separated into two main categories (a) creating a list of valid SIP addresses [1–7], and (b) when SPIT automation is required [8–10].

Sending ambiguous requests to proxies. A proxy server needs to decide the next step destination of a call or a message. The decision is made by looking at a specific part of the header of the request message or by contacting a service which is implemented by the SIP registrar server. If the header request/URI part does not offer adequate information then the proxy service is used. The proxy server offers an answer via a 485 SIP message. This kind of response message can contain a contact header field with a list of new URIs available for trial. For example if there are many URIs which contain a specific sequence of characters, then the 485 message would contain all the possible URIs with this string. The worst case arises where a special character can evoke a 485 message with all the registrars of the service's URI database. Therefore, this is an important weakness because a SPITter can easily collect a set of valid URIs.

Listening to a multicast address. Every participant in an SIP session needs to be aware of the current location of other UAs. The physical location and the IP address of a user are discovered during the session initiation phase. In this phase, the proxy servers answers user or server REGISTER requests by consulting the location service. In order for the location service to have a valid URI database, it should be aware of any user change of address. This is achieved by listening to a specific multicast address where the UAs send a REGISTER message which contains their current IP addresses. The UAs listening to that address are therefore informed of users' location. On the other hand, if a SPITter is listening to that multicast address, he is in the position to assemble both a set of the addresses of the registered users and their registration.

Population of "active" addresses. Registration is a main function in the SIP infrastructure. The registration is accomplished when the REGISTER messages link the user's URI with the machine it currently uses and the IP address it happens to be using at that exact point in time. Therefore, both of the aforementioned weaknesses combined facilitate the determination of online users by SPITters. This vulnerability allows SPITters to send a smaller amount of SPIT messages and a larger number of successful SPIT since it can forward messages to users that are currently logged on.

Contacting a redirect server with ambiguous requests. As proxy servers can handle a great amount of messages; redirect servers are employed to balance the load of proxy servers. A redirect server is designed to refuse any other request than CANCEL. But it can collect a possible list of alternative locations of UAs and forward it to the requestor. The SPITter can therefore send random requests to redirect servers and as a result collect a list of URIs in order to send SPIT messages. This attack could be more effective if conducted in conjunction with sending ambiguous requests to proxies (first vulnerability).

Throwaway SIP accounts. The creation of many accounts in various domains, by the same user, can prove to be a great vulnerability which can be exploited by SPITters. This is because the SPIT traffic can not be easily associated with a specific SIP account, using conventional anti-SPIT countermeasures, such as Black Lists. This vulnerability is due to the SIP specification [7] which clearly states that a user can be registered in a SIP domain by simply sending a REGISTER message. Thus, a SPITter is free to issue multiple REGISTER messages and as result to create multiple accounts in a SIP domain.

Misuse of stateless servers. Open Relay Servers, which exist in the email domain, are used by spammers to forward their spam emails. The SIP specification states that there can be stateless servers which offer the same service as Open Relay Servers in that they forward any call to its destination. SPITters can exploit it for hiding their IP address and location. Stateless proxies are not frequently used by SIP domains but are useful because they easily handle the rare flood of request messages.

Anonymous SIP service and Back-to-Back User Agents. The anonymous SIP option permits the forwarding of any call to a recipient without disclosing the caller identity. SPITters use this option to overcome various detection methods, such Black Lists and header content analysis. In this scenario a Back-To-Back User Agent (B2BUA) acts as a concatenated User Agent Server (UAS) and User Agent Client (UAC) [7]. The B2BUA collects and parses every request of a dialog [7]. Thus, a SPITter can replace the contact information of every message with its own before forwarding the message. The recipient upon answering to his request would engage in a malicious dialog.

Sending messages to multicast addresses. The via header is one of the most important SIP header fields. The content of the via field shows the route that should be followed by the response message and the path the request has followed towards the target UA. The Maddr tag is a possible property of this specific field and indicates a multicast address. A SPITter who obtains such an address is able to initiate calls to a set of users belonging to the multicast address.

Exploitation of forking proxies. A forking proxy server is a SIP proxy server which can send an INVITE message to multiple recipients simultaneously. These kinds of servers help SPITters to collect a list of active SIP URI's faster, by using the vulnerability of sending ambiguous requests, which might result in a return of unambiguous new addresses. After a SPITter sends an ambiguous INVITE request, it can initiate simultaneous calls towards all the addresses returned by the location service. In this case it is obvious that a SPITter using one single request can automate bulk SIP calls generation utilizing a statefull proxy.

Exploitation of messages and header fields structure. A way of sending SPIT is through bulk unsolicited messages. Actually this could be achieved by (a) using the MESSAGE method, or (b) hiding the message in various SIP header fields of the SIP protocol message bodies. A SPITter is capable of sending a SIP message, initiating a request, or spoofing the SIP responses transferred in the network. A request message might use the INVITE and the ACK methods, which both include a message body, which in turn might convey SPIT traffic. This body might include any media file that eventually will be delivered to the recipient. Furthermore, the MESSAGE method can be used to encapsulate SPIT traffic. This option might exploit the fact that any SIP flow outside a dialog does not require the authentication of the caller UA. On the other hand, the response messages include data fields that can be manipulated by a SPITter. Table 19.1 illustrates the main response messages that can be used for SPIT content encapsulation:

Finally both request and response messages contain header fields that can conceal information which can be rendered by a UA. This kind of information can help SPITters to send large messages since the attributes of these header fields allow (dynamic length, extent to multiple lines) or to perform other

TABLE 19.1

Response Messages and Their Possible SPIT Content

Response messages	Possible SPIT content
Provisional (1xx)	The message bodies might include session descriptions
180 Ringing	The message body might contain textual information, or an audio file, or even animations
182 Queued	It might contain a reason phrase, illustrating further details about the status of a call, or a message body that play an audio file, from an on-hold music palette.
183 (Session Progress)	The Reason-Phrase or message body of it could provide additional details.
200 OK	It might contain returned information that depends on the method used in the request
300 Multiple Choices	It might include a message body
380 Alternative Service	It might contain a message body in which the alternative services are described.
480 Temporarily Unavailable	It might contain a reason phrase indicate a more precise cause, such as why the callee is unavailable
484 Address Incomplete	It might contain a reason phrase
488 Not Acceptable Here	It might contain a message body with a description of media capabilities
606 (Not Acceptable)	It might contain a message body

action as to trigger the UA to execute the body. The main header fields of SPITters' interest and the way they can be exploited are:

Subject: Might be parsed by the user's software and displayed to the user's terminal.

From: Allows a text to be displayed to the user.

Alert-Info: Specifies an alternative ring back tone that could be used as a pre-recorded audio SPIT message.

Call-Info: Some parameters of the specific header, i.e., purpose, icon, info and card, could be used for sending SPIT messages.

Contact and To: The display name flag could be altered to provide a SPIT message.

Retry After: The optional textual comment parameter might be used to convey a SPIT message.

Error-Info: The pointer to additional information, in relation to the error status response, could be set by an attacker to point to a SPIT message.

Warning: The warning text that this header contains could be replaced by a SPIT message.

Content-Disposition: Even if this header field indicates the way that a message should be interpreted by a UA, it could also be used as a SPIT message.

Content-Type: This field denotes the media type of the message body. Therefore, it could be used as a SPIT message. It could also trigger the receiver's UA to execute the message body.

Priority: This field indicates the urgency of a request. A SPITter could exploit it in order to facilitate the delivery of the SPIT message.

There are two more vulnerabilities; one due to the protocol's optional recommendations and the other due to the interoperability with other protocols. The first vulnerability is owed to the exploitation of registrar servers. The registrar server is in charge for updating the location information of the UAs by handling REGISTER requests. A server answers to queries of UAs and discloses the location of registered users. Therefore, a SPITter can issue an attack to registrar in order to collect a set of URIs. This attack is successfully accomplished because the user, who initiates a dialog to a registrar, is not compulsory registered or authenticated to a domain since the SIP specification does not oblige the domain to register each user. Thus, the impersonation of a legitimate user is easily acquired. The second vulnerability is the exploitation of particular domains' address resolution procedures. The SIP proxy servers and the afterward communication inherits the vulnerabilities which are introduced by the communication supplementary protocols, such as DNS queries, ARP, RARP, etc. These protocols can be vulnerable to several attacks, such as spoofing or man in the middle [8–10]. Therefore, a malicious intermediate could impersonate a legitimate user, spoof messages, deliver SPITs by using the victims' identities and create lists of other possible victims.

19.4 SPIT Identification Criteria

A list of SPIT identification criteria are analyzed in this section. These may be applied as SPIT detection rules on both sides of a SIP communication and they can be proactive or reactive depending on the point they are applied (SPITter's or target's domain). We propose two generic categories of SPIT identification criteria:

- **SIP Message criteria:** This category includes the criteria that are related to attributes of SIP messages and they can be based on (a) call and message patterns or (b) headers' semantics.

- **SIP User Agent criteria:** This category includes the criteria that are related to attributes of a SIP User Agent and they can examine the origin of the call or message as well as the relationship of the SIP participants.

19.4.1 SIP Message Criteria

19.4.1.1 Call-Messages Patterns

This category includes criteria that analyze specific call or message characteristics or patterns, in order to determine whether a call (message) is possibly a SPIT call.

- *Path traversal:* A call or a message might pass through many intermediates before reaching its final destination. This path is denoted in the *via* header. Thus, if in the via header a SPIT domain is recognized, the call or the message may be a potential SPIT.

- *Number of calls-messages sent in a specific time frame:* It analyzes the number of calls (messages) made in a specific time period by a user. If this number is above a specific pre-defined threshold then the call (message) is characterized as a possible SPIT call.

- *Static calls' duration:* If the calls initiated by a single user have a static duration, then the user is a potential SPITter who possible used an automated script in order to initiate the calls.

- *Receivers' address patterns:* If the receivers' addresses follow a specific pattern (e.g., alphabetical SIP URI addresses), then the call (message) is flagged as potential SPIT.

- *Small percentage of answered/dialed calls:* It indicates the number of successful call completions from this caller per a pre-defined time period, which is relative to the number of failed ones.

- *Large number of errors:* When a user sends a large number of INVITES and the SIP protocol returns a large number of error messages (e.g., 404 Not Found) is a sign of a potential SPIT attack.

- *Size of SIP messages:* In this case a set of SIP messages sent by a user to other users is analyzed. If those messages have a specific size then it is very possible to be sent by a "bot" software, and therefore the call is characterized as SPIT.

19.4.1.2 SIP Headers' Semantics (SIP Message Oriented)

This category includes criteria that identify a SPIT call or message through a semantic analysis of the contents of the SIP messages (e.g., Bayesian filtering). These particular criteria are further categorized, according to the different parts of SIP messages that could be used. These are: (a) a message's headers,

(b) a message's body, and (c) the reason phrases of a message. In addition, we have identified three possible types of SPIT that could be injected in a SIP message, namely: (a) text SPIT injected in a header field, (b) media SPIT carried in the message body, and (c) hyperlink SPIT injected in a header field.

Tables 19.2 through 19.4 depict the specific SIP header fields that can be used for a detailed semantic analysis, so as to detect a SPIT call or message alongside with the type of the SPIT that could be sent.

Table 19.3 presents the type of messages containing a body, which could be used in order to make a SPIT call (or message). They are categorised in terms of request and response messages. SPIT contained in the message body can be text, media or hyperlink.

TABLE 19.2
SIP Headers That Could Be Used for SPIT

Header fields	SPIT type	Request messages	Response messages
Subject	Text	✓	✓
From	Text	✓	✓
Call-Info	Hyperlink	✓	✓
Contact	Text	✓	✓
To	Text	✓	✓
Retry After	Text	✓	✓
Alert-Info	Hyperlink	✓	✓
REPLY TO	Text	✓	—
Error-Info	Hyperlink	—	✓
Warning	Text	—	✓

Header fields related to SIP messages' bodies (not carrying SPIT "directly")

Content-Disposition	Displayed message body	✓	✓
Content-Type	Displayed message body	✓	✓

TABLE 19.3
Request–Response Messages That Could Be Used for SPIT

	INVITE
Request messages	ACK
	180 Ringing
	183 Session Progress
	200 OK
Response messages	300 Multiple Choices
	380 Alternative Service
	488 Not Acceptable Here
	606 Not Acceptable

TABLE 19.4
Reason Phrases That Could Be Used for SPIT

182 Queued
183 Session Progress
200 OK
400 Bad Request
480 Temporarily Unavailable
484 Address Incomplete

Table 19.4 presents the reason phrases of response messages that could be used for SPIT purposes. Reason phrases may contain text or hyperlink type of SPIT.

19.4.2 SIP User Agent Criteria

These criteria examine the characteristics of a SIP session, meaning the SIP addresses of the sender/caller (i.e., SIP URI or IP address), as well as the domain the session was initiated in. They can be used in traditional white and black listing techniques.

- *Caller SIP URI*: It detects and analyzes the SIP URI of the sender of a call/message, so as to determine if he/she is a potential SPITter or not.

- *Caller IP address:* It analyzes the IP address of the sender/caller so as to characterize him/her as a SPITter.

- *Caller domain:* It analyzes the identity of the domain of the caller (sender), which is determined either by SIP URI of the caller, or through DNS lookup from the IP address. If the identity of the domain is a well-known SPIT source, then the call or the message is characterized as potentially SPIT.

- *Caller/callee relationship:* It examines whether the caller/sender is trusted by the callee/receiver. Typical examples include whether a caller is known to the callee (inclusion in address book, previous calls have been established, marked as SPITter by the callee).

19.5 Anti-SPIT Mechanisms

19.5.1 Anti-SPIT Mechanisms Description

SPIT is likely to influence the future use and adoption of the VoIP technology. In order for SPIT to be efficiently managed the following three progressive steps are necessary: (a) prevention, the avoiding of SPIT altogether,

(b) detection, the ability to identify a SPIT call or message, and (c) handling, the dealing with a detected SPIT call or message.

Certain general frameworks from the e-mail spam paradigm have been considered as potential candidates for SPIT avoidance [8]. Some of them appear to be basic building blocks of the anti-SPIT architecture that have been proposed in the literature. We shall first provide a brief description of the general anti-SPIT frameworks before proceeding to classify them according to the three aforementioned progressive steps.

Black and white lists. White lists are made up of trusted users that are not categorised as SPITters. An end-user accepts the calls, or messages, initiated by any of the members in his/her white list. On the contrary, black lists contain any/all potential initiators of SPIT calls. These calls are therefore to be blocked.

Content filtering: This method is based on filters that scan the content of messages. They appear to be inappropriate as anti-SPIT, since real-time filtering is hard to accomplish. Nevertheless, this technique could be used for the detection of instance messaging SPIT (similar to e-mail spam).

Challenge–response. Communication becomes possible only once the caller has replied correctly to a challenge sent by the callee. This approach aims at preventing SPIT by operating a distinction between humans and 'bots'. Other such mechanisms include Turing tests and computational puzzles.

Consent-based. Here, communication is not achieved unless the callee explicitly consents.

Reputation-based. Trust is the central notion of this approach. When a callee receives a request for communication the level of trust of the caller should be determined. This is accomplished through direct estimations or second-hand reputations. If the trust level is above a predefined threshold then the communication is permitted, otherwise it is rejected.

SIP addresses management: One of the SPITters' main goals is to collect as many valid addresses as possible. So, to prevent SPIT it is important for end-users to protect their addresses (i.e., URIs) from being collected. For this to be done, two different approaches are possible: address obfuscation and use of multiple addresses.

Charging-based. This approach obliges SPITters to pay for every unsolicited bulk call (messages), as a result, the cost born by the SPITter is increased in terms of financial or computational resources.

As more attention is drawn to the SPIT problem, a number of anti-SPIT frameworks begin to emerge. They are commented upon briefly in the sequel.

19.5.1.1 SPIT Prevention Using Anonymous Verifying Authorities (AVA)

The so-called AVA system [11] illustrates how the prevention of voice spam may be achieved by extending the call setup procedure. This method is founded upon a "call me back scheme" and the use of two new entities in the IP infrastructure, namely: (a) the mediator, and (b) the AVA. Through the exchange of information between these two entities, SPIT is mitigated by anonymously blocking unwanted calls. This way the caller is not aware of the existence of the callee, provided that the call failed to be established.

19.5.1.2   SPIT Mitigation through a Network Layer Anti-SPIT Entity

The anti-SPIT entity [12] is based on an approach which detects and mitigates SPIT by using a sniffing-oriented network-level entity. This entity filters and analyzes the network traffic so as to detect a SPIT call by means of educated guesses. These guesses are based on a list of criteria, which contribute to different extent toward the final decision. The mechanism only takes the SIP packets into account and ignores the rest of the network traffic. Finally, handling actions are enforced based on the results of the SPIT analysis. These actions rely on the policies adopted by the domain to which the specific user belongs, and the end-user's preferences.

19.5.1.3   SPIT Detection Based on Reputation and Charging Techniques

This approach proposes a SPIT detection mechanism [13] based on two different techniques, namely reputation (i.e., trust networks) and payments at risk. The reputation-based technique is based on the notion of trust, in other words, SPIT detection relies solely on the callee's trust toward the caller. On the other hand, the charging based technique aims at reducing SPIT by increasing its cost. This is achieved by imposing a charge to each message sent by the caller. The specific technique functions alongside simultaneously with other anti-SPIT frameworks such as authentication modules and white lists.

19.5.1.4   DAPES

The DAPES system [14] determines in real time depending on whether a call is identified as being SPIT or not. The main characteristics of DAPES are the following: firstly, all messages are sent through proxies that serve as authenticators, secondly, all outbound proxies have certificates granted by Trusted CA, thirdly, all communication between proxies has to be encrypted and fourthly, the sender has to be authenticated by its domain proxy. In order for all of the above to be effective, certain characteristics have to be supported. Current systems are therefore unable to effectively implement the abovementioned infrastructure.

19.5.1.5   Progressive Multi Grey-Leveling

Progressive multi grey-leveling (PMG) [15] determines and allocates a grey-level for each caller that enables it to determine whether a call qualifies as

SPIT or not. This mechanism is not based on the feedback of other users with regards to a specific caller but on the last calls made by that caller itself. In practice, the grey level of each user is calculated by means of the addition of a long-term and short-term level. If a summary is greater than a threshold than the user is considered as being a SPITter and all his calls are consequently blocked. However, a caller who is classified as a SPITter is bound to loose such status since the grey-level of the caller is not constant.

19.5.1.6 Biometric Framework for SPIT Prevention

This technique [16] offers an approach based on identity-management since a SPITter changes his identity frequently. The specific technique proposes that personal detail (such as biometric data) of all users be recorded, so as to create a unique link between every user's identity and its biometric data. So as to enable the functioning of this infrastructure, each user must be registered to authenticated servers when using VoIP. This technique requires an effective communication between servers that can enable the exchange of user credentials (based on biometric data).

19.5.1.7 RFC 4474

This proposal [17] focuses on identity-management and aims at resolving the problem of user authentication by using PKI and certificate authorities. So as to enable the functioning of this RFC, two new sip-header fields are required: (a) identity, containing a signature that is used to verify the caller's identity (b) identity-info, for transmitting a reference to the signer's certificate. Despite the fact that this infrastructure had not been initially proposed for SPIT, it can nevertheless be easily implemented to SPIT.

19.5.1.8 SIP SAML

The Security Assertion Markup Language (SAML) [18] is used for the expression of security assertions, such as authentication, role membership, or permissions. A SIP-authentication service is proposed that authenticates the user via certain asserted features. Each VoIP message includes caller identity information along with a reference to a SAML assertion which has various features of the caller and the caller's domain certificate. This infrastructure tries to avoid the frequent change of SPITter addresses through identity control.

19.5.1.9 DSIP

Differentiated SIP [19] is an extension to the SIP protocol which classifies users into three distinct categories. These categories are deduced from the e-mail context: the white list, made up legitimate callers, the black list comprises the SPITters and the grey-list contains the callers who have yet to be categorised. DSIP usually employs human verification test for the uncategorised callers.

Once the grey-list callers have succeeded the test, the communication with the callee is established.

19.5.1.10   Voice Spam Detector

The voice spam detector system [20] is a combination of anti-SPIT techniques that is founded upon reputation and trust. The basic characteristics of a VSD system are the following: (a) Presence filtering that depends on the current status of the callee, (b) the traffic patent filter evaluates the incoming calls received from a specific caller or ship domain so as to ensure that they do not exceed the predetermined threshold, (c) the black and white list that is based on the black and white list filtering, (d) Bayesian learning, every call is evaluated, on the basis of trusted information that is made available by third party entities, with regards to the behavior of the participating entities. The existence of trusted information is based on the presumption that all participating entities have established prior calls, (e) social networks and reputation: this technique is used so as to accept a call that is based on social relationships already established by the user in its VoIP environment.

19.5.1.11   VoIP SEAL

VoIP SEAL [21] is a system that is divided into two main stages. The first stage uses various models which give a score between −1 and 1, the higher the score, the higher the probability of the call being SPIT. Moreover, two thresholds exist: On the one hand, if the score is inferior to the lower threshold, the call moves to a second stage that usually entails a CAPTCHA test. On the other hand, if the score is higher than the superior threshold the call is rejected. The main characteristic of VoIP SEAL is the adoption of a modular architecture that facilitates adding or updating of certain modules so as to take defence measures against SPIT.

19.5.2   Anti-SPIT Mechanisms Classification

Classification of anti-SPIT countermeasures in prevention, detection and handling is useful because it provides information about the scope. For example, it might be useful to combine prevention with handling mechanism, or detection with handling mechanism.

In the following prevention refers to the mechanisms applied to avert SPIT, i.e., avoid a SPIT call reaching the callee domain or phone. This avoids communication and processing overheads, conserves costs and resources, and minimizes the annoyance of the end-user. Detection means the identification of SPIT when a SPIT call is actually under transit to the callee, or under processing to the callee's proxy server or SIP-phone. The issue here is to mitigate or avoid any annoyance, and if possible, to minimize the overheads and to conserve resources. Finally, handling means any reaction to the SPIT. Table 19.5 illustrates this classification.

TABLE 19.5
Classification of the Anti-SPIT Mechanisms

Mechanism	Prevent	Detect	Handle
AVA [5]	✓		
Anti-SPIT Entity [6]		✓	✓
Reputation Charging [7]	✓	✓	
DAPES [8]	✓		
PGM [9]		✓	
Biometrics [10]	✓	✓	
RFC 4474 [11]	✓		
SIP SAML [12]	✓	✓	
DSIP [13]	✓	✓	
VoIP SEAL [14]	✓	✓	✓
VSD [15]	✓	✓	

19.6 Anti-SPIT Mechanisms Evaluations

19.6.1 Assessment Criteria

An assessment of the proposed anti-SPIT mechanisms requires the definition of various qualitative and quantitative criteria. We have based our research on the following criteria:

Percentage of SPIT calls avoided. How many SPIT call attempts have been identified and handled by the anti-SPIT mechanism?

Reliability. The precision of making the right adjustments about SPIT calls and callers, in terms of false positive and negative rates.

Latency. Due to the real-time nature of VoIP, quick decisions regarding SPIT detection are a major requirement, especially when legitimate calls are analyzed. In such a case, trustworthy users should not tolerate large delays.

Human interference. This metric represents the transparency of the anti-SPIT mechanisms to the end-user.

Resource overhead for the SIP provider. SIP providers should estimate the required resources for the implementation of the mechanism. This quantitative criterion seems essential for providers, since the number of calls that should be analyzed per unit time might be enormous, when the number of registered users increases.

Vulnerabilities. This parameter refers to the capability of a SPITter to bypass any of the anti-SPIT countermeasures.

Privacy risk. This criterion is associated with the collection, manipulation, and dissemination of private data. We assume that the end-user consents for the collection and manipulation of his/her private data, and has authorized specific legal entities for these purposes.

Scalability. This is an important criterion, since VoIP networks grow fast. Scalability should be considered when authentication is involved, since PKI and CA might need to establish complex cross-certification chains, or when reputations and assertions are used.

Adoption. This parameter corresponds to the success of the anti-SPIT countermeasure, and depends on the effort it takes for an end-user, or a provider, to begin using it.

Availability. It denotes the increase in the availability of network, computing, memory, or human resources when preventing, detecting or handling SPIT countermeasures apply.

More specifically, in the description of VoIP SEAL mechanism [21], it is mentioned that the particular mechanism is intrusive for the end-user, as there is an interactive part requiring user's feedback. In our analysis, we do not estimate how intrusive this mechanism is or not, but we emphasize that there are information in the description of the mechanism that could help someone value accordingly the particular criterion, namely human interference. Finally, regarding the vulnerabilities criterion we must emphasize that we only focus on the vulnerabilities considerations that the authors take into account, whether these are few or many.

19.6.2 Compliance of SPIT Mechanisms to Assessment Criteria

Table 19.6 presents which mechanisms takes into account the SPIT identification criteria we defined. For this purpose we only took into consideration an abstract description of each mechanism. Furthermore, we do not consider whether the mechanisms meet the criteria well or not, but we rather provide the mere existence of each criterion in the mechanisms' description. For example, in the description of reputation/charging mechanism, the use of black and white lists requires the existence of a way to identify and handle users, either by SIP URI, IP address or even domain of origin. However, as something like that is not explicitly mentioned, we put the appropriate negative value in the table. Furthermore, the table can be used as a reference to choose the appropriate mechanism for anti-SPIT mechanisms in a given context. For example the call and message patterns might be highly cost demanding, in terms of data gathering and analysis, and thus mechanisms that focus on, and better fulfill the other criteria might be of preference. Finally, the table can be read as a concentrated area of further research directions regarding anti-SPIT countermeasures. Some of the questions that one can answer using the table include how can a particular mechanism contribute in terms of prevention, detection or handling of SPIT, and which combinations of techniques should someone use in order to fight SPIT more effectively, etc.

TABLE 19.6

Compliance of SPIT Mechanisms to Identification Criteria

Anti-SPIT mechanisms	Call and message origin			Path of message	Number of calls	Calls/messages patterns				Size of SIP message	Caller/callee relationship
	Caller SIP URI	Caller IP address	Caller domain			Calls duration	Sequential call numbers	Dialed/answered calls	Large number of errors		White/black lists
AVA	✓	—	—	—	—	—	—	—	—	—	—
Anti-SPIT Entity	✓	✓	—	—	✓	✓	✓	—	✓	—	✓
Reputation/Charging	—	—	—	—	—	—	—	—	—	—	✓
DAPES	✓	✓	✓	✓	—	—	—	—	—	—	—
PMG	✓	—	—	—	✓	—	—	—	—	—	✓
Biometrics	—	—	—	—	—	—	—	—	—	—	—
RFC 4474	✓	—	✓	—	—	—	—	—	—	—	—
SIP SAML	✓	—	✓	—	—	—	—	—	—	—	—
DSIP	✓	—	✓	—	—	—	—	—	—	—	✓
VoIP Seal	—	—	—	—	—	—	—	—	—	—	✓
VSD	✓	✓	✓	✓	✓	—	—	—	—	—	✓

Finally it must be mentioned, that although many mechanisms check the content of the SIP messages, for example the headers *FROM, VIA*, etc, none of them check the appropriate the messages' bodies and reason phrases for carrying the actual SPIT message. This explains why the corresponding column in the table has only negative values.

19.7   Anti-SPIT Mechanisms and Legal Issues

Protection of individual rights has—or should have—an impact on the choice and design, implementation as well as on the legal assessment of anti-SPIT measures and mechanisms. The deployment of anti-SPIT mechanisms raises a lot of issues relating to everyone's right to respect to his private life and his correspondence. The confidentiality of communications is guaranteed explicitly by several national, supranational and international legal instruments [among them: the Fourth Amendment (USA), the EU e-Privacy Directive (Art. 5), the European Convention for the Protection of Human Rights and Fundamental Freedoms (Art. 8)]. Both the EU e-Privacy Directive and the European Convention for Human Rights prohibit any form of interception, e.g., a third party acquiring access to the content or traffic data (data processed for the conveyance of a communication or the billing thereof) to private communications between two or more correspondents. Such interceptions are acceptable only on the basis of some fundamental criteria, i.e., legal basis, need for such a measure in a democratic society and conformity of the measures adopted with legitimate aims such as national security, public safety or the economic well-being of the country, prevention of disorder or crime, protection of health or morals, protection of the rights and freedoms of others.

Electronic communications are combining, in the most regular cases, both the notions of "private life" and "correspondence" [5]. Interception, opening, reading, delaying reception of communications or impeding the sending of messages have been considered by the Courts to be an intrusion into the right of correspondence, which includes not only confidentiality but also the right to send and receive correspondence. In this institutional context anti-SPIT mechanisms have to respond to a double challenge: they have to prevent successfully the invasion into the receiver's privacy while respecting the privacy and other fundamental rights of communicating parties, even of potential SPITters.

Filtering and withholding of received communication constitutes an interference with the freedom of communications. Filtering and screening of communications content for the purpose of detecting SPIT, without the consent of the communicating parties, should be considered as intruding fundamental rights and therefore unlawful. In any case defining SPIT by content is related with the risk of infringing freedom of expression and introducing a kind of censorship. Especially some techniques of communication screening, like blacklisting, can raise questions in relation to the freedom of expression and freedom of information [5]. Filtering can also result in the blocking of legitimate information (the so-called false positive), which has as consequence the infringement of the freedom of expression [2,5].

Blocking spam by technical means could intrude on the right to informational self-determination of the communicating parties. By definition some anti-SPIT modules need information from the user, such as his preferences, to be able to work [22]. Mechanisms based on the analysis of behavior and reactions (such as the challenge–response mechanism), the caller's characteristics reputation or the use of black/white lists imply the collection and use of personal data, i.e., any information relating to an identified or identifiable individual. Phone numbers allow indirect identification of subscribers through the use of reverse directories as well as through electronic communications services providers. Indirect identification is also possible through IP addresses, which can be traced back to a computer and through the provider consequently to a subscriber. Even if the link between subscribers and users is, in the case of IP address, less strong than by e-mail addresses and phone numbers, most IP addresses can be tied to a log-in and may qualify as personal data [2].

This challenge–response mechanism can have as result the processing of personal data of SPITters to the extent that these data can be considered as personal. The authentication and reputation manager mechanisms are based on the calling party reputation as well as on the enhancing identity management of the SIP users and consequently they may imply the use of personal data. In the case of black/white lists, the SIP URL of the users is stored and consequently it could be used for indirect identification. The list of "friends" and "non-friends" are also stored by their URIs, revealing a user's personality and life profile. In the aforementioned cases the processing of personal data should be based either on the explicit consent of the user, as far as his data are processed, or on the protection his right to protect his privacy, which regularly overrides the interests of the SPITter. Filtering tools may not be in compliance with the existing data protection legislation. Subscribers should keep the control over the information concerning them by having the possibility to opt out of SPIT detecting and the possibility to decide, what kind of spam should be filtered out.

19.8 Conclusions

The SIP protocol raised significant concerns, as to whether SPIT will be equivalent to the current spam prevalence. In order to address and evaluate these concerns, we addressed existing vulnerabilities of SIP for SPIT and proposed a SPIT identification framework. Additionally, we provided a macroscopic view of SPIT management techniques and mechanisms, alongside an extensive list of evaluation criteria that can be used for self-assessment against SPIT. Moreover, we presented a high-level and theoretical evaluation of the existing anti-SPIT mechanisms, and, finally we addressed the legal issues that are raised by the deployment of these candidate anti-SPIT mechanisms.

References

[1] McAfee Avert Labs. 2007. Top 10 threat predictions for 2008.

[2] Asscher, L. V., and J. van Erve (IViR -Institute for Information Law). 2004. Regulating spam–directive 2002/58 and beyond. Amsterdam, 1–74.

[3] Posegga, J., and J. Seedorf. 2005. Voice over IP: Unsafe at any bandwidth? In *Proceedings of EURESCOM Summit 2005: Ubiquitous Services and Applications.* April.

[4] Kabel, J. 2003. Spam: A terminal threat to ISPs?. *Computer und Recht International* 1:1–5.

[5] Data Protection Working Party, Opinion 2/2006 on privacy issues related to the provision of e-mail screening services (WP 118).

[6] Sury, U. 2005. Datenschutzgerechte nutzung von intrusion detection— Organisatorische und juristische implikationen. *DuD* 29(7): 393–98.

[7] Rosenberg, J. 2002. SIP: Session Initiation Protocol. *RFC 3261.* June.

[8] Rosenberg, J., and C. Jennings 2008. The Session Initiation Protocol (SIP) and Spam. *RFC 5039.* January.

[9] Kuhn, D., T. Walsh, and S. Fries. 2005. Security considerations for voice over IP systems. Special Publication No. 800-58, NIST, USA.

[10] Sisalem, D., J. Kuthan, and S. Ehlert. 2006. Denial of service attacks targeting a SIP VoIP infrastructure: Attack scenarios and prevention mechanisms. *IEEE Network Journal* 20(5): 26–31. September–October.

[11] Croft, N., and M. Olivier. 2005. A model for spam prevention in voice over IP networks using anonymous Verifying Authorities. In *Proceedings of the 5th Annual Information Security South Africa Conference (ISSA 2005).* South Africa. July.

[12] Dantu, R., and P. Kolan. 2005. Detecting spam in VoIP networks. In *Proceedings of Steps to Reducing Unwanted Traffic on the Internet Workshop (SRUTI '05).* USA. July.

[13] Dritsas, S., J. Mallios, M. Theoharidou, G. F. Marias, and D. Gritzalis. 2007. Threat analysis of the session initiation protocol regarding spam. In *Proceedings of the 3rd IEEE International Workshop on Information Assurance (WIA 2007)*. USA. April.

[14] Srivastava, K., and H. Schulzrinne. 2004. Preventing spam for SIP-based instant messages and sessions. Technical Report, University of Columbia.

[15] Dongwook, S., A. Jinyoung, and S. Choon. 2006. Progressive multi gray-leveling: A voice spam protection algorithm. *IEEE Network Journal* 20(5): 18–24. September–October.

[16] Baumann, R., S. Cavin, and S. Schmid. 2006. Voice over IP—Security and SPIT. Swiss Army, FU Br 41, KryptDet Report, University of Berne, September.

[17] Peterson, J., and C. Jennings 2006. Enhancements for authenticated identity management in the session initiation protocol. *RFC 4474*. August.

[18] Tschofenig, H., R. Falk, and J. Peterson. 2006. Using SAML to protect the session initiation protocol. *IEEE Network Journal* 20(5): 14–17. September–October.

[19] Madhosingh, A. 2006. The design of a differentiated SIP to control VoIP spam. Technical Report, Computer Science Department, Florida State University.

[20] Niccolini, S. 2007. SPIT prevention: State of the art and research challenges. NEC Europe, Germany: Network Laboratories.

[21] Dantu, R., and P. Kolan. Detecting spam in VoIP networks. In *Proceedings of Steps to Reducing Unwanted Traffic on the Internet Workshop (SRUTI '05)*. USA. July 2005.

[22] Rohwer, T., C. Tolkmit, M. Hansen, M. Hansen, J. Moeller, and H. Waack. Abwehr von spam over internet telephony. (SPIT-AL), Kiel 2006. 1–30.

20

Anonymity in SIP

L. Kazatzopoulos, K. Delakourides and G. F. Marias
Athens University of Economics and Business

CONTENTS

20.1 Introduction 481
20.2 Motivation 482
20.3 Generalized Anonymity Architectures 484
20.4 Proposals for Anonymity in SIP 487
20.5 A New Proposal for Anonymity in SIP 488
20.6 Conclusions 495
     References 495

20.1 Introduction

In 1993, Peter Steiner published a cartoon in the *The New Yorker**, which illustrated a dog communicating over the Internet. The caption of this cartoon appears to be extremely prophetic for the Internet community, since it asserted that "On the Internet, nobody knows you are a dog". Actually, as far back as the early 1990s, this assumption was somehow valid, mainly because the communication intermediates or relays, did not care, worry, or ask if the end-user was actually a dog or a human. Nowadays, end-users, mainly humans, should employ technical and procedural means to defend against attackers that maliciously survey or spy on the Internet using network traffic analysis tools. This will protect the personal freedom and privacy, achieving digital dignity, and, moreover, defend confidentially in business, as well as in human relationships. That is why today we should say "On the Internet, nobody should know you are a dog".

In this scope, privacy and anonymity when communicating over the Internet gained substantial consideration in the technical, procedural and legal domain. For every new service that is launched and massively adopted in the Internet, privacy issues arise immediately. The same applies for VoIP services, and especially for SIP, which currently prevails in this new market. There are various reasons why an end-user wishes to maintain its anonymity when

---

* *The New Yorker*, July 5, 1993 issue (Vol. 69, no. 20) page 61.

communicating using SIP. Firstly, a caller might wish to conceal his identity from displaying in the receivers' phone, a feature that is usually used in mobile phone calls. On the other hand, a callee might want to be unlinkable from her/his personal preferences and direct marketing campaigns.

In its original specification, SIP supports anonymity, since the originator of a call could remain "anonymous" to the callee, and for that reason default values are used when the user agent initiates a call. This feature supports caller anonymity against the callee, but not to the entire set of SIP realms, since practically the user agent server of the serving domain requires strong authentication of the caller. Additionally, using tunnelling techniques, and especially end-to-end S/MIME encryption, selective anonymity can be supported. This option enables caller's privacy within the set of intermediate relays and the serving domains, if authentication is not required, but not against the callee. Finally, if network analysis tools are used in the network, then a malicious third parry can track the locations, using the caller's address-of-record fields. In such a case it could link address-of-records to physical locations, using data mining techniques, and finally with people, since there would be only a few people that make phone calls from particular residential addresses during a day. So, the question is: is total anonymity possible in SIP, and how could this be applied to shield the identity, or the character of a dog or a human?

20.2 Motivation

According to Justice Louis Brandeis, the right to privacy is "the right to be left alone"[†]. Additionally, Alan Westin identifies privacy as "the desire of people to choose freely under what circumstances and to what extent they will expose themselves, their attitude and their behavior to others"[‡]. Nowadays, we can define privacy in different domains, not vertical, but overlapped:

- Physical privacy—such as DNA searching.

- Information privacy—the unsanctioned invasion of privacy by the government, corporations or individuals in order to identify, or even handle, our personal information such as our age, address, market profiles, daily communications, or even sexual preference.

- Context privacy—each individual's fundamental right not to be linked with places, people, locations and preferences in his daily life because of surveillance cameras, sensor networks and RFID systems.

[†]L. Brandeis, S. Warren. 1890.
[‡]A. Westin. 1967.

Privacy is sometimes related to anonymity. According to Rana and Sharma [1] and the *Oxford English Dictionary*, anonymity is defined as "the state of being anonymous" which in turn is described as "nameless, having no name; of unknown name". This definition is vague, since in real-world implementations it should be clear which identity should be hidden and from whom. To better define anonymity, Pfitzmann and Kohntopp introduced the most common definition of anonymity used in the information and information community [2]: "anonymity is the state of being not identifiable within a set of subjects, the anonymity set". The anonymity set is a sensible metric since it associates the sender or receiver anonymity with a set of people and their actions. For instance the receiver anonymity set is the set of people who could have received a message intercepted by an attacker. Obviously the cardinality of the anonymity set is a measure of anonymity. A user is k-anonymous, or has k-anonymity, if he/she is one of at least k users within a specific anonymity set associated with a particular action. Recently, Serjantov and Danezis defined an information theoretic measure of anonymity [3]: each member of the anonymity set is assigned a probability equal to the likelihood that the member performed the anonymous action. Shannon's entropy is then used to quantify the level of uncertainty, or anonymity, that the members of the anonymity set achieve collectively with respect to their action.

To apply the anonymity set concept in SIP we should discriminate roles and actions. Even if various servers, intermediate proxies, and end-entities contribute on SIP, the set of actions, or service building blocks, that they contribute is actually very restricted. Subscription, registration, location (or redirection), call forwarding (or routing), call setup initiation-termination, and, optionally, authentication. Some additional, but proprietary services, such as anti-spitting, could also be reported. This set of actions normally is performed by the entities belonging into two district sets of service providers: those of the callee and those of the caller. Thus, if we consider a model in which an attacker wishes to reveal the identity of the calling parties, we can then define four legitimate parties in a SIP session: the caller, the callee, the service provider of the caller, and the service provider of the callee. In this direction we can define some privacy protection classes:

- Caller's absolute anonymity, in which the caller does not expose its identity to, or otherwise its identity cannot be exposed by, any other entity, or the attacker.

- Caller's anonymity only to the callee, in which the identity of the caller should be revealed only to the callee.

- Caller's anonymity only to her/his provider, in which the identity of the caller should be revealed only to the his/her provider.

- Caller's anonymity only to callee's provider; same as above, but for the peer's provider.

Except the first privacy class, the other three are not disjoint, and may coexist. In the following sections we will see how the existing SIP anonymity proposal and specifications deal with these four classes. We should mention here that the potential attacker might be one of the service providers or the callee, depending on the privacy protection class. For instance, the attacker might be a callee that aims to expose the name of any caller that wishes not to display his/her name to the peer party.

To support the above privacy classes, any anonymity architecture should make an attacker unable to distinguish between the occasions when a callee transmits or receives an SIP message and the occasions when she/he does not [4]. Additionally, it should take into account some of the idiosyncrasies of the SIP, such as:

- The SIP messages should not be delayed, e.g., for mixture.

- The sequence of SIP messages should not be violated.

- The traverse path of the SIP messages might be predetermined, according to service agreements between local, regional and national operators.

Moreover, any anonymity architecture should protect the physical location of the end-user. No one in the system, neither the system itself, will know from which point a user is connected. Even if the relation of the transmitted or received SP messages with a particular callee is not possible, the anonymity system should prevent attackers from linking the messages with physical locations. This will avoid the provable exposed conditions [4], whereas an attacker can prove the identity of the sender to others. For instance consider a user who decided to use anonymous SIP features. The UAC uses a meaningless URI, such as sip:thisis@anonymous.invalid [5]. If this meaningless URI is always used for this particular user, then it is possible to intercept SIP traffic, and connect this URI with different "Addresses-of-Record" (AoR). Then, using commercial or open source tools, the attacker will link these AoRs with physical locations, and then with end-users' identities.

20.3 Generalized Anonymity Architectures

To enhance or provide privacy in the internet services several privacy enhancement technologies (PET) have been proposed. Chaum's Mixes [6], Stop-and-Go Mixes and MixeNets [7], Crowds [4], Hordes [8], Onion Routing [9], and Mist [10] are some of the privacy preserving techniques.

Mixes [6] introduced the notion of anonymous digital communication. The Mix system provides unlinkability of sender and receiver. This ensures that while an attacker is able to determine that the sender and receiver actually sends or receives messages, they cannot determine with whom they are communicating. The system consists of special mix nodes which store, mix, and then forward the messages in transit. The sender predetermines the route of the message through one or more mix nodes using a well-defined protocol. A public key cryptography protocol is also used to ensure that any message cannot be tracked by an attacker as it passes through the mix network. In its simplest form (called a threshold mix) a mix node waits until it collects a number of messages as input. It uses its private key to reveal the address of the next mix node (or final destination) and reorders the received and buffered messages by some metric before forwarding them. In that sense, an omnipresent attacker cannot trace a message from its source to its destination without the collusion of the mix nodes. To provide a mix-network routing protocol, Kesdokan et al. introduced the Free Route and Mix Cascade concepts [7]. The former gives autonomy to the sender for dynamically choosing the trust path of the mix-nodes, while in the latter the routing paths are predefined. Mix networks introduce delays due to buffering and mixing and different padding patterns for mixing real with dummy traffic. Continuous Mixes try to avoid the delay issue by introducing fixed delay distributions for buffering and mixing. Mixes became subject to several attacks, such as timing attacks [11], statistical analysis of message distribution [12], statistical properties of randomly constructed routes [13,14], and packet flow correlation attacks [15,16].

Crowds [4] is a network that consists of voluntarily collaborating nodes. It is based on the idea that our anonymity can be protected better when we move ourselves within a crowd. According to [4], Crowds' web servers are unable to learn the true source of a request because it is equally likely to have originated from any member of the crowd. Even collaborating crowd members cannot distinguish the originator of a request from a member who is merely forwarding the request on behalf of another. In Crowds each user (browser) is represented in the system by a jondo process. Each message that needs anonymity enters into the Crowd node, its presence is announced via the local jondo, and is sent to another randomly chosen jondo with probability p, or to the actual server with probability $1 - p$. When the server (recipient jondo) receives the message it answers using the same, forward, path. Crowds can effectively face trace back attacks, and they can mitigate collusion attacks if the users randomly select the set of forwarding jondos.

Onion Routing [9] is an overlay infrastructure for providing anonymous communication over a public network. It supports anonymous connections through three phases: connection setup, data exchange, and connection termination. In the setup phase the initiator creates a layered data structure, called an onion, which implicitly defines the route path through the network. An onion is recursively encrypted using public key cryptography. The number of encryptions is equal to the number of the onion routes through which this structure should be delivered and processed towards the destination. The outer cryptographic control information refers to the first onion router in the path, while the inner cryptographic control information refers to the last onion router in the path (i.e., the predecessor of the destination). Each onion router along the route uses its public key to decrypt the entire onion that it receives. This operation exposes the embedded onion, and as a result, the identity of the next onion router. After decrypting a cortex each onion router then pads the embedded onion to maintain a fixed size, and sends it to the next onion router. Once the onion reaches the destination, all the inner control data are displayed as plaintext. This establishes the anonymous end-to-end connection, and then data can be sent in both directions. As data are routed through the anonymous end-to-end connection, each onion router removes one layer of encryption, so the data arrives as plaintext at the recipient. This layering occurs in the reverse order (using different algorithms and keys) for data moving backward. Connection tear-down can be initiated by either end, or in the middle if needed. All the messages (onions and real data) that transferred through the Onion Routing network illustrate identical sizes. The messages arrive at an onion router at fixed time intervals. They are mixed to avoid correlation by potential attackers. Additional cover traffics the semi-permanent connections between onion-routers and thus deludes external eavesdroppers. Onion Routing effectively resists traffic analysis.

Hordes [8] is an anonymity infrastructure that combines elements from Onion Routing and Crowds. It is the first protocol that uses multicast transmission, when the destination answers to the sender. It includes two phases, the initiation and the transmission phase. In the first phase, Hordes borrows the jondos idea from Crowds, and a public key scheme is used to add authentication services. In the first phase, called initiation phase, the sender sends a join request message to a proxy server. The proxy authenticates the sender, it returns a signed message that contains the multicast address of jondos, and informs the multicast group for the new entry. In the second phase, or transmission phase of a message, the sender selects a subset of jondos for the forwarding path and a multicast group address for the backward path. When data message is scheduled for transmission, the sender chooses a jondo member of the forwarding subset and sends this message to this peer as an encrypted onion data structure. The chosen jondo sends this message to another, randomly chosen, jondo with probability p, or to the receiver with probability $1 - p$, using encryption layers as well. The receiver replies on the backward path, and for that reason it sends an acknowledgment as plaintext message to the multicast group.

For the most of the aforementioned approaches, which are oriented mostly for e-mail and asynchronous web communication, there are some problems when directly adapting this approach to other types of Internet services, such as SIP. The first issue is related to latency. SIP call setup requests, encapsulated e.g., via the INVITE method, require somehow immediate respond, either positive or negative. This feature is not supported directly by the PET mechanism already presented. Additionally, these PETs do not support bidirectional communications, excluding the Onion Routing, a characteristic that is essential for every VoIP mechanism, such as SIP. Moreover, anonymity should be semantically supported. In that sense, the PET mechanism should support unlinkability of location that calls are initiated (or terminated) from SIP-registered URIs, or physical addresses (e.g., IP addresses). Most of the previously mentioned approaches support anonymity (of sender or receiver) in transit and do not have extensions to support the transactions unlinkability feature. Finally, TOR, the second generation Onion Routing service, might be immune to this location tracing attack, basically because it supports rendezvous points and location-hidden services.

A promising privacy system that overcomes these drawbacks is Mist. Mist [10] handles the problem of routing a message though a network while keeping the sender's location private from intermediate routers, the receiver and potential eavesdroppers. The system consists of a number of routers, called Mist routers, ordered in a hierarchical structure. According to Mist, special routers, called "Portals", are aware of the user's location, without knowing the corresponding identity, while "Lighthouse" routers are aware of the user's identity without knowing his/her exact location. We will discuss Mist in more detail in the next section.

When practical issues arise, proxy servers offer anonymity services on the World Wide Web. For instance the Anonymizer.com provides proxy services via rewriting URLs such that a link to e.g., http://www.you.gr might be rewritten as https://anonymity.proxy.net/www.you.gr. SSL encryption is used to ensure confidentiality of the connections between the end-user and the proxy server.

Another theoretical model for ensuring anonymity is the k-anonymity concept [17,18] which was earlier introduced in the context of relational data privacy. It addresses the question of how a data holder can release its private data with guarantees that the individual subjects of the data cannot be identified while the data remain practically useful [19]. Regarding LBSs and mobile clients, location k-anonymity refers to the k-anonymity usage of location information. A subject's location is considered k-anonymous if and only if the location information sent from a mobile client to LBS is indistinguishable from the location information of at least $k-1$ other mobile clients [20]. The location perturbation is an effective technique for supporting location k-anonymity and dealing with location privacy breaches exemplified by the location inference attack scenarios. If the location information sent by each mobile client is perturbed by replacing the position of the mobile client with a coarser grained spatial range such that there are $k-1$ other mobile clients within that range $k > 1$, then the adversary will have uncertainty in matching the mobile client to a known location-identity association or an external observation of the location-identity binding. This uncertainty increases with the increasing value of k, providing a higher degree of privacy for mobile clients.

20.4 Proposals for Anonymity in SIP

In the original specification of SIP the anonymity feature was juvenilely supported. To enable anonymity, the UAC should use in the "From" header field the display name "Anonymous", along with a syntactically correct, but otherwise meaningless URI (e.g., sip:an@anonymous.user). Additionally, tunneling encryption is suggested for anonymity. This is achieved by encrypting the header fields, and producing an outer new "From" header field that includes the "Anonymous" value in the display name subfield. This end-to-end encryption is not immune to location tracing attacks, since statistical analysis of sniffed data might prove the identities of the communicating parties.

A more recent draft RFC discussed privacy issues in SIP [21]. This RFC introduces guidelines for the creation of messages that do not reveal personal identity information, and a new "privacy service" logical role for intermediaries is defined to answer some privacy requirements. Additionally, an Internet draft proposal suggested the extension of the SIP that enables parties in an SIP session to be identified by different types of party information, which are authenticated by a trusted entity [22]. These trusted entities are delegated by end-users, and include the logic to reveal the identity of the calling or called party to peer entities. Trusted peers might receive information that identifies an end-user, if these entities are supposed to provide the same level of privacy, i.e., to reveal party information to other trusted peers. In addition to suppressing the delivery of party information, this draft proposal provides extensions to enable a party to obtain IP address privacy.

Sipanon is another SIP anonymity proposal that introduces an architecture two user agents (a User Agent Client and User Agent Server) that are coupled back-to-back (B2BUA) [23]. A message from the anonymous user's UAC is received by an Anonymizer's UAS. This message is then anonymized. The "From" header is changed and sent out from the Anonymizer's UAS to the remote UAC. The Anonymizers maintain a store of mappings between "real" SIP addresses and "anonymous" ones. This approach is very close to Chaum mixes solutions, and, thus, it inherits the corresponding drawbacks and vulnerabilities, such as clocking attacks.

20.5   A New Proposal for Anonymity in SIP

In this section we introduce the use of the Mist architecture in order to enable anonymity in SIP.

20.5.1   Mist at a Glance

Mist provides several advantages that led us to this decision. The key point of the Mist architecture is the distribution of knowledge. The "Lighthouse" routers are aware of the user's identity without knowing his/her exact location while the "Portal" routers are aware of the user's location, without knowing the corresponding identity. Furthermore, due to the decentralized Mist

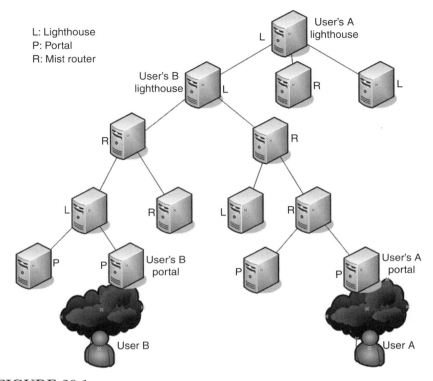

FIGURE 20.1

The Mist hierarchical structure.

architecture, a possible collusion between the aforementioned Mist routers is extremely difficult since the routers are unaware of each other's identity. It should be noted also that the architecture is applicable to a variety of network environments since it uses a general purpose protocol that can enable privacy on IP networks despite the underlying technology (i.e., Ethernet-based fixed networks, WLANs, 2/3G Mobile systems).

In short, the Mist architecture consists of a number of routers, called "Mist routers" ordered in a hierarchical structure. A typical structure is shown in Figure 20.1.

The leaf nodes of hierarchy are called Portals. Portals are the gateways that interface the virtual world with the physical one. In other words, they are connection points where users can connect to the Mist system. Let us assume that user A requires a network connection that ensures privacy and data confidentiality. User A has to register himself to the Mist system. His mobile device locates and interfaces directly with one of the available portals in the surrounding physical space. The portal, upon receiving a registration request, replies with a list of its ancestral Mist routers that exist at a higher level within the Mist hierarchy and are willing to act as a Lighthouse for the user. A Lighthouse is a Mist router that acts as a point of contact for user A. Users that intend to communicate with user A have to contact his Lighthouse.

Following Lighthouse selection, a virtual circuit (Mist circuit) must be established between user A and the corresponding Lighthouse. This process, called "Mist Circuit Establishment," aims to entitle user A Lighthouse to authenticate user A without revealing his/her physical location while hiding, at the same time, from the portal, user's identity and the designated Lighthouse. Furthermore, the Mist circuit applies a hop-to-hop handle-based routing technique for packet transmission between source and destination nodes and in combination with data encryption manages to conceal from the intermediary nodes information related to the identities and location of the communicating parties.

In order to establish a Mist circuit, user A generates a "Mist Circuit Establishment" packet and transmits it to the corresponding portal, without informing the portal of the selected Lighthouse. The portal, upon receiving the packet, assigns a special number, called handle ID, to the communication session with user A.

Thereafter, the portal encloses the assigned Handle ID to the received packet and forwards it to its Mist router ancestor. As the packet propagates through the Mist hierarchy, each Lighthouse router attempts to decrypt the payload using his private key. If the decryption fails, the particular router infers that it is not the recipient of this packet and thus, forwards it to the next router on the hierarchy. This process is repeated on each of the intermediate Mist routers until the packet reaches its final destination. In the case that decryption of the payload is successful, this indicates that the user selected the current Mist router to act as his Lighthouse. Finally, the Lighthouse answers back to user A and confirms the registration. From this point, a secure circuit is established through which user A can communicate securely with his Lighthouse. Note that even though the Lighthouse of user A can infer that his/her physical location is underneath Mist router "Y", it is very difficult to determine the exact position.

Following circuit establishment, the Lighthouse undertakes the role of representing the end user. An issue that has to be addressed is the detection of the user's Lighthouse. A public directory (i.e., LDAP server) or a Web server can be used for that purpose.

Let us assume now that user B intends to communicate with user A (Figure 20.2) and both of them have previously established a Mist circuit with LighthouseB and LighthouseA, respectively. User B transmits to his Lighthouse a packet indicating that he wants to set up a connection with user A. The LighthouseB verifies that the originator of the message is user B, locates the Lighthouse of user A and performs the initialization procedure for connection establishment with the Lighthouse of user B. As soon as the communication path is established, user A and user B are able to communicate. Note that the intermediate routers are unaware of the two ends of the communication. Moreover, it is impossible for user B to determine the location of user A and vice versa.

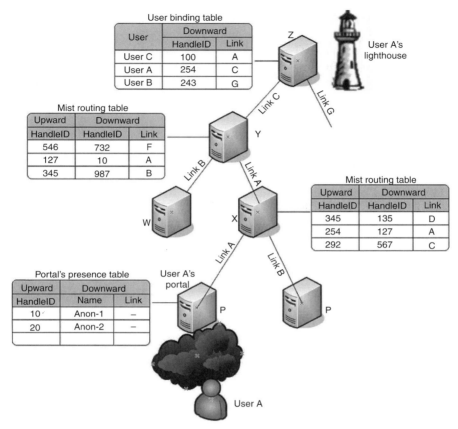

Mist routing table (upper left)

Upward	Downward	
HandleID	HandleID	Link
546	732	F
127	10	A
345	987	B

User binding table

User	Downward	
	HandleID	Link
User C	100	A
User A	254	C
User B	243	G

Mist routing table (right)

Upward	Downward	
HandleID	HandleID	Link
345	135	D
254	127	A
292	567	C

Portal's presence table

Upward	Downward	
HandleID	Name	Link
10	Anon-1	–
20	Anon-2	–

FIGURE 20.2

The Mist circuit establishment.

20.5.2 Applying MIST in SIP

By taking a closer look into the SIP protocol, someone can note that the home server (registrar, redirect or proxy server) has knowledge both for the user's ID and her current location. Our goal would be to distribute this knowledge to more than one entity. By doing so, it will be difficult for eavesdroppers or the home server itself to inference the user IDs.

Since on SIP, the user registers to the home server (with her ID) and the latter is the one that all SIP entities refer to in order to locate the former, which we will consider being the user's Lighthouse. Furthermore, we will consider as portals all the remote SIP servers that the user is connected to in order to establish communication through the SIP. Based on the above, we could say in general that each SIP server acts as Lighthouse for the users that have registered to it whilst as portal for the remote users that at some point can connect to the former.

In order to apply Mist to SIP, there should be some small modifications on the SIP.

The SIP location service is a LDAP directory that keeps the current position of each registered user. However on the RTTMist, the SIP location service is not required. The Mist user's binding table will be used instead. The latter keeps routing information about the Mist communication circuit with each user.

Furthermore, we consider that:

- Mist hierarchy has been deployed to the SIP network. As we mentioned earlier a Mist hierarchy considers that all Mist servers are ordered in a hierarchical structure.

- Public key infrastructure (PKI)

  - A PKI has been established and has created and distributed to Mist servers a pair of keys. Furthermore the keys are accessible from every Mist server.

  - Each user holds a pair of keys issued to his nickname

20.5.2.1   Registration Phase

Let assume that Alice has a 3G handset that allows her to place calls using an SIP-based phone service. When she triggers this service for the first time, her handset initializes the SIP registration routine and connects to the first available register SIP server. During this registration phase Alice is prompted for nickname and password. She is clever enough to hide her real identity by giving false data. She is recognized by the nickname "Mother" to all the friends that would like to call her in the future. From a Mist point of view, the register SIP server considers to be her Lighthouse. The Lighthouse will be the point of contact for the other SIP users in order to get in touch with her.

20.5.2.2   Mobility Issues

Alice is visiting a friend in the other part of town and she wants to be reachable but not traceable. She connects to the first available SIP server. From the Mist point of view, this is considered to be her portal. The next step is to set up a Mist circuit between Alice portal and Lighthouse. Note that the portal does not have to send to Alice a list with the available lighthouses since Alice knows that her lighthouse will be the SIP server where she was registered to (i.e., the registrar server of the SIP protocol). Thus the portal, instead of informing Alice about the available lighthouses, retrieves from the Mist LDAP directory information about the Mist network so Alice will be able to establish a Mist circuit with her Lighthouse. Accordingly, Alice's handset locates Alice Lighthouse and determines the available Mist circuit through which the Lighthouse is reachable. An open issue is the size of the data volume that the handset receives. It is impossible to receive all Mist links and try to determine the route to its Lighthouse.

20.5.2.3   Mist Circuit Establishment

Her handset encrypts a predefined message with the public key of the Lighthouse and forwards it to the portal. The latter follows the aforementioned Mist procedure and routes the packet to her Lighthouse. The Lighthouse upon receiving the packet stores the Mist circuit information to the user binding table. At this point, the Mist circuit has been established. The Lighthouse has knowledge of Alice's "name" (her pseudonym) but not her current position (just a way to get in touch with her).

20.5.2.4   Making a VoIP Call

Bob wants to call Alice. Note that both users should have established a Mist circuit with their Lighthouses and be able communicate through them. Bob is aware that Alice's nickname is "Mother". The procedure is illustrated in Figure 20.4.

1) Bob (who has registered with the nickname "Father") creates a MIST packet towards its SIP Lighthouse and encapsulates a SIP INVITE request for the user "Mother". He sends this up to the Mist hierarchy.

2) LighthouseBob receives the packet, determines the destination user and looks up in the Mist registry for the corresponding Lighthouse.

3) LighthouseBob creates a Mist packet towards the LighthouseAlice and encapsulates the INVITE Request that he received.

4) LighthouseAlice receives the packet, which is the SIP redirect server, determines that the called person is "Mother" and looks in the binding table to locate her.

5) Upon retrieving the Mist routing information, it creates a Mist packet with the SIP INVITE request and sends it to her through the Mist circuit.

6) Alice receives the packet, determines that it is an INVITE request from her friend Bob (she knows that his nickname is "Father").

7) Alice creates a SIP redirect packet to inform Bob about her current location, encrypts this message with Father's public key, and encapsulates everything in a Mist packet towards her Lighthouse. Note that the public key of Bob is based on his nickname. This is to enforce Bob's anonymity.

8) LighthouseAlice upon receiving the packet, determines that destination is LighthouseBob, encapsulates the content of Alice packet to a Mist packet and send it to LighthouseBob.

9) LighthouseBob forward the packet to Bob.

10) Bob, upon receiving, creates a SIP packet to acknowledge. At this point, Bob knows Alice's remote current address.

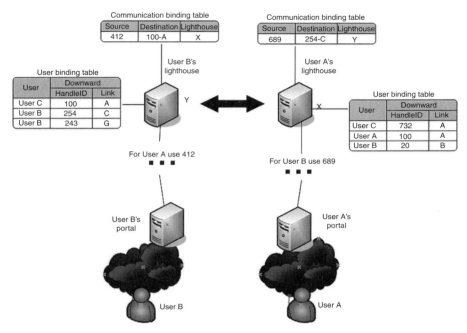

FIGURE 20.3
Establishment of a Mist communication path.

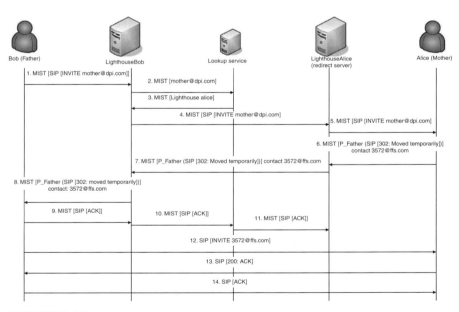

FIGURE 20.4
The message sequence chart when applying Mist in SIP.

11) Therefore the next step is to send directly to her an SIP INVITE request.

12) They both acknowledge. The SIP circuit is established.

Through the aforementioned procedure, the two users can place phone calls through the Mist without revealing their position. Therefore, it will be difficult for eavesdroppers to trace them and inference about their identity.

20.6 Conclusions

In many Internet applications, end-user privacy is essential. This avoids the identification of sexual, political, social and economic preferences of the individual and upgrades the digital dignity. The VoIP paradigm, and the SIP as its dominant architecture, is one of the Internet services that is expected to prevail. Solving the privacy concerns might accelerate its success and broadness. This section addresses these concerns and suggests a solution that might be useful for VoIP providers, network operators, or local network administrators. The proposed solution is based on the Mist privacy enhancement architecture. The research in this area is ongoing. Several issues are still open, and the proposed system should be further evaluated against privacy threats, vulnerabilities and potential attacks.

References

[1] Rana, S., and J. Sharma. 2006. *Frontiers of Geographic Information Technology.* Berlin: Springer; New York: Heidelberg Publishers.

[2] Pfitzmann, A., and M. Khntopp. 2004. Anonymity, unobservability, and pseudonymity: A proposal for terminology, draft v0.21, September.

[3] Serjantov, A., and G. Danezis. 2002. Towards an information theoretic metric for anonymity, in Paul Syverson and Roger Dingledine, editors, *Privacy Enhancing Technologies.* LNCS, San Francisco, CA.

[4] Michael K. Reiter, and Aviel D. Rubin. 1998. Crowds: Anonymity for web transactions. *ACM Transactions Information Systems Security* 1 (1):66–92.

[5] Rosenberg, J., H. Schulzrinne, G. Camarillo, A. Johnston, J. Peterson, R. Sparks, M. Handley, and E. Schooler. 2002. SIP: Session Initiation Protocol. *Request for Comments 3261, IETF.* June.

[6] Chaum, D. 1981. Untraceable electronic mail, return addresses, and digital pseudonyms. *Communications of the ACM* 4 (2): February.

[7] Kesdogan, D., J. Egner, and R. Buschkes. 1998. Stop-and-go MIXes providing probabilistic security in an open system, *2nd International Workshop on Information Hiding*. Lecture Notes in Computer Science 1525:83–98.

[8] Levine, B.N., and C. Shields. 2002. Hordes: A multicast-based protocol for anonymity. *Journal of Computer Security* 10 (3):213–40.

[9] Reed, M.G., P.F. Syverson, and D.M. Goldschlag. 1998. Anonymous connections and onion routing. *IEEE Journal on Selected Areas in Communications* 16 (4):482–94.

[10] Al-Muhtadi, J., R. Campbell, A. Kapadia, D. Mickunas, and S. Yi. 2002. Routing through the mist: Privacy preserving communication in ubiquitous computing environments. In *Proceedings International Conference of Distributed Computing Systems*, July.

[11] Øverlier, L., and P. Syverson. 2006. Locating hidden servers. In *Proceedings IEEE Symposium on Security and Privacy*, 100–14. May.

[12] Danezis. G. 2003. Statistical disclosure attacks. In *Proceedings Security and Privacy in the Age of Uncertainty*, IFIP Conference Proceedings, 250:421–26.

[13] Wright, M., M. Adler, B. Levine, and C. Shields. 2002. An analysis of the degradation of anonymous protocols. In *Proceedings ISOC Network and Distributed System Security*, 38–50. February.

[14] Shmatikov, V. 2004. Probabilistic analysis of an anonymity system. *Journal of Computer Security* 12 (3–4):355–77.

[15] Back, A., U. Moller, and A. Stiglic. 2001. Traffic analysis attacks and trade-offs in anonymity providing systems, In *Proceedings of the 4th International Workshop on Information Hiding*, LNCS 245–57.

[16] Serjantov, A., and P. Sewell. 2003. Passive attack analysis for connection-based anonymity systems. In *Proceedings of the 8th European Symposium on Research in Computer Security*, LNCS, 116–131.

[17] Samarati, P., and L. Sweeney. 1998. Protecting privacy when disclosing Information: k-anonymity and its enforcement through generalization and suppression. In *Proceedings IEEE Symposium. Research in Security and Privacy*, May.

[18] Samarati, P. 2001. Protecting respondent's privacy in microdata release. *IEEE Transactions Knowledge and Data Engineering* 13 (6):1010–27.

[19] Sweeney, L. 2002. k-Anonymity: A model for protecting privacy. *International Journal of Uncertainty, Fuzziness, and Knowledge-Based Systems* 10 (5):557–70.

[20] Gruteser, M., and D. Grunwald. 2003. Anonymous usage of location-based services through spatial and temporal cloaking. In *Proceedings*

ACM International Conference Mobile Systems, Applications, and Services (Mo-biSys '03).

[21] Peterson, J. 2002. A Privacy Mechanism for the Session Initiation Protocol. *Request for Comments 3323, IETF.* November.

[22] Marshall, W., K. Ramakrishnan, E. Miller, G. Russell, B. Beser, M. Mannette, K. Steinbrenner, D. Oran, F. Andreasen, J. Pickens, P. Lalwaney, J. Fellows, D. Evans, K. Kelly, and M. Watson. 2001. SIP Extensions for Caller Identity and Privacy. *Internet Draft, IETF.* November.

[23] Castleman, M. 2001. Sipanon: A SIP Anonymizer CS W3998. Columbia University.

21

Secure Intelligent SIP Services

Handoura Abdallah
Ecole Nationale Supérieure des Télécommunications de Bretagne

CONTENTS

21.1 Background 499
21.2 Introduction 500
21.3 Implementing Intelligent Network in VoIP Applications with SIP 501
21.4 Telecommunications Services Security with SIP 510
21.5 Conclusion 518
    References 518

21.1 Background

Intelligent network must become the standard of telephony services. However, the transition to Internet-based telephony services also provides an opportunity to create new services more rapidly and with lower complexity than in the existing public switched telephone network (PSTN) [1]. The Session Initiation Protocol (SIP) is a signaling protocol that creates, modifies and terminates associations between Internet end systems. The SIP uses DNS procedures to allow a client to resolve a SIP URI into the IP address, port, and transport protocol of the next hop to contact. Generally, these are the problems of mapping the name of a destination into an address, and to find the best route to the destination in a combined IP and PSTN network. Two protocols are being developed by the Internet Engineering Task Force (IETF) to solve these problems. The Telephony Routing over IP (TRIP) protocol solves the gateway location problem by distributing routing information between entities on the IP network. The tElephony NUmbering Mapping (ENUM) provides a solution to the terminal location problem based on DNS.

The intelligent network is a basis to establish and commercialize the services of the telecommunication network. Selling services and information through a network does not define solely a considerable increase of the sum of the information flowing through the network, but also a question of confidentiality and integrity. With the rapid evolution of wireless systems, the need for security is increasingly indispensable. This necessitates the advanced methods of cryptography and authentication, management and distribution of keys between the different entities on the system, while respecting constraints imposed by wireless systems, such that the capacity of the radio interface and resources of the terminal that creates a bottleneck that strangles this type of system.

This chapter describes the problem of implementing intelligent network services in VoIP applications with SIP (signaling protocol), TRIP (routing protocol over IP) and ENUM and proposes a method based on the integration of signaling protocol SIP on IN to realize a procedure of client authentication and data confidentiality.

21.2   Introduction

The architectural model of Internet telephony is rather different than that of the traditional telephone network. The basic assumption is that all signaling and media flow over an IP-based network, either the public Internet or various intranets. IP-based networks present the appearance at the network level that any machine can communicate directly with any other, unless the network specifically restricts them from doing so, through such means as firewalls. This architectural change necessitates a dramatic transformation in the architectural assumptions of traditional telephone networks. In particular, in a traditional network a large amount of administrative control, such as call-volume limitation, implicitly resides at every switch, and thus additional controls can easily be added there without much architectural change, whereas in an Internet environment an administrative point of control must be explicitly engineered into a network, as in a firewall; otherwise end systems can simply bypass any device which attempts to restrict their behavior. In addition, the Internet model transforms the locations at which many services are performed. In general, end systems are assumed to be much more intelligent than in the traditional telephone model; thus, many services which traditionally had to reside within the network can be moved out to the edges, without requiring any explicit support for them within the network. Other services can be performed by widely separated specialized servers which result in call setup information traversing paths which might be extremely indirect when compared with the physical network's actual topology.

Most of the services and service features of ITU-T Q.1211 can be provided by IETF's draft signaling standards for Internet Telephony, SIP [2] and, for some specialized features, its Call Control extensions [3].

ENUM provides the capability to translate an E.164 telephone number into a set of URIs using DNS. This capability has different uses depending on the applications being used and, in the case of voice services, the technology available at the source and destination of the communication.

For voice services, ENUM allows easy end-user identification and interworking between terminals on PSTN and IP-based networks. It may also allow for the implementation of more advanced services.

TRIP was engineered as a tool for inter-domain exchange of telephone routing information. However, it can also be used as a means for gateways and soft switches to export their routing information to a Location Server (LS), which may be co-resident with a proxy or gatekeeper. With these protocols is it possible to implant a voice services that will be exploited by an IP client or PSTN?

Some engineers have developed VoIP systems based on the SIP protocol; Astol, that interconnects Internet and Fax, Cisco Systems developed routers of SIP messages that interconnect to PSTN, Eurotelstat developed equipment-based SIP that enables call transfer, videoconference and voice messaging.

However, the growing globalization and liberalization of the telecommunications market necessitates a more global IN infrastructure to satisfy multinational subscribers. A lot of these services are offered on current, but specialized, systems. The IN concept is to provide these services on a coherent and stable basis [2]. Some security functions have already been introduced in current systems, but the end-user is constrained by private lines, proprietor equipment, and software algorithms.

There is a major difference between todays limited service and the original goals at the introduction of IN. IN originally offered open public access to user groups to communicate across open yet secure networks.

Therefore, the security services provided on the current network will not be sufficient for IN. New security functions need to be publicly usable, economically feasible and insured.

21.3 Implementing Intelligent Network in VoIP Applications with SIP

21.3.1 Interworking SIP and IN Applications

The IN functional architecture identifies several entities such as the service switching function (SSF), the service control function (SCF), and the call control function (CCF) that model the behavior of the IN network. The CCF represents the normal activities of call and connection processing in traditional switch, for two clients. The SSF models additional functionality required for interacting with a service logic program executed by the SCF. The CCF and the SSF are generally co-located in specific switching centers, known as SSP, while the SCFs are hosted in dedicated computers known as SCP. Communication between the different nodes uses the INAP protocol.

Accessing IN services from VoIP network requires that one or more entities in these networks can play the role of an SSP and communicate with existing SCPs via the INAP protocol.

To realize services with SIP it is important that the SIP entity is able to provide features normally provided by the traditional switch, including operating as an SSP for IN features. The SIP entity should also maintain call state and trigger queries to IN-based services, just as traditional switches do.

The most expeditious manner for providing existing IN services in the IP domain is to use the deployed IN infrastructure as much as possible.

The creation of a service by SIP is possible by the three methods INVITE, Bye and options and fields of header SIP Contact plus, Call-Disposition Replace and Requested-by that are extensions of SIP specified by IETF [4]. Some services are already specified by IETF [5] by using methods and fields of already quoted header: Call, Hold, Transfer of call, Return of call, Third party control.

The key to programming Internet telephony services with SIP is to add logic that guides behavior at each of the elements in the system [3]. In an SIP proxy server, this logic would dictate where the requests are proxied to, how the results are processed, and how the packet should be formatted.

The basic model for providing logic for SIP services is shown in Figure 21.1. The figure shows an SIP server that has been augmented with service logic, which is a program that is responsible for creating the services [18]. When requests and responses arrive, the server passes information up to the service logic. The service logic makes some decisions based on this information, and other information it gathers from different resources, and passes instructions back to the server. The server then executes these instructions [3].

The separation of the service logic from the server is certainly not new. This idea is inspired by the IN concept and also exists in the Internet. This separation enables rapid development of new services [14,17].

The mechanisms used in these environments can provide valuable insight into a solution for SIP IN integration. SIP does not operate the call IN model directly to access IN services. The machine of the entity SIP is bypassed with the layer IN such that the acceptance of call and the routing is executed by the basic states. The service machine is accessed by IN layers with the call IN model [7]. The model of service programming with SIP consists of an add-on SIP server and an IN layer, that manages the interconnection with the IN called SIP Intelligent Network (SIN) [8]. This operation necessitates

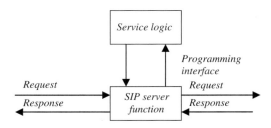

FIGURE 21.1

Model for programming SIP services. (From Schweizer, L. 2001. Scripts et APIs pour la gestion de serveurs SIP. www.tcom.ch; 23/12/2001. With permission.)

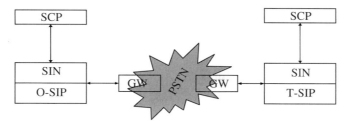

FIGURE 21.2
Interconnection SIP-IN.

the definition of a correspondence between the call IN and call SIP models, it is to tell a correspondence between the state machine of the SIP protocol (SIP defines the header Record_Route that allows to order the SIP server to function in mode with states until liberation of the call) and the state machine of IN. A call will be processed by the two machines, the state machine SIP processes the initiation of call and the final reply deliverance, and the IN layer communicates with the intelligent point SCP to provide services during call processing [8]. Figure 21.2 illustrates the integrated SIP-IN.

Similarly to the machine of states IN, one defines the quoted calling and called the entity (O-SIP) and (T-SIP) these entities correspond, respectively, to the O-BCSM and T-BCSM of the model IN.

In the basic SIP system, the SIP proxy with control of intelligent call is designed to interconnect with IN, this IN is realized by the utilization of significant control calls called 'state' that can synchronize with the model of call IN (BCSM). According to RFC 2543 (SIP) one can define calls of the state machine of the SIP client and SIP server [8,9]. The 11 PICs of O_BCSM come into play when a call request (SIP INVITE message) arrives from an upstream SIP client to an originating SIN-enabled SIP entity running the IN call model. This entity will create an O_BCSM object and initialize it in the O_NULL PIC. The next IN PICs—O_NULL, AUTH_ORIG_ATT, COLLECT_INFO, ANALYZE_INFO, SELECT_ROUTE, AUTH_CALL_SETUP, CALL_SENT, O_Alerting and O_Active, can all be mapped to the SIP 'calling' state. Figure 21.3 provides a visual mapping from the SIP protocol state machine to the originating half of the IN call model. Note that control of the call shuttles between the SIP protocol machine and the IN O_BCSM call model while it is being serviced.

The SIP calling protocol state has enough functionality to absorb the seven PICs. From the proxy SIP server, its initial state could correspond to the O_NULL PIC of the O_BCSM. Its processing state could correspond to the AUTH_ORI_ATT, COLLECT_INFO, ANALYSE_INFO, SELECT_ROUTE, AUTH_CALL_SETUP, CALL_SENT and O_Alerting PICs. Its success, confirmed and complete states could correspond to the O_Active PIC. Figure 21.4 provides a visual mapping from the SIP protocol state machine to the terminating half of the IN call model.

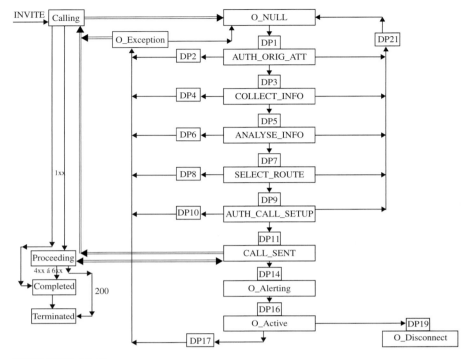

FIGURE 21.3

Mapping from SIP to O_BCSM.

When the SIP server of termination receives the message INVITE, it creates the T_BCSM object and initials it to the PIC T_NULL, this operation is realized by the 'proceeding' state that orders the five PICs: T_NULL, AUTH_TER_ATT, SELECT_FACILITY, PICs of the T_BCSM. Its calling state could correspond to the present_Call PIC. Its call processing state could correspond to the T_Alerting. Its complete state could correspond to the T_Active PIC. The service-level call flows for VoIP communication where interworking between the PSTN and IP-based networks are necessary to complete a call.

21.3.2 ENUM Call Flows for VoIP Interworking

ENUM (E.164 Number Mapping, RFC 3761) is a system that uses DNS (RFC 1034) in order to translate certain telephone numbers, into URIs (RFC 2396), e.g., sip:user@ domain. ENUM exists primarily to facilitate the interconnection of systems that rely on telephone numbers with those that use URIs to route transactions. E.164 is the ITU-T standard international numbering plan, under which all globally-reachable telephone numbers are organized. A telephone number cannot be routed in accordance with the traditional DNS resolution procedures standardized for SIP, which rely on SIP URIs.

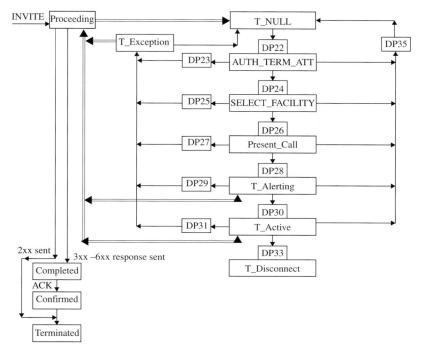

FIGURE 21.4
Mapping from SIP to T_BCSM.

ENUM provides a method for translating E.164 numbers into URIs, including potential SIP URIs. ENUM uses the domain 'e164.arpa' [5] to store the mapping. Numbers are converted to domain names using the schema defined in Figure 21.5. The E.164 number must be in full form, including the country code [5].

In the next section we describe the problem of locating terminals using E.164 numbers, and the problem of selecting a suitable gateway for calls from an IP telephony network to PSTN. Generally, these problems relate to mapping the name of a destination into an address, and finding the best route to the destination in a combined IP and PSTN network [16]. The Telephony Routing over IP (TRIP) protocol solves the gateway location problem by distributing routing information between entities on the IP network.

21.3.3 TRIP for Exporting Phone Routes

The number entered by the user, usually given as an E.164 number, an URI or e-mail, must be mapped to at least one routing address. For calls from IP telephony terminals to other IP telephony terminals, the destination host address must be found. For calls over the network boundary to the PSTN, a gateway must be located. In the opposite direction, from the PSTN to the IP

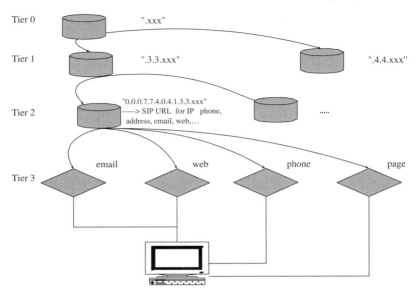

FIGURE 21.5
Different ENUM layer. (From Faltstrom, P. 2000. E.164 number and DNS;
Network Working Group, RFC 2916. With permission.)

network, a gateway must also be selected. The TRIP framework [10] divides
the problem into three subproblems:

1. Given a phone number corresponding to a specific host on the IP network, determine the IP address of the host.

2. Given a phone number corresponding to a terminal on the PSTN, determine the IP address of a gateway capable of completing calls to that
   phone. This is called the gateway location problem.

3. Given a phone number corresponding to a user of a terminal on the
   PSTN, determine the IP address of an IP terminal owned by the same
   user [10].

As the usage of ToIP grows and the number of gateways increase, the management of gateways and routes between the IP and PSTN becomes more
complex. In the situation where the IP address approaches the size of the
PSTN address, a large part of the calls will pass through one or even several
gateways on their path. The caller must locate a gateway for calls from the
IP network to the PSTN, the gateway is able to complete calls to the desired
destination. There may be several available gateways, and selecting the most
suitable one is not a trivial process. Currently the gateway must be selected
by the user or by the signaling servers.

TRIP uses the concept of Internet Telephony Administrative Domains
(ITADs) and location server (LS) nodes which exchange information with

other location servers. The location servers are administered by a single ITAD. TRIP distributes information between ITADs and contains functions for inter-domain synchronization of routing information. The decision about which gateway to use depends on many factors, including their availability, remaining call capacity, and cost for terminating to a particular number. For the proxy to do this adequately, it needs to have access to this information in real-time. This means there must be some communication between the proxy and the gateways to convey this information.

In TRIP, the routing tables are exchanged when they change updates sent and exchange the open message and send the 'keep alive' message frequently between peers. After connection, the LS exchange their entire routing table and incremental update messages are sent as the routing table changes. When synchronized, the TRIP routing table of all internal peers are identical. The routes external or internal or local are stored in the Telephony Routing Information Base (TRIB). The LS collect information and use it to reply to queries about routes to destinations. Any directory access protocol can be used, for example LDAP or DNS.

21.3.4 Implementing Intelligent Network Services in VoIP Application

The SELECT_ROUTE PIC allows selection of the pass route to the next server. The route can be according to the INVITE request, a URI or E.164 number. The ENUM standard described an architecture based on the DNS system corresponding to an E.164 number into the communication services identifier, with an order of priority (e-mail, URL of web site, address SIP server, etc). The ENUM protocol therefore finds the various addresses of the user from a simple telephone number. The user can also personalize who can join, with just an E.164 number. It is easy to add or to modify this supplementary information without changing the number used for the access. Thereby, the protocol is considered as a technical bridge of correspondence between the Internet and the PSTN, thus allowing the interoperability between these two systems [10]. DNS stores information in different types of records. The NAPTR record is used for identifying available ways of contacting a specific node identified by that name. ENUM defines a new service named E.164 to URI, termed 'E2U'. Before all, it is necessary to define and record the type of application in the DNS server. The following shows some example NAPTR records with the E2U service:

```
$ORIGIN 2.2.7.3.5.3.8.9.6.1.2.e164.arpa.
 IN NAPTR 10 10 "u" "sip+E2U" "!^.*$!sip:handoura@sip.tn".
 IN NAPTR 10 10 "u" "mailto+E2U" "!^.*$!mailto:handoura@enig.tn".
 IN NAPTR 10 10 "u" "http+E2U" "!^.*$!http://hand.enig.tn".
 IN NAPTR 10 10 "u" "tel+E2U" "!^.*$!tel:+216-98-353722".
 IN NAPTR 10 10 "u" "ldap+E2U" "!^+216(.*)$!ldap://ldap.tn/cn=01!"
```

And this shows specific SRV records added to your DNS for SIP:

```
IN A @IP or
IN SRV 0 1 5060 server.example.com
```

The sources of location information in the network (SIP client, phone, PSTN, IP) are:

- SIP REGISTER or SIP NOTIFY for client SIP registration.

- TRIP for location gateway or server.

- DNS A, SRV, NAPTR records for SIP server or ENUM services registration.

- LDAP or TRIB for local data bases.

Figure 21.6 shows our proposal allowing interconnection between PSTN, IN and IP.

Intelligent service gateway (ISG) is a gateway presenting the processor realizing the different transitions between the SIP state machine O, T and the O,T BCSM model. It is the SIN, presented in the preceding paragraph.

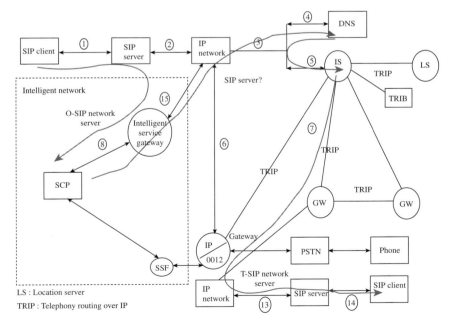

FIGURE 21.6

Interconnection PSTN-IN-IP and Services (in gray) invocation by a SIP client.

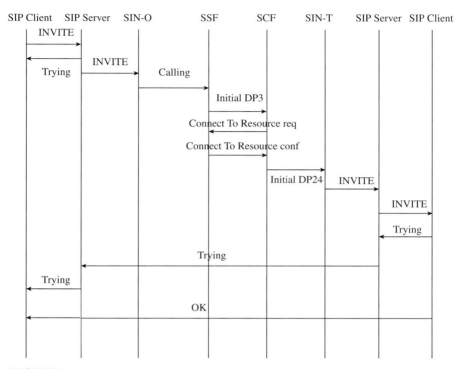

FIGURE 21.7
Diagram of to access IN services by SIP.

Figure 21.7 illustrates the following functions:

- Communication between SIP client and SIP client: classical communication over IP, when the TRIP can localize the GW through the LS or DNS: $1 + 2 + 3 + 4 + 5 + 7 + 13 + 14$.

- Communication SIP client and PSTN client: in this communication the NAPTR records with the E2U service (sip+E2U or tel+E2U) and makes the conversion SIP-PSTN or PSTN-SIP: $1 + 2 + 3 + 4 + 5 + 7 + 10 + 11$.

- To access IN services by a PSTN client: classical connection: $11 + 10 + 12 + 9$.

- To access IN services by a SIP client: this is the contribution of this work. Figure 21.6 illustrates this access.

The SIP client connects to the SCP as it is already shown. If the asked service is accepted, the IN broadcasts the request to the SIP client caller, the SIP client caller will be found with DNS or TRIP through the LS. The diagram to access IN services by SIP client is described in Figure 21.7.

21.4   Telecommunications Services Security with SIP

21.4.1   Threats of the Intelligent Network

As all elements of a network can be distributed geographically and that elements of system IP are totally opened, several security threats can rise and attack these different elements. The interconnection IN is the process to execute a demand of service IN on the part of the user through at least two different autonomous areas. Typically, each area represents a system IN separated with different legal entities and entities of resource IN. The IN service is only to be executed, if the two different areas cooperate by exchanging management, control, and services of data on the basis of a legal contract (co-contract of operation). In this competition, each area applies these clean mechanisms to provide the integrity, the availability, and the confidentiality of service user. If the two distribution operators cooperate, they have to apply the same totality of mechanisms to exchange data between their SCPs and SDPs.

The main security problems posed by multimedia communications and telephony over IP (in an IN under IP system) are as follows:

- Imitation of attack: unauthorized users can try to access IN services. For example, in the case of IMR service a user not recorded can try to view video services.

- Simulating attack: an authorized user can try to avoid security policy and illegitimately obtain access to sensitive services. For example, a user with common access privileges can try to act as an administrator of service IN.

- Denial of Service attack: an adversary can try to block users to access service IN. For example, by sending a great number of requests to the system simultaneously.

- Communication to spy and alter: an adversary can try to spy and/or to modify the communication between a legitimate user and elements of service IN.

- Lack of responsibility: if IN is not capable of verifing the communication between users and its elements of service, then it will not be possible to make users responsible for their actions.

This list is not exhaustive, in practice one can be confronted with other problems of security, not belonging to the area of application (for example, problems linked to the policy of security, to the security of management systems, to the placement in action of the security, to the security of the implementation, to the operational security or to the processing of security incidents). As the technological evolution of software increases in a manner not estimable, similarily the technological connection tools to the system and the attack passive and active also become inestimable. If the system and its components are not sufficiently secured, many threats can occur. However, it is necessary always to consider:

- What is the probability of a threat (occurrence; likelihood)?

- What are the potential damages (impact)?

- What are the costs to prevent a threat?

Depending on these suppositions, the cost and the efficiency of security mechanisms should be implemented. The potential of the threat depends on the implementation of IN and the specific service IN. They depend also on the implementation of security mechanisms (e.g., PIN, strong authentication, placement of authentication, management of key, etc).

Manufacturers as well as the standardization groups carry out analysis of risks in order to improve the security of systems IN. Although a lot of improvements have already been realized concerning the security of access to SCP and SMP, new services and new architectural concepts necessitate supplementary improvements. Figure 21.8 presents areas of different threats.

Intelligent networks are distributed by nature. This distribution is realized not only to the superior level where the services are described as a collection of service feature (SF), but also to lower levels, where the functional entities (FE) can be propagated on different physical entities (PE). As communication

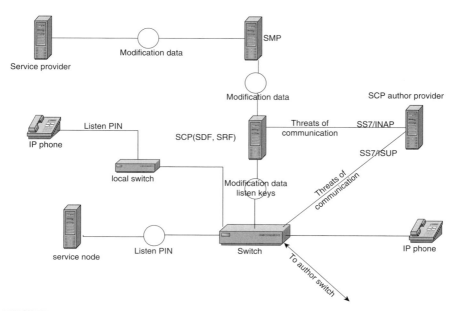

FIGURE 21.8
Areas of potential threats.

of various elements of the network are realized on the open and uncertain environments, security mechanisms need to be applied.

A responsible system can prevent these threats of security by using the various mechanisms. The authentication has to be applied in order to prevent unauthorized users gain access to the distribution of services.

21.4.2   Secure the Intelligent Network with SIP

SIP communication is susceptible to several types of attack. The simplest SIP attack permits assailants access to information on users identities, services, media and topology of distribution. This information can be employed and executed in other types of attack. The modification of attack occurs when an assailant intercepts the effort, covers the signal and modifies the SIP message in order to change a certain service characteristic. For example, this type of attack can be employed for diverted flow of signaling by forcing a particular itinerary or for changed recording of user or modifing a profile of service [11]. The attack type depends on the type of security (or lack of security) employed (the type of authentication, etc.). These attacks can also be employed for the denial of service.

The main two mechanisms of security employed with SIP are authentication and data encryption. The authentication of data is employed to authenticate the sender of message and insure that certain critical message information was not modified in transit. It has to prevent an assailant from modifying and/or listening to SIP requests and reply. SIP employs Proxy-Authentication, Proxy-Authorization, authorization, and WWW-Authentication in the letterhead, similar to these of HTTP, for the authentication of the terminal system by means of a numerical signature. Rather, proxy-par-proxy authentication can be executed by using protocols of authentication of the transport layer or the network layer such that TLS or IPSEC.

Data encryption is employed to insure the confidentiality of communication SIP, allowing only the destined client to decipher and read data. It is usual to use encryption algorithms such as the DES (Data Encryption Standard) and Advanced AES (Advanced Encryption Standard). SIP endorses two forms of cryptography: end-to-end and hop-by-hop. The end-to-end encryption provides confidentiality for all information (some letterhead and the body of SIP message) that needs to be read by intermediate servers or proxy. The end-to-end encryption is executed by the mechanism S/MIME (Figure 21.10). On the contrary, the hop-by-hop encryption of whole SIP message can be employed in order to protect the information that would have to be accessed by intermediate entities, such as letterhead From, To and Via. Securing such information prevents malevolent users determining the calls or accessing the itinerary information. The hop-by-hop encryption can be executed by external security mechanisms to SIP (IPSEC or TLS).

As we conclude, it is necessary to consider the efforts of standardization processing in the improvement of security mechanisms for SIP. The important issue is the problem in agreement on the chosen security mechanism between two SIP entities (user agents and/or proxy) that want to communicate by applying a level of 'sufficient' security. For this reason, it is very important to define how an SIP entity can select an appropriate mechanism to communicate with a next proxy entity. One of the proposals for an agreement security mechanism that allows two agents to exchange their clean security aptitudes of preferences in order to select and apply a common mechanism is one where a client initiates a procedure, the SIP agent includes in the first request sent to the neighbor proxy entity, the list of its mechanisms of security sustained. The other element replies with a list of its clean security mechanisms and its parameters. The client then selects the preferred common security mechanism and then makes use this chosen mechanism (e.g., TLS), to contact the proxy by using the new mechanism of security [12].

With this technique, another problem is identified. It is the problem of the assertion and validation of the user identity by SIP server. The SIP protocol allows a user to assert its identity by several manners (e.g., in the letterhead); but the information of identity requested by the user is not verified in the fundamental SIP operation. On the other hand, an IP telephony client could have initially required the identity of a user in order to provide a specific service and/or to condition the type of service to the user identity. The model of SIP authentication could be a way to obtain such identity; however, the user agents do not always have the necessary key information to authenticate with all the other agents. A model is proposed by Profos [11] for 'confirmed identity' and is based on the concept of a 'confirmed area'. The idea is that when a user authenticates its clean identity with a proxy, the proxy can share this authenticated identity (the confirmed identity) with all the other proxies in the confirmed area. A confirmed area is a totality of proxies that have a mutual configuration of security association. Such association of security represents a confidence between proxies. When a proxy in a confirmed area authenticates the identity of the author of a message, it adds a new letterhead to the message containing the confirmed identity of the user. Such an identity can be employed by all other proxies belonging to the confirmed area.

Using this mechanism the client UAC, is capable of identifying himself to a proxy UAS, to an intermediate proxy or to a registration proxy. Therefore, the SIP authentication is applied only to the end-to-end or end-to-proxy communication; the authentication proxy-by-proxy would have to count on other mechanisms such as IPsec or TLS.

The procedure of authentication is executed when the UAS, the proxy intermediate, or the necessary recording proxy for the call of the UAC is authenticated before accepting the call or the recording. At the start of the message the UAS sends a request of SIP message (e.g., INVITE). On receipt of this message, the UAS, proxy, or proxy of recording decides that authentication is necessary and sends the client a specific SIP error message for the request of authentication. This message of error represents a challenge. When the message error 401 (Unauthorized) is sent by UAS and recording, the message error 407 (Proxy Authentication Required) is sent by proxy sever. The UAC receives the error message, calculates the reply, and includes it in a new

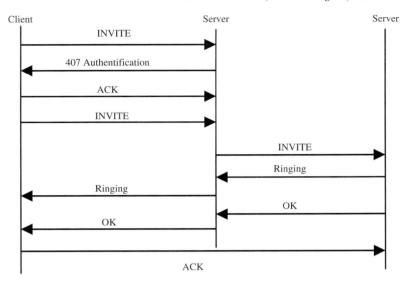

FIGURE 21.9
Authentication SIP.

message of SIP request. Figure 21.9 shows the sequence of message for the case of request of authentication by the proxy server. Noting that, the UAC sends a message ACK immediately after the message of error is received. This message closes the first transaction; then the second message INVITE opens a new transaction.

21.4.3 Application of Secure VoIP

The VoIP service allows users connected to an Internet supplier (ISP) to realize calls in PSTN. This scenario has the advantage that the ISP already has a security report from its client. It is 'standard', that the ISP offers this service to the client in addition to Internet access. An SIP proxy server in the system ISP will be configured as a server band of proxy for client SIP in the system ISP. This proxy server dispatch of calls to the proxy server of the Internet telephony service provider (ITSP), will select and contact the gateways appropriate to SIP.

In this scenario, a possible realization of a security mechanism for the calls is as follows: The proxy server ISP employs the proxy authentication procedure of SIP to authenticate the user to call. The ISP authenticates one of its clients. Once the user is authenticated by the proxy server, the proxy verifies if the user is authorized to make a call. If it is the case, the proxy contacts the proxy server ITSP by sending INVITE.

VoIP is essentially an IP system and therefore suffers the same threats inherent with all IP systems [13].

Research and development has been employed to solve the problem of the security protocols for specific applications on the IP system (the layer SSL

```
INVITE sip:xy@domain SIP/2.0
Via: SIP/2.0/UDP
To: xy <sip:xy@domain>
From: yx <sip:yx@domain>
Call-ID:
CSeq: 1 INVITE
Contact: <sip:yx@domain>
Content-Type: multipart/signed;boundary=…;
micalg=sha1;protocol=application/pkcs7-signature
Content-Length:
Content-Type: application/pkcs7-mime;
smime-type=envelopeddata; name=smime.p7m
Content-Disposition: attachment;handling=required;filename=smime.p7m
Content-Transfer-Encoding: binary
<envelopedData object encapsulating encrypted SDP attachment not
shown>
Content-Type: application/pkcs7-signature;name=smime.p7s
Content-Disposition: attachment;handling=required;filename=smime.p7s
Content-Transfer-Encoding: binary
<signedData object containing signature not shown>
```

FIGURE 21.10
S/MIME encryption in SDP.

that is mainly employed in e-commerce and (PGP), employed for e-mail). There are already security protocols that could be employed for VoIP. However, the exigencies in the resource and security characteristics of voice circulation (and the multimedia data circulation of a general manner) are different from traditional IP circulation. In real-time application, VoIP is highly sensitive to the distribution and delay of packets. The delay is acceptable for an approximate quality in attacking a speech, that does not exceed 150 ms. Delay Codec, delay of serialization, delay of propagation, etc. all contribute to general packet delays, which could be variable and depends on the dynamics of the system. Authentication and algorithms of key exchanges are by nature an intensive calculation and when employed for VoIP add delay. More specifically, the public key cryptography (employed for the key exchange) necessitates very high calculation. The choice has to be made such that the security protocol does not affect the packet that processes the delay beyond acceptable limits, with a negative impact on the quality of the transmission of voice. An option is to employ the end-to-end encryption, so that the calculation demand is largely distributed. However, the problems occur with system management, the key exchange, investment and the service cost. The other option is to employ security services to the edge of the routers, the gateway, etc.

With already quoted techniques and demands in terms of security, we suggest a physical structure for implementing a security platform of network and intelligent services in Figure 21.11 and Figure 21.12.

SAO: Operator Server authorization.
SOS: Service server authorization.
SAC: Service Access Code

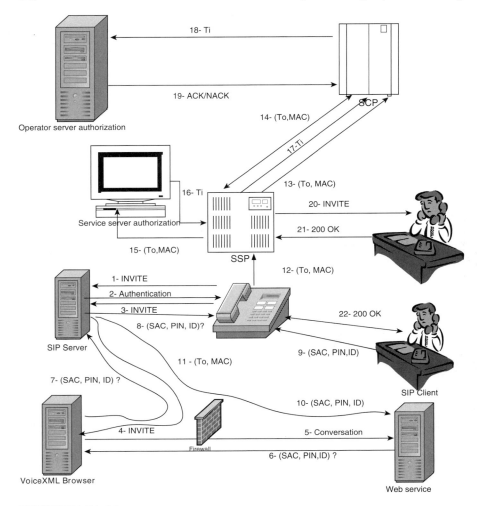

FIGURE 21.11
Service telecommunication security with SIP.

PIN: Personal Identification Number
ID: User ID
To: Operator Ticket
MAC: Message Authentication Code
Ti: Service Ticket

F1: SIP INVITE with authentication and ID
F2: 407 Authentication Proxy
F3: ACK
F4: INVITE a session with a service

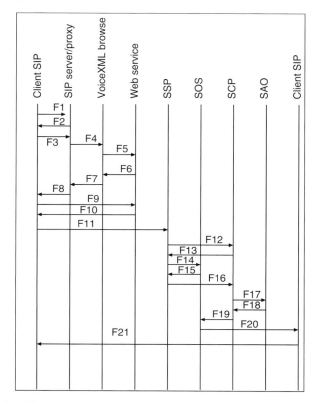

FIGURE 21.12
Diagram of intelligent network secure.

F5: http://service1.dialog.vxml [9]
F6: Give your identifiers (SAC, PIN, ID)
F7: Give your identifiers (SAC, PIN, ID)
F8: Give your identifiers (SAC, PIN, ID)
F9: Here is my identifiers
F10: calculation of To and MAC
F11: To and MAC transmitted to SSP
F12: then to SCP
F13: validation or no
F14: If validated ; send to SOS
F15: calculation of Ti
F16: Ti to SCP
F17: verification of Ti by SOA
F18: Validation or not of Ti
F19: If validated, interrogate SSP.
F20: INVITE
F21: OK.

21.5   Conclusion

The approach of integration of SIP and IN permits the use of intelligent services already deployed from same states described in the BCSM model by IP client. It is very easy to APIs (Application Programming Interfaces) implementation for access to services used in the traditional PSTN model.

Nevertheless there are some problems to resolve, such as the choice of SCP to put in communication with the SIP server and also the security problem in IN or more precisely, the authentification of the caller and the callee.

With the introduction of intelligent functionality in telecommunications systems, the necessity for security functions is emphasized. Integrating the application of this functionality will be important to give a clear view to each user that information and orders are processed correctly, and that information is processed solidly in the system. Of course, there is a long list of functionality demands for services of security, but the main demand list is given from the user viewpoints. Therefore, in order to market IN as a technical and economic success, we have to make good work not only in our laboratory development and in the adminstration of telecommunications, but also to establish a level of human comprehension.

References

[1] Schulzrinne, H. 1998. *Signalling for Internet Telephony*, Columbia University, Dept. of Computer Science Technical Report Number CUCS-005-98

[2] Schulzrinne, H. 2001. The Session Initiation Protocol (SIP). Columbia University, New York, http://www.cs.columbia.edu/hgs. May 2001.

[3] Johnston, A., A. Sparks, R. Donovan, S. Summers, K. 2002. SIP service examples. draft-ietf-sip-service-examples-03.txt, June.

[4] Fondation internet nouvelle génération (FING). 2001. Principes et conditions de mise en oeuvre du standard ENUM en France. http:// www.art-telecom.fr/publications/index-cp-enum.htm. 18 June 2001.

[5] Huston, G. 2002. ENUM—Mapping the E. 164 Number Space into the DNS. The Internet Protocol Journal. Volume 5, number 2. Page 13–23, June.

[6] Chapron, J., and B. Chatras. 2001. An analysis of the IN call model suitability in the context of VoIP. *Computer Networks* 35:521–35.

[7] El ouahidi, B. 2000. Internet/Telecommunication integration: Towards IN-capable SIP network; Revue internationale: Calculateurs parallèles réseaux et systèmes repartis. *Hermès-Editions, Edition spéciale ingénierie des services de télécoms* 12 (2): 259–280.

[8] VK. Gurbani. 2002. Interworking SIP and intelligent network (IN) applications; draft-gurbani-sin-02.txt, June 2002.

[9] EEEurescom Project P916 Supporting of H323 by IN, Providing IN functionality for H323 telephony calls, October 2000

[10] Nicklas, B. TRIP, ENUM and Number Portability, Networking Laboratory. HUT, Finland: Helsinki University of Technology.

[11] Profos, D. Security requirements and concepts for Intelligents networks. *Ascom Tech AG Bielstrasse* 122: 4502 Solthurn.

[12] Jennings, C., J. Peterson, and M. Watson. 2002. Private Extensions to the Session Initiation Protocol (SIP) for Asserted Identity within Trusted Networks. draft-ietf-sip-asserted-identity-01 June 2002.

[13] Ranganathan, M.K., and L. Kilmartin 2003. Performance analysis of secure session initiation protocol based VoIP networks. *Computer Communications* 26:552–65.

[14] Van Leekwijck W., and D. Brouns. 2000. Siplets, la programmation de services de téléphonie sur IP avec Java. *Revue des Télécommunications d'Alcatel* - 2e trimestre.

[15] Faltstrom, P. 2000. E.164 number and DNS; Network Working Group, RFC 2916.

[16] Lind, S., and AT & T. 2002. ENUM call flows for VoIP interworking. Proceedings of the fifty-second Internet Engineering Task Force, salt Lake City, Utah, December 9–14.

[17] Schweizer, L. 2001. Scripts et APIs pour la gestion de serveurs SIP. www.tcom.ch (accessed December 23, 2001).

[18] Rosenberg, J. Lennox, J. Schulzrinne, H. 1999. Programming internet telephony services, Columbia University technical report CUCS-010-99.

22

SIP Security and Quality of Service Performance

Johnson Ihyeh Agbinya
University of Technology

CONTENTS

22.1 Introduction 521
22.2 QoS Performance of SIP 522
22.3 SIP Security 548
    References 559

22.1   Introduction

Session Initiation Protocol (SIP) is a new application level protocol which was developed to support the exchange of information that is required for the establishment of media sessions in IP networks. It supports mobility at the application layer through its capability to establish and tear down sessions. Its main function is for creating, establishing and tearing down media sessions between one or more SIP end points. The end points have valid SIP addresses and form part of the information that is exchanged during SIP signaling. The end points may reside either in IPv4 or IPv6 domains, and Application Level Gateways (ALG) have been developed for the translation of the addresses inside the SIP messages.

SIP targets the enabling of emerging IP services including Voice over IP (VoIP), Internet telephony, Presence, IPTV and instant messaging (IM). Although SIP does not carry the media itself, it works closely with other protocols such as Real-Time Transport Protocol (RTP), which is used to transport the media, RTP Control Protocol (RTCP) which is used for the management and control of the data stream, Real-Time Streaming Protocol (RTSP), session description protocol (SDP) is used for specifying the parameters of the media sessions (equivalent to the H.245 Call Control in H.323 Protocol), TCP and UDP.

Securing SIP by itself is not easy since each session that is established has to be secured. In a highly mobile network environment when either the terminal or the network is mobile, the problem is compounded. Securing SIP means securing the dynamic links, signaling and data involved.

The traditional PSTN is relatively secure and provides a level of quality of service (QoS) that subscribers are used to. Hence when SIP is used in IP networks, subscribers will demand, as a minimum, the same levels of security and QoS. However, SIP has not matured enough to support that form of security and QoS. No similar solutions exist for providing the security mechanisms that subscribers are familiar with over PSTN and Internet. There are however several proposals that have been published that are aimed at addressing the identified security concerns in SIP. This chapter provides a summary of the QoS performance of SIP, SIP security issues and the proposed solutions.

22.2   QoS Performance of SIP

SIP was not developed originally with QoS considerations in mind. SIP is never used in isolation. It has to work in agreement with other QoS determining protocols such as DiffServ and real-time protocols such as RTP, RTCP, RTSP and RSVP. The IETF has provided several QoS solutions in IntServ, RSVP and DiffServ. The RSVP is an end-to-end QoS solution but it is relatively hard to design due to its high cost. RSVP cannot provide guaranteed QoS in a wireless environment because of the dynamic links and significant bit error rates (BER) that result. Therefore, the QoS performance of SIP is intricately woven to the QoS performances of these protocols. The QoS performance of SIP alone can be measured by how fast it establishes and maintains sessions through its signaling mechanisms under TCP (UDP) and either IPv4 or IPv6. This section summarizes these two approaches to the QoS performance of SIP. To handle QoS, extension to SIP proxy server is normally undertaken [1].

22.2.1   Measuring Session Based QoS Performance

The key QoS performance measures for SIP sessions include communication time, handoff delay when mobile nodes are involved in and between neighboring subnets, data volume exchanged during the SIP session time, jitter and BER. For real-time traffic, these parameters refer to end-to-end performance. In non-real-time traffic, guaranteed bit rates are used. An SIP session normally consists of several ON and OFF periods. Most data types manifest these times.

Three data types are considered, audio, data and video. For real-time traffic such as VoIP, the ON time refers to talk periods and the OFF time refers to silence periods. In the case of data, the ON time refers to the download time and the OFF time refers to data reading time. When video is considered, only the ON time is necessary.

To provide a means for comparing different implementations and SIP based platforms, this section provides a set of parameters and expressions. Based on these definitions, in one flow of a multimedia TCP session, the following parameters apply in defining SIP QoS performance following the example of Hassan [1]. Consider that arrival time for the traffic can be modeled as a Poisson process with the following parameters [1]:

D: Session rate per second (session/sec)
N: Session load (number of simultaneous sessions)
T: Average mean duration of one session (sec)
ρ: Average rate of one session (kbps)
L: Average rate of one flow (kbps)
N_p: Average number of ON–OFF patterns in one session
T_{on}: Average duration of ON period (sec)
T_{off}: Average duration of OFF period (sec)
R_{on}: Average rate in ON period (kbps)
Q: Average file size in ON period (kb)

The ON period for data transfer is a function of the data rate and the average file size used during the ON period. This is given by the expression

$$T_{on} = \frac{Q}{R_{on}} \qquad (22.1)$$

Several authors have modelled the ON and OFF times. The ON and OFF durations for Web and FTP sessions have Pareto distributions.

Normally the download rate is a function of the quality of the network or the level of activity in the network. At some point, the download speed will be very high and at other times could grind to a very low value when the network is busy. Hence the average download rate is used. The average download rate can be calculated based on a statistical average of per minute rates accumulated over a length of time. Consider that a session consists of several ON and OFF patterns. That being the case, the average data transferred during a session is given by the product of the data transfer rate and the number of ON and OFF patterns:

$$\alpha = Q * N_p \qquad (22.2)$$

Let the number of active sessions be given by the expression

$$N = D * T \qquad (22.3)$$

and the duration T of one session is

$$T = N_p(T_{on} + T_{off}) \qquad (22.4)$$

The data flow rate has an average value L given by

$$L = N * d \qquad (22.5)$$

The session data transfer rate can therefore be estimated with the expression:

$$\rho = R_{on} * \frac{T_{on}}{T_{on} + T_{off}} \qquad (22.6)$$

Apart from the level of activity in a network which determines the return trip time (RTT), the rate at which packet acknowledgments is performed impacts the average rate of a TCP connection. The approach used by Hassan [1] is adopted here. It considers that acknowledgment is provided for every two packets received. This leads to the expression:

$$T(Q) = \text{RTT} * \log Q \tag{22.7}$$

When long TCP connections are involved, this expression should be substituted with the maximum of TCP rate permitted by the source.

In the following sections we discuss the QoS performance of SIP from various delay perspectives.

22.2.2   QoS and SIP

Quality of service is a term that has been loosely used in telecommunications to refer to network [22] parameters that impact the performance of the system. In telecommunications, QoS refers mostly to the network performance variables such as delay, BER, packet loss rate, interference, bandwidth, system capacity and throughput. In a dynamic network environment the definition includes the velocity of the nodes, the length of the links between nodes and the characteristics of the links. In a multimedia situation, it includes the quality of the data such as the quality of the audio content, the mean opinion scores in VoIP systems and for IPTV where video is involved to the quality of the video (images), compression quality and the frame rate. Thus, the definition of QoS includes quite a lot of system parameters and is dependent on the application in question. In this chapter, a subset of these system parameters that typically affect SIP performance is considered. This includes delay in its various forms in SIP networks including set up delay, signaling delay, call set up delay, queuing delay, message transfer delay, session setup delay, handoff delay, disruption time and throughput.

The SIP-T signaling mechanism is used to facilitate the interconnection of PSTN with carrier-class VoIP network. Its major attributes are the promises for scalability, flexibility, and inter-operability with PSTN. It also provides the call control function of a Media Gateway Controller required for set up, tearing down and managing VoIP calls in carrier-class VoIP networks. The performance of the SIP-T signaling system has a direct impact in optimizing network QoS. The QoS performance attributes include the queue size, mean of queuing delay and the variance of queuing delay. To represent the SIP-T signaling system, Wu [2] assumed an M/G/1 queue with non-pre-emptive priority assignment. Formulae for queuing size, queuing delay and delay variation for the non-pre-emptive priority queue from queuing theory, respectively were presented and readers are urged to check this work. The number of SIP-T signaling messages in the queue is called the queuing size and the queuing delay is the average delay of messages in the queue. Kueh [3] also evaluated the performance of SIP-based session setup over satellite-UMTS (S-UMTS) taking into account the larger propagation delay over a satellite link as well as the contribution of the UMTS radio interface. In the following paragraphs we discuss similar works in more details.

22.2.2.1 Queuing Delay Performance

Queuing delay is a result of queuing size and is one of the several sources of delay in SIP operations. This was analyzed by Wu and Wang [2] using imbedded Markov chain and semi-Markov processing while queuing delay and delay variation was analyzed using the Laplace-Stieltjes transform (LST) [4]. Queuing size is defined as the number of SIP-T messages in the system. Table 22.1 and Table 22.2 give the comparative views between the mathematical and simulation results including buffer size, mean queuing delay and standard deviation of queuing delay for error probability of a SIP-T message, $p_e = 0.004$. Mean queuing delay and standard deviation of queuing delay vary slowly as a function of the queue size for SIP-T messages arrival rate of less that 450. However, these values increase dramatically when the arrival rate of SIP-T messages exceeds 450. This phenomenon is due to the SIP-T message arrival rate approaching the processing capability of the system for heavy traffic intensity [2]. It can be observed that the theoretical estimates are in excellent agreement with simulation results. Therefore we can determine the cost to performance planning and design compromises needed to meet the requirements of a carrier class VoIP network.

22.2.2.1.1 Call setup delay Call setup delay consists of the post-dialing delay, answer-signal delay and call-release delay. The post dialing delay occurs when signaling takes place through the network to the receiver. These delays happen during the lifetime of a SIP call. A measure of these delays was

TABLE 22.1
Mathematical Results

SIP-T message queue size (messages/s)	Mean (ms)	Standard deviation (ms)	Buffer size
50	0.49	0.49	0.12
100	0.62	0.68	0.23
150	0.78	0.89	0.37
200	1.00	1.14	0.52
250	1.29	1.46	0.72
300	1.71	1.90	0.97
350	2.37	2.57	1.33
400	3.57	3.75	1.92
450	6.39	6.41	3.17
500	25.00	18.57	8.20

Source: Wu, J.S., and P.Y. Wang. 2003. The performance analysis of SIP-T signaling system in carrier class VoIP network. *17th International Conference on Advanced Information Networking and Applications (AINA'03), IEEE.* 39. March.

TABLE 22.2
Simulation Results

SIP-T message queue size (messages/s)	Mean and 95th percent (ms)	Standard deviation (ms)	Buffer size	Sample size (SIP-T message)
59.2	0.49 ± 0.01	0.47	0.12	4925
100.7	0.61 ± 0.01	0.65	0.25	10074
150.4	0.79 ± 0.01	0.88	0.39	15044
201.8	0.98 ± 0.02	1.10	0.54	20178
251.4	1.27 ± 0.02	1.40	0.76	25138
301.1	1.70 ± 0.02	1.91	1.03	30115
348.9	2.36 ± 0.03	2.57	1.40	34891
399.5	3.62 ± 0.04	3.98	1.97	39955
450.7	6.49 ± 0.06	6.67	3.49	45067
502.5	21.08 ± 0.18	20.12	8.85	50251

Source: Wu, J.S., and P.Y. Wang. 2003. The performance analysis of SIP-T signaling system in carrier class VoIP network. *17th International Conference on Advanced Information Networking and Applications (AINA'03), IEEE.* 39. March.

undertaken by Cursio and Landan [5] using a 3G network emulator. The results account for local, national, international and overseas Intranet LAN calls.

Post-dialing delay (PDD) is also called post-selection delay or dial-to-ring delay. This is the time between when the caller clicks the button of the terminal to call another caller and the time the caller hears the callee terminal ringing.

Answer-signal delay (ASD) is the time between when the callee picks-up the phone and the time the caller receives indication of this. This delay is obviously a function of the depth and extent of the network infrastructure between the caller and callee. In a local call scenario, this time is relatively short. For international and long distance calls is can be significant.

Call-release delay (CRD) is the time elapsed between when the releasing party (the caller) hangs-up the phone and the time the same party can initiate/receive a new call. Table 22.3 is a summary of these delays:

TABLE 22.3
3G SIP Calls vs. Intranet Calls Delay Results

PDD	24 ms	38 ms	153 ms	24 ms	62 ms
ASD	23 ms	31 ms	147 ms	237 ms	45 ms
CRD	11 ms	30 ms	138 ms	230 ms	50 ms

Source: Curcio, I.D.D., and M. Lundan. 2002. SIP call setup delay in 3G networks. *Seventh International Symposium on Computers and Communications (ISCC'02),* 835.

The values in Table 22.3 are relative to the point of call and the destination and therefore are subject to wide variations. For example, a SIP call between Europe and Australia will result to significantly larger PDD, ASD and CRD. Tests in the case of bandwidth limitations were also carried out by Curcio and Lundan [5]. The call success rate was always 100%, therefore bandwidth limitation does not prevent a successful SIP call. The PDD for 2 kbps bandwidth was less that 1 sec. This is significantly large compared with the PDD for 5 kbps bandwidth which was approximately 420 ms. PDD decrease with increasing bandwidth. At a bandwidth of 254 kbps, the PDD was around 50 ms. In realistic 3G network configurations, the bearer allocated for signaling is a few kbps [5] requiring very small time. ASD values recorded by Curcio and Lundan [5] were constantly 45 ms for channels of at least 5 kpbs, but increases for very narrow channels (2 kpbs to 166 ms). This could be attributed to message queuing and congestion at lower channel bandwidths and the possibility of retransmissions [5]. The CRD could not be measured, because the media packets queued up in the simulator and blocked the channel for a long time [5].

Globally, SIP signaling delays are well within the Grade of Service (GoS) bounds proposed by the ETSI TIPHON QoS classes [6].

Losses in the air interface and narrow channels have great impact on the overall SIP call setup time. They result in large call setup times. However, it is expected that UDP/IP header compression and SIP message compression algorithms will reduce the SIP call setup delay over 3GPP networks significantly.

22.2.2.1.2   Message transfer delay   Since the INVITE method is considered to be the most important method in SIP, because it is used to establish a session between participants and normally contains the description of the session to be setup, it is interesting to determine its performance in general. This was done by Kueh [3] for message transfer delay at block error rates of 0, 10 and 20%. It was found that the message transfer delay increases as the message size (and hence the session description) increases. This is expected. It is found that the transfer delay of the INVITE request is substantially reduced compared to when no link-layer retransmission is employed; whereby the delay reduction increases as the message size and the block error rates increase. This is because without retransmission at the link-layer, a segment that is lost means that the whole message cannot be recovered at the receiver side and thus, the whole message needs to be retransmitted at the session-layer, according to the SIP reliability mechanism.

The tests also show that the delay is lower with the unsolicited and solicited status report option set, compared to only having the solicited feedback. This is because by incorporating unsolicited feedback on top of solicited, the missing protocol data units (PDU) can be recovered faster since retransmissions of missing PDUs can be performed before polling; also the reduction in delay is greater at a higher block error rate (BLER).

22.2.2.1.3   Session setup delay   Handley and Jacobson [7] have shown that session setup delay and call blocking probability for a simple call setup sequence consists of the sending of the INVITE request and the 180 ringing response. Two status report trigger settings present a bound on the delay. Tgood status report trigger settings range between 0.5 and 10 s, and Tbad equal to 0.5, 2 and 4 s. Comparing both schemes, it can be seen that when the channel is good, i.e., for a low Tbad value or a high Tgood value, there is hardly any difference in performance, but as the channel gets worse, combining both unsolicited and solicited feedback options gives a lower delay and blocking probability.

SIP signaling is transactional-based and generous in size. Therefore merely transport of packets over the radio interface is not sufficient. Consideration needs to be given to the delay values and the errors in the process. When passing over the error-prone wireless link plus a larger satellite propagation delay, the session establishment delay can be rather large. At the moment SIP over Satellite is not advisable. From a previous study [3], it was shown that with the presence of RLCAM, the session setup performance can be substantially improved. Also it was shown that the combination of unsolicited and solicited status report triggers gives better performance than just solicited alone in terms of delay and blocking probability in a more hostile environment.

22.2.2.2   Mobility and Handoff Delay

The mobility of a user can be described under three distinct terms which are roaming, micro mobility and macro mobility. Roaming is the mobility of a user in the absence or independent of Internet access. Roaming is initiated or triggered when the mobile user gains Internet access. Semantically, this is not different from roaming in cellular networks. Micro and macro mobility are the changes in Internet access with a domain (intradomain) and between domains (interdomain), respectively. If ad hoc networking is in existence within the domain, micro mobility results to handoff between nodes in the same domain or subnet. Otherwise, in interdomain or macro mobility, handoff is required for movement of nodes between two neighboring domains or subnets.

Two approaches are used for handling mobility when IP services are involved. Mobility is handled at the network layer using mobile IP or in the application layer using SIP. The most prominent problem with mobility management at the network layer with mobile IP is the so-called triangular routing that is inherent in the protocol. Triangular routing leads to more delay in the routing process. Secondly, because mobile IP depends on tunneling to forward packets to the mobile node's foreign network; it increases overheads specifically in terms of the encapsulation using the new care-of-address. This increases the IP header and is not suitable for narrow band wireless links because it consumes the limited capacity. Although the SIP-based approach to mobility using UDP offers several advantages over mobile IP by removing the triangular routing problem and the overhead associated with tunneling, it suffers from several drawbacks. The most prominent is the handoff delay and handoff disruption time. Disruption can occur if the new SIP session is not completely created within the overlapping coverage regions of the two nodes involved in handover. The time required to acquire DHCP address renewal while in the overlap region can be significant and can cause disruption in calls (VoIP) and silence periods during handoff. To understand the mechanisms that lead to the delays and the performance of mobility management at the network and the application layers, this section summaries and compares the two mobility management schemes.

22.2.2.2.1 Mobility management using mobile IP The well-known mobile IP triangular routing scheme is shown in Figure 22.1. Two agents (routers) are involved. The home agent (HA) is at the home network (subnet A) and the foreign agent (FA) is at the foreign network (subnet B). The correspondent node (CN) seeks to establish communication with a mobile terminal originally attached to the HA. In this mobility management scheme, while the mobile node is within the intersecting region of the two neighboring subnets, it must acquire a new address using DHCP from the foreign agent. The FA

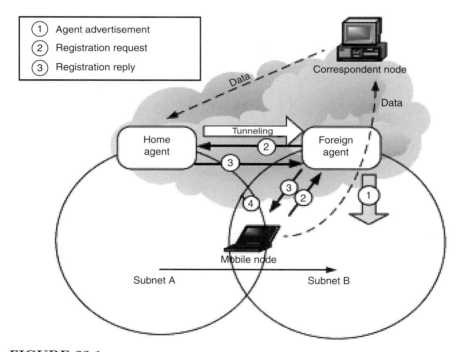

FIGURE 22.1

Mobile IP mobility management scheme. (From Jung, J.W., Mudumbai, R., Montgomery, D. and Kahng, H.K. Performance evaluation of two layered mobility management using mobile IP and session initiation protocol. *Global Telecommunication Conference 2003 (GLOBECOM'03)*, 3, 1190–1194, IEEE 2003. With permission.)

advertises its availability using an agent advertisement (1) to all the nodes within its coverage region. The mobile node upon hearing this advertisement sends a request for registration message (2) to the FA. This message contains in part the identities of the mobile node and of its HA. If a DHCP address is available, it is issued to the mobile node and it is registered at the subnet B. The FA also informs the HA of the new registration (2). Data sent by the CN for the mobile node originally through the HA can therefore now be sent to the mobile node at the foreign network (subnet B). To do this tunneling is used. From this point on, the mobile node can start to communicate with the CN directly through the foreign agent (Figure 22.1).

22.2.2.2.2 Mobility management using SIP The initial design of SIP did not consider mobility management of the end nodes as an issue and hence mobility management using SIP is an after thought. However, to support real-time applications, it is essential to address mobility of nodes to ensure there is no disruption of communication due to lost attachment and poor link quality.

When mobile nodes move, they make, break and re-establish connections with their correspondent nodes. This mobility affects the performance of SIP. Using IPv6 SIP can support terminal mobility without increasing the SIP payload but incurs delay penalties resulting from the desire to handoff. In terminal mobility, the terminal changes access between two nodes or subnets to maintain communication. This change could be as a result of better received signals from the new node or better link conditions to the new node. Which ever one of this is involved, there are delays to be considered during handoff.

In the mobility management scheme described in Figure 22.2, during an active communication session the mobile node obtains a new address through a DHCP server and sends a new session invitation (SIP re-INVITE) to the correspondent node (1). This invitation message contains details of the identity of the mobile node, its new IP address and the new subnet it is attached to. Usually this is done through updating the session description. The correspondent node therefore knows where to send data to the mobile node first by issuing a SIP OK message (2) and sending data (3). A re-registrar message is sent to SIP server to complete the signaling.

There are two crucial problems with SIP mobility management scheme. Firstly, SIP does not offer a solution for the mobility management scheme for long-term TCP connections. Secondly, disruption can occur during handoff when the mobile node is within the overlap region if the acquisition of the new SIP session is not completed before the node moves out of the overlap region. Because the mobile node must acquire in this case a new address through DHCP, there are delays associated with completing the acquisition of the new IP address. This handoff delay can be significant and is a function of the implementation of SIP and the mobility schemes. Some versions of the DHCP can cause a delay of about 2 sec during address renewal. This delay is reduced to about 0.1 sec if duplicate address detection scheme is removed from the DHCP. Delays in SIP mobility-based communication can further be distinguished and analyzed.

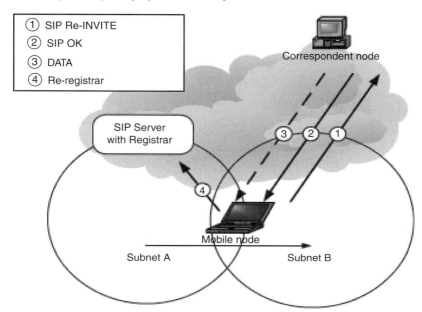

FIGURE 22.2

Mobility management scheme using SIP. (From Jung, J.W., Mudumbai, R., Montgomery, D. and Kahng, H.K. Performance evaluation of two layered mobility management using mobile IP and session initiation protocol. *Global Telecommunication Conference 2003 (GLOBECOM'03)*, 3, 1190–1194, IEEE 2003. With permission.)

Figure 22.3 shows that in trying to establish sessions, SIP operations lead to delays associated with the signaling.

In the scenario described in Figure 22.3, the SIP proxy is assumed to have Anike's location address (host_b.telkom.com) and Johnson therefore invites Anike through it. The sequence of messages sent is as follows. Johnson sends an INVITE message through the proxy server (1), and the proxy server relays the INVITE message to Anike at host_b.telkom.com (2). While Anike is being alerted a 180 Ringing message is sent back to Johnson via the proxy server (3, 4). After Anike has answered to the call, 200 OK response is sent back to Johnson to notify him that Anike accepts the call (5, 6). Then Johnson sends an ACK (7) message to Anike to confirm a session establishment. Since Johnson knows the contact address of Anike, anike@host_b.company.com, via 200 OK and Johnson can send an ACK to Anike directly without going through the proxy server. When Johnson wants to terminate the call, Johnson sends a BYE request to Anike (8) which is confirmed by another 200 OK (9) issued by Anike.

Apart from establishing sessions which causes delays, two types of terminal mobility can be identified. They are the pre-call terminal mobility and mid-call

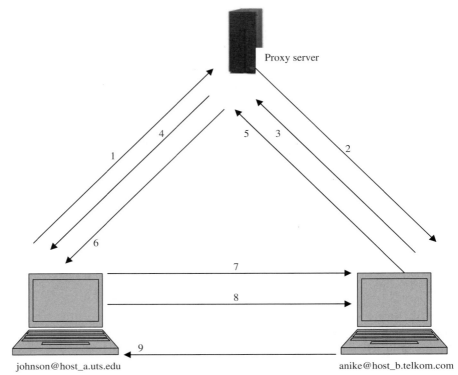

FIGURE 22.3
Basic SIP procedure. (From Nakajima, N., Dutta, A., Das, S., and Schulzinne, H. Handoff delay analysis and measurement for SIP based mobility in IPv6. *International Conference on Communications (ICC'03)*, 26, 1085–1089, May 2003. With permission.)

terminal mobility. The pre-call terminal mobility ensures that a correspondent node (CN) can reach a mobile node (MN), as in Figure 22.4 [8]. The pre-call terminal mobility scenario is akin to mobile IP. When an SIP-enabled MN moves to a foreign network (Visited Network) it registers and has a new contact address with the SIP proxy server (1). Therefore an INVITE request from a CN through the SIP server (2) is used by the server to notify the new contact address of the MN to the CN. This provides the means for the CN to send an INVITE request to the MN directly (4) with an OK message sent back by the MN (5).

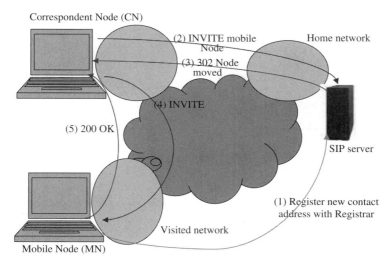

FIGURE 22.4
Pre-call mobility in SIP.

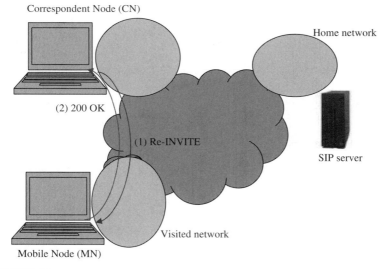

FIGURE 22.5
Mid-call mobility in SIP.

The second type of mobility is the mid-call mobility. It allows a node to maintain an ongoing session with its peer during handoff as in Figure 22.5 [8]. As shown in Figure 22.5 an MN sends a re-INVITE request with a new IP address to CN (1), and the CN sends packets to MN directly at the new point of attachment to the network (2).

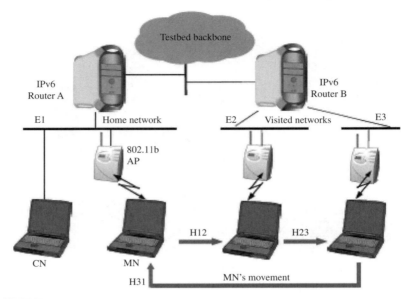

FIGURE 22.6

Experimental test bed for handoff measurement in IPv6 mobile networks. (From Nakajima, N., Dutta, A., Das, S., and Schulzinne, H. Handoff delay analysis and measurement for SIP based mobility in IPv6. *International Conference on Communications (ICC'03)*, 26, 1085–1089, May 2003. With permission.)

22.2.2.3 Handoff Delay

Handoff duration is composed of two broad delays, which are link layer establishment and signaling delays. Link layer establishment delay is normally negligible compared with signaling delay. Handoff delay as a result of mobility was measured by Nakajima et al. [8] using a small network test bed of three mobile nodes, one correspondent node, a home network and a visited network consisting each of a router within an IPv6 domain. The network used by Nakajima [8] is shown as Figure 22.6.

Total handoff delay (τ) is defined [8] as the time between detachment from the old access point and establishment of communication with a new correspondence node (CN). It consists of three components:

1. The time for switching lower layer medium to access network (τ_1).

2. The time for detecting a new router and a new link (τ_2).

3. The time for recovery of communication with a CN after detecting a new link (τ_3).

Therefore $\tau = \tau_1 + \tau_2 + \tau_3$. Nakajima [8] concentrated on measuring τ_3. There are two main factors which contribute to delay τ_3; Duplicate Address Detection (DAD) and router selection. The purpose of DAD is to confirm the uniqueness of the IPv6 address on the link. DAD imposes a delay between

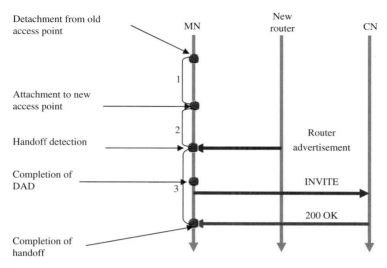

FIGURE 22.7

Handoff process in SIP in IPv6 environment. (From Nakajima, N., Dutta, A., Das, S., and Schulzinne, H. Handoff delay analysis and measurement for SIP based mobility in IPv6. *International Conference on Communications (ICC'03)*, 26, 1085–1089, May 2003. With permission.)

receiving a router advertisement (RA) and sending a packet out of the interface with auto configured IPv6 address.

Nakajima [8] also measured the handoff delay of SIP terminal mobility in the IPv6 testbed. Two different scenarios were considered:

(a) SIP mobility without kernel modification

(b) SIP mobility with kernel modification.

Measurements have been performed for scenarios (a) and (b). Table 22.4 shows the handoff delay τ_3, which is related to signaling.

H12 is the handoff time between mobile nodes 1 and 2, H23 is the handoff time between mobile nodes 2 and 3 and H13 is the handoff delay between mobile nodes 1 and 3 as shown in Figure 22.7.

Table 22.5 shows the results of another handoff delay related to voice communication using UDP.

From Table 22.4 and Table 22.5, modified kernel has reduced handoff delay. Although the delay Figures are unacceptable for real-time multimedia communications. For example a delay of 420 ms is likely to result to echo if other delays in the network add to it. This could lead to packets being dropped or played back with significant delays and hence render the playback unsuitable for listening. It was shown by Nakajima et al. [8] that by integrating MIPL MIPv6 in the network for SIP mobility, the handoff delay for signaling was observed to be about 2 ms and the handoff delay for media UDP packet is less than 31 ms. This is acceptable in most voice communications. The results

TABLE 22.4
Handoff Delay of Signaling

Handoff case	(a)	(b)
H12	38290.0 ms	171.4 ms
H23	3932.2 ms	161.6 ms
H31	1934.7 ms	161.1 ms

Source: From Nakajima, N., Dutta, A., Das, S., and Schulzinne, H. Handoff delay analysis and measurement for SIP based mobility in IPv6. *International Conference on Communications (ICC'03)*, 26, 1085–1089, May 2003. With permission.

TABLE 22.5
Handoff Delay of Media UDP Packet

Handoff case	(a)	(b)
H12	38546.3 ms	420.8 ms
H23	4187.7 ms	418.6 ms
H31	1949.4 ms	408.4 ms

Source: From Nakajima, N., Dutta, A., Das, S., and Schulzinne, H. Handoff delay analysis and measurement for SIP based mobility in IPv6. *International Conference on Communications (ICC'03)*, 26, 1085–1089, May 2003. With permission.

showed that MIPL MIPv6 with modified kernel outperforms the SIP mobility with modified kernel. Hence application layer mobility based on SIP has potential for real-time applications, provided MIPL and MIPv6 and other similar speed up processes are integrated.

22.2.2.4　Disruption Time

The amount of disruption time for processing handoff and interconnections is a major concern in VoIP services. Handoff between terminals or nodes could disrupt speech communications if speech packets are lost or delayed too much. Lost packets lead to lost words and sentences and hence makes the communication annoying and uncomfortable. Delayed packets also lead to long delays between sections of words or between words and hence lead to the same uncomfortable speech communication. Hence handoff has the potential to lead to an annoying listening experience in VoIP and IPTV services. Roaming users are interested in staying connected with the network while moving from one network to another network with multiple network interfaces. Hence an efficient mobility management scheme is necessary for handling micro-mobility, macro-mobility and roaming. The objective for the scheme should be to reduce handoff delay and disruption effects. Micro-mobility (*intradomain*

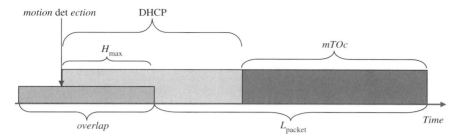

FIGURE 22.8
Handoff delay and packet loss instances.

mobility) protocols aim to handle local movement (e.g., within a domain) of mobile hosts without interaction with the mobile IP enabled Internet [9]. A macro-mobility (or *interdomain* mobility) protocol based on mobile IP manages mobility between domains [9]. Kwon et al. [4] found that the disruption for handoff of the mobile IP approach is smaller than that of the SIP approach in most situations, however, SIP shows shorter disruption when the mobile node (MN) and correspondent node (CN) are close because the need for handoff is reduced in such situations. While the SIP-based approach offers several advantages over a corresponding MIP-based solution for typical UDP-based VoIP streams, it continues to suffer from several drawbacks. These include the absence of a mobility management for long-term TCP connection. Second, it can cause call disruption if the new SIP session is not created completely while the mobile terminal is in the overlapped area. As opposed to a mobile node using mobile IP, a mobile node using SIP-mobility always needs to acquire an IP address via DHCP which can be a major part of the overall handoff delay.

Several delays or disruption factors [10] affect the handoff delay and packet loss over wireless networks (Figure 22.8). These are:

1. τ_{mTOc}: The one-way delay from the mobile node to correspondent node.

2. τ_{hTOn}: The delay in sending a message between the new foreign agent and HA.

3. τ_{DHCP}: IP address renewal time through DHCP

4. $H_{\max}$ is the maximum time for seamless handoff. It is the time difference between when motion is detected and the time when the node escapes from the old cell.

5. τ_{overlap} is the time over which the mobile node is within the overlap area of two adjacent cells.

6. L_{packet}: Packet loss time during handoff process. This is the length of time over which the node will lose packets when a handoff process is on going.

7. $\tau_{\mathrm{overtime}} = (H_{\max} - \tau_{\mathrm{DHCP}})$: The time difference between $H_{\max}$ and τ_{DHCP}.

TABLE 22.6
Disruption Time vs. τ_{mTOn}

	Disruption time (msec)		
τ_{mTOn}	MIP	SIP	MIP-SIP
10	42	100	41
20	70	99	71
30	90	107	95
40	110	95	95
50	131	100	98
60	151	100	100
70	170	92	91
80	191	93	92
90	210	93	92
100	230	94	94

Source: From Jung, J.W., Mudumbai, R., Montgomery, D., and Kahng, H.K., Performance evaluation of two layered mobility management using mobile IP and session initiation protocol. *Global Telecommunication Conference 2003 (GLOBECOM'03)*, 3, 1190–1194, IEEE 2003. With permission.

Tests were performed by Jung [10] to assess some of these disruption times and the results are reflected in Table 22.6, Table 22.7 and Table 22.8. Three factors are considered. Table 22.6 illustrates the handoff disruption time as the delay, τ_{hTOn} increases. Here τ_{mTOc} is set at 30 ms.

Table 22.7 shows the disruption time as the delay τ_{mTOc} increases. Here τ_{hTOn} is assumed to be 30 ms. As SIP mobility only depends on the distance between the MN and CN, the disruption time of SIP-mobility increases according to the delay τ_{mTOc}.

In Table 22.6, τ_{mTOc} and τ_{hTOn}, are set at 30 and 100 ms, respectively. Also the H_{max} is fixed at 500 ms. Because the handoff in SIP over mobile IP does not need IP renewal time, its disruption time is the same as that in Disruption Time versus τ_{hTOn}. However as the SIP-based approach always needs a new IP address, its disruption time increases proportional to the rate of increase for $\tau_{overtime}$. Simulation results show that the proposed approach out-performs the existing approach in most cases. This shows that VoIP can be supported over wireless Internet.

22.2.3 Handoff Delay Disruption of SIP in IP Services

The performance of SIP in IP services can better be appreciated if we adopt similar metrics and measures for comparing them. One such approach uses the so-called "shadow registration" proposed by Kwon et al. [4]. In shadow registration, the mobile node establishes security association with the neighboring AAA server a priori before handoff is initiated. This can be done in

TABLE 22.7

Disruption Time vs τ_{mTOc}

τ_{mTOc}	Disruption time (msec)		
	MIP	SIP	MIP-SIP
10	90	49	48
20	90	70	59
30	90	90	91
40	90	120	91
50	90	125	91
60	90	143	91
70	90	180	91
80	90	200	90
90	90	205	90
100	90	230	90

Source: From Jung, J.W., Mudumbai, R., Montgomery, D., and Kahng, H.K., Performance evaluation of two layered mobility management using mobile IP and session initiation protocol. *Global Telecommunication Conference 2003 (GLOBECOM'03)*, 3, 1190–1194, IEEE 2003. With permission.

TABLE 22.8

Disruption Time vs. τ_{overtime}

τ_{overtime}	Disruption time (msec)		
	MIP	SIP	MIP-SIP
0	0.1	0.1	0.21
0.1	0.2	0.2	0.21
0.2	0.21	0.3	0.21
0.3	0.21	0.4	0.21
0.4	0.21	0.5	0.21
0.5	0.21	0.6	0.21
0.6	0.21	0.7	0.21
0.7	0.21	0.8	0.21
0.8	0.21	0.9	0.21
0.9	0.21	1	0.21
1	0.21	1.1	0.21

Source: From Jung, J.W., Mudumbai, R., Montgomery, D., and Kahng, H.K., Performance evaluation of two layered mobility management using mobile IP and session initiation protocol. *Global Telecommunication Conference 2003 (GLOBECOM'03)*, 3, 1190–1194, IEEE 2003. With permission.

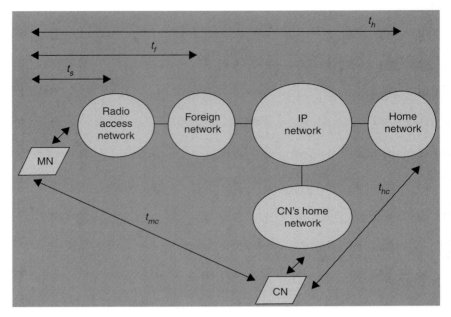

FIGURE 22.9

Handoff delay measurement architecture. (From Kwon, T.T., Gerla, M., Das, S.K., and Das, S. Mobility management for VoIP services: mobile IP vs SIP. *IEEE Wireless Communications,* 9, 66–75, IEEE October 2002. With permission.)

a distributed manner with many known AAA servers or selectively with only the neighboring server. This section does not discuss how this is done.

22.2.3.1 Handoff Delay Disruption of SIP in VoIP Services

Handoff delay for SIP-based VoIP services was measured by Kwon et al. [4] for both mobile IP and SIP using the set up illustrated in Figure 22.9. Following Kwon et al. [4], the following variables are defined:

1. t_s, the time for message transfer in the wireless link between the mobile node and the foreign agent router (RFA) or DHCP server.

2. t_f, the delay between the mobile node and the foreign AAA server (AAAF). This is the time required to send a message in the foreign network.

3. t_h, the delay between the mobile node and the communicating entities in its home network (its home registrar (HR), AAAH or HA). It is therefore the time for the mobile node to send messages to its home network. Generally, $t_s < t_f < t_h$.

4. t_{mc}, the delay between the mobile node and the correspondent node.

5. t_{hc}, the delay between the home network and the correspondent node. These delays are shown in Figure 22.9.

6. t_{ARP}, the time required for the ARP (address resolution protocol).

7. t_{no}, the time to send a message between the new foreign agent (NFA) and the old foreign agent (OFA).

It is assumed in this diagram that a simple VoIP application, unaware of mobility, is involved in the QoS parameter measurements and the mobile node caches the address of the callee's correspondent node (CN). Furthermore, it is assumed that there was no a priori Internet connection. Hence the mobile node must initiate the connection when it requires a VoIP application. Therefore the initial delay is the sum of the time required for Internet connection and VoIP signaling. These times are different for the MIP and SIP situations.

22.2.3.1.1 Initial registration and session setup The time required by mobile IP for initial registration and session setup is given by three components:

(a) the sum of the delay for router solicitation message and router advertisement messages ($2t_s$) as a round trip delay, plus

(b) the round trip registration message with the HA ($2t_h$) and

(c) the time for the simple VoIP application (SVA) to initiate a VoIP service request with the correspondent node. This time consists of the request and the OK response from the CN ($2t_{mc}$). For mobile IP, the total time is:

$$t_{\text{MIP\_init}} = 2t_s + 2t_h + 2t_{mc} \qquad (22.8)$$

When SIP is used instead of MIP, the time required to acquire a new address through the DHCP ($4t_s$) and for the ARP (t_{ARP}) need to be added to the time for the MIP case. The total time is:

$$t_{\text{SIP\_init}} = 4t_s + t_{ARP} + 2t_h + 2t_{mc} \qquad (22.9)$$

When the same networks and nodes are used, the time required to initiate the IP service using SIP is distinctly higher than for MIP.

22.2.3.1.2 Intradomain handoff delay To compute the handoff delay for the case of MIP, the mobile node detects a new access point and then the new IP subnet. It then sends the router solicitation message to the RFA and receives a response in terms of the router advertisement message. This round trip time is ($2t_s$). Since intradomain does not involve AAA, the registration time is also ($2t_f$) and the total intradomain handoff is:

$$t_{\text{MIP\_init}} = 2t_s + 2t_f \qquad (22.10)$$

If SIP is used instead of MIP, the time to acquire a new IP address using a DHCP ($4t_s$) and the ARP time (t_{ARP}) are needed. The remaining time is for the mobile node to register for the simple VoIP application with the foreign registrar $(2t_f)$. The total time for intradomain handoff in this case is:

$$t_{\text{SIP\_init}} = 4t_s + t_{\text{ARP}} + 2t_f \qquad (22.11)$$

Generally in the example given in Kwon et al. [4] t_s is about 10 ms and t_f is about $(t_s + 2)$ ms.

22.2.3.1.3 Interdomain handoff delay

In interdomain, two sub networks are involved. Therefore the delays involve communication and signaling between the elements of the two sub networks and the mobile node. Using MIP, for handoff to take place, the mobile node must first detect the new IP sub network, select it and initiate handoff. The delays involved include the time for the mobile node to initial router solicitation and router advertisement messages $(2t_s)$. The next delays are the time for the mobile node to register with the NFA $(2t_h)$, cache the message from the home agent $(2t_h - t_s)$. During the message caching process two signaling flows take place: route optimisation and smooth handoff. As the NFA caches the registration message, it sends a binding update message to the OFA. The OFA upon receiving the binding update message (t_{no}) starts sending packets destined to the MN to the NFA $(t_s + t_{no})$ thus starting the interdomain handoff. This result to the total time for starting the interdomain handoff or smooth handoff is:

$$h_{\text{MIP\_smh}}(2t_s + (2t_h - t_s) + t_{no} + (t_s + t_{no})) = 2t_s + 2t_h + 2t_{no} \qquad (22.12)$$

During the route optimization, other delays are involved in the handoff process including the time when the OFA updates its binding cache with the care of address (CoA) of the mobile node $(t_h - t_s))$ and sends a binding warning message to the HA of the mobile node. The HA then sends the binding update message to the correspondent node (t_{hc}). Lastly, the correspondent node will then send the packets for the mobile node to the NFA which are then subsequently delivered to the mobile node with time (t_{mc}). The total time for route optimization is:

$$\begin{aligned} h_{\text{MIP\_ropt}} &= (2t_s + (2t_h - t_s) + t_{no} + (t_h - t_s) + t_{hc} + t_{mc}) \\ &= 3t_h + t_{no} + t_{hc} + t_{mc} \end{aligned} \qquad (22.13)$$

After route optimization, handoff is triggered. It is essential to understand the two signaling flows involved in route optimization in terms of the time packets from the correspondent node reaching the mobile node and the time the packets forwarded by the OFA reach the mobile node as well. There is a blackout period until packets from the OFA arrive at the mobile node. This blackout period is about $2t_s + 2t_h + 2t_{no}$ after which the VoIP communication resumes with a disruption until packets from the CN reach the mobile node. The MIP interdomain handoff time becomes

$$t_{\text{MIP\_inter}} = 3t_h + t_{no} + t_{hc} + t_{mc} \qquad (22.14)$$

Note that when the mobile node is located in its home network $t_{mc} = t_h + t_{hc}$. With SIP the interdomain handoff time occurs after

(a) the times for acquiring new IP address through DHCP and an ARP $(4t_s + t_{\text{ARP}})$ and

(b) the time for the mobile node to send SIP REGISTER message to the home registrar (HR) which is $(2t_h)$.

At this point Internet connectivity is enabled. Then the mobile node re-invites the CN by sending to it an INVITE message.

(c) This takes the time $(2t_{mc})$. Hence the interdomain handoff time is:

$$t_{\text{SIP\_inter}} = 4t_s + t_{\text{ARP}} + 2t_h + 2t_{mc} \qquad (22.15)$$

In general the choice of t_{ARP} is between 1 and 3 ms and t_{no} is about 5 ms. The disruption time increases as the delay distance between the mobile node and CN increases. Since in the case of MIP the mobile node the disruption time is $2t_s + 2t_h + 2t_{no}$ and t_f is about $(t_s + 2)$ ms, the disruption time is about 54 ms, which is large enough (for speech sampled at 8 KHz) to drop more than 432 contiguous speech samples which is a significant portion of the speech communication and makes it annoying to listen to. However, MIP outperforms SIP because the mobile terminal starts to receive speech after only about $2t_s + 2t_h + 2t_{no}$ seconds and the terminal can start to playback some of the speech until the overall handoff is completed. In both interdomain and intradomain handoffs, MIP performance is superior. The intradomain disruption times for both MIP and SIP are relatively flat with increasing delay distance between the mobile node and home network and also always smaller than the interdomain disruption times. The interdomain disruption times increase linearly for both MIP and SIP. At smaller delay separation, the MIP outperforms the SIP performance but only after some time, SIP is advantageous as its disruption time becomes smaller than MIP over larger delays.

For both MIP and SIP the interdomain handoff disruption time also increases with the delay separation between the CN and mobile node and their intradomain handoff disruption times remain flat with MIP always out performing SIP.

Over low bandwidth wireless links, the disruption times resulting from inter and intradomain handoffs increase linearly as the wireless link delay increases and the intradomain disruptions are always smaller (MIP smaller then SIP). This is shown in Figure 22.10 from Kwon et al. [4].

22.2.3.1.4 Disruption time with shadow registration When shadow registration is included in the operation of MIP and SIP, the disruption times change. Since AAA resolution for the mobile node can be performed within the foreign network (AAAF), the disruption time for the MIP case is:

(a) a function of the router solicitation and advertisement message exchange time $(2t_s)$, the registration message time from the mobile node to the NFA

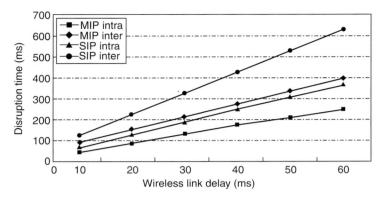

FIGURE 22.10

Disruption time as a function of wireless link delay for MIP and SIP. (From Kwon, T. T., M. Gerla, S. K. Das, and S. Das. 2002. Mobility management for VoIP services: Mobile IP vs SIP. *IEEE Wireless Communications* 9: 66–75.)

(b) the reply time to the mobile node $(2t_f - t_s)$ and

(c) the route optimization signaling flow time $(t_h - t_s) + t_{no} + t_{hc} + t_{mc}$

Therefore the total disruption time is given by the sum:

$$t_{MIP\_inter\_sdw} = 2t_s + (2t_f - t_s) + (t_h - t_s) + t_{no} + t_{hc} + t_{mc}$$
$$= 2t_f + t_h + t_{no} + t_{hc} + t_{mc} \qquad (22.16)$$

In the case of SIP, the REGISTER message is handled locally also in the foreign network DHCP acquisition and the ARP consumes the time $(4t_s + t_{ARP})$. Furthermore, the AAA resolution and REGISTER is performed in the foreign network with time $(2t_f)$. This is followed by the mobile node re-inviting the CN by sending the re-INVITE message with time $(2t_{mc})$. The total disruption time is therefore given by the expression:

$$t_{SIP\_inter\_sdw} = 4t_s + t_{ARP} + 2t_f + 2t_{mc} \qquad (22.17)$$

Note that the disruption time now has t_f replacing t_h. Hence shadow registration has decreased the disruption time in both cases of MIP and SIP. Therefore shadow registration is useful when the mobile node is far from the home network.

22.2.3.2   Performance of SIP in Fax over IP

The performance of Fax over IP (FoIP) needs to rival its performance in legacy fixed line at user premises. Fax over the PSTN is very robust and reliable. However, the reach and the popularity of the Internet combined with the fact that its use is almost free, is a major driver for VoIP networks and for similar data applications that exist on the PSTN. Fax transmission has special requirements. Firstly while the loss of a packet during a human conversation does not affect a voice call a great deal, it can easily affect a fax call. This is because fax transmission requires far more signaling and handshaking than a regular telephone call. This includes negotiating details such as speed, paper size, and delivery confirmation. Apart from the signaling in a fax call, the sending and receiving of fax documents are mostly done and interpreted by automated fax machines. Therefore errors in either signaling or the actual transmission result in lengthy recovery times.

A study was performed by Choudhary [11] on the performance of FoIP networks. The performance parameters for FoIP are significantly different compared with VoIP. In fax transmissions, no roaming or handoffs are required. However, end-to-end delays, link utilization and the performance of the wireless links are essential.

22.2.3.2.1 Experimental network models With three network models, Choudhary [11] measured the link utilization of the inter-proxy-server link (IPSL), the link utilization of the auxiliary link (AUXL), the average end-to-end delay of the SIP signaling packets, the average end-to-end delay of fax data packets, the average SIP call setup time, and the average fax call setup time.

1. In Network Model 1 calls are generated from T.38 gateway, and all messages are sent to the SIP proxy server on its network. The path between the T.38 gateway and the SIP proxy server contains an IP router. Although this is not necessary, it is however done to maintain compatibility with the next two network models where IP routers play an important role. The SIP proxy server of the sender's network communicates with that of the recipient's network and initially sends SIP messages to set up a call. Once that is done, the T.38 gateway starts fax transmission. The originating T.38 gateway starts sending fax packets to the SIP proxy server. That is, the SIP proxy server routes and interprets SIP messages. It also routes IFP packets without actually interpreting them. This scenario is likely if the SIP proxy server is implemented on a router within the network. In such a case, it interprets SIP messages, translating them and maintaining the state for them. All other messages are not interpreted by it but merely routed. This model is shown in Figure 22.11 [11].

2. The Network Model 2 is the same network as model 1 but a different setup. It not only includes the components and links of the first model, but also has a direct link between the IP routers (the so-called proxy-bypass approach). That is, all the SIP signaling is carried out on the path that traverses the two SIP proxy servers. However, once the call is setup, all the fax data packet transmission is done through IP routers and the proxy-bypass links only. This mimics a network where signaling travels on separate links, and data is sent across another set of links. This is possibly because data does not need to go through SIP proxy servers or other network entities. This frees up resources at such entities and segregates signaling from data transmission, much like PSTN

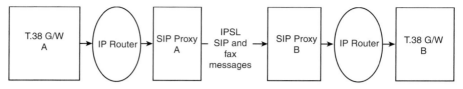

FIGURE 22.11

SIP proxies routing SIP and fax PACKETS. (From Choudhary, U., Perl, E., and Sidhu, D. Using T.38 and SIP for real-time fax transmission over IP networks. *26th Annual IEEE Conference on Local Computer Networks (LCN'01),* 74, IEEE November 2001. With permission.)

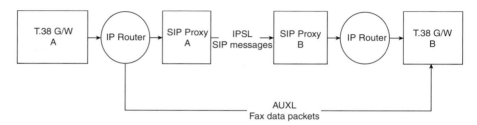

FIGURE 22.12

Separate paths for SIP messages and fax images. (From Choudhary, U., Perl, E., and Sidhu, D. Using T.38 and SIP for real-time fax transmission over IP networks. *26th Annual IEEE Conference on Local Computer Networks (LCN'01),* 74, IEEE November 2001. With permission.)

networks where the Signaling System #7 (SS7) links are distinct from the trunks. Figure 22.12 gives a view of this model.

3. Network Model 3 uses the same network as network model 2. However, not all SIP messages travel between the SIP proxy servers in this case. In general, all SIP terminals are configured to know their network's SIP proxy server. Hence, to set up a SIP call, they contact their SIP proxy server first not having to know themselves the entire route to a receiver. They only need to know how to route SIP calls they initiate to their designated SIP proxy servers. The SIP proxy server then handles all the signaling for the rest of the call. The AUXL link is used for fax data packets as is in network model 2.

Table 22.9 shows the performance measures recorded by Choudhary [11].

In terms of link utilization values on IPSL, network model 3 is the most suitable. In terms of just link utilization values on AUXL, network model 2 is only marginally better. In terms of end-to-end fax data packet delay, network models 2 and 3 are on average equally good. When just packet end-to-end delay is considered, network model 3 is the best. The average SIP call setup time and average fax call setup for network model 3 is the best.

TABLE 22.9
Experiment Results

	Model 1	Model 2	Model 3
Average link utilization of IPSL (Mbps)	77.50	9.223	0.9546
Average link utilization of AUXL (Mbps)		69.03	75.96
Average end-to-end fax data packets delay (sec)	0.2023	0.2017	0.2017
Average end-to-end signaling delay (sec)	0.2014	0.2015	0.2009
Average SIP call setup times (sec)	3.4023	3.4042	3.4020
Average fax call setup time (sec)	4.2086	4.2106	4.2060

Source: From Choudhary, U., Perl, E., and Sidhu, D. Using T.38 and SIP for real-time fax transmission over IP networks. *26th Annual IEEE Conference on Local Computer Networks (LCN'01)*, 74, IEEE November 2001. With permission.

SIP appears to be a powerful and useful signaling protocol supporting mobility for wireless IP networks but it has inherent weaknesses and dangers. These weaknesses and dangers are discussed in Section 22.2.4.

22.2.4 Effects of Security Protocols on Performance of SIP

SIP has been in use for setting up secure calls in VoIP. For this, SIP employs several security protocols such as TLS, DTLS or IPSec, combined with TCP, UDP or SCTP, as a security protocol in VoIP (Figure 22.13). These security mechanisms introduce additional overheads into SIP performance. From a privacy point of view VoIP delivered over an IP network is vulnerable to message spoofing and eavesdropping. Impersonation of a legal user could also result to false accounting and avoidance of payment. Hence many vendors resort to encrypting VoIP messages. SIP therefore assumes an additional role of authenticating users who send and receive VoIP messages with the potential to negatively affect the performance of SIP. An obvious effect of a security protocol on the performance of SIP is the call setup delays. Traditionally SIP was not developed with its own security features but the standard recommends that existing security mechanisms could be used instead. Figure 22.13 illustrates how SIP uses some of the security mechanisms in the protocol stack.

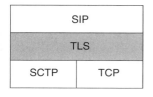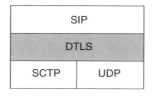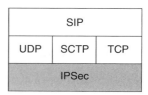

FIGURE 22.13
Protocol stack for SIP security.

Two factors are responsible for the increased call setup delay. These are authentication of messages and security handshake. A security channel is established during the handshake. During this process a key is generated and messages are exchanged. This results in additional setup delay. Also during authentication the SIP messages are encrypted resulting in increased packet size or payload, causing additional delay.

There are two types of handshakes, full handshake and abbreviated handshake. During full handshake for normal session initiation, the user proxy or servers authenticate themselves and also create an encryption session key. During abbreviated handshake (Phase 2 rekey) a previous security channel could be reused for session resumption. This is called abbreviated security handshake because the information of the previous security channel is still available. Fewer messages are exchanged during the abbreviated handshake since the authentication process may be skipped and consequently the delay involved is smaller as well.

Furthermore, additional setup delays are introduced by the processes required for establishing confidentiality and integrity of messages. These processes include encryption and adding of message authentication code (MAC) to the message. For example SHA-1 MAC process results in 25 bytes of extra data on top of the original message. In addition every encryption step introduces its own processing delay which worsens if multiple levels of encryption and decryption are involved.

Different security mechanisms have different effects on SIP particularly in terms of delays and overheads. This section summarizes the performance of SIP when combined with IPSec, TLS and DTLS within the domains of transport protocols such as TCP, UDP and streaming control transport protocol (SCTP). Noticeable performance degradation was observed during SIP call setup up [12]. A relatively large call setup delay is observed with the combination of TLS and SCTP than in other combinations (e.g., TLS/TCP or TLS/UDP). The best performance is achieved with the combination of UDP and other security mechanisms (IPSec/UDP and DTLS/UDP). However, there is no congestion control involved with SIP over IPSec/UDP or DTLS/UDP. A high level of congestion increases the call setup delay with TLS/TCP. At low network congestion there is no significant difference between the call setup times for the combinations of IPSec/TCP and TLS/TCP. In general the overhead associated with message authentication is very small when TLS and IPSec are involved.

22.3   SIP Security

SIP security threats fall into one of four different categories. The first set of threats result from the vulnerabilities of the SIP protocol. These threats take advantage of the flaws in the definition of SIP in terms of its characteristics, functionality and description. The second set of threats result from its interoperation with other protocols. This relates to SIP interfacing with other protocols. The third group is threats that exploit the operational details of SIP including mandatory, non-mandatory and/or optional requirements of SIP. The fourth type of threat result from the inherent security risks associated with the system under which SIP is used. These four categories of threats are discussed well elsewhere [13].

SPam over Internet Telephony (SPIT) is emerging as a major security problem brought to the fore because of the reduced cost of VoIP services compared with the traditional voice calls over PSTN. The deterrent posed by the high cost of voice calls over PSTN has been removed and in doing so a new problem has emerged (SPIT) or has it? From experience, there is no significant spam through short messaging service (SMS) which suggests that SPIT may not be as major a problem as anticipated. SPIT consists of two components which are signaling and media. Three different forms of SPIT can be identified. These are

(a) Voice or call SPIT. Call SPIT is defined as 'a bulk unsolicited set of session initiation attempts in order to establish a multimedia session' [13].

(b) Instant message SPIT. This is defined as an unsolicited set of bulk instant messages.

(c) Presence SPIT is defined as an unsolicited set of bulk presence requests made with the objective of the initiator to gain membership of the 'address book of a user or potentially of multiple users' [13].

From these definitions SPIT threats are more potent that spam threats because three sources of SPIT attack can be mounted compared with just a single direct spam attack through bulk e-mails.

22.3.1 Threats from Vulnerability of SIP

SIP is not an easy protocol to secure. Its use of intermediaries, its multi-faceted trust relationships, its expected usage between elements with no trust at all and its user-to-user operation make security far from trivial. Security solutions are needed that are deployable, without extensive coordination, in a wide variety of environments and usages. In order to meet these diverse needs, several distinct mechanisms applicable to different aspects and usages of SIP are required. Although security and privacy should be mandatory for IP telephony architecture, most of the attention during the initial design of the IETF IP telephony architecture and its SIP has been focused on the possibility of providing new dynamic and powerful services with simplicity. Less attention was paid to security features.

IP networks using SIP can replace PSTN provided the same basic telephone service with comparable level of QoS and network security are part and parcel of the IP network. The following security characteristics should therefore be guaranteed [24]:

1. High service availability.

2. Stable and error-free operation.

3. Protection of the user-to-network and user-to-user network traffic.

SIP security should cover the overall communication chain. This includes the source, destination, links, signaling, packet headers and messages. In addition, protocols which interface with SIP should also be secured.

Unlike traditional mobile phone text messages which are protected by the network and therefore are forced to remain unmodified from sender to destination, SIP text messages can be modified in transit because they are not secured.

SIP messages may contain information a user or server wishes to keep private. The headers can reveal information about the communication patterns and content of individuals, or other confidential information. The SIP message body may also contain user information (media type, codec, address and ports, etc) that should not be revealed.

22.3.1.1 Messages not Encrypted

Traditionally SIP messages are not encrypted and are therefore open to interception in the links or at any points in the network. The 'To' and 'Via' fields of the headers are visible so that SIP messages can be routed normally. This makes it possible for an attacker to send spoofed INITIATE messages with deceitful IP addresses. Furthermore, it is possible to use a spoofed BYE message to end SIP sessions prematurely. While this might require capturing SIP messages and holding them for use, it is not a hard task.

There are several methods that can therefore be used to protect SIP messages. They include using security mechanisms such as IPSec, TLS/SSL and S/MIME. Another approach is to deploy firewalls at the perimeter of the network to protect and provide application layer security as well as authentication. This also protects against transport layer and protocol attacks. Securing SIP header and overall information can be implemented for the following reasons:

1. Maintaining private user and network information in order to guarantee some form of user privacy.

2. Avoiding SIP sessions being established or charged by someone faking the identity of another person.

22.3.2 Threats in SIP Communication Chain

Threats exist at all levels of SIP operations, at the servers, proxies, the SIP addresses, messages and requests and the message headers:

1. Sending ambiguous requests to proxies.

2. Exploitation of forking proxies.

3. Listening to a multicast address.

4. Population of 'active' addresses.

5. Sending messages to multicast addresses.

6. Contacting a redirect server with ambiguous requests.

7. Throwaway SIP accounts.

8. Misuse of stateless servers.

9. Anonymous SIP servers and back-to-back user agents.

10. Exploitation of messages and header fields structure.

22.3.2.1 Threats from SIP's Optional Recommendations

This category of threat exploits the registrar servers. A registrar server is responsible for receiving REGISTER requests and updating location information into the location service for the domain it controls. It is the front end to the location service for a domain. As such, it could query the location service to gather information for specific registrations. The registrar server could be fooled. A SPITter could issue a dictionary attack, or submit a query with special characters to a registrar and get a list of URIs. The authentication of a user to any proxy server, including the registrar is not defined as a mandatory requirement. Therefore this form of attack can be mounted. More specifically, the SIP specification does not dictate proxy servers to authenticate each user. Therefore, without any authentication between a UA and a proxy server several vulnerabilities might occur, including impersonation of a legitimate user.

22.3.3 Attacks and Threat Models

SIP security mechanisms must be defined and combined properly to obtain a trusted network scenario. Attackers on the network may be able to modify packets (perhaps at any compromised intermediary). Attackers may wish to steal services, eavesdrop on communications or disrupt sessions. SIP communications are susceptible to several types of attacks:

1. The simplest attack in SIP is snooping, which permits an attacker to gain information on users' identities, services, media and network topology. This information can be used to perform other types of attacks.

2. Modification attacks occur when an attacker intercepts the signaling path and tries to modify SIP messages in order to change some service characteristics. This kind of attack depends on the kind of security used.

3. Snoofing is used to impersonate the identity of a server or user to gain some information provided directly or indirectly by the attacked entity.

4. Finally, SIP is especially prone to denial of service attacks that can be performed in several ways, and can damage both servers and user agents. This attack technique may cause memory exhaustion and processor overload.

Although the security mechanisms provided with SIP can reduce the risks of attacks, there are limitations in the scope of the mechanisms that must be considered. These limitations are discussed later in the chapter.

The following security protection requirements should be considered as priority. They are to first ensure the protection of the registration process by ensuring confidentiality and authentication. Providing confidentiality is a harder requirement than authentication. End-to-end authentication ensures that the man-in-the middle security breaches are reduced. This can also limit denial of service attacks. This should be done for both random access and for repeat services. In addition to end-to-end authentication, end-to-end message confidentiality will create the required confidence in subscribers for SIP-based services. One open source of problem is the possibility of numerous INVITE messages that could be unsolicited. To limit this, SIP security should ensure that the INVITE messages are actually from valid IP addresses.

22.3.4   Security Vulnerabilities and Mechanisms

The security mechanisms in SIP can be classified as end-to-end or hop-by-hop protection. The first provides long term and long reach security protection. The second provides intermediate and short-distance protection.

22.3.4.1   End-to-End Security Mechanisms

The end-to-end mechanism involves the caller and/or callee SIP user agents and is realized by features of the SIP protocol specifically designed for this purpose (e.g., SIP authentication and SIP message body encryption). Hop-by-hop mechanisms secure the communication between two successive SIP entities in the path of signaling messages. SIP does not provide specific features for the hop-by-hop protection and relies on network-level or transport-level security. Hop-by-hop mechanisms are needed because intermediate elements may play active roles in SIP processing by reading and/or writing some parts of the SIP messages. End-to-end security cannot apply to these parts of messages that are read and written by intermediate SIP entities.

22.3.4.2   Security Mechanisms

Two main security mechanisms are used with SIP: authentication and data encryption. Data authentication is used to authenticate the sender of the message and to ensure that critical message information is not modified in transit. This is to prevent an attacker from modifying and/or replaying SIP requests and responses. Data encryption is used to ensure confidentiality of SIP communications, so that only the intended recipient decrypt and read the data. This is usually done using encryption algorithms such as Data Encryption Standard (DES) and Advanced Encryption Standard (AES).

Four fundamental security services are required for the SIP protocol: preserving the confidentiality and integrity of messaging, preventing replay attacks or message spoofing, providing for the authentication and privacy of the participants in a session, and preventing denial-of-service attacks. Entities within SIP messages separately require the security services of confidentiality, integrity, and authentication. Rather than defining new security mechanisms specific to SIP, SIP re-uses wherever possible existing security models derived from the HTTP and SMTP space.

Full encryption of messages provides the best means to preserve the confidentiality of signaling as it can also guarantee that messages are not modified by any malicious intermediaries. However, SIP requests and responses cannot be naively encrypted end-to-end in their entirety because message fields such as the Request-URI, Route, and VIA need to be visible to proxies in most network architectures so that SIP requests are routed correctly. Note that proxy servers also need to modify some features of messages (such as adding VIA header field values) in order for SIP to function. Therefore the trust levels on proxy servers must be high. For this purpose, low-layer security mechanisms for SIP are recommended. Low-layer security mechanisms encrypt the entire SIP requests or responses on the wire on a hop-by-hop basis. They also allow endpoints to verify the identity of proxy servers to whom they send requests.

SIP components also need to identify one another in a secure manner. When an SIP endpoint provides the identity of its user to a peer UA or to a proxy server, that identity should be verifiable. A cryptographic authentication mechanism is used in SIP to address this requirement. An independent security mechanism for the body for an IP message provides an alternative means of end-to-end mutual authentication. It also provides a limit on the degree to which user agents must trust intermediaries.

22.3.4.2.1 Network layer security mechanisms In transport or network layer security, the signaling traffic is encrypted, guaranteeing message confidentiality and integrity. Often, certificates are used for establishing lower-layer security. The certificates may also be used to provide a means of authentication.

Two popular alternatives for providing security at the transport and network layer are, respectively, TLS [14] and IPSec [15]. IPSec is a set of network-layer protocol tools that collectively can be used as a secure replacement for traditional IP. IPSec is most commonly used in architectures in which a set of hosts or administrative domains have an existing trust relationship with one another. IPSec is usually implemented at the operating system level in a host, or on a security gateway that provides confidentiality and integrity for all traffic it receives from a particular interface (as in a VPN architecture). IPSec can also be used on a hop-by-hop basis. In many architectures IPSec does not require integration with SIP applications; IPSec is perhaps best suited to deployments in which adding security directly to SIP hosts would be arduous. UAs that have a pre-shared keying relationship with their first-hop proxy server are also good candidates for IPSec use. Any deployment of IPSec for SIP would require an IPSec profile describing the protocol tools that would be required to secure SIP.

22.3.4.2.2 Transport layer security mechanisms TLS provides transport-layer security over connection-oriented protocols (for the purposes of this document, TCP); 'tls' (signifying TLS over TCP) can be specified as the desired transport protocol within a via header field value or a SIP-URI. TLS is most suited to architectures in which hop-by-hop security is required between hosts with no pre-existing trust association. For example, Alice trusts her local proxy server, which after a certificate exchange decides to trust Bob's local proxy server, which Bob trusts, hence Bob and Alice can communicate securely. TLS must be tightly coupled with a SIP application. Note that transport mechanisms are specified on a hop-by-hop basis in SIP, thus a UA that sends requests over TLS to a proxy server has no assurance that TLS will be used end-to-end. The TLS_RSA_WITH_AES_128_CBC_SHA cipher suite [16] must be supported at a minimum by implementers when TLS is used in a SIP application. For purposes of backwards compatibility, proxy servers, redirect servers, and registrars should support TLS_RSA_WITH_3DES_EDE_CBC_SHA [16]. Implementers may also support any other cipher suite.

The most commonly voiced concern about TLS is that it cannot be compliant with UDP; TLS requires a connection-oriented underlying transport protocol, which for the purposes of this document means TCP.

It may also be necessary for a local outbound proxy server or registrar to maintain many simultaneous long-lived TLS connections with numerous UAs. This introduces scalability concerns, especially for intensive cipher combinations. Maintaining redundancy of long-lived TLS connections especially when a UA is solely responsible for their establishment is cumbersome.

TLS only allows SIP entities to authenticate servers to which they are adjacent. TLS offers strictly hop-by-hop security. Neither TLS, nor any other mechanism allows clients to authenticate proxy servers to which they cannot form direct TCP connection.

22.3.4.2.2.1 SIP URI scheme The SIP URI scheme adheres to the syntax of the SIP URI, although the scheme string is 'sips' instead of 'sip'. However, the semantics of SIP are very different from the SIP URI. SIP allows resources to specify that they should be reached securely.

A SIP URI can be used as an address-of-record for a particular user—the URI by which the user is canonically known (on their business cards, in the 'From' header field of their requests and in the 'To' header field of register requests). When used as the Request-URI of a request, the SIP scheme signifies that each hop over which the request is forwarded, until the request reaches the SIP entity responsible for the domain portion of the Request-URI, must be secured with TLS. Once it reaches the domain in question it is handled in accordance with local security and routing policies. When used by the originator of a request (as in the case when a SIP URI is employed as the address-of-record of the target), SIP dictates that the entire request path to the target domain be secured.

The SIP scheme is applicable to many of the other ways in which SIP URI's are used in SIP today in addition to the Request-URI i.e., inclusion in addresses-of-record, contact addresses (the contents of contact headers, including those of register methods), and route headers. In each instance, the SIP URI scheme allows these existing fields to designate secure resources. The manner in which a SIP URI is dereferenced in any of these contexts has its own security properties which are detailed elsewhere [17].

The use of SIP in particular entails that mutual TLS authentication should be employed, and the cipher suite TLS_RSA_WITH_AES_128_CBC_SHA. Certificates received in the authentication process should be validated with root certificates held by the client. Failure to validate a certificate should result in the failure of the request.

Note that in the SIP URI scheme, transport is independent of TLS, and thus 'sip:anike@tut.com;transport=tcp' and 'sip:anike@tut.com;transport=sctp' are both valid (although note that UDP is not a valid transport for SIP). The use of 'transport=tls' has consequently been deprecated, partly because it was specific to a single hop of the request. This is a change since RFC 2543. Users that distribute a SIP URI as an address-of-record may elect to operate devices that refuse requests over insecure transports.

22.3.4.2.2.2 HTTP authentication SIP provides a challenge capability, based on HTTP authentication, which relies on the 401 and 407 response codes as well as header fields for carrying challenges and credentials. Without significant modification, the reuse of the HTTP digest authentication scheme in SIP allows for replay protection and one-way authentication.

22.3.4.2.2.3 HTTP digest One of the primary limitations in using HTTP digest in SIP is that the integrity mechanisms in digest do not work very well for SIP. Specifically, they offer protection of the Request-URI and the method of a message, but not for any of the header fields that UAs would most likely wish to secure.

The existing replay protection mechanisms described in RFC 2617 also have some limitations for SIP. The next-nonce mechanism, for example, does not support pipelined requests. The nonce-count mechanism should be used for replay protection.

Another limitation of HTTP digest is the scope of realms. Digest is valuable when a user wants to authenticate themselves to a resource with which they have a pre-existing association, like a service provider of which the user is a customer (which is a common scenario and thus digest provides an extremely useful function). By way of contrast, the scope of TLS is interdomain or multi-realm, since certificates are often globally verifiable, so that the UA can authenticate the server with no pre-existing association [18].

22.3.4.2.3   Application layer security mechanisms   Encrypting entire SIP messages end-to-end for the purpose of confidentiality is not appropriate because network intermediaries (like proxy servers) need to view certain header fields in order to route messages correctly, and if these intermediaries are excluded from security associations, then SIP messages will essentially be non-routable. However, S/MIME allows SIP UAs to encrypt MIME bodies within SIP, securing these bodies end-to-end without affecting message headers [19]. S/MIME can be used to provide end-to-end confidentiality and integrity for the bodies of messages, as well as mutual authentication. Furthermore, S/MIME can be used to provide some form of integrity and confidentiality for SIP header fields with SIP message tunneling.

The most significant defect of the S/MIME mechanism is the absence of a prevalent public key infrastructure for end users. If self-signed certificates (or certificates that cannot be verified by a participant in a dialogue) are employed, the SIP-based key exchange mechanism is open to a man-in-the-middle attack. Thus an attacker can potentially inspect and modify S/MIME bodies [19]. The approach is for the attacker to intercept the first exchange of keys between the two communicating parties, extract the existing CMS-detached signatures from the request and response, and put a different CMS-detached signature containing the attacker's certificate (which appears to be a certificate for the correct address-of-record). Both parties will think they have exchanged keys with each other, when in fact they have the public key of the attacker.

SSH is open to the similar man-in-the-middle attack on the first exchange of keys; however, it is widely accepted that while SSH is not perfect, it improves the security of connections. The use of key fingerprints can provide some help in SIP applications, just as for SSH. For example, if two end points use SIP to establish a voice communications session, each party reads off the fingerprint of the key they received from the other party which may be compared against an original. It is however more difficult for the man-in-the-middle to emulate the participants voices than their signaling (this practice is used with the Clipper chip-based secure telephone).

The S/MIME mechanism allows UAs to send encrypted requests without preamble if they possess a certificate for the destination address-of-record on their key ring [18]. However, it is possible that any particular device registered for an address-of-record will not hold the certificate that has been previously employed by the device's current user, and that it will therefore be unable to process an encrypted request properly, which could lead to some avoidable error signaling. This is especially likely when an encrypted request is forked.

The keys associated with S/MIME are most useful when associated with a particular user (an address-of-record) rather than a device (a UA). When users move between devices, it may be difficult to transport private keys securely between UAs.

Another difficulty with the S/MIME mechanism is that it can result in very large messages, especially when SIP tunneling is used. For that reason, it is recommended that TCP should be used as a transport protocol when S/MIME tunneling is employed [18].

22.3.4.2.3.1 SIP URIs Using TLS on all segments of a request path requires that the terminating UAS must be reachable over TLS (perhaps registering with a SIP URI as a contact address) [20]. This is the preferred use of SIP. Much valid architecture, however, use TLS to secure part of the request path, but rely on a different mechanism for the final hop to a UAS. Thus SIP cannot guarantee that using TLS will be truly end-to-end. In essence since many UAs will not accept incoming TLS connections, therefore even those UAs which do support TLS may be required to maintain persistent TLS connections.

Ensuring that TLS is used for all of the request segments up to the target domain is complex. The possibility that noncompliant cryptographically authenticated proxy servers along the route that are compromized, may choose to disregard the forwarding rules associated with SIP. Hence malicious intermediaries could, for example, retarget a request from a SIP URI to a different SIP URI in an attempt to downgrade security.

Alternatively, an intermediary might legitimately retarget a request from a SIP to a SIP URI. Recipients of a request whose Request-URI uses the SIP URI scheme thus cannot assume on the basis of the Request-URI alone that SIP was used for the entire request path (from the client onwards) [20].

To address these concerns, it is recommended that recipients of a request whose Request-URI contains a SIP or SIP URI inspect the 'To' header field value to see if it contains a SIP URI (though note that it does not constitute a breach of security if this URI has the same scheme but is not equivalent to the URI in the 'To' header field). Although clients may choose to populate the Request-URI and 'To' header field of a request differently, when SIP is used this disparity could be interpreted as a possible security violation, and the request could consequently be rejected by its recipient. Recipients may also inspect the 'Via' header chain in order to double-check whether or not TLS was used for the entire request path until the local administrative domain was reached. S/MIME may also be used by the originating UAC to help ensure that the original form of the 'To' header field is carried end-to-end.

If the UAS has reason to believe that the scheme of the Request-URI has been improperly modified in transit, the UA should notify its user of a potential security breach.

As a further measure to prevent downgrade attacks, entities that accept only SIP requests may also refuse connections on insecure ports. End users will undoubtedly discern the difference between SIP and SIP URIs, and they may manually edit them in response to stimuli. This can either benefit or degrade security. For example, if an attacker corrupts a DNS cache, inserting

TABLE 22.10

Experimental Results (Threads and Throughput)

Active threads	1	2	3	4	5	6
Single thread throughput (s^{-1})	27.8	15.9	10.6	7.9	6.4	5.3
Total throughput (s^{-1})	27.8	31.7	31.9	31.7	31.8	31.7

Source: From Salsano, S., Veltri, L., and Papalilo, D. SIP security issues: The SIP authentication procedure and its processing load. *Network IEEE,* 16(6), 38–44, November/December 2002. With permission.

a fake record set that effectively removes all SIP records for a proxy server, then any SIP requests that traverse this proxy server may fail. When a user, however, sees that repeated calls to a SIP AOR are failing, they could on some devices manually convert the scheme from SIP to SIP and retry. Of course, there are some safeguards against this (if the destination UA is truly paranoid it could refuse all non-SIP requests), but it is a limitation worth noting. On the bright side, users might also divine that SIP would be valid even when they are presented only with a SIP URI [20].

22.3.4.3 SIP Processing Cost

Starting from the SIP-based telephony service scenario, eight procedures/ scenarios have been identified to compare their processing costs [21]. As reported, Table 22.10 and Table 22.11 provide the basic procedure/scenarios which are:

1. SIP call setup with no authentication, where the proxy server is stateless and always uses UDP communication.

2. SIP authentication which corresponds exactly to the call flow has been considered in order to see the difference between the use of UDP or TCP between the proxy and tester UAS. The motivation for considering TCP-based SIP communication is to obtain an incremental analysis toward a TLS-based SIP communication setup. Scenarios (5 through to 8) replicate (1 through to 4) as far as authentication and UDP/TCP/TLS are employed, by considering a call stateful proxy server.

Table 22.10 reports the call throughput for the procedure/scenario (1) for different numbers of active threads. Starting from two threads in parallel, the capacity of the server is saturated: each of the N threads takes 1/N of the server capacity, so the server throughput is 1/N of the maximum throughput.

In Table 22.11 the results of the evaluation by Salsano [21] are reported. The third column reports the theoretical maximum throughput in terms of calls per second that the proxy server can accommodate. This includes all the processing that the proxy performs for a call from its setup to its termination or tear down. The two rightmost columns are the most important ones and report the throughput values converted into relative processing cost. In the

TABLE 22.11

Experimental Results

	Procedure/scenario	Total throughput (s^{-1})	Relative processing cost	
1	No authentication, stateless server, UDP	34.8	100	
2	Authentication, stateless server, UDP	19.6	177	
3	Authentication, stateless server, TCP	19.4	180	
4	Authentication, stateless server, TLS	19.2	182	
5	No authentication, call stateful server, UDP	21.0	166	100
6	Authentication, call stateful server, UDP	14.6	239	144
7	Authentication, call stateful server, TCP	13.8	253	152
8	Authentication, call stateful server, TLS	13.6	256	154

Source: From Salsano, S., Veltri, L., and Papalilo, D. SIP security issues: The SIP authentication procedure and its processing load. *Network IEEE,* 16(6), 38–44, November/December 2002. With permission.

first row a reference value of 100 has been assigned to procedure/scenario (5). The results show that the introduction of SIP security accounts for nearly 80% of the processing cost of a stateless server and 45% of a call stateful server. This increase can be explained with the increase in the number of exchanged SIP messages and with the actual processing cost of security. Salsano [21] estimated that 70% of the additional cost identified was for message processing and 30% for actual security mechanisms [23,24].

References

[1] Hassan, H., J.M. Garcia, and O. Brun. 2006. In *Proc. 2nd International Conference on Communication and Information Technologies, 2006, ICTTA'06* 2, Damas, Syria, pp. 3251–56. April 24–28.

[2] Wu, J.S., and P.Y. Wang. 2003. The performance analysis of SIP-T signalling system in carrier class VoIP network. In *17th International Conference on Advanced Information Networking and Applications (AINA'03), IEEE.* 39. Xi'an, China. March.

[3] Kueh, V.Y.H., R. Tafazolli, and B. Evans. 2003. Performance evaluation of SIP based session establishment over satellite-UMTS. *Vehicular Technology Conference 2003 (VTC 2003-Spring). The 57th IEEE Semi-Annual, IEEE* 22–25 2: 1381–85. Jeju Island, Korea. April.

[4] Kwon, T.T., M. Gerla, S.K. Das, and S. Das. 2002. Mobility management for VoIP services: Mobile IP vs SIP. *IEEE Wireless Communications* 9: 66–75.

[5] Curcio, I.D.D., and M. Lundan. 2002. SIP call setup delay in 3G networks. *Seventh International Symposium on Computers and Communications (ISCC'02)* 835.

[6] ETSI Telecommunications and Intranet Protocol Harmonization Over Networks (TIPHON), 2000. End to End Quality of Service in TIPHON System; Part 2: Definition of Quality of Service (QoS) Classes, TS101 329-2 v.1.1.1, July.

[7] Handley, M., and V. Jacobson. 1998. SDP: Session Description Protocol, *IETF, RFC 2327.* April.

[8] Nakajima, N., A. Dutta, S. Das, and H. Schulzinne. 2003. Handoff delay analysis and measurement for SIP based mobility in IPv6. *International conference on communications (ICC'03)* 26: 1085–89. May.

[9] Campbell, A.T., J. Gomez, S. Kim, A.G. Valko, and C Wan. 2000. Design, implementation, and evaluation of cellular IP. *IEEE Personal Communications* 7(4): 42–49. August.

[10] Jung, J.W., R. Mudumbai, D. Montgomery, and H.K. Kahng. 2003. Performance evaluation of two layered mobility management using mobile IP and session initiation protocol. *Global Telecommunication Conference 2003 (GLOBECOM'03) IEEE*3: 1190–94, San Francisco, USA.

[11] Choudhary, U., E. Perl, and D. Sidhu. 2001. Using T.38 and SIP for real-time fax transmission over IP networks. *26th Annual IEEE Conference on Local Computer Networks (LCN'01)* 74 IEEE November.

[12] Eun-Chul Cha, Hyoung-Kee Choi, and Sung-Jae Cho. 2007. Evaluation of security protocols for the session initiation protocol. In *Proceedings IEEE* 611–16.

[13] Dritsas, S., J. Mallios, M. Theoharidou, G. F. Marias, and D. Gritzalis. Threat analysis of the session initiation protocol regarding spam. In *Proceedings of IEEE International Conference on Performance, Computing, and Communications Conference, IPCCC 2007,* New Orleans, Louisiana, USA, pp. 426–33. April 11–13.

[14] Dierks, T., and C. Allen. 1999. The TLS protocol version 1.0. *RFC 2246.* January.

[15] Kent, S., and R Atkinson. 1998. Security architecture for the internet protocol. *RFC 2401.* November.

[16] Chown, P. 2002. Advanced Encryption Standard (AES) cipher suites for Transport Layer Security (TLS). *RFC 3268.* June.

[17] Rosenberg, J., and H Schulzrinne. 2002. SIP: Locating SIP Servers. *RFC 3263.* June.

[18] Rosenberg, J., H. Schulzrinne, G. Camarillo, A. Johnston, J. Peterson, R. Sparks, M. Handley, and E. Schooler. 2002. RFC 3261 SIP: Session Initiation Protocol. http://www.faqs.org/rfcs/rfc3261.html.

[19] Duanfeng, S., L. Qin, H. Xinhui, and Z. Wei. 2004. Security mechanisms for SIP-based multimedia communication infrastructure In *Proceedings International Conference on Communications, Circuits and Systems ICCCAS 2004,* 1: 575–78. June 27–29.

[20] Elthea Trevolee Lakay. 2006. SIP-based content development for wireless mobile devices with delay constraints, MSc Thesis, Department of Computer Science, University of the Western Cape, Bellville, Cape Town, South Africa.

[21] Salsano, S., L. Veltri, and D. Papalilo. 2002. SIP security issues: The SIP authentication procedure and its processing load. *Network IEEE* 16(6): 38–44. November/December.

[22] Agbinya, J.I. 2005. QoS functions and theorems for moving wireless networks. In *Proceedings IEEE International Conference on Information Technology and Application.* Sydney Australia 4–7 July.

[23] Tivoli, E.L., and J.I. Agbinya. 2005. Communication cost of SIP signalling in wireless networks and services. In *Proceedings ICT 2005 12th International Conference on Telecommunications,* 3–6 May, Cape Town, South Africa.

[24] Tivoli, E.L., E.L, and J.I. Agbinya. 2005. Security issues in SIP signaling in wireless networks and services. In *Proceedings Fourth International Conference on Mobile Business (ICMB 2005).* 11th–13th July, Sydney, Australia.

23

A Conceptual Architecture for SPIT Mitigation

Yacine Rebahi
Fraunhofer Institute for Open Communication Systems

Stelios Dritsas
Athens University of Economic and Business

Tudor Golubenco
Iptego

Benjamin Pannier
eleven GmbH

Johan Fredrik Juell
Telio Telecom AS

CONTENTS

23.1 Introduction 563
23.2 Related Work 565
23.3 SPIT Emergence and Persistence: Potential Reasons 565
23.4 Architecture 567
23.5 Conclusion 580
    Acknowledgment 580
    References 580

23.1   Introduction

In general, spam refers to any unsolicited information that is received, often with topics related to money, pornography, or health. A European Union study in 2001 estimates that the worldwide cost of spam to Internet subscribers could be in the vicinity of US\$10 billion per year. A recent study from the Ferris Research Institute [1] estimates that in 2007 spam will cost US\$100 billion worldwide, with US\$35 billion in the U.S. alone. The same study also mentions that the percentage of spam messages sent daily is greater than 75%.

Spam poses several challenges to both Internet users and regulatory agencies. It is typically anonymous, indiscriminate and global. With these characteristics spam has become a popular vehicle for promotions that may

563

be illegal, unscrupulous or use tactics that would not be commercially or legally viable outside the virtual environment. A report to the U.S. Federal Trade Commission (FTC) [2] estimates that roughly half of all unsolicited commercial e-mail contains fraudulent or deceptive content. Some of the key issues raised by spam include privacy, illegal/offensive content, misleading and deceptive trade practices and burdensome financial and resource costs.

As the spam information is without interest for the majority of people to whom this information is sent, the spammers try to utilize widely used carrier tools that have extremely low cost in order to guarantee high gains. A few years ago, fax was the main tool used for sending spam, however the current favorite tools utilized for transmitting spam information are e-mails and phone calls. Considering Voice over IP (VoIP) spam (also known as spam over Internet telephony, SPIT), the costs for delivering spam messages are roughly three orders of magnitude cheaper than traditional circuit-based telemarketer calls. For the future, IP telephony seems to be the adequate means for sending spam because of the low cost of the Internet connection and the convergence of data and voice provided by the corresponding VoIP protocols. The Ferris Research Institute [1] predicted in April 2005 that VoIP spam would be a common problem within a year.

Currently, there are very little telemarketing calls across international borders, largely due to the large cost of making international calls. This is one of the reasons why local "do not call lists", e.g., the U.S. national list of numbers that telemarketers cannot call, has been effective. This law only affects U.S. companies, but since most telemarketing calls are domestic, it has been effective. Unfortunately (and fortunately), the IP network provides no such boundaries, and calls to any VoIP destination are possible from anywhere in the world. This will allow for international spam at a significantly reduced cost. International spam is likely to be even more annoying than national spam, since it may arrive in languages that the recipient does not even speak.

VoIP has been already established in the communication world. While previously it has been based on the H.323 [3] and the session initiation protocol (SIP) standard [4], the support for the former is decreasing [3]. As the new emerging IP telephony standard, SIP will certainly be the target of spam attacks. As a consequence, identifying SIP spam in advance and the mechanisms to deal with, is a crucial task before the problem arises or becomes serious. The different forms of SIP spam can be categorized in the following way [5],

- *Call spam*: This is the case of unsolicited messages for establishing voice or video session. The spammer proceeds to relay his message over the real time media. This form is the usual method used by telemarketers.

- *IM spam*: This form is similar to e-mail spam, unsolicited IMs whose content contains the message that the spammer is seeking to convey. The SIP MESSAGE request will be used here but also some other messages such as INVITE with text or HTML bodies.

- *Presence spam*: Another form of spam similar to IM spam is called presence spam. The latter is generated by unsolicited presence (subscribe) requests sent to get on the buddy list of a user and then IMs will be sent to this user or some other form of communications will be initiated.

E-mail spam protection is a topic of already intensive research, with solutions available based on simple white/blacklist or advanced Bayesian filters that have a high rate in detecting this kind of spam. Until now, close to no work was done in the area of detection and prevention of VoIP spam in general and specifically, SIP. While some of the experience gained from the work dedicated to current web and communication systems will be surely reused for the protection of VoIP systems, further work needs to be dedicated to examine the specific possibilities of spam calls on VoIP systems. The main difference between e-mail and VoIP spam lies in the fact, that voice is an "instant and real-time" communication tool, while e-mail is a relayed one. This allows e-mail spam detectors to analyze messages in before they are received by the client, while this is not possible with voice communication. Hence, new and innovative solutions are needed to counter the VoIP spam threat.

23.2   Related Work

In order to fight the SPIT phenomenon, several anti-SPIT mechanisms have been proposed so far. Some of them appear to be the basic blocks-modules of more sophisticated anti-SPIT architectures that have been proposed in the literature [6–9]. Nowadays, the most important techniques are: black, white and gray lists [10,11], content filtering, challenge/response schemes, consent-based approaches, reputation-based mechanisms [12,13], SIP addresses management and charging-based mechanisms [14].

In fact, the above mentioned anti-SPIT mechanisms were evaluated based on some quantitative and qualitative criteria [15]. Accordingly, it is argued that the effectiveness and applicability of the proposed anti-SPIT mechanisms are judged as not adequate especially that they fail to address the majority of the evaluation criteria. More specifically, most of them try to satisfy only some straightforward criteria of anti-SPIT identification and handling, while they ignore more sophisticated ways of handle SPIT such as the usability, and the overall adoption parameters.

23.3   SPIT Emergence and Persistence: Potential Reasons

In this section, we provide an overview of the major issues behind the persistence of the current spam activities and which will certainly be driving the emergence of SPIT in the future.

23.3.1 Fake Identities and Anonymity

On of the main reasons behind the persistence of spam activities is the difficulty in determining the origins of calls and messages. This can be achieved by using spoofed identities or anonymity services.

The identity claimed by a spammer as well as the claimed domain might be some fake data or represent some information related to another domain that has nothing to do with spam.

The VoIP technologies also allow the users to utilize aliases and anonymity services. Anonymity is a way of protecting private data such as SIP URI and phone number. When anonymity is employed, the recipient is unable to identify the origin of the incoming calls. This service that can be used for protecting the private users data, can unfortunately be used for carrying out SPIT activities. This can be achieved simply by using the anonymity service when generating SPIT requests.

One can note that using strong authentication mechanisms in the current VoIP infrastructures will help efficiently in reducing the amount SPIT.

23.3.2 Insecure Third-Party Relays

A way for SPITters to hide their identities is to take advantage of SIP open relay servers. A SIP open relay is a SIP proxy that accepts unauthorized third party relays of SIP messages even they are not destined to its domain. These relays could be used by SPITters to route large volumes of unwanted SIP messages without being detected.

SIP open proxies are another kind of third-party relay, however, they are not specifically used for SIP, they could also route some other protocols requests.

SPITters are able to locate accessible third-party SIP servers by using free available automated tools. By relaying instant messages or calls through several open relay SIP proxies at the same time, it is possible to flood the SIP network with large amounts of SPIT calls in a very short time before being detected.

The problem generated by using open relays can be addressed on the one side by forcing the SIP users to utilize their outgoing proxies and on the other side by enhancing security between the different providers.

23.3.3 Automated SPIT

There are hardware or software that generate calls and send messages automatically. Any SIP client that can be configured to generate a huge number of simultaneous calls can be used for this purpose. The SIP client SIPp [16] is a good example of these automats. These automats may be used alone or with other equipment such as Interactive Voice Response (IVR) to deliver pre-recorded voice messages to the chosen recipients.

Let us mention that an efficient way to deal with the SPIT automat problem is to use any mechanism that challenges the SIP request sender as automats will certainly fail to answer correctly to the challenges. Another mechanism is to analyze the audio stream generated by an audio file that a SPITter wants to relay over once the SIP session is established. This will be discussed later.

23.3.4 Human SPITters

Telemarketing is a method for advertising products and offering services. It is usual that companies and organizations use professional telemarketers or call centers to make phone calls and send messages to potential customers on their behalf. The phone calls here could be PSTN or VoIP based. We mentioned previously that PSTN was the main tool used by spammers to carry out their activities and VoIP was the potential carrier for spam in the future because of the convergence of voice and data provided by this technology. Although, this spam category (generated by human beings) is also annoying and frightening, its impact is very limited if we compare it with the impact of the spam generated by automats.

On the other side, the use of white and black lists (if they are well implemented) can be an efficient way to mitigate this kind of SPIT activity.

23.4 Architecture

The objective of this section is to define a common architectural view for the anti-SPIT solutions. This generic vision follows a modular approach and hence will ensure the following:

- Incorporation of new detection mechanisms in response to the emergence of new SPIT forms activities.

- Seamless replacement of SPIT detection mechanisms without harming the overall architecture.

- Flexibility of the suggested solution to be adjusted to the providers requirements. This means that for a provider whose customers accept with difficulty to be challenged, this provider can use another technique (for instance, audio analysis) as a substitute.

The architecture we are proposing is depicted in Figure 23.1. Our design approach lies in the introduction of two basic layers:

- *Detection layer*: the main task of this layer is to test the received SIP requests against some predefined rules describing the spam behavior. This layer consists of one module or several ones that operate in general in a separate fashion. The results of these tests is passed to the Decision point which will integrate the received results from the different modules and decide whether the SIP request (or the call) under consideration is SPIT or not. Some concrete examples of modules will be discussed later on.

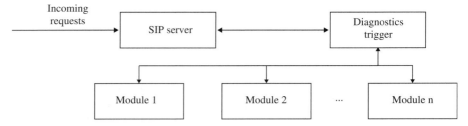

FIGURE 23.1

The generic architecture.

- *Decision layer*: the main task of the Decision point is to receive the results of the tests performed by the detection modules, as mentioned earlier, combine them, and classify the intercepted requests as SPIT or not. The Decision point could be either an independent component or just a functionality collocated with another component in our architecture. One way for combining the test results assumes that a score is assigned to each test [17]. By combining these scores together in an appropriate way, a real number is generated and compared to a pre-defined threshold. If this number is greater than the threshold, the SIP request is classified as SPIT. In the opposite case, the request is classified as non-SPIT. Another way of combining the tests results is to use them in a logical sequence. In other words, if the detection modules are organized in a logical sequence, and if the test corresponding to the first module is negative, a next test is performed, otherwise, the received request is classified as SPIT. For this kind of combination, a simple script can be used. A third way would be to build a decision tree depending on the used modules. The tree could be described in a scripting language where the scripting engine traversals the tree, calls every anti-SPIT module until a final decision is made. This solution adds more flexibility in handling the results of each module and opens the ability of interaction between the different modules. But this also adds more complexity to the whole system and makes it more difficult to add or remove modules to the system.

23.4.1   Anti-SPIT Module Examples

In Section 23.3, we have identified four types of problems that are behind the persistence of e-mail spam and which will certainly be the reasons behind the emergence and probably the persistence of SPIT. As a response to these issues, a non-exclusive list of SPIT detection modules is defined. These modules can be used as blocks for building potential anti-SPIT solutions.

23.4.1.1   Authentication Module

The authentication module simply aims at enhancing domain verification for the anti-SPIT solutions. This module tries to identify senders by the set of

SIP servers they use to send their requests. As a consequence, a SIP message apparently issued from iptel.org could not be sent by a user outside iptel.org. The solution adopted here is described in RFC 4474 [18]. We assume in this technique that there is a trustful third party that is able to issue certificates, which could be in particular acquired by the SIP proxies in the network. To be more concrete, this technique works as follows:

Before being able to establish a call, the caller needs to authenticate himself against the SIP proxy of his domain.

- If the authentication succeeds, the SIP proxy in the caller domain computes a hash (using its private key) on some header fields of the SIP message and inserts the result as a new header field (called Identity header) in the same SIP message. The SIP proxy also creates a second header field (called Identity-info) and inserts the URL of the certification authority in it. After that, the SIP message is forwarded further.

- When the SIP proxy on the callee side receives the SIP message, it uses the URL within the Identity-info field to retrieve the public key used by the SIP proxy of the originating domain. This key is used to verify the signature that the originating SIP proxy has inserted in the forwarded SIP message.

23.4.1.2 Proxy Check Module

In spite of its high reliability, the authentication module described earlier requires a trusted third party for issuing certificates which could make this module less attractive. In addition, encryption and decryption can generate a lot of processing that can add delays to communications. For these reasons, we suggest here another module that has the same spirit, but which is less complex and does not require third party nor encryption. Unfortunately, this module is less reliable than the one mentioned previously.

One mode of operation in SIP networks is to use the proxy of the domain not only for incoming calls, but also for all outgoing calls. If users want to make an outgoing call, they don't send their INVITE directly to the proxy of the target domain. Instead, they send it to the proxy of their own domain which then authenticates the user through the digest mechanism. This mode is so common that it may as well be called the standard mode of operation. The SIP specification calls it the SIP trapezoid.

This can be exploited for detection of SPIT. If an INVITE is received from a host that is not known to be a proxy of the domain of the caller, it is more likely to be SPIT and should be tested further. The Via header provides a list of all hosts that the request has traversed so far. Since it is used to route a response back, the addresses in the Via cannot be tampered with or the response will never arrive. Although, in order to successfully establish a call, this response has to arrive.

A simple version of the test for an outgoing proxy is to merely count the number of Via values present. If the request passed through a proxy, there will be at least two. Of course a SPITter (to differentiate from someone spewing saliva or tobacco juice) can easily deal with this by simply adding a second Via. On the other hand, many operators use SBCs or B2BUAs which by storing the path a response has to take internally and are able to remove the Via list.

Therefore, the test has to check the IP addresses given in the Via list. In most cases, comparing them with the addresses returned by the standard procedure for finding the SIP proxies for the domain is sufficient. However, some operators use different addresses for outgoing proxies so that the last hop may not be in this list of addresses. Comparing earlier entries in the Via list does not work reliably either, since it is possible to add additional entries in that list before the entry that will lead the response to the originating host.

The proxy check module can be more efficient if a list reporting available open relay proxies is built. So, any suspicious information present in the SIP message header of the intercepted SIP request can be checked against this list. If this information matches any entry in the list, the call will simply be dropped. The mentioned list can be built through community efforts. This means that the SIP users who are aware of the open relay servers menace can coordinate their efforts in order to build a list with these proxies that could be used by the SIP providers for blocking any SIP requests going through these servers. If we observe the e-mail case, such community efforts are already available for use [19].

In the above discussion, we provided both advantages and disadvantages of such modules. As a consequence, the corresponding implementation has to be optimized in order to avoid that such modules harm more than help in mitigating SPIT activities.

23.4.1.3   White/Black Lists Module

A SPIT black list is a list of SIP URIs and IP addresses of known SPIT users and servers. In practice, black lists are used to block all SIP requests coming from certain SIP sources identified as being used to send spam.

White lists are the opposite of black lists. They involve trusted SIP URIs and IP addresses that are always allowed to establish calls and send instant messages, no matter what the content is. White lists allow calls and instant messages coming from a trusted provider to come through, however they do not provide a solution for blocking SPIT. These lists require constant maintenance to be very effective. If they are not properly maintained, the risk of losing legitimate calls and instant messages is high.

Although their simplicity, the white/black lists can be very efficient if they are implemented correctly. In this section, we will discuss how the white/black listing technique can be handled in order to be efficient.

23.4.1.3.1   Building white/black lists   To build white and black lists, different techniques (which could be combined) are possible:

- *Using an interface*: this interface will allow the user to manage his lists and update them. A web interface could be the best choice.

- *Call rate monitoring on the callee side*: the system can observe the calls issued by a user towards a given destination. If the number of calls goes beyond a certain threshold, the call is automatically added to the user's white list. To get the agreement of the user, the system should first notify this user, however this is not easy when the user is often using a hardphone. Another option is that the user can delete any entry from his lists whenever he does not agree on adding any.

- *Call rate monitoring on the caller side*: the purpose behind this is to build some traffic patterns and make the SPIT analysis based on these patterns. We can for instance monitor continuously the arriving calls from a given user/host/domain targeting one or more end users. If we see that the number of the arriving calls deviates from the usual average, the sender will be classified as spammer.

23.4.1.3.2 White/black lists classification The white/black lists can be categorized as follows:

- Local lists: this type of list simply means that to each user, a white list and a black list will be bound. The way these lists are managed was described earlier.

- Global lists: there will also be global white and black lists besides the individual ones. If for a given sender a minimal number of users have opted to put this sender's URI on their white or black list, this URI will automatically advance to the global black or white list. The minimal number (or threshold) described earlier is policy dependant. Taking different values of the threshold has to be investigated and special care will be dedicated to the study of the generated false alarms if any.

Global lists can cause problems if the users preferences have to be taken into account. Let us for instance assume that the above mentioned threshold is set to five. In this case, if six users opt for not receiving calls or instant messages related to job advertisements, this means that all the other users will be prevented from receiving this information even if they are not against it. As a consequence, global and local lists have to be handled dependently.

The use of global lists might raise conflict issues. Imagine the situation where a given URI is on both global white and black lists, this yields to an undefined situation.

23.4.1.3.3 Scored global lists In this part, another alternative for white/ black global lists is discussed. While the global white and black lists can be effective in fighting SPIT, some drawbacks can affect their performance. In the previous paragraph, we reported briefly some of the disadvantages of using white and black lists. Another drawback that we have not mentioned yet is the fact that if the provider policy is to add a user to the global white list when a given number of users have reported his calls as not spam, an attacker could create a series of legitimate calls (to his own accounts, for example) in order to get on the white list.

Similarly, a user can be mistakenly blacklisted if some users repeatedly report him to create spam either intentionally (e.g., friends who don't get along anymore) or unintentionally (e.g., wrong button pressed). As a consequence, the entrance policies for the white/black lists are difficult to define and maintain.

These drawbacks come from the fact that the white/black lists offer very limited information about the entries, making it easy for a SPITter to work around them once it understands the policies. Therefore, the white/black lists can be improved by attaching a score to each entry. This score represents the probability of a user being a SPITter based on its past activity. Every time a callee marks a call as SPIT, it will raise the score of the caller, and every time a callee marks a call as non-SPIT, it will lower the score of the caller.

A basic example of such a score is "the number of times the user was reported as SPITter" minus "the number of times the user was reported as non-SPITter". Even with this simple score definition, an attacker would need to create more legitimate calls than spam ones in order to obtain a favorable score. More advanced scoring systems could limit the number of positive reports per user and per day, weight the score to the number of reporters or make use of a reputation system.

In addition, external modules can be used to update the score of one user. For example, if one user creates many calls to different users in a short period of time, or if a user fails a Turing machine test, its score is dynamically raised. The score value that it is added or subtracted can depend on the reporter module's weight, number of users, reputation of users.

23.4.1.4   Machine Learning Module

This module can be used to automatically filter SPIT based on some machine learning techniques. The SIP message can be considered as a block of text to which a learning algorithm (decision tree, probabilistic approach, etc.) could be applied. Because of the similarities between e-mail and SIP instant messaging, this technique can be applied successfully to detect SPIT instant messages. As for SIP calls, one can restrict himself to the SIP message header because of the very useful data that it involves.

There are two very 'opposite' approaches for how such a module should work, and the machine learning module could be optimized for each of these. Even multiple machine learning modules could be used in collaboration with other SPIT prevention modules.

23.4.1.4.1   Detect suspicious traffic   The SIP message header contains related information to the From field, the To field, the Subject field and some other data that might be valuable for detecting SPIT. The learning algorithm,

TABLE 23.1
Some Potential SPIT Patterns Coming within the From Field

%yacine@domain_name. com	invalid URI: starts with %
1237@hotmail.com	invalid URI: starts with numbers
Yacine@yahoo	invalid URI: short domain name
Fokus.fraunhofer.de	invalid URI: no @
@yahoo.com	invalid URI: too short

when applied to the SIP message header, may deal particularly with the following situations

- The From field (see Table 23.1).

- The Subject field: Is the Subject field empty? If not, which words are there: pills, money, or some other good indicators?

- Is this request the first one from this user?

- Number of blank lines in the SIP header.

As SPIT is sent in bulk, a big number of the recipients' URIs or IP addresses will come from old lists that are no longer in operation. Also, SPITters might use 'dictionary attack' or trying out local parts 'after the @' in order to build a list of their targets. As a consequence, the SIP messages sent by a SPITter (to differentiate from someone spewing saliva or tobacco juice) will present various weaknesses that a learning algorithm can exploit to detect the corresponding SPIT activity.

With other respects, the above criteria are defined according to our observation of the current e-mail case. The learning algorithms need concrete data about the SIP messages sent by the SPITters. Unfortunately, this is still early because the SPIT problem is expected to become more serious and hence detectable, as VoIP deployment goes further.

Another way of defining machine learning algorithms is to log the calls or the instant messages that we are sure are SPIT. For instance, some spam tools create calls which fail a challenge/response test. These calls are logged and a set of attributes is stored. This can be specific SIP headers, IP ranges, etc.

The machine learning module recognizes a pattern so that when next time this tool is used, calls could force a challenge/response or even drop the calls completely.

23.4.1.4.2 Detect safe calls The idea behind this approach is to allow a call to quickly progress to the endpoint based on an assumption that this is almost guaranteed a non-spam pattern. This is not necessarily a direct inverse as the first architecture. Completely different attributes could detect a known safe caller than in the first architecture. A module as described here has its logical place in the test hierarchy much earlier than the former, since this will reduce the need for further checking.

Some VoIP endpoint vendors have launched a new series of products. Its early software version is partly buggy and resembles a SPIT tool. After the first few passed challenge/response tests, the machine learning module will know the minor details from a SPIT call. Future calls with this product will then not need to be sent through the challenge/response module.

A SPIT detection module for SIP instant messaging could easily benefit from the first approach. As instant messages are in general short, containing little more than SIP headers and the message as payload, the time needed for inspection and calculations will not reduce the user experience to an unacceptable level.

SPIT detection for voice calls could benefit from the second approach. As most calls will be deemed 'safe', there will be little or non-general delay for all calls. And there is an obvious goal to reduce the amount of challenge/responses to a minimum.

The challenge for any machine learning technique is to reach an acceptable level of false positives and false negatives, and as a result these tools will need to be monitored and configured periodically. In addition, an expert/administrator must provide the different attributes, which the module will use to categorize the different calls. A too large set of attributes will slow down the throughput. A too small set of attributes will probably not detect SPIT. Attributes being subsets of other attributes, or being indirectly related to each other, will affect the calculation of the results. Since the output of any machine learning technique is statistical, it is essential to get these numbers right.

A proper study on the different attributes to be used is an important early phase if any machine learning based SPIT tool has to be made.

Let us mention that in the context of this chapter, the machine learning module is intended to be used to help decide whether the challenge/response test is needed, or even drop the intercepted call.

23.4.1.5   Challenge/Response Module

Challenge/response (C/R) is a common authentication/registration technique whereby an entity is prompted to provide a valid answer or some information in order to have access to computational resources. A well-known challenge/response protocol is user authentication via pass phrase, where a service provider is asking for a pass phrase (challenge) and the valid answer (response) is the correct password. Since C/R is widely used to prevent spam, it can also be used to prevent SPIT.

The idea behind a C/R spam preventing system is that the spammer is not willing to consume time or resources to respond to a challenge, while people who are interested in contacting with the callee will be willing to spend a minute or part of their resources.

There are two main C/R categories that can be considered for the anti-SPIT scenario. The first one is about economic solutions, so called computational puzzles. These puzzles prompt the SIP client to perform a calculation that limits the rate of incoming SPIT calls and increases their establishment cost. The second one is about puzzles, which can tell humans and computers apart.

These are puzzles, which are considered easy for a human and difficult for an automated process to solve.

The main advantage of the C/R mechanism is that it can utilize any Turing test implementation. Everyone who wishes to implement such a mechanism can use visual, audio or commonsense challenges. Moreover, a single implementation can support multiple options according to the capabilities of the devices and the preferences of the caller. So, if a caller holds a device without a screen, the challenge could be audio. On the other hand, if the caller is hearing impaired or is a non-native speaker then this caller must be submitted to a visual challenge. This conclusion can be exported by reading and processing the SIP messages' headers.

The disadvantage of this mechanism is that the recipient domain is responsible to produce and judge the challenge test. Therefore, this domain is vulnerable to Denial of Service attacks. If a spammer, who 'owes' a lot of zombie PCs, starts calling a large number of subscribers of a specific SIP provider, then the provider would create a large number of tests and consume a lot of his resources.

The other disadvantage of this technique is that the caller has to support this technique on his client. Otherwise he will get some undesired messages whenever he tries to call.

23.4.1.6 Audio Analyzer

The Audio Analyzer module is intended to be used for fighting SPIT generated by automats. This implies that the audio message which is heard by the receiver is always the same for the same SPIT wave. So a calculated checksum of the audio data that are sent by the SPITter is the same. In a normal VoIP communication, the audio checksums of two different calls are never the same. If a checksum is seen more than once the probability of SPIT is very high for this signature (so also for the call).

The audio signature must be calculated before the call reaches the receiver. For that reason, all calls from unknown callers will be handled by an Audio Analyzer. The latter will receive the call, calculate the checksum and redirect the call to the receiver if the audio checksum was not seen before.

The analyzer has similarities with an Interactive Voice Response (IVR). It answers a call and plays a pre-recorded audio for human callers. This audio informs the caller of a short delay while processing his call. This does not only inform human callers, it also simulates a human receiver for SPITters, which are waiting for an audio answer before sending their messages. While playing the information the Audio Analyzer starts recording the caller audio. The recorded audio data will be feed to a checksum calculating algorithm and the checksum is checked against a database. If the counter of the checksum is unknown this call is potentially not SPIT. Otherwise the call is very likely to be SPIT.

The heart of the Audio Analyzer module is the calculation of the checksum itself. But also very important is how the algorithm is fed. The algorithm must be able to identify the same or even slightly different audio streams as the same. This is not that difficult when the input of the audio streams is always the same. But in the VoIP environment, the input of two calls where the caller streams always the same message could result in different audio data. Due to different compressions, loss of packets, background noise or other things the audio data could be slightly or greatly different. The algorithm must take care of such possibilities. Also important is that comparing different checksums against each other or finding a checksum in a list must be very fast. In addition, the checksum algorithm must not produce false positives in a way that two different input data results in the same checksum. After discussing the requirements for such algorithms, we would like to list some concrete ways that SPITters can use to confuse these algorithms.

- Changing the audio playback starting time.

- Changing the audio playback speed.

- Pitching the audio data.

- Adding background noise or even sound.

The mind of the Audio Analyzer module is the Checksum Database (see Figure 23.2). It receives the requests with the checksum from the Audio Analyzer, looks for the checksum in its memory, increments the counter and sends the reply back to the analyzer. All checksums will be held in RAM in a list or hash tree where the search can be performed very quickly. Each time, a checksum is found or created, the counter of that checksum will be incremented and the modification date is changed. Checksums that are older than a certain amount of time and the corresponding counter is one, can be removed from the list because the mass criteria was not reached. Checksums with counter greater than one could also be removed if they were not updated for a much greater period of time (the recommendation is a time period greater than one month). Deleting old entries will depend on the traffic volume. The more traffic is checked, the earlier checksums can be removed. The suggested procedure

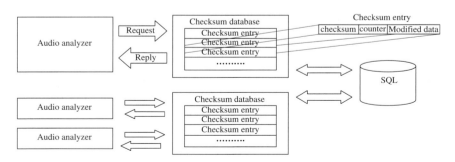

FIGURE 23.2
Checksum storage procedure.

is to reserve a certain amount of RAM for the list of checksums; when that amount is reached the delete procedure should start and delete 10–15% of old checksums.

23.4.2 Case Studies

In Section 23.4, we suggested a generic architecture to face the SPIT problem. In this section, however, we provide some case studies to show how our architecture works in a practical way. In fact, different possible combinations of the detection modules are investigated and accordingly, some more complete anti-SPIT solutions are built. Although different combinations are investigated, a unique logic is behind the investigated use cases. Indeed, we start by checking whether the call or the message is coming from a reliable source. If it is the case, we check whether this source has already been seen which means that the corresponding information is stored either in the white or black lists. If not, we need to verify whether the intercepted request is originated by an automat or a human being. To achieve this, a challenge/response mechanism (or another mechanism in the same spirit) is used. At the end and if the request is issued by a human being, the session will be established and it is up to the recipient to classify this request as SPIT or legitimate.

23.4.2.1 Case Study 1

The current case study is depicted in Figure 23.3. The involved components have already been discussed previously and are not going to be discussed here again. We will restrict ourselves to the description of how the corresponding architecture is used for filtering SPIT.

Upon receiving a SIP request, the SIP proxy forwards this request to the Decision point. The latter, in turn, forwards it further to the first detection module in this solution, namely, the authentication module. The latter will assure that the caller is exactly who is claiming to be. If the operation triggered by this module fails, the call will be dropped. If this operation succeeds or if one of the domains (inbound or outbound) is not implementing this technique, the call will be treated as a normal SIP session, so it will be submitted to the next anti-SPIT test, which is represented by the white/black lists module. This module analyzes the corresponding URI information and compares it to the entries stored in the black lists repository. If the request comes from

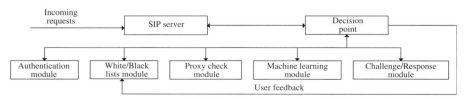

FIGURE 23.3

First case study architecture.

a domain listed in the black list, the request will be rejected without any further processing. This means that the white/black list module will inform the Decision point about that, which in turn informs the SIP server about the decision. As a consequence, the SIP server will simply reject the request under consideration.

If the domain from which the SIP request was issued is not on the black list, a second test is performed. Here, the domain information of the intercepted SIP request is checked against the information stored in the white list. If the request comes from a domain listed in the white list, the request will be forwarded without any further processing. In this case, the white/black list module will inform the Decision point, which in turn informs the SIP server about the decision. The SIP server will simply process the considered request.

If the caller is not on the white/black lists, the request is passed to another test. The latter, called machine learning module, will examines the content of the SIP message and some other related information as describe in Section 23.4. In fact, this module will help in distinguishing between potential SPIT requests and potential legitimate calls, so only potential SPITters will be challenges using the challenge/response module.

At this stage, the system reaction has to take into account the following.

- The SPITters are not using proxies: the callers will be challenged, if they fail, the calls will be dropped. An option is to blacklist the corresponding IP addresses for some time (this will help in case these IP addresses are legitimate which is the case where botnets are used). The drawback of this is the fact that the user to whom the IP address belongs will be penalized and could not make calls for some time.

- The SPITters are using open relays: here if a list with open relay information exists already, for each received call we check whether an open relay is used. If it is the case, the call will be challenged and dropped if the test fails.

- The SPITters are using their outgoing proxies: if the test fails, the call can be dropped or the user can be challenged again using some sort of scoring mechanisms. However, the calls in this kind of situation will never be blocked.

In our architecture, the C/R module uses reverse Turing tests and not computational puzzles based on hash functions. This is because the VoIP communication should be able to be established between low CPU power devices (e.g., mobile or handheld phones) and the delay of solving hash-based computational puzzles may be unacceptable by a legitimate caller.

If the C/R test is successful, the request will be forwarded to the callee. At this stage, the callee will answer the call or check the instant message, however, it will have also the opportunity to classify this call (resp. message) as SPIT or legitimate. The user feedback mechanism depicted in Figure 23.3 will allow the callee to insert the URI or the IP address used by the caller either in the black list, so further calls or messages from these user will be blacked, or in the white list, so all further calls or messages from this user will be accepted without any anti-SPIT detection.

23.4.2.2 Case Study 2

The audio analysis framework is another solution for dealing with SPIT activities. This solution follows the same spirit as the first case, however, the authentication module and the C/R module are replaced respectively, with proxy check module and Audio Analyzer module. In fact, on the one side, both the authentication module and the proxy check module are used to enforce the authentication process and prevent the use of fake identities, however, they work differently. On the other side, the C/R module and the Audio Analyzer module have the same scope, i.e., the detection of SPIT activities handled by automats. However, the detection techniques used in both are different. The authentication module filters the SIP signalling, however the Audio Analyzer monitors the RTP session. The audio analysis based solution is depicted in Figure 23.4.

The first detection entity in this framework is the check proxy module. Its way of operating is described in detail in Section 23.4. If the check proxy does not report any suspicious behavior, the SIP request will be tested further through the white/black lists module and the machine learning module. The manner this is achieved was already discussed in Section 23.4. If these tests are successful, the request is directed to the last test, namely the Audio Analyzer module.

Contrary to the other modules which take their decision based on the SIP session, the Audio Analyzer takes its decision based on the RTP session. For this reason, all modules whose decision rely on the SIP protocol need to be addressed before the Audio Analyzer. If the latter is used, the Decision point must also handle the rejection of calls differently. In this case, a rejection must also close the RTP session.

If the audio analysis test is successful, the request will be forwarded to the callee. At this stage and similarly to the first case study, the callee has the possibility to classify the request as SPIT or legitimate.

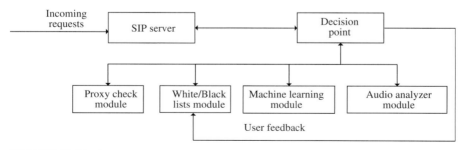

FIGURE 23.4
Second case study architecture.

23.5 Conclusion

In this chapter, we started by identifying the issues behind the persistence of the current spam that will drive the emergence of SPIT. Based on the corresponding results, we have presented a generic modular architecture for detecting SPIT. The aim of combining different classifiers is to achieve excellent discrimination between spam and legitimate SIP requests. The modular architecture we have adopted would allow the proposed anti-SPIT solutions to be updated, upgraded and adjusted to the needs of the providers without any difficulty. In this architecture, we have identified two main layers. A detection layer where the SPIT detection intelligence is distributed over different modules, and a decision layer where the results received from the detection layer are gathered and accordingly a decision is made. Finally, two case studies are considered to show how the overall architecture works.

Acknowledgment

First of all, I would like to express my deep and sincere gratitude to the SPIDER project [20] members for their important support throughout this work. I am also grateful to Prof. Dimitris Gritzalis, Athens University of Economics and Business, for his detailed and constructive comments. Many thanks go in particular to the editors for inviting the authors to contribute to this book.

References

[1] Ferris Research. 2007. http://www.ferris.com/, January.

[2] Federal Trade Commission. 2007. http://www.ftc.gov/spam/, January.

[3] H323 versus SIP: A Comparison. 2007. http://www.packetizer.com/voip/h323_vs_sip/, March.

[4] Rosenberg, J. 2002. SIP: Session Initiation Protocol. *RFC 3261*.

[5] Rosenberg, J. 2006. The Session Initiation Protocol (SIP) and spam. Draft-ietf-sipping-spam-02. March 6.

[6] Niccolini, S. 2007. Signalling to prevent SPIT (SPITSTOP) reference scenario. Draft-niccolini-sipping-spitstop-00. January.

[7] Niccolini, S. 2007. SIP extentions for SPIT identification. Draft-niccolini-feedback-spit-03.

[8] Schwartz, D. 2006. SPAM for Internet Telephony (SPIT) prevention using the Security Assertion Markup Language (SAML). draft-schwartz-sipping-spit-saml-01.txt. June.

[9] Croft, N., and M. Olivier. 2005. A model for spam prevention in voice over IP networks using anonymous verifying authorities. In *Proceedings of the 5th Annual Information Security South Africa Conference (ISSA 2005)*. South Africa. July.

[10] Mathieu, B. 2006. SPIT mitigation by a network-level anti-spit entity. In *Proceedings of the 3rd Annual VoIP Security Workshop*. Germany. June.

[11] Dongwook, S., A. Jinyoung, and S. Choon. 2006. Progressive multi gray-leveling: A voice spam protection algorithm. *IEEE Network* 20(5): 18–24. September/October.

[12] Rebahi, Y., D. Sisalem, and T. Magedanz. 2006. SIP spam detection. In *Proceedings of the International Conference on Digital Telecommunications*. France. August 29–31.

[13] Dantu, R., and P. Kolan. 2005. Detecting spam in VoIP networks. In *Proceedings of Steps to Reducing Unwanted Traffic on the Internet Workshop (SRUTI '05)*. USA. July.

[14] Baumann, R., S. Cavin, and S. Schmid. 2006. Voice over IP-security and SPIT. Swiss Army, FU Br 41, KryptDet Report, University of Berne, September.

[15] Marias, J., S. Dritsas, M. Theoharidou, J. Mallios, and D. Gritzalis. 2007. SIP vulnerabilities and antispit mechanisms assessment. In *Proceedings of the 16th IEEE International Conference on Computer Communications and Networks (ICCCN '07)*. Hawaii: IEEE Press.

[16] SIPp. 2007. http://sipp.sourceforge.net, March.

[17] Niccolini, S. 2006. *SPIT prevention: State of the Art and Research Challenges*. Germany: Network Laboratories, NEC Europe.

[18] Peterson, J. and C. Jennings. 2006. Enhancements for authenticated identity management in the session initiation protocol. *RFC 4474*.

[19] Community efforts in the email case. 2007. http://www.stayinvisible.com/index.pl/proxy_list, March.

[20] The SPIDER project. 2007. http://www.projectspider.org. Deliverable 2.2 (Spit detection and handling strategies for VoIP infrastructures). September.

24

Towards a Fraud Detection Framework in VoIP Networks

Yacine Rebahi
Fraunhofer Institute for Open Communication Systems

Thomas Magedanz
Technical University of Berlin

Dorgham Sisalem
Tekelec

CONTENTS

24.1 Introduction 583
24.2 Background 584
24.3 The Fraud Detection Framework Specification 589
24.4 Conclusion 597
     Acknowledgment 598
     References 598

24.1 Introduction

The openness and low cost structure of Voice over IP (VoIP) services has helped Application Service Providers to attract large numbers of subscribers over the last few years. However, as evidenced by various press reports the same reasons have unfortunately also attracted attackers and malicious users. Thus, misuse of services and fraud are currently considered as one of the biggest obstacles to VoIP providers. Fraud detection has been an active area of research and development in the banking and credit card industry. In the VoIP area, there is still hardly any research or products that can assist providers in detecting anomalous behavior. Hence, providers refer to general monitoring tools and manual search methods for identifying fraud attempts. To fill this gap, we suggest a complete solution for automatic fraud detection that alarms providers when suspicious behavior is detected. These tools can be used by VoIP providers to protect their services against losses due to fraud and identify previously uncollected revenue sources.

24.2   Background

24.2.1   VoIP Overview

In contrast to the closeness of the traditional PSTN networks, in which the introduction of novel services requires a long cycle process, the simplicity and openess of the Internet architecture holds the promise of rapid deployment of innovative services and efficient integration of various technologies such as web, gaming, messaging with the basic voice service. Based on this vision, VoIP technologies have already been considered for a few years as the killer application. These big expectations were coupled with huge investments in fast Internet infrastructure, start-up companies working on new technologies and services as well as research and development. Further, a great effort was spent in various standardization groups like IETF [1], ITU-T [2], 3GPP [3] in order to standardize the necessary protocols and interfaces for realizing VoIP services. Although, various VoIP products based on protocols such as H323 [4,5] and MEGACO [6] exist already in the market, the session initiation protocol (SIP) [7] has been establishing itself as the most widely accepted protocol for initiating and controling various kinds of interactive communication applications in the Internet. While primarily designed for VoIP, SIP is now currently finding wide acceptance as the basis for instant messaging applications, online gaming and conferencing.

The most important SIP operation is the invitation of new participants to a call. To achieve this functionality we can distinguish different SIP entities, such as proxy, redirect, User Agent (UA), registrar, location service, application server and Back-to-Back User Agents (B2BUA).

In SIP, a user is identified through a SIP URI in the form of "user@domain". This address can be resolved to a SIP proxy that is responsible for the user's domain. To identify the actual location of the user in terms of an IP address, the user needs to register his IP address at the SIP registrar responsible for his domain. As shown in Figure 24.1, when inviting a user, the caller sends his invitation to the SIP proxy responsible for the user's domain, which checks in the registrar's database the location of the user and forwards the invitation to the callee. The callee can either accept or reject the invitation. The session initiation is then finalized by having the caller acknowledging the receipt of the callee's answer. During this message exchange, the caller and callee exchange the addresses at which they would like to receive the media and what kind of media they can accept. After finishing the session establishment, the end systems can exchange data directly without the involvement of the SIP proxy.

24.2.2   Fraud in General

Fraud refers in general to any illegal use of resources and services by a fraudster. Usually, if fraudulent activities against an institution are perpetrated by

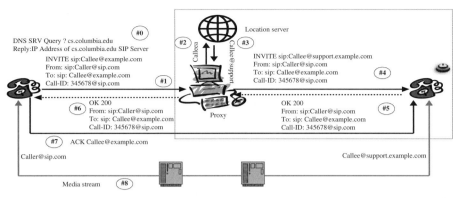

FIGURE 24.1
SIP architecture.

a person who is employed by this institution, the term misuse is used. If this person is external to this institution, the term used is fraud. This chapter will make no difference between the two terminologies. Fraud is a problem that affects all providers and is an important factor in their annual revenue losses. Beside the financial impact, it might also constrain new service deployment and adversely affect the customers perception.

Fraud activities can fit into two main categories. The non revenue fraud happens when a service is used with the intention of avoiding the costs but without the intention of making money. It includes providing no-cost services to friends or private usage. However, revenue fraud intends financial benefits as in call selling or premium rate services fraud.

Fraud is a general menace being faced by many of fields including, credit cards, e-commerce, insurance and telecommunications. VoIP is an emerging technology and as with most existing Internet applications, it will certainly be the target of fraud and misuse in the future. As a consequence, in today's financial environment, fraud detection has become a necessity in order to maintain a successful business. To realize how fraud detection and prevention is important, it is worth mentioning that for the high data traffic of 400000 transactions per day, a reduction of 2.5% of fraud triggers a saving of $1 million per year.

24.2.3 Fraud in VoIP

In traditional telecommunication networks, various experts estimate that fraud accounts for annual losses at an average of 3–5% of the operators' revenue and is still increasing at a rate of more than 10% yearly. Hence, with the openness of VoIP technology, one can expect an even higher threat of fraud and higher losses of revenue. Actually, there has already been various press reports about identity stealing and misuse of services in the Internet, which makes fraud the biggest threat to VoIP providers currently.

In traditional telecommunications, service fraud can take different shapes including:

- *Call selling*: sale of high tarif calls below their market value with the intent of evading payment to the operators.

- *Surfing*: use of another person's service by illegally obtaining his access information.

- *Ghosting*: obtaining cheap or free access to the network through technical means of deceiving the network, e.g., manipulating a switch or database.

- *Premium rate service*: this includes setting a premium rate number which receives a certain sum of money from the originating provider for each call destined to that number. There have been cases where office cleaners were organized to call these numbers and leave the phone off the hook all night or where the fraudster subscribes to an operator under a false identity and generate these calls.

- *Roaming fraud*: in this case, the attacker might obtain fraudulently a subscription from an operator and use it in high value activities (call selling to expensive locations, Data/3G session, etc) in the network of another operator. Currently, this kind of fraud needs a few days to be detected, so huge amounts of money can be lost by the providers.

It is common that these techniques are combined together in order to maximize the profits. While all of these fraud scenarios can be seen as direct threats to VoIP networks, the VoIP technology also adds additional risks.

1. *Service plan misuse*: in general VoIP services are offered as flat rate services. Operators calculate these plans based on average usage scenarios. While such services are intended for personal use only, some subscribers offer this service to other people—family and friends—resulting in high usage and high losses to the operator.

2. *Credit card fraud*: in this case, fraudsters use the toll free service of a VoIP provider to find out the PIN numbers of a stolen credit card. That is, while sitting in Nigeria for example, the fraudster would call a premium service in the U.S. and that would require the credit card and PIN number and keep trying this until the right number was found.

3. *Identity theft*: while the access control scheme of the VoIP protocols (based of MD5 hash technology) is fairly robust it can be attacked. This would result in revealing a user's password to the attacker who would then misuse the user's service.

These are just examples of possible fraud scenarios that have become public. In fact, there are different techniques that scammers can use to reach their goals:

- Eavesdropping on public transactions to obtain personal data. This can be done by looking over the shoulder of a person performing a transaction or by sniffing on VoIP transactions packets especially if encryption is not used.

- Using viruses and Trojan horses to get private data stored in computers.

- Phishing, impersonate a trusted organization or a VoIP provider and ask customers for some private in order to fix a feigned problem.

- Spam: most spam activities require you to take advantage of the good deals they are offering. If some personal data is given to them, this information will be used for carrying out their attacks.

- Browsing social networks, for instance web sites of public domains where personal data is posted.

- Equipment stealing or hacking.

24.2.4 State-of-the-Art

Fraud plagues all kinds of business activities ranging from credit card misuse, Internet transactions, telecommunications services and most recently VoIP services. The most obvious approach to prevent such misuse is to reduce the possibility of identity theft and the possibility for a fraudster to misuse the system with false or stolen personal data. Such protection measures include PIN numbers for credit cards or encrypted passwords and are usually standardized by industrial groups. Such prevention measures are outside the scope of this chapter. Instead, we will focus more on the fraud detection issue. That is, we will investigate solutions for identifying fraud as quickly as it happens, i.e., once fraud prevention has failed. When looking at the literature dealing with fraud detection, especially in the credit card and telecommunications areas, we can identify two major approaches, namely rule based and artificial intelligence based detection.

24.2.4.1 Rule Based Detection

With this approach, a set of criteria that each transaction must meet is defined. These rules are usually based on past experience, users' identity, price limit, etc. The fraud detection system scans the transaction information, checks it against these rules and decides to reject or accept the transaction. Depending on the field under consideration, these criteria can maintain information such as stolen credit card numbers and bad shipping addresses in case of financial transactions, IP addresses, logging in from unknown hosts and e-mail addresses in the case of Internet services, and changing personal data and establishing a large number of calls to unusual destinations in case of telecommunication services. In the literature, various rule-based schemes can be found such as the Protocol Data Analysis Tool (PDAT) [8]. While the rule-based technique is widely used in diverse fields, it can be very difficult to manage as fraud activities can be complex and can not always be defined by simple rules. In addition, fraud attacks are evolving, thus, static rules will be unable to detect them. With other aspects, when this technique is used, its corresponding scalability is under question. The more data the system must process, the poorer the performance.

24.2.4.2 Artificial Intelligence Based Detection

Artificial intelligence is another paradigm that can be used to detect fraud. In this case, very large databases are analyzed and some models (or profiles) are built. These models describe the user's normal behavior in term of performed transactions. Any suspicious behavior showing a significant deviation from this norm is then considered as a fraud attempt. Peer group analysis, break point analysis, activity profiling, Neural networks, Decision trees, Bayesian networks, and Knowledge Discovery in Databases (KDD) [9] are examples of intelligent algorithms that allow models to be built for normal behavior.

Peer group analysis [10] is a new tool for monitoring behavior over time in data mining situations. This tool particularly detects individual accounts that begin to behave in a way different from accounts to which they have previously been similar to. Each account is selected as a target account and is compared with all other accounts in the database, using either external comparison or internal criteria summarizing earlier behavior patterns of each account. Based on this comparison, a peer group of accounts most similar to the target account is chosen. The behavior of the peer group is then summarized at each subsequent time point, and the behavior of the target account is compared with the summary of its peer group. The targeted accounts showing a behavior different from their peer group summary behavior are flagged as suspicious. This technique might suffer from the fact that defining adequate metrics to measure the correlation between different accounts is very difficult or even impossible, thus the false alarm rate could be very high.

Break point analysis [11] is a tool that identifies changes in spending behavior based on the transaction information in a single account. Recent transactions are compared with previous spending behavior to detect features such as rapid spending and an increase in the level of spending, features that would not necessarily be captured by other detection. This technique, if it is used alone, might also suffer from false alarms.

The basic idea of activity profiling [12,13] is that the past behavior of a user or a standard operation such as logins, program execution, etc is accumulated in order to construct a pattern of the expected behavior of this user or application. The profile serves as a signature/description of normal activity for its respective subject. The observed behaviors are in general characterized in terms of metrics (which might be some statistical variables) and models. A metric is a variable reflecting both user activities and resource usage, obtained through observations. The model determines whether a new observation is consistent (or not) with respect to previous observations by applying, for instance, comparison rules, Hidden Markov models or Baye's rules. The profiles can be generated manually or automatically from profile templates.

Neural networks have been of great interest for a lot of communities over the last few years, and are being successfully applied across a large range of problem domains namely, finance, medicine, engineering and computer science. A neural network [14] is a computational model developed to be an imitation of the human brain with several computational units working in parallel. In other words, a neural network is a system of elementary decision units that can be adapted by training in order to recognize and classify arbitrary patterns. The interaction of a high number of elementary units makes it possible to learn arbitrary complex tasks. These units are connected by links which are assigned some weights representing the output of the unit. Some of the units are connected to the external environment and are dedicated as input or output units of the network. The network learns from examples, applying a function to each unit's input giving its output. The weights are adjusted so that the computational result conforms to the wanted result, given by the example. The network structure can either be feed forward or recurrent depending on whether the units are connected in a noncyclic manner or not. They can also be single layer or multi-layer networks. The latter adds hidden layers between the input and the output layer, to make more complex computations possible. A major drawback with neural networks is the construction time.

A decision tree [9,10] takes as input an object or situation described by a set of properties (attributes) and outputs a yes/no decision. The internal nodes in the tree correspond to a test of the value of one of the attributes, and the branches from the node are labeled with the possible values of the test. Each leaf specifies the goal predicted to return if that leaf is reached. Decision trees are implicitly limited to make decisions about a single object and thus limited to propositions, though they are fully expressive within the class of propositional languages. There are several algorithms that solve problems like keeping the representation small, eliminating false positives and negatives and handling noise (incorrect or missing values) and over fitting (finding meaningless regularities in large sets of hypothesis). Even if these algorithms are applied, decision trees are still vulnerable to noisy data, but the noise can be tolerated. If the prediction accuracy is high the false alarm rate is low, which is good, but it requires two large and disjoint data sets—the training data and the test data.

Unfortunately, these predictive statistical models suffer from problems such as threshold values determination, false positives and false negatives, which reduce considerably their usability especially in scenarios where a high diagnostic quality is required.

24.3 The Fraud Detection Framework Specification

24.3.1 Special Requirements in the VoIP Anti-Fraud Fight

As described in Section 24.2.4, there has been a fair amount of research efforts dedicated fraud detection either in the Internet based commercial transactions or in the telecommunications field. While VoIP offers a similar general service to traditional telecommunication services and is based on similar infrastructures (such as e-commerce), VoIP services are, however, based on a different set of signaling protocols, access technologies and business plans. Hence, extrapolating state-of-the-art fraud detection schemes to VoIP is unfortunately not straightforward and requires at least the extension of these approaches to take the specific nature of VoIP into account.

Further, the techniques developed to face the fraud problems in telecommunication networks and credit card services are usually designed with the assumption of large and centralized databases and the availability of sufficiently large hardware and software resources. This has lead to the design of often inefficient, unflexible and very complex schemes that are not adequate for the distributed and open nature of VoIP services.

In this chapter, we suggest a conceptual fraud detection framework for detecting fraud and misuse in VoIP environments. Here, for clarity purposes, the framework under consideration will be called *SAFER*. In the latter, the above-mentioned anti-fraud techniques will be improved in terms of performance and applicability to the VoIP environment. Moreover, the following new themes will be addressed.

1. *New vulnerabilities consideration*: The emerging VoIP technologies also brought new threats and vulnerabilities. These threats are especially crucial in cases where operators are interfacing with each other and to the users through servers connected to the public Internet. A great amount of work can be found in the literature about monitoring and fraud detection. However, this work mainly targets financial transactions through web applications or misuse of services provided by mobile operators and does not take into account the very specific characteristics of the VoIP protocols such as SIP, DNS/ENUM, RTP/RTCP and many others. As a consequence, the SAFER framework will not only investigate approaches for combating fraud attempts as seen today in the telecommunications world, it will also aim at identifying other attack scenarios that target the VoIP protocol and application architecture.

2. *Multi-layered adaptive fraud detection*: The VoIP anti-fraud solutions to be developed within this framework will on the one side investigate the optimal approach for applying state-of-the-art algorithms to VoIP usage scenarios. On the other side, they will also investigate the use of different techniques (rule based, intelligence based) in conjunction with each other in order to provide the best detection rate.

3. *Resource effective solutions*: On the one hand, the design of the SAFER detection framework and algorithms will take the specific characteristic of VoIP protocols and services into account. On the other hand, it has to accommodate the needs of the VoIP providers—often small enterprises themselves—in terms of low cost solutions and open and distributed architecture. Thereby, the design of the fraud detection algorithms has to

target resource efficient solutions with relatively low CPU and memory consumption.

4. *Impact on the service quality*: The anti-fraud techniques to be implemented for protecting the VoIP infrastructures cannot be deployed in a real world if their impact on the service under consideration is negative. The issues that can be a handicap for the developed techniques to be used are in particular, high complexity, expensive computation, high false alarm rate, and refusing transactions due to a fear of potential fraud.

5. *Inter-provider collaboration*: The fraud activities might go beyond one provider to another. In the case of fraud roaming, the attacker might obtain fraudulently a subscription from a provider and use it in high value activities (e.g., call selling to expensive locations, gain access to personalised services, etc) in the network of another provider. Currently, this kind of fraud needs a few days to be detected, so huge amounts of money can be lost by these providers. Roaming fraud is an example that motivates the issue of collaboration between different providers. In fact, whenever a provider detects a suspicious activity from a user who has a subscription with another provider, the latter has to be informed without delay in order to compare the user current behavior to the one stored in the subscription issuing provider and also to this provider policies. A more elaborate method is to exchange users' information between providers whenever there is an agreement between them. This technique cannot be achieved, however, without addressing legal issues such as privacy and data exchange.

24.3.2 Steps Towards the Anti-Fraud Framework

A major step in the process of providing any public service is to ensure the security of this service. This involves securing access to the servers and databases to avoid malicious hackers and instaling firewalls and fraud detection solutions. When considering services such as e-mail, e-commerce or traditional telecommunication services, there are various solutions addressing the aspect of fraud detection. A novel service such as VoIP bears great similarity to traditional telecommunication services, thus it could also utilize some of those available solutions. However, to provide a truly secure VoIP infrastructure one needs to address a broader range of scenarios that accommodate the specific characteristics of VoIP signaling protocols as well as the open and distributed nature of VoIP. To provide effective counter measures against fraud in the SAFER framework, we will be investigating the following points and provide the appropriate solutions in the form of specifications and implementations.

1. **Placement analysis**: As a first step in protecting the VoIP system, a detailed look into the used signaling protocols, mainly SIP, and VoIP

usage scenarios will be taken to determine possible ways a fraudster can misuse the service. Based on this analysis, not only the scope of the detection schemes will be defined but also the underlying information models and data visualization possibilities will be decided.

2. **Detection analysis**: The core work of the SAFER framework will be dedicated towards the specification and evaluation of efficient and highly accurate fraud detection algorithms. In this context the following approaches will be investigated.

    (a) *Rule-based detection techniques*: Within this category, the providers define a set of criteria that each call or transaction must meet. These rules can be based on past experiences, price limits, IP addresses, SLA agreements, etc. This approach works well with the information in the users profiles structured in two data levels: the customer data (long term profile) and the behavior data (usage characteristics in a short time frame). For instance, customer data is often used in conjunction with call detail data when trying to identify fraud activities. The system will automatically check the monitoring results against these rules and decide to reject or accept the transaction. For instance, the rule-based techniques are able to identify anomalies such as a user logging in from an unknown host or from a risky IP address, establishing a large number of calls to unknown destinations and changing some personal information. In fact, there are several rule-based mechanisms such as PDAT (Protocol Data Analysis Tool) [8] that can be adapted to our context and applied to detect fraud. However, the SAFER framework also aims at developing new rule-based tools that are more appropriate to the VoIP area and which take its specificities.

    (b) *Artificial intelligent techniques*: The diversity and the heterogeneity of the fraud scenarios require the use of intelligent detection techniques. The latter have to be extremely flexible in order to cope with the large number of existing fraud scenarios and to face the new ones. These issues might make the rule-based techniques difficult to manage as the later require precision and rigor for each fraud possibility. This category uses some mathematical patterns defining the normal behavior of the users and attempts to identify any deviation from this model. Such deviations can be a sign of ongoing attacks. The system will analyze the data provided by large and historical databases, build the model and apply it to real time calls and transactions. Many approaches of intelligent data analysis exist already: activity profiling, Machine Learning, Neural networks and Knowledge Discovery in Databases (KDD). These methods are quite general and have been used in different fields. Within the SAFER framework, these techniques will be investigated with regard to their applicability to the VoIP fraud detection context. On the other side, some new intelligent algorithms will be developed where the VoIP specificities are taken into account. Let us note that it is generally difficult to be certain when using statistical analysis alone that a fraud has been perpetrated. Rather, the analysis should be regarded as alerting us to the fact that an observation is abnormal, or likely to be more fraudulent than others. The different algorithms to be developed here will be evaluated with regard to their complexity and the level of achieved false positive and false negative values.

To be more precise, our framework will consider mature detection techniques from both categories and combine them in an appropriate way, which will certainly lead to more efficiency.

3. **Adaptation phase**: Fraud attacks change often in order to escape the detection techniques. For these reasons, the SAFER framework will also offer the ability to uncover and respond to the changing fraud patterns. This will be mainly achieved through intelligent techniques and the self-learning models discussed in Section 24.2.4. These techniques will use the past fraud transactions data in order to predict potential fraud cases. If we take neural networks as an example, usually different phases are considered. An identification phase where the detection algorithm analyzes the data and informs about any suspicious behavior. At this stage, some parameters are needed and could be setup randomly. Based on the amount of the phase error calculated in each fraud identification step, the parameters are corrected. This can be achieved iteratively and according to some specific rules until the phase error is reduced to zero. In SAFER, appropriate adaptive algorithms will be investigated. Of particular interest is the assessment of the correctness of the estimations in terms of false alarms and the speed at which new variations of a known fraud scenario can be detected.

24.3.3 Functional Architecture

Based on the research work dedicated to analyzing detection and prevention mechanisms, the SAFER fraud detection framework will be developed. In designing this framework, we aim at achieving the following.

- Highly accurate fraud detection.

- Low complexity system.

- Highly scalable and resource efficient detection.

- Customisable and adaptable fraud detection system.

- Advanced visualisation for human operators.

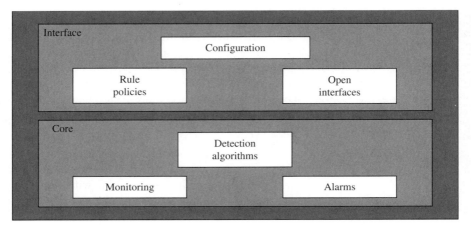

FIGURE 24.2
Functional architecture of the SAFER fraud detection framework.

On the other side, this framework will provide the following components, see also Figure 24.2.

- **The fraud detection core**: This part of the system provides the core components that include the parts of the system that can be reused for different versions of the system and different usage scenarios.
  - **System monitoring**: The monitoring activities will deal in particular with the following:
    * Call start and end time, corresponding dates, average call duration, number of calls to/from a different provider, number of weekdays calls (originated or received), number of calls destined to PSTN networks and which ones.
    * The IP addresses used, are they known or not, are they anonymous or open proxy.
    * The errors detected.
    * The same account is being used simultaneously from different IP addresses.
    * A given account is being accessed from an usual location.
    * Unauthorized calls: problems with username and password or both.
    * Users credit monitoring (the user credit is still positive?).
    * Call detail records (CDRs) monitoring.
    * Number of transactions received from the same IP address.
    * Collect monitoring information and alarms from other systems such as intrusion detection systems or SNMP traps.
  - **Fraud detection modules**: This component provides the fraud detection engine. Based on a modular architecture, this component

will include different fraud detection algorithms and mechanisms. New detection algorithms can be added to the system in a pluggable manner. Which algorithms and modules to use will be controlled through a logic provided through the configuration and policy interfaces.

— **Detection and alarm**: Based on the collected monitoring information, and the misuse signatures and scenarios identified by the threat analysis step and the provider's policies and rules, the detection and alarm component can decide whether fraud is being attempted. Whenever a fraud attempt is discovered, an alarm will be raised. These alarms will provide the provider with sufficient information so as to determine the appropriate reaction. The framework under consideration will provide the provider with the following information:

  * Suspected fraud method.
  * Level of certainty, e.g., to which level the system is certain that the attempted call is a fraud attempt.
  * Fraudster identity including his user VoIP information, IP number and destination.
  * A log of information that have lead the system to its decision such as the number of call attempts, the destination of the numbers, the number of log attempts.

The system we are intending to implement will not take preventive measures by itself. This would require deep involvement in the signaling process and access to the operator's user database. As our system will be designed to act as an independent component in the VoIP network, realizing such interaction is not planned. However, an open interface will be provided through which the operator can access the alarms generated by the system and use its established infrastructure for taking the preventive measures.

• **Interface components**: This part provides the necessary interface to enable the communication of SAFER with other systems as well as providing the operator with various status information and configuration possibilities.

  — **Provider and user policies**: The work on placement and detection analysis will result in the specification of certain detection mechanisms that can be utilized by the fraud detection system. One problematic issue with fraud is, however, that there is no simple classification. While one operator might allow for certain behavior others might consider it as a fraud attempt. Therefore, the VoIP providers deploying a fraud detection system must also have the ability to specify certain rules and policies describing which requests and usage scenarios should be prevented as well as the actions to invoke after detecting certain irregularities. These policies will enhance the detection mechanisms and ensure that the users' wishes are taken into account.

- **Configuration interface**: It is obvious from the state-of-the-art discussion, that not all detection and prevention mechanisms will be useful in each scenario. Thus, the fraud detection system will be configurable with different detection modules. Hence, to achieve a high detection rate and rapidly adjust to new fraud methods and scenarios, the SAFER system will provide for rich configuration possibilities enabling the operators to chose the appropriate detection schemes and enhance the system with new mechanisms in a rapid manner.

- **Distributed communication**: For fraud detection systems to be effective, the system needs to be constantly updated. To achieve this, communication methods will be defined to update the system from different source, e.g., from operator input or cooperating partner networks. The SAFER peer-to-peer technology is designed as follows:

  * Each secured VoIP provider network is protected by an instance of the SAFER system that is responsible for fraud detection at the operator.

  * If one network has identified a certain fraud scenario, this scenario is expressed as a policy and is distributed to other networks.

- **Open interface**: The SAFER system under consideration will provide an open interface over which the results of the analysis can be exported. This would enable the operator to process the results of the SAFER system in an automated manner, e.g., process the data, add it to some database or web site, or use it to trigger other management processes.

With other respects, the general anti-fraud defence architecture of the SAFER framework is depicted in Figure 24.3.

- In this architecture, each secured VoIP provider network is protected by a general fraud guard, that delivers general fraud management features, e.g.,

  - Monitoring the service activity by collecting data from the network, other monitoring systems, SNMP traps and intrusion detection systems [15,16] and possibly from SAFER systems in other networks domains.

  - Analysing the collected data using various detection algorithms as well as a set of operator defined policies and rules.

  - Inform the operator about the results of the analysis. This can be visualized over a GUI in the form of alarms. Further, the detection

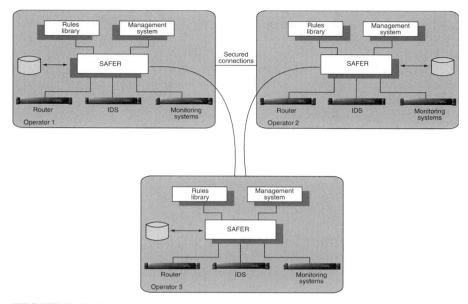

FIGURE 24.3

Anti-fraud defence architecture.

results can be delivered to the operator's operation and management systems. These systems can then take the needed measures to counter the fraud attempt, e.g., disconnect the caller or flag the caller as fraudster and reject future calls from him. The exact actions to take and the policies defining these depend on the operator's preferences.

- To enhance fraud prevention, communication methods will be defined to distribute fraud information into cooperating networks. Hence, if one network has clearly identified certain users/domains as fraudulent sources, it will inform other VoIP networks of these fraud sources.

- To enhance effectiveness, inter-provider connections need to be secured, e.g., by TLS. Hence, requests over unsecured connection will be more crucially investigated.

24.4 Conclusion

In this chapter, we have defined a framework that could be used by VoIP providers to detect fraud and misuse in their networks. We started by describing some fraud scenarios that may occur with traditional telecommunications and the risks that can be added when migrating to the VoIP paradigm. We have also discussed the current state-of-the-art fraud especially in areas such as credit card and Internet transactions, and telecommunication services. The strengths and the weaknesses of the discussed techniques were identified and their applicability to the VoIP area was investigated. After describing the different steps planned to build the SAFER framework, a functional architecture of this framework was provided. In addition to the description of the different functional entities, an overall architecture specifying how the VoIP providers can collaborate to defend their networks against fraud was presented.

Acknowledgment

First of all, I would like to express my deep and sincere gratitude to all who have closely followed this chapter. Special thanks to Dr. Dorgham Sisalem and Prof. Thomas Magedanz who contributed as co-authors. Many thanks go in particular to the editors for inviting the authors to contribute to this book.

References

[1] The Internet Engineering Task Force. 2007. http://www.ietf.org/, January.

[2] ITU Telecommunication Standardization Sector (ITU-T). 2007. http://www. itu.int/ITU-T/, January.

[3] The 3rd Generation Partnership Project (3GPP). 2007. http://www. 3gpp.org/, January.

[4] The H323 Protocol. 2007. http://www.itu.int/rec/T-REC-H.323-200606-I/en, March.

[5] International Telecommunication Union. 2003. Packet based multimedia communication systems, Recommendation H. 323, July.

[6] Groves, C., M. Pantaleo, T. Anderson, and T. Taylor. 2003. Gateway Control Protocol Version 1, *RFC 3525*.

[7] Rosenberg, J., H. Schulzrinne, G. Camarillo, A. Johnston, J. Peterson, R. Sparks, M. Handley, and E. Schooler. 2002. SIP: Session Initiation Protocol, *RFC 3261*.

[8] Senator, T. E. 2000. Ongoing management and application of discovered knowledge in a large regulatory organization: A case study of the use and impact of NASD Regulation's Advanced Detection System (ADS),

In *Proceedings of the 6th ACM SIGKDD International Conference on Knowledge Discovery and Data Mining*, New York, 44–53.

[9] http://kdd.ics.uci.edu/databases/kddcup99/kddcup99.html.

[10] Chau D. H. and C. Faloutsos. Fraud detection in electronic auction. http://www.cs. cmu.edu/dchau/papers/chau fraud detection.pdf.

[11] http://retailindustry.about.com/library/uc/02/uc fraud1.htm.

[12] Hollmen, J. 2000.User profiling and classification for fraud detection in mobile communications networks, PhD thesis, Helsinki University of Technology.

[13] Jianyun Xu, Andrew H. Sung, and Qingzhong Liu. 2006. Tree based behavior monitoring for adaptive fraud detection, *18th International Conference on Pattern Recognition (ICPR'06)*.

[14] Olusegun, O. I. 2005. Telecommunications fraud detection using baysian networks, Master thesis, Cape Town University, South Africa, May.

[15] Denning, D. E. 1987. An Intrusion Detection Model, *IEEE Transactions on Software Engineering* 13(2), 118–31.

[16] Manikopoulos, C. and S. Papavassiliou. 2002. Network intrusion and fault detection: A Statistical Anomaly Approach. In *Proceedings of the IEEE Communications Magazine*, 76–82.

Index

A

ACK responses, 194–195
Adapter peers, 8, 11–12
Address-dependent filtering behavior, 281–282
Ad-hoc networks, 11
Advanced Encryption Standard (AES), 553
Advanced media services
building, 16–25
developing, using open source tools, 23–25
experimental results, 33–40
introduction to, 15–16
Agent-based solutions, 144
ALEX (address list extension), 302–304
ALEX cache, 304
ALEX-item, 302, 304
Alternative Network Address Types (ANAT), 47, 59
Anonymity, 481–495
generalized architectures for, 484–487
introduction to, 481–482
Mist, 488–495
motivation, 482–484
proposals for, 487–495
SPIT and, 566
Anonymous communication, 11
Anonymous SIP, 463
Anonymous verifying authorities (AVA), 469–470
Answer-signal delay (ASD), 526
Anti-SPIT mechanisms, 468–477, 565;
see also Spam over Internet Telephony (SPIT)
anonymous verifying authorities, 469–470
architecture, 567–579
audio analyzer, 575–577, 579–580
authentication module, 568–569
biometric framework, 471
case studies, 577–580
challenge/response module, 574–575
classification, 472–473

compliance of, to assessment criteria, 474–476
DAPES, 470
description, 468–472
DSIP, 471–472
evaluation, 473–474
legal issues, 476–477
machine learning module, 572–574
module examples, 568–577
network-layer, 470
progressive multi grey-leveling, 470–471
proxy check module, 569–570
reputation and charging techniques, 470
RFC 4474, 471
SAML, 471
scored global lists, 571–572
voice spam detector, 472
VoIP SEAL, 472
white/black lists module, 570–571
Any source multicast (ASM), 138–140
Application layer mobility, 402–403, 406–408
Application layer multicast, 141–142
Application layer security mechanisms, 556–557
Application layer solutions, in heterogeneous wireless networks, 260–262
Application Level Gateways (ALGs), 44
Application servers, 17, 20–25
Areas of Influence (AoIs), 234
Artificial intelligence, fraud detection using, 588–589, 592–593
Asterisk, 23, 24–25
Attacks, see Security attacks
Attend command, 332
Atypes, 59–61
Audio analyzer, 575–577, 579–580
Audio conferences, 128–129
Authentication, 236–238, 552–553, 555
Authentication module, 568–569
Authentication protocols, 449, 449–450
Authenticity, 448

AVA, *see* Anonymous verifying
   authorities (AVA)
Availability, 448

B

Background validation sub-phase, 304
Back-to-Back User Agents, 463, 584
Backward compatibility, 4, 51
Bamboo network, 208–209
Basic support protocol (BSP), 74
Bi-directional tunneling, 144
Binding tables, 279
Biometric framework, for SPIT
   prevention, 471
Black lists, 469, 567, 570–571
Block rate error (BLER), 427
Bootstrapping mechanism, 10
Broadband convergence network (BcN)
   heterogeneous subnetwork
      technologies in, 349–350
   QoS-guaranteed real-time
      multimedia service
      provisioning on, 352–363
Bye Delay attacks, 446
ByeDrop attacks, 446
Bye signaling attack, 446–447, 452

C

Call Control eXtensible Markup
   Language (CCXML), 18–19
Call control function (CCF), 501–502
Call Programming Logic (CPL), 24
Call-release delay (CRD), 526
Call selling, 586
Call Session Control Function (CSCF),
   411
Call setup delay, 525–527
Call SPIT, 460, 549, 564
Call Whisper feature, 324
CAPTCHA test, 472
CCANCEL attack, 376–377
CCXML, *see* Call Control eXtensible
   Markup Language (CCXML)
Challenge-response communication, 469
Challenge/response module, 574–575
Channel validation, 303–304
Charging-based communication, 469
Chord, 5, 6, 205–207, 222
CIPID (Contact Information in
   Presence Information Data
   Format), 121
Client–server infrastructure, 3
Client–server networks, 205
Client–server paradigm, 4, 5
Codenomicon SIP test tool, 189

Communication, anonymous, 11
Communication chain, threats in,
   550–551
Compliance engine, 187–188
Compliance testing, 174
   vs. interoperability testing, 175–176
Conference control, 326–327
Conference event states, 137–138
Conferences; *see also* Group conference
      management
   infrastructure-assisted, 145–147
   loosely coupled, 132, 326
   multimedia, 323
   narrowcasting, 323–343
   peer-managed, 147–152
   privacy, 323–325, 327
   SIP conference control, 326–327
   tightly coupled, 132–134, 326
   WebEx, 325
Conferencing, 325–328
   architecture, 326
   media mixing and distribution,
      334–335
   policy configuration, 333
   policy evaluation, 333–334
   practical, 339–340
   SIP model, 326
   system design and implementation,
      332–339
Confidentiality, 447, 552
Connection admission control (CAC),
   347, 348
Connectivity checks, 300
Consent-based communication, 469
Content filtering, 469
Continuous media, 16
Continuous media services (CM), 17, 30
Control messages, MSIP, 310–311
Convergence, 342–343
CPL, *see* Call Programming Logic
   (CPL)
Credit card fraud, 586
CRESS diagrams, 377–378
Cross-layering, 28, 29, 36
Crowds, 485, 486
Cyberlink, 168–169

D

DAPES system, 470
Data delivery, 82
Data encryption, 552–553
Data Encryption Standard (DES), 553
Deafen command, 331–332
Decision layer, 568

Denial of service attacks, 510, 552
Detectionl layer, 567
DHT, *see* Distributed hash tables
 (DHTs)
DHT algorithm, 7
DHT-PeerID, 8
DHT protocols, 6
DIA, *see* Digital Item Adaptation (DIA)
Differentiated SIP (DSIP), 471–472
Digital authentication, 236–238
Digital Item Adaptation (DIA), 28
Digital Objects (DOs), 234
Digital Video Transport System
 (DVTS), 37
Disruption time, 536–538, 543–544
Distributed hash tables (DHTs), 4–5,
 141
 SIP-based telephony systems and,
  213–214
Domain Name Service (DNS), 44,
 233–234
dSIP protocol, 6–8, 218–220
 other features of, 220
 P2P overlay structure, 219–220
 peer functions and behavior, 219
 routing mechanisms, 7–8
 use of SIP messages in, 220
Dual-stack user agents, 50–51
Dynamic Host Configuration Protocol
 (DHCP), 404

E
Eavesdropping, 439, 586–587
Edge proxies, 296–298
Electronic Numbering (ENUM), 234
E-mail spam protection, 565
Empirix Hammer Call Analyzer, 186
Endpoint independent mapping
 behavior, 279
Endpoint independent filtering
 behavior, 281
End-to-end authentication, 552
End-to-end QoS session negotiation,
 352–356
End-to-end security mechanisms, 552
ENUM call flows, 504–505
European Telecommunications
 Standards Institute (ETSI),
 408
European Telecommunications
 Standards Institute Test
 Specification (ETSI TS),
 179–180

Event notification, 87
 for exchanging narrowcasting control
  status information, 340–342
 Session Initiation Protocol
  (SIP)-Specific Event
  Notification, 88–94, 102–115
Event states, conferencing, 137–138
Extended finite state machine (EFSM),
 377–378

F
Fake Busy attacks, 446
Fake identities, 566
Fast/Gigabit Ethernet, 359–361,
 364–365
Fault tolerance, 11
Fax over IP (FoIP), 544–547
File Transfer Protocol (FTP), 44
Filtering behavior, 281–282
Finite state machine (FSM), 176–177,
 191–193, 374, 375–378
Firewalls, 229
First-contact problem, 10
Flooding attacks, 441–444, 452
Fore validation sub-phase, 304
Forking proxies, 463
Fraud, 584–585
 artificial intelligence based detection,
  588–589
 rule based detection, 587–588, 592
Fraud detection framework, 583–598
 background, 584–589
 functional architecture, 593–597
 introduction to, 583
 special requirements for VoIP,
  589–591
 specification, 589–597
 steps toward, 591–593
Full distribution conferences, 134–135

G
Gaming, 130
General Packet Radio Service (GPRS),
 403–404
Ghosting, 586
GL Communications, 186
Globally unique identifier (GUID), 7, 8
Global University Phone System
 (GUPS), 35–36
Globe Project, 233
Green Networks, 234
Group conference management,
 123–153; *see also* Conferences;
 Conferencing
 application domain for, 128–130

audio conferences, 128–129
concepts and technologies, 132–144
conferencing event states, 137–138
gaming, 130
infrastructure-assisted conferences,
145–147
introduction to, 123–124
introductory scenarios, 124–128
large-scale conference, 127–128
mobile group members, 142–144
multicast, 138–142
peer-managed conferences, 147–152
presence and instant messaging,
130–131
service models, 132–135
SIP extensions used in, 135–138
three-way conference, 125–127
videoconferencing, 129–130

H
Hairpinning behavior, 282, 283
Half-Quarter Video Graphics Array
(QVGA), 129
Handle-DNS proxy (HDP), 241, 246–247
Handles, 229, 235–236
authentication and registration,
237–238
routing, 238–239
Handle-SIP, 235–236
Handle System, 229, 233–236
implementation, 240–247
Handoff delay, 534–536
analysis, 415–424
interdomain, 542–543
intradomain, 541–542
mitigations, 424–428
Handoff delay disruption, 538–547
initial registration and session setup,
541
with shadow registration, 543–544
of SIP in IP services, 538–547
in VoIP services, 540–541
Handoff latency, 84
Handoff mechanism, 399–400; *see also*
Vertical handoffs
Handoff operations, 82
Handoff period, 264–265
Hard handoffs, 400
Hash values, 5
HCPN, *see* Timed hierarchical CPN
Header fields, 463–465, 466–468
Header overhead, 83–84
Heterogeneous subnetwork technologies,
349–350

Heterogeneous wired networks, 347,
363–370
Heterogeneous wireless networks
application layer solutions, 260–262
handoff management, 257–262
network layer solutions, 257–259
QoS provisioning in, 363–370
Q-SIP/SDP for, 347–371
transport layer solutions, 259–260
vertical handoffs in, 253–272
Hierarchical mobile SIP (HMSIP), 262
Hierarchical Petri Net (HPN), 380–382
Hole punching, 284–286
Home networking systems
introduction to, 159–160
middleware implementation, 168–169
redesign issues, 165
with SIP, 159–170
UPnP, 160–162, 165–168
Home/office intranets, 347
Hordes, 486
Horizontal handoff, 400
Host identification, mobility and,
233–240
Host identity protocol (HIP), 233
Host mobility, SIP-based, 71–74
HTTP authentication, 555
HTTP digest, 449, 450–451, 555–556
HTTP protocol, 23

I
IBM rational software architect, 378,
380
ICE, *see* Interactive Connectivity
Establishment (ICE)
Identity management, 10, 471
Identity theft, 437, 444, 586
IEEE 802.11e WLAN, 361–362, 365–366
IEEE 802.3 Fast/Gigabit Ethernet
switch, 359–361
IETF protocol, 3, 116
Infrastructure-assisted conferences,
145–147
Instant message SPIT, 460, 549, 564
Instant messaging, 130–131
Integrated Services Digital Network
(ISDN), 124
Integrity, 448
Intelligent network, 499–518
background, 499–500
implementing, in VoIP applications,
501–509
introduction to, 500–501
securing, with SIP, 511–514
threats to, 510–512

Intelligent network (IN) applications,
    501–504
Interactive Connectivity Establishment
    (ICE), 298–302
Interactive Voice Response (IVR), 566
Inter-domain mobility, 229, 229–232
Inter-domain Multicast Gateway (IMG),
    147
International Telecommunication Union,
    123–124
Internet
    SIP and future of, 31–32
    telecommunications revolution and,
        435
Internet Engineering Task Force
    (IETF), 5, 44, 404
Internet Protocol (IP), 43; *see also* IPv6
Internet Protocol (IP) networks, 15
Internet Protocol Television (IPTV),
    127
Internet Telephony Administrative
    Domains (ITADs), 506–507
Interoperability standards, 3
Interoperability testing, 174–175,
    175–176
INVITE client model, 384
INVITE client transaction, validation
    of, 190–196
INVITE-contained URI lists, 137
INVITE flooding attack, 391–392
INVITE message, 5, 59–61, 163–164,
    354–355
INVITE method, 486
INVITE reflection syndrome, 443–444
INVITE server model, 385
IP/MPLS backbone network, 366–368
IP Multimedia Subsystem (IMS), 16,
    30–31, 43
    SIP and, 131
    third generation, 115–117
    wireless networks and, 254
IP Security (IPSec), 449–450, 553–554
IP Telephony Administrative Domain
    (ITAD), 58–59
IP telephony services
    handoff delay disruption of SIP in,
        538–547
    P2P architecture for SIP-Based,
        212–215
    quality of service issues, 57–58
IP/(T-)MPLS subnetwork, 351–352
IPv4, 57–58
IPv6
    address scheme, 46
    blank SDP procedure, 61–67

business and service considerations
    with VoIP migration, 47–58
introduction to, 43–46
migration scenarios, 53–56
problems for SIP-based architectures
    associated with, 46–47
quality of service issues, 57–58
security issues, 58
service engineering
    recommendations, 56–57
SIP and, 43–68
SIP internetworking, 49–51
SIP support, 48–49
technical issues raised from
    migration to, 58–67
telephony routing optimization,
    59–61
transport migration, input from,
    51–52
Iterative routing, 218

J
JAIN, *see* Java for Advanced Intelligent
    Network (JAIN)
Jain/Java Call Control (JCC), 23
JAIN-SIP proxy, 240
Jain SLEE, 23
Java for Advanced Intelligent Network
    (JAIN), 22–23
JCC, *see* Jain/Java Call Control (JCC)
Join header, 135–136
JXTA, 9

K
K-anonymity concept, 487
Kazaa architecture, 203

L
Large-scale conferences, 127–128
Latency, 84
Lightweight Directory Access Protocol
    (LDAP), 404
LinkBit, 185–186
Link-layer (subnetwork-layer) mobility,
    403–404
Location data, 121
Location service (LOC), 3, 5, 12
Loosely coupled conferences, 132, 326

M
MACHINE, 38, 40
Machine learning module, 572–574
Macro-mobility protocols, 404–405
MANET, *see* Mobile ad hoc network
    (MANET)

Man-in-the-middle attacks, 445–446,
    552, 556
Mapping refresh, 279–281
Media distribution, 334–335
Media filters, 323
Media mixing, 125, 334–335
Media privacy, 323–325, 328–332
Media Server Control API, 18, 20–21
Media Server Control Markup Language
    (MSCML), 19
Media Server Control (MEDIACTRL),
    19–20
Media servers, 17
    control mechanisms, 17–21
Media Service Component Model,
    332–333
Media Session Markup Language
    (MSML), 19
Message flow
    MSIP, 311–322
    session, 312
Message injection attacks, 439–441
Message transfer delay, 427
Micro-mobility protocols, 405
Mid-call mobility, 409, 412, 533–534
Mid-call operation, 73–74
Middleware, home networking, 168–169
Mist, 487, 488–495
    applying, in SIP, 491–495
    architecture, 489
    circuit establishment, 493
    hierarchical structure, 489
    mobility issues, 492
    overview, 488–491
    registration phase, 492
    VoIP calls, 493–495
Mobile ad hoc network (MANET), 143
Mobile devices, 227
Mobile group members, 142–144
Mobile IP (MIP), 256, 404–405
    motility management using, 529–530
    vs. SIP, 81–85
Mobile multicast, 144
Mobile router (MR), 74
Mobility
    application layer, 406–408
    host identification and, 233–240
    inter-domain, 229, 229–232
    link-layer (subnetwork-layer),
        403–404
    mid-call, 409, 412, 533–534
    network-layer, 404–405
    pre-call, 409, 411, 531–532
    session, 229–232

transport layer, 405–406
    unicast-based, 143–144
    user-controlled, 240
Mobility delay, 528–534
Mobility management, SIP, 71–85,
    397–429, 530–534
    application layer, 402–403
    comparison of schemes, 407
    handoff delay analysis, 415–424
    handoff delay mitigations, 424–428
    host mobility, 71–74
    mobility support, 408–410
    network layer, 401–402
    network mobility, 74–81
    overview, 399–400
    protocol analysis, 81–85
    protocols, 403–408
    quantitative analysis, 408–424
    session, 229
    subnetwork layer, 401
    system architecture for performance
        evaluation, 410–415
    transport layer, 402
Mobility management, using mobile IP,
    529–530
Mobility protocols, 400, 403–408
    application-layer, 406–408
    link-layer (subnetwork-layer),
        403–404
    network-layer, 404–405
    transport-layer, 405–406
Mobility support, 82–83
Mobility technologies, 227–248
    abstraction layer, 235–236
    authentication and registration,
        236–238
    implementation, 240–247
    introduction to, 228–229
    routing, 238–239
Modeling and simulation, 373–393
    CRESS diagrams, 377–378
    finite state machine, 375–378
    formal methodologies and tools for
        SIP, 375–382
    general network simulators for SIP,
        390–393
    introduction to, 373–375
    methods, 384–386
    NS-2 and, 390–392
    P2P simulators, 393
    Petri Net cluster, 380–382
    simplified SIP modeling, 385–386
    simulation and analysis, 389–390
    of SIP through times hierarchical
        CPN and CPN tools, 382–390

timed SIP model, 386–388
Unified Modeling Language (UML),
    378–380
whole SIP modeling, 384–385
Modification attacks, 551
MPEG-21/SIP-based cross-layer, 28, 29
MSCML, *see* Media Server Control
    Markup Language (MSCML)
MSIP, *see* Multipoint Session Initiation
    Protocol (MSIP)
MSML, *see* Media Session Markup
    Language (MSML)
Muffle command, 342
Multicast, 138–142
    any source, 138–140
    application layer, 141–142
    mobile, 144
    source specific, 139–141
Multicast addresses, 462, 463
Multimedia conferencing, 323
Multimedia service provisioning,
    347–349
Multiplayer games, 130
Multipoint extensions, 307–322
    introduction to, 307–308
Multipoint Session Initiation Protocol
    (MSIP), 307–322
    architecture, 308
    client behavior, 309
    controlling the session and session
        updates, 311, 318–321
    control messages, 310–311
    entities, 308–309
    general message format, 310
    joining a session, 311, 317–318
    message flow, 311–322
    reflector behavior, 309
    registration, 310
    registration phase, 312–314
    server behavior, 309
    session establishment, 310
    session establishment phase, 317
    session initiation, 310
    session initiation phase, 314–317
    terminating the session, 311, 321–322
Mute, 324, 330–331
Muzzle command, 342

N
Narrowcasting, 323–343
    architectural and interface
        refinement, 342
    attend command, 332
    commands, 330–332
    concept, 328–332

convergence, 342–343
deafen command, 331–332
event notification framework, for
    exchanging control status
    information, 340–342
interfaces, 335–338
introduction to, 323–325
media privacy and, 328–332
mute command, 330–331
parallel conversation streams and,
    339–340
privacy control, 327
select command, 332
system design and implementation,
    332–339
system performance, 338–339
NAT, *see* Network address translators
    (NATs)
NAT traversal, 277–304; *see also*
    Network address translators
    (NATs)
    ALEX, 302–304
    generic techniques, 284–292
    hole punching, 284–286
    for media flows: ICE, 298–302
    problem statement, 282–284
    relaying, 286–289
    signaling layer, 292–295
    SIP-specific techniques, 292–302
    STUN, 289–291
    TURN, 291–292
NEMO, *see* network mobility
Nesting impact, 84–85
Network address, 279
Network address translators (NATs), 10,
    277–278; *see also* NAT
    traversal
    ALEX, 302–304
    behavior, 278–282
    deterministic properties, 282
    filtering behavior, 281–282
    hairpinning behavior, 282
    mapping refresh, 279–281
    network address and port
        translation, 279
Network layer security mechanisms,
    553–554
Network layer solutions, in
    heterogeneous wireless
    networks, 257–259
Network mobility, 401–402, 404–405
    SIP-based, 74–81
Network models, 545–547
Network–Node Interface (NNI)
    signaling, 351–352, 356–359

Networks
    ad-hoc, 11
    Bamboo, 208–209
    chord, 205–207
    client–server, 205
    home, 159–170
    intelligent, 499–518
    IP, 15
    overlay, 204
    Pastry, 208–209
    peer-to-peer. *see* Peer-to-peer (P2P)
        networks
    telecommunications, 28–31
    wireless. *see* Wireless networks
    WLANs, 253, 255–256, 361–362,
        365–366, 414–415
Network Simulation version 2 (NS-2),
    390–392
Network simulators, 390–393
New Generation Networks (NGN), 30
NGN, *see* New Generation Networks
    (NGN)
NNI, *see* Network–Node Interface (NNI)
    signaling
Notification mechanism, 87
    event list—notification extension,
        102–115
    Session Initiation Protocol
        (SIP)-Specific Event
        Notification, 88–94, 102–115
NOTIFY mechanism, 107–115

O
OMA presence service, 117–120
183 Session Progress message, 360
Onion routing, 485–486
On-line transcoding, 33–34
OpenDHT, 9
Open Relay Servers, 462
Open SIP Express Router (OpenSer),
    23–25
Open source tools, for advanced media
    services, 23–25
OpenWengo NG UA, 304
OPNET, 392
Optical transport network (OTN), 352
Optional recommendations, 551
Overlay networks, 204

P
P2P File-sharing simulator, 393
P2P-PROXY, 8–9, 12
P2PSim, 393
P2P SIP, 3
    architecture for, 216–218

background, 4–6
concepts and high-level description,
    215–216
dSIP protocol, 218–220
early development of, 210–215
field of application, 10–11
goal of, 222–223
hierarchical architecture, 217–218
introduction to, 203
issues in, 10
node-level operations, 210–211
P2P simulators for, 393
protocol development and
    implementation, 218–222
recursive and iterative routing, 218
RELOAD protocol, 220–221
resource lookup algorithm, 222
SOSIMPLE approach, 210–212
standardization, 11, 215–218
user operations, 211–212
P2P SIP location service (P2P-LOC),
    8–9, 12
P2P-UA, 8
Parsing, 439–441
Partial presence, 120
Pastry network, 208–209
PATH message, 357
Peer-IDs, 219–220, 222
Peerism, 393
Peer-managed conferences, 147–152
    based on SSM, 150–152
    distributed point-to-point model,
        148–150
Peer-to-peer algorithm, 222
Peer-to-peer (P2P) networks; *see also*
    P2P SIP
    background, 4–5
    infrastructures, 7
    integrating into SIP, 3–4, 6–9
    for SIP-based IP telephony system,
        212–215
    structured, 4–5
    technologies, 204–209
    unstructured, 4–5
    in VoIP, 3
Peer-to-peer (P2P) simulators, 393
Performance issues, 10
Petri Net cluster, 380–382
Phishing, 587
Phone routes, 505–507
PIDF, *see* Presence Information Data
    Format (PIDF)
PIDF-diff, 120
Point-to-Point Protocol (PPP), 405
Port-dependent filtering behavior, 282

Port places, 381
Port translation, 279
Post-dialing delay (PDD), 526
Post Office Protocol (POP), 404
Pre-call mobility, 409, 411, 531–532
Pre-call operation, 72–73
Premium rate service, 586
Presence event package, 96–97
Presence features, 120–121
Presence information, 87
    definition, 95
    example: SIP REGISTER sesion
        state machine, 177–178
    instant messaging and, 130–131
    operations, 95–97
    overview, 94–95
    walkthrough example, 97–99
Presence Information Data Format
        (PIDF), 99–102, 119–121
    contents, 100–101
    examples, 101–102
    overview, 99–100
Presence protocol, 87
Presence server (PS), 119
Presence service, 87
    OMA, 117–120
    operations, 95–96
    reference architecture, 116–117
Presence SPIT, 460, 549, 565
Privacy control, 323–325, 327
Privacy enhancement technologies
        (PET), 484–487
Privacy issues, 437, 459; *see also*
        Security
Private Branch Exchanges (PBXs),
        128–129
Private Conversation feature, 324–325
Processing costs, 558–559
Progressive multi grey-leveling (PMG),
        470–471
ProLab SIP Test Manager, 187
Protocol analyzer tools, 185–186
Protocol fuzzers, 188–189
Protocol testing, 173–183
    compliance testing, 174
    ETSI TS 102 027 technical
        specification for SIP IETF
        RFC 3261, 179–180
    interoperability testing, 174–175
    interoperability testing vs.
        compliance testing, 175–176
    methods, 176–183
    SIP interoperability test events,
        180–183
    SIP testing, 178–180

SIP testing tools, 183–189
    standard test specification, 193–195
    state machine approach, 177,
        191–193
    validation of INVITE client
        transaction, 190–196
PROTOS test-suite c07-Sip, 189
Provisioning and Monitoring of Optical
        Services (PROMISE), 36
Proxy check module, 569–570
Proxy servers, 3, 5, 8–9, 229, 350–351,
        461
PR-SCTP, SIP over, 269–270
Public key cryptography, 485
Public Switched Telephone Network
        (PSTN), 130, 408, 435

Q
QoS-aware customer network
        management (Q-CNM),
        362–363
QoS-aware Station (QSTA), 360
QoS-extended SIP (Q-SIP)/Session
        Description Protocol (SDP)
    background and related work,
        349–352
    interaction for end-to-end QoS
        session negotiation, 352–359
    introduction to, 347–349
QoS-guaranteed real-time multimedia
        service provisioning, 347–349,
        365–366
    connectivity establishment for, at
        Fast/Gigabit Ethernet,
        364–365
    connectivity establishment for, at
        IEEE 802.11e WLAN, 365–366
    connectivity establishment for, at
        IP/MPLS backbone network,
        366–368
    experimental implementations and
        performance evaluation of,
        363–370
    in heterogeneous convergence
        networks, 363–370
    resource reservation and CAC in
        IEEE 802.11e WLAN, 361–362
    resource reservation and CAC on
        IEEE 802.3 Fast/Gigabit
        Ethernet switch, 359–361
    RSVP-TE UNI/NNI signaling,
        356–359
    session and connection management
        in, 353
    session establishment for, 352–363

Quality of service (QoS), 524–525
Quality of service (QoS) performance
    call setup delay, 525–527
    disruption time, 536–538
    effects of security protocols on
        performance of SIP, 547–548
    in fax over IP, 544–547
    handoff delay disruption, 538–547
    measuring session based, 522–524
    message transfer delay, 427
    mobility and handoff delay, 528–536
    queuing delay performance, 525–527
    session setup delay, 528
    of SIP, 522–548
QualNet, 392
Queuing delay performance, 525–527

R

Radio signal strength (RSS), 256
Real-time multimedia service
        provisioning, *see*
        QoS-guaranteed real-time
        multimedia service
        provisioning
Real-Time Transport Protocol
    (RTP), 46
Recrusive routing, 218
Redirect servers, 351
REFER method, 136–137
REGISTER message, 5, 7–8, 9, 59–61,
    163
Registrar servers, 3, 351, 551
Registration, 236–238
Re-INVITE operation, 136, 151–152
Relaying, 286–289, 291–292
RELOAD protocol, 220–221
Remote subscription, 144
Reputation-based communication, 469
Requests for comments (RFCs), 374
ReSerVation Protocol with Traffic
    Engineering (RSVP-TE), 348,
    351–352
Resource identifiers (IDs), 5
Resource-IDs, 219–220, 222
Resource list attributes, 104–107
Resource List Meta-Information (RLMI)
    document, 104, 108, 111
Resource list server (RSL), 103–115, 119
    definition, 103
    operation, 103–104
    walkthrough example, 107–115
Resource lookup algorithm, 222
RESPONSE message, 164
Response routing, 292–295
RESV message, 357

RFC 3261, 5
RFC 4474, 471
Roaming, 229, 586
Robustness tools, 188–189, 195–196
Routing, 238–239
RPID (rich presence information data
    format), 120
RSVP-TE messages, 356–359
RTP Proxy, 23, 25
RTSP, SIP and, 21–22
Rule-based detection, 587–588, 592

S

SAFER framework, 591–593
Scalable Video Coding (SVC), 129
Scored global lists, 571–572
SCTP, *see* Stream Control Transmission
    Protocol (SCTP)
SDP, *see* Session Description Protocol
    (SDP)
Secure Multipurpose Internet Mail
    Extensions (S/MIME), 450,
    556–557
Secure VoIP, 514–517
Security, 10, 435–453, 548–559
    anonymity, 481–495
    application layer, 556–557
    authenticity, 448
    availability, 448
    confidentiality, 447, 552
    fraud detection framework, 583–598
    HTTP authentication, 555
    HTTP digest, 449, 450–451, 555–556
    integrity, 448
    intelligent SIP services, 499–518
    introduction to, 435–436
    IP Security, 449–450
    with IPv6 migration, 58
    mechanisms and services, 448–452,
    552–559
    network layer security mechanisms,
    553–554
    requirements, 447–448
    SIP URIs, 557–558
    for SPIT, 457–478, 563–580
    telecommunications services, 510–517
    threats and vulnerabilities, 436–437,
    548–551
    transport layer, 554
    Transport Layer Security (TSL), 450
Security Assertion Markup Language
    (SAML), 471
Security attacks, 551–552
    denial of services, 510, 552
    eavesdropping, 439

flooding attacks, 441–444, 452
on intelligent network, 510–512
man-in-the-middle attacks, 445–446,
    552
message injection attacks, 439–441
modification attacks, 551
parsing, 439–441
signaling attacks, 444–447
against the SIP, 437–447
snoofing, 552
snooping, 551
social attacks, 447
Security protocols, 547–548
Security restrictions, 448
Select command, 332
Server farms, 11
Servers
    application, 17, 20–25
    media, 17, 17–21
    proxy, 229, 350–351
    redirect, 351
    registrar, 351, 551
    SIP, 17, 162–163, 350–351
    stateless, 462
    STUN, 289–290
Service control functions (SCF), 501
Service level agreements (SLAs), 240,
    352, 354
Service Logic Execution Environment
    (SLEE), 23
Service-Oriented Architecture (SOA),
    16, 25–27, 30
    characteristics of, 25–26
    SIP convergence and, 26–27
Service plan misuse, 586
Service switching function, 501
Session Description Protocol (SDP), 28,
    46, 348
Session Initiation Protocol (SIP); *see
    also* P2P SIP
    advanced media integration, 15–40
    advantages of, 16
    anonymity in, 481–495
    anonymous, 463
    application-layer mobility and,
        406–408
    blank SDP procedure, 61–67
    communication chain, 550–551
    event notification, 87–94, 102–115
    extension of, 3–4, 6–8
    fundamentals, 5–6
    future of Internet and, 31–32
    group conference management with,
        123–153

home networking systems with,
    159–170
implemnting intelligent network
    services in, 499–518
IMS and, 131
infrastructure, 34–36
integrating P2P into, 3–4, 6–9
introduction to, 3–4, 228–229,
    397–399, 421–422
IPv6 and, 43–68
message flow, 5–6
vs. Mobile IP, 81–85
mobility technologies, 227–248
modeling and simulation, 373–393
with multicast, 138–142
multipoint extensions for, 307–322
narrowcasting in, 323–343
over PR-SCTP, 269–270
presence information, 87, 94–102
processing costs, 558–559
QoS performance of, 522–548
RTSP and, 21–22
S/MIME, 450
SOA and, 26–27
SUBSCRIBE request, 103–115
third generation wireless networks,
    115–121
URI scheme, 554–555
vertical handoffs and, 253–272
Session Initiation Protocol
    (SIP)-Specific Event
    Notification, 88–94, 102–115
    definitions, 88–89
    introduction to, 102–103
    operation overview, 89
    resource list server, 103
    walkthrough example, 90–94,
        107–115
Session mobility, 229, 229–232
Session setup delay, 528
Session Traversal Utilities for NAT
    (STUN), 289–291
Short message service (SMS), 549
Signaling attacks, 444–447
Signaling overhead, 83
Signal-to-interference ration (SIR), 256
Simulation, *see* Modeling and simulation
Simulation tools, 187–189
SIP, *see* Session Initiation Protocol
    (SIP)
SIP address management, 469
SIP analyzer, 186
SIP application servers
    application servers, 20–25
    Java-based approach, 22–23

SIP Application Server (SMPC), 36
SIP-based host mobility, 71–74
    mid-call operation, 73–74
    pre-call operation, 72–73
SIP-based mobility management, 71–85,
            397–429, 530–534
    application layer, 402–403
    comparison of schemes, 407
    handoff delay analysis, 415–424
    handoff delay mitigations, 424–428
    host mobility, 71–74
    mobility support, 408–410
    network layer, 401–402
    network mobility, 74–81
    protocol analysis, 81–85
    protocols, 403–408
    quantitative analysis, 408–424
    subnetwork layer, 401
    system architecture for performance
            evaluation, 410–415
    transport layer, 402
SIP-based network mobility
            (SIP-NEMO), 74–81
    re-invitation operation, 78–79
    re-registration operation, 76–78
    route optimization, 79–81
    system components, 75–76
Sipbomber, 187
SIPCAT, 33–36
SIP-CMI (SIP-based continuous media
            integration), 16
SIP conference model, 326
SIP devices
    compliance and interoperability
            testing, 173–197
    protocol testing, 173–183
    validation of INVITE client
            transaction, 190–196
SIP Express Router, 187, 240
SIP headers, 466–468
SIP interoperability test events
            (SIPITs), 180–183
SIP invitations, 351
SIP location databases, 351
SIP location service, 8–9
SIP messages
    criteria, 465–466
    unencrypted, 550
    use of, in dSIP, 220
SIP Messenger tool, 187–188
SIP modeling, *see* Modeling and
            simulation
Sipp, 187
SIPPEER, 9
SIP requests, 292–295

SIPSAK, 188
SIP/SDP session layer signaling,
            350–351
SIP security, *see* Security
SIP servers, 17, 162–163, 350–351
SIP Servlet API, 23, 24
SIP signaling, 139
SIP testing, 178–180
SIP testing tools, 183–189
    protocol analyzer tools, 185–186
    robustness tools, 188–189
    simulation tools, 187–189
SIP traffic, efficient transport of, over
            SCTP, 267–272
SIP URIs, 5–6, 7, 8, 557–558, 584
SIP URI scheme, 554–555
SIP user agents, 350
    type discovery, 59–61
Skype, 3, 10–11, 203
SLEE, *see* Service Logic Execution
            Environment (SLEE)
Snoofing, 552
Snooping, 551
SOA, *see* Service-Oriented Architecture
            (SOA); Service-Oriented
            Architecture (SOA)
Social engineering, 447
Social networks, 587
Soft handoffs, 400
SOSIMPLE approach, 210–212, 215
Source specific multicast (SSM), 139,
            140–141
Spam, 458, 563–564, 587
Spam over Internet Telephony (SPIT),
            457–478, 549; *see also*
            Anti-SPIT mechanisms
    anti-SPIT mechanisms, 468–477
    automated, 566–567
    background, 459–461
    call, 460, 549, 564
    definitions, 459–460
    emergence and persistence of,
            565–567
    identification criteria, 465–468
    instant messaging, 460, 549, 564
    introduction to, 457–459, 563–565
    motivation, 460–461
    presence, 460, 549, 565
    vulnerability analysis, 461–465
Spirent Protocol Tester (SPT), 188
SPIT, *see* Spam over Internet Telephony
            (SPIT)
SSH, 556
S-SIP handoff scheme, 263–264

SSM, peer-managed conferences based on, 150–152
Standardization, 11
Stateless servers, 462
State machine transitions, 194–195
Stream Control Transmission Protocol (SCTP), 221, 256
  retransmission mechanism, 268–269
  transporting SIP traffic over, 267–272
Streaming services, 17
Structured P2P systems, 4–5
STUN, *see* Session Traversal Utilities for NAT (STUN)
STUN binding requests, 299–301
STUN clients, 289–290
STUN server, 289–290
Subnetwork layer, mobility management at, 401
SUBSCRIBE request, 102–115, 117
Substitution transition, 380
Surfing, 586
Synchronous digital hierarchy (SDH), 352

T
TCP-Migrate, 405–406
Telecommunications networks, service convergence platform of, 28–31
Tele-immersion, 37
Tele-Immersion For Applications Supporting New Interactive Srevices (TIFANIS), 37
Telemarketing, 564, 567
Telephony Routing Information Base (TRIB), 507
Telephony routing optimization, 59–61
Third generation (3G) cellular communications, 253
Third generation (3G) wireless networks, 115–121, 203
  IMS, 115–117
  OMA presence service, 117–120
  presence features, 120–121
Third Generation Mobile System, 116
Third Generation Partnership Project (3GPP), 3, 16, 115–117, 254–255
Third-party relays, insecure, 566
Three-way conferences, 125–127
Tightly coupled conferences, 132–134, 326
Timed hierarchical CPN
  definition of, 382–384

modeling and simulation of SIP through, 382–390
  simplified SIP modeling, 385–386
  simulation and analysis, 389–390
  timed SIP model, 386–388
  whole SIP modeling, 384–385
Timed presence, 120–121
Timers test, 195
TISPAN architecture, 43
TLS, 552, 554, 557
Toll fraud, 437, 444
Translating Relaying Internet Architecture Integrating Active Directories (TRIAD), 234
Transmission Control Protocol (TCP), 256
Transmission Reception Subsystem (TRS), 37
Transmission routes, 84
Transport addresses, 279, 299, 302–303
Transport layer mobility, 402, 405–406
Transport layer security mechanism, 554
Transport Layer Security (TLS), 20, 450
Transport layer solutions, in heterogeneous wireless networks, 259–260
Transport MPLS (T-MPLS), 352, 367
Traversal Using Relays around NAT (TURN), 291–292
TRIP framework, 505–507
Trojan horses, 587

U
UDDI, *see* Universal Description Discovery Integration (UDDI)
UltraGrid, 37–38, 39–40
UMTS network, 410–415
  mobile host moving to, 412–414
UMTS Release 5, 410–411
Unauthorized access, 436–437, 510
UNI, *see* User–Network Interface (UNI) signaling
Unicast-based mobility, 143–144
Unified Modeling Language (UML), 378–380
Uniform resource identifiers (URIs), 5–8, 229, 326, 557–558, 584
Universal Description Discovery Integration (UDDI), 26
Universal Mobile Telecommunication System (UMTS), 403–404
Universal Plug and Play (UPnP), 159, 160–162
  discovery functionalities, 165–168

Unstructured P2P systems, 4–5
UPDATE message, 355
User agent capability, 120
User Agent Client (UAC), 16–17
User agents, 3, 5, 8–9, 162, 350
    Back-to-Back, 463, 584
    dual-stack, 50–51
    SPIT and, 468
    type discovery, 59–61
User Agent Server (UAS), 16–17
User-controlled mobility, 240
User Datagram Protocol (UDP), 221,
    256
User–Network Interface (UNI) signaling,
    351–352, 354, 356–359

V
Validation tables, 303
Vertical handoffs, 253–272, 400
    handoff initiation based on user
        mobility, 265–266
    handoff periods, 264–265
    management, 257–262
    simulations results and discussions,
        266–267
    S-SIP, 263–264
    using SIP, 262–267
Video
    high definition, 37–38
    real-time transport of high-quality,
        33
Videoconferencing
    examples of high-end, 36–40
    over IP, 129–130
VideoLAN, 33–34
Viruses, 587
Voice-over-Internet Protocol (VoIP),
        3–4, 10–11, 374, 435–436, 564
    fraud in, 585–587

overview, 584
quality of service issues, 57–58
secure, 514–517
threats and vulnerabilities,
    436–437
TRIP framework for, 505–507
Voice-over-Internet Protocol (VoIP)
    internetworking, ENUM call
    flows for, 504–505
Voice-over-Internet Protocol (VoIP)
    services
    handoff delay disruption of SIP in,
        540–541
    implement intelligent network
        services in, 501–509
Voice spam detector, 472
Voice SPIT, 549
VoiceXML, 18–19
VoIP SEAL, 472

W
WDSL, *see* Web Service Description
    Language (WDSL)
WebEx conferences, 325
Web Service Description Language
    (WDSL), 26
Web services, 16
Weighted round-robin (WWR), 359
Whisper Coaching, 325
White lists, 469, 567, 570–571
Wireless local area networks (WLAN),
        253, 255–256
    IEE 802.11e, 361–362
    IEEE 802.11e, 365–366
    mobile host moving to, 414–415
Wireless networks
    heterogeneous, 253–272
    third generation, 115–121
Wireshark, 185, 193–194